Renewable Polymers

Scrivener Publishing
3 Winter Street, Suite 3
Salem, MA 01970

Scrivener Publishing Collections Editors

James E. R. Couper	Ken Dragoon
Richard Erdlac	Rafiq Islam
Norman Lieberman	Peter Martin
W. Kent Muhlbauer	Andrew Y. C. Nee
S. A. Sherif	James G. Speight

Publishers at Scrivener
Martin Scrivener (martin@scrivenerpublishing.com)
Phillip Carmical (pcarmical@scrivenerpublishing.com)

Renewable Polymers
Synthesis, Processing, and Technology

Edited by

Vikas Mittal

Chemical Engineering Department,
The Petroleum Institute,
Abu Dhabi, UAE

Scrivener

For general information on our other products and services or for technical support, please contact our Customer Care Department within the United States at (800) 762-2974, outside the United States at (317) 572-3993 or fax (317) 572-4002.

Wiley also publishes its books in a variety of electronic formats. Some content that appears in print may not be available in electronic formats. For more information about Wiley products, visit our web site at www.wiley.com.

For more information about Scrivener products please visit www.scrivenerpublishing.com.

Cover design by Russell Richardson

Library of Congress Cataloging-in-Publication Data:

ISBN 978-0-470-93877-5

Printed in the United States of America

10 9 8 7 6 5 4 3 2 1

Contents

Preface   xiii
List of Contributors   xv

1. **Polymers from Renewable Resources**   **1**
    V Mittal
    1.1   Introduction   1
    1.2   Naturally Renewable Methylene Butyrolactones   4
    1.3   Renewable Rosin Acid-Degradable Caprolactone
          Block Copolymers   6
    1.4   Plant Oils as Platform Chemicals
          for Polymer Synthesis   7
    1.5   Biosourced Stereocontrolled Polytriazoles   9
    1.6   Polymers from Naturally Occurring Monoterpene   10
    1.7   Polymerization of Biosourced
          2-(Methacryloyloxy)ethyl Tiglate   11
    1.8   Oxypropylation of Rapeseed Cake Residue   12
    1.9   Copolymerization of Naturally
          Occurring Limonene   13
    1.10 Polymerization of Lactides   14
    1.11 Nanocomposites Using Renewable Polymers   19
    1.12 Castor Oil Based Thermosets   19
    References   22

2. **Design, Synthesis, Property, and**
    Application of Plant Oil Polymers   **23**
    Keshar Prasain and Duy H. Hua
    2.1   Introduction   24
    2.2   Triglyceride Polymers   25
          2.2.1   Formation and Copolymerization
                  of Monoglycerides and Diglycerides   25
          2.2.2   Copolymerization of Fatty Acids   33

2.2.3 Polymerization of Functionalized
Triglycerides   39
2.3 Summary   65
References   65

3. **Advances in Acid Mediated Polymerizations**   **69**
Stewart P. Lewis and Robert T. Mathers
3.1 Introduction   70
3.2 Problems Inherent to Cationic Olefin
Polymerization   72
3.3 Progress Toward Cleaner Cationic Polymerizations   75
3.3.1 Improvements Resulting from Initiator
System Design   75
3.3.1.1 Progress in Homogeneous Initiator
Systems   76
3.3.1.2 Developments in Heterogeneous
Initiator Systems   134
3.4 Environmental Benefits via New
Process Conditions   158
3.5 Cationic Polymerization of Monomers
Derived from Renewable Resources   161
3.6 Sustainable Synthesis of Monomers for Cationic
Polymerization   163
References   164

4. **Olive Oil Wastewater as a Renewable Resource
for Production of Polyhydroxyalkanoates**   **175**
*Francesco Valentino, Marianna Villano, Lorenzo Bertin,
Mario Beccari, and Mauro Majone*
4.1 Polyhydroxyalkanoates (PHAs): Structure,
Properties, and Applications   175
4.2 PHA Production Processes Employing
Pure Microbial Cultures   177
4.3 PHA Production Processes Employing
Mixed Microbial Cultures   178
4.3.1 The Acidogenic Fermentation Stage:
Key Aspects   182
4.3.2 The Mixed Microbial Culture (MMC)
Selection Stage   185
4.3.3 The PHA Accumulation Stage   188
4.4. Olive Oil Mill Effluents (OMEs) as a Possible
Feedstock for PHA Production   197

| | 4.4.1 | Olive Oil Production | 197 |

4.4.1   Olive Oil Production   197
4.4.2   Chemical and Physical Characteristic
        of OMEs   198
4.4.3   Wastewater Treatment and Disposal
        Alternatives   201
4.4.4   Biological Wastewater Treatment   203
4.5   OMEs as Feedstock for PHA Production   206
4.6   Concluding Remarks   211
References   212

5.   **Atom Transfer Radical Polymerization (ATRP) for
     Production of Polymers from Renewable Resources   221**
     Kattimuttathu I. Suresh
5.1   Introduction   221
5.2   Atom Transfer Radical Polymerization (ATRP)   222
      5.2.1   General Considerations   224
      5.2.2   Kinetics of ATRP   226
      5.2.3   Macromolecular Architecture   227
      5.2.4   Choice of Reaction Medium   227
5.3   Synthetic Strategies to Develop Functional
      Material Based on Renewable Resources -
      Composition, Topologies
      and Functionalities   227
      5.3.1   Use of Functional Initiators   228
      5.3.2   Modified Processes   230
5.4   Sustainable Sources for Monomers
      with a Potential for Making Novel
      Renewable Polymers   231
      5.4.1   Plant Oil Derived Monomers –Fatty Acid
              Acrylates/Methacrylates   231
      5.4.2   Monomers Prepared Through
              Derivatization of Vegetable Oils   232
      5.4.3   Block Co-Polymers Based on
              Renewable Monomers   234
      5.4.4   Cardanol – the Phenolic Lipid   235
      5.4.5   Catalyst for ATRP of Vegetable
              Oil Monomers   237
      5.4.6   Rosin Gum   238
      5.4.7   Miscellaneous Monomers   239
5.5   Conclusions and Outlook   241
References   242

6. Renewable Polymers in Transgenic Crop Plants   **247**
Tina Hausmann and Inge Broer
  6.1   Natural Plant Polymers   248
       6.1.1   Starch   249
       6.1.2   Cellulose   254
       6.1.3   Rubber   261
  6.2   De Novo Synthesis of Polymers in Plants   269
       6.2.1   Fibrous Proteins   270
           6.2.1.1   Silk   270
           6.2.1.2   Collagen   274
           6.2.1.3   Elastin   276
       6.2.2   Polyhydroxyalkanoates   279
       6.2.3   Cyanophycin   285
  6.3   Conclusion   289
  References   291

7. Polyesters, Polycarbonates and Polyamides Based on Renewable Resources   **305**
Bart A. J. Noordover
  7.1   Introduction   306
  7.2   Biomass-Based Monomers   307
       7.2.1   Monomers from Saccharides   308
       7.2.2   Monomers from Vegetable Oils and Other Sources   309
  7.3   Polyesters Based on Renewable Resources   311
       7.3.1   Polyesters from Saccharide-Derivatives   311
       7.3.2   Aliphatic DAH-Based Polyesters   313
       7.3.3   Aromatic DAH-Based Polyesters   327
       7.3.4   Polyesters Based on Furan Monomers   330
       7.3.5   Vegetable Oil-Based Polyesters   331
  7.4   Polycarbonates Based on Renewable Resources   332
       7.4.1   Polycarbonates Based on 1,4:3,6-dianhydrohexitols   333
       7.4.2   Other Biomass-Based Polycarbonates   343
  7.5   Polyamides Based on Renewable Resources   344
       7.5.1   Linear, Aliphatic Polyamides   344
       7.5.2   Fatty Acid-Based Polyamides   346
       7.5.3   Other Biomass-Based Polyamides   347
  7.6   Conclusions   349
  References   350

8. **Succinic Acid: Synthesis of Biobased Polymers from Renewable Resources**   355
    Stephan Kabasci and Inna Bretz
    8.1 Introduction   355
         8.1.1 General   355
         8.1.2 Biotechnological Production of
                Succinic Acid   356
         8.1.3 Chemical Conversion   358
    8.2 Polymerization   359
         8.2.1 Polyesters   360
         8.2.2 Polyamides   365
         8.2.3 Poly(ester amide)s   370
    8.3 Conclusions   371
    References   372

9. **5-Hydroxymethylfurfural Based Polymers**   381
    Ananda S. Amarasekara
    9.1 Introduction   381
    9.2 5-Hydroxymethylfurfural   382
         9.2.1 Preparation of 5-Hydroxymethylfurfural
                from Hexoses   382
                9.2.1.1 Dehydration of Fructose Using
                         Acid Catalysts   385
                9.2.1.2 Dehydration of Fructose Without
                         a Catalyst   386
                9.2.1.3 Dehydration of Fructose in
                         Ionic Liquids   386
                9.2.1.4 Inulin in Ionic Liquids   388
                9.2.1.5 Dehydration of Glucose in
                         Ionic Liquids   388
                9.2.1.6 Cellulose in Ionic Liquids   389
    9.3 5-Hydroxymethylfurfural Derivatives   393
         9.3.1 Monomers Derived from
                5-Hydroxymethylfurfural   393
         9.3.2 Synthesis of 5-Hydroxymethylfurfural
                Derivatives   394
                9.3.2.1 Synthesis of
                         2,5-bis(hydroxymethyl)furan   394
                9.3.2.2 Synthesis of 2,5-bis(formyl)furan   394
                9.3.2.3 Synthesis of 2,5-furandicarboxylic
                         Acid   395

9.3.2.4 Synthesis of 5,5′(oxy-bis(methylene))
bis-2-furfural   396
9.3.2.5 Synthesis of
2,5-bis(aminomethyl)furan   397
9.3.2.6 Synthesis of
2,5-bis(chloromethyl)furan   397
9.3.2.7 Synthesis of 2,5-furandicarboxylicacid
Dichloride and Related Compounds   398
9.4 Polymers from 5-Hydroxymethylfurfural
Derivatives   398
9.4.1 Furanic Poly Schiff Bases   398
9.4.2 Furanic Poly Esters   400
9.4.3 Furanic Polyamides   410
9.4.4 Furanic Polyurethanes   414
9.4.5 Furanic Polybenzoimidazoles   416
9.4.6 Furanic Polyoxadiazoles   417
9.4.7 Poly(furalidine bisamides)   418
9.4.8 Poly(furylenethylenediol)   418
9.4.9 Miscellaneous Furanic Polymers
Derived from HMF   420
9.5 Conclusion   421
References   422

10. Natural Polymers—A Boon for Drug Delivery   429
Rajesh. N, Uma. N, and Valluru Ravi
10.1 Introduction   429
10.2 Acacia   429
10.3 Agar   431
10.4 Alginate   433
10.5 Carrageenan   436
10.6 Cellulose   438
10.7 Chitosan   440
10.8 Dextran   444
10.9 Dextrin   445
10.10 Gellan Gum   447
10.11 Guar Gum   448
10.12 Inulin   451
10.13 Karaya Gum   452
10.14 Konjac Glucomannan   453
10.15 Locust Bean Gum   454
10.16 Locust Bean Gum   455

10.17 Pectin   455
10.18 Psyllium Husk   457
10.19 Scleroglucan   458
10.20 Starch   460
10.21 Xanthan Gum   462
References   465

Index   **473**

Preface

It has been a long-term desire to replace polymers from fossil fuels with the more environmentally friendly polymers generated from renewable resources. Although a lot of effort has been devoted to the development of such polymers, in most cases however, the polymers from renewable resources were costly and did not match the properties of the fossil fuel-based polymers. Consequently, in the last decades, their importance and use declined due primarily to the rise of the fossil-based raw resources from which a large number of low cost polymers could be synthesized. These polymers had also, in general, superior properties than the polymers obtained from renewable resources. However, with the recent advancements in synthesis technologies and the discovery of new functional monomers, research in natural polymers has surged and has shown strong potential in generating better property polymers from renewable resources. Thus, this book aims to present these research successes by underlining the wide potential of the renewable polymers for large number of applications.

Chapter 1 introduces the reader to various polymers derived from renewable resources. Different polymer systems like naturally renewable methylene butyrolactones, renewable rosin acid-degradable caprolactone block copolymers, plant oils derived polymers, biosourced stereocontrolled polytriazoles, monoterpene based polymers, biosourced 2-(Methacryloyloxy)ethyl tiglate based polymers, rapeseed cake residue based polymers, limonene derived polymers, polylactides, castor oil based thermosets are described. Chapter 2 specifically demonstrates the design, synthesis, properties and applications of plant oil based polymers. The authors present reactions of triglyceride, a major component in natural oils, which can be directly polymerized under cationic conditions or functionalized to various moieties followed by either free-radical or cross-linking polymerization reactions to generate a wide spectrum of polymers having soft, rubbery, elastic, or toughening properties. Chapter 3 presents an elaborate review of acid mediated polymerization techniques for the generation of green polymers. Progress in both homogenous and heterogeneous initiator systems are described. Chapter 4 presents the production of polyhydroxy-alkanoates (PHA) from olive oil based wastewater. The advantage of olive oil mill effluents OMEs to be easily fermentable into vola-tile fatty acids and other suitable substrates for PHA production are demonstrated. Chapter 5 describes the use of atom transfer radical polymerization techniques for the production of polymers from renewable resources. The general considerations of ATRP along with a few case studies are reviewed. Chapter 6 describes the renewable polymers derived from transgenic crop plants. The authors review the great progress made in the production of bio-polymers in plants by gene technology owing to the careful selec-tion of production compartments like chloroplast or endoplasmic reticulum (ER), production organs like seeds or tubers as well as the selection and combination of genes involved in the synthesis. Chapter 7 provides an overview of a range of biomass-based poly-mers (like polycarbonates, polyesters and polyamides) prepared through polycondensation chemistry. An overview of the biopoly-mers (e. g. polyesters, polyamides and poly(ester amide)s) based on succinic acid and its derivatives is provided in Chapter 8. Recent progress in the synthesis, characterization, and physical properties studies of the poly-Schiff-base, polyester, polyamide, polyurethane, polybenzoimidazole, and polyoxadiazole type furanic polymers from 5-Hydroxymethylfurfural (HMF) derived monomers are discussed in Chapter 9. Chapter 10 reviews the recent efforts and approaches exploiting the natural materials in developing drug delivery systems.

It gives me immense pleasure to thank Scrivener Publishing and John Wiley & Sons for kind acceptance to publish the book. I dedi-cate this book to my mother for being constant source of inspira-tion. I express heartfelt thanks to my wife Preeti for her continuous help in co-editing the book as well as for her ideas to improve the manuscript.

Vikas Mittal
July 12, 2011

List of Contributors

Ananda S. Amarasekara is an Associate Professor of Chemistry at Prairie View A&M University, Texas, USA, and he received his PhD in organic chemistry from City University of NewYork. His research interests are in biomass based chemicals, polymeric materials, and fuels. Dr. Amarasekara has authored or co-authored over 70 peer reviewed research publications.

Lorenzo Bertin is Assistant Professor at the Department of Civil, Environmental, and Materials Engineering of Alma Mater Studiorum University of Bologna (Italy). His research activity is mainly focused on the valorization of organic wastes by means of anaerobic biotechnological processes, carried out under both methanogenic or acidogenic, and by the extraction or biotechnological production of added value natural molecules.

Mario Beccari is a Senior Professor at the Faculty of Mathematical, Physical and Natural Sciences of Sapienza University of Rome (Italy). He is a Member of the Scientific Council of the Department of Earth and Environment of CNR and a Member of the Scientific Council of the Institute of Italian Encyclopaedia. He is the co-Editor of the *Encyclopaedia of Hydrocarbons* (published by ENI and the Institute of Italian Encyclopaedia). His scientific activity is focused on the following areas: hydrocarbon oxidation processes, evaporation desalination processes, advanced processes of treatment/disposal of wastewaters and wastes. He is author of about 150 scientific papers, many of them published in international journals.

Inge Broer is a molecular biologist and holds the Professorship for Agrobiotechnology and Risk Assessment for Bio- and Gene Technology at the University of Rostock in Germany. Her research is focussed on the usage of transgenic plants for a sustainable agriculture involving the production of vaccines and biodegradable polymers in transgenic plants, as well as the environmental influence on transgene expression and risk assessment on the plants produced in the group.

Inna Bretz studied chemistry and chemical education at the University of Karaganda, Kazakhstan, and at the Technische Universität Berlin, where she prepared her PhD thesis on poly (lactic acid). She joined Fraunhofer UMSICHT in 2006 as a senior scientist working on polymer chemistry using bio-based raw materials.

Tina Hausmann is currently pursuing her PhD in the Agrobiotechnology and Risk Assessment for Bio- and Gene Technology department at the University of Rostock in Germany. Her research interests are the optimization of cyanophycin in commercial tobacco lines and the usage of cyanophycin as feed additive and its storage in silage.

Duy H. Hua is a University Distinguished Professor at the Department of Chemistry, Kansas State University. His research interests include design, synthesis, and application of renewable polymers, nanomaterials, design, synthesis, and bio-evaluation of anti-Alzheimer, anti-norovirus, and anti-cancer compounds. Hua has received numerous awards including the Alumni Achievement Award from Southern Illinois University at Carbondale, Commerce Bank Distinguished Graduate Faculty Award, and Higuchi-University of Kansas Endowment Research Achievement Awards.

Stephan Kabasci is a chemical engineer and prepared his PhD at the Technische Universität Dortmund. He has been business unit manager of "Renewable Resources" at Fraunhofer UMSICHT since 2004. His main research areas are biogas technology and the production of polymers and plastic materials from renewable resources.

Stewart P. Lewis is a polymer scientist, an inventor, a part-time professor, and an entrepreneur running his own research for hire business (Innovative Science Corporation). He invented systems for the aqueous polymerization of isobutene using chelating diboranes and discovered that sterically hindered pyridines react with carbocations that are paired with weakly coordinating anions. Recently he has invented a number of novel polymerization systems for the preparation of high molecular weight grades of isobutene polymers in a more sustainable manner as well as additional methodologies for aqueous carbocationic polymerization.

Mauro Majone is Associate Professor of Chemical Engineering at the Department of Chemistry, Sapienza University of Rome (Italy). Main research interests are in the field of biological and chemical/physical processes for treatment of wastes and wastewaters, remediation of polluted soils and groundwater, environmental and industrial biotechnologies. He is author of more than 100 papers on international scientific journals and books with peer review and more than 100 communications to scientific conferences and other publications.

Robert T. Mathers obtained his PhD from The University of Akron in Polymer Science. After two years of postdoctoral research at Cornell University in the Department of Chemistry and Chemical Biology, he joined Pennsylvania State University where he is an Associate Professor of Chemistry. His research interests focus on polymerization methods that integrate renewable resources, such as monoterpenes and plant oils, with catalysis. Robert recently served as coeditor for the *Handbook of Transition Metal Polymerization Catalysts* (Wiley) and *Green Polymerization Methods: Renewable Starting Materials, Catalysis, and Waste Reduction* (Wiley-VCH).

Bart A. J. Noordover received his PhD degree in Polymer Chemistry from the Eindhoven University of Technology, the Netherlands. His research interests include step-growth polymerization, molecular, thermal and mechanical characterization as well as structure-property relations of engineering plastics, elastomers and coatings. He currently holds a position of Researcher at the Laboratory of Polymer Chemistry at the Eindhoven University of Technology.

Keshar Prasain is now perusing his PhD in organic chemistry at Kansas State University under Duy H. Hua. His research focuses on synthesis and characterization of triglyceride based polymers and synthesis of laccase inhibiting compounds. He has received graduate classroom and department research awards from the Department of Chemistry, Kansas State University.

N. Rajesh is an Assistant Professor in the Department of Biochemistry at CSI Holdsworth Memorial Hospital and College, Mysore, India. He obtained his PhD in Polymer Science from the University of Mysore. He has 15 research papers to his credit in international journals along with 20 publications in proceedings of national & international conferences. His research interest centers on the fields of biopolymers for sustained drug delivery and transdermal drug delivery systems.

Valluru Ravi is a Lecturer in the Department of Pharmaceutics at JSS College of Pharmacy, Mysore, India. He is pursuing his PhD in Pharmacy from the University of Mysore. He has 12 research papers to his credit in international journals along with 15 publications in proceedings of national & international conferences. His research interest is on biopolymers for use in drug delivery systems.

Kattimuttathu I.Suresh is a Senior Scientist at the Organic Coatings & Polymers Division of the Indian Institute of Chemical Technology (CSIR), India and is involved in basic and applied research in the broad areas of Polymer Science and Technology. Specific research interests are in the area of polymer synthesis, structure–property relationship studies and polymers from renewable resources. He is recipient of CSIR-DAAD (Germany) fellowship and the Marie Curie fellowship of the European Commission.

N. Uma is an Assistant Professor in the Department of Biochemistry at BGS International Foundation for Health Science, Mysore, India. She is pursuing her PhD in Chemistry from the University of Mysore. She has 4 research papers to her credit in international journals along with 7 publications in proceedings of national & international conferences. Her research interest centers on the fields of metal complexes materials and medicinal chemistry.

Francesco Valentino is a PhD student in chemical engineering at the Department of Chemistry, Sapienza University of Rome (Italy). Main research interests are in the field of biochemical processes for production of added value materials from wastewater and simultaneous treatment of wastewaters.

Marianna Villano is a postdoctoral researcher at the Department of Chemistry of Sapienza University of Rome (Italy). She earned a PhD in Industrial Chemical Processes from the same University in 2011. Her main research interests are in the field of biological processes for polyhydroxyalkanoates production and bioenergy generation using mixed microbial cultures.

Polymers from Renewable Resources

V. Mittal

Chemical Engineering Department,
The Petroleum Institute,
Abu Dhabi, UAE

Abstract

In the modern world, the importance and use of macromolecular materials based on renewable resources declined owing to the rise of the fossil based raw resources from which a large number of low cost polymers could be synthesized. These polymers had also, in general, superior properties than the polymers obtained from renewable resources. The high price of renewable resourced-based polymers also caused limitations in their use. However, changing world scenarios have once again thrusted the macromolecular materials from renewable resources to the forefront of research and application. Reasons for such a paradigm shift are: dwindling amounts of fossil fuels, increasing awareness regarding the environment and increasing price of fossil fuels. Recent advances in the synthesis and properties of polymers from renewable resources with good properties and potentially low costs have kindled the hopes that the replacement of the conventional fossil-based polymers with these renewable polmers will be permanent and commercially economic.

Keywords: Naturally renewable methylene butyrolactones, rosin acid, plant oils, polytriazoles, monoterpene, rapeseed, limonene, lactides, nanocomposites, thermosets

1.1   Introduction

Macromolecular materials based on renewable resources have found applications in various human activities for centuries. In the modern world, their importance and use declined owing to the rise of the fossil based raw resources from which a large

Vikas Mittal (ed.) Renewable Polymers, (1–22) © Scrivener Publishing LLC

number of low cost polymers could be synthesized. These polymers had also, in general, superior properties than the polymers obtained from renewable resources. The high price of renewable resources based polymers also caused the declination in their use. However, changing world scenarios have once again thrusted the macromolecular materials from renewable resources to the forefront of research and application. Reasons for such a paradigm shift are: dwindling amounts of fossil fuels, increasing awareness regarding the environment and increasing price of fossil fuels. Apart from that, recent advances in the synthesis and properties of polymers from renewable resources with good properties and potentially low costs have kindled the hopes of the replacement of the conventional fossil based polymers with these materials. A lot of research effort has been ongoing in order to further explore the novel renewable recourses as well as to modify the currently existing polymers derived from renewable polymers. In a recent comprehensive review on the subject [1], Gandini also underlined the importance associated with these materials but also pointed towards more efforts to be done in order to realize the true commercialization of such materials. There are many polymers which are more abundant in nature than and have also received maximum research attention like cellulose, chitin, starch etc. These materials find uses either in the unmodified state or after suitable bulk or surface modifications. The hydroxyl groups in polysaccharides are commonly used to modify these materials and to add required chemical functionality. As the OH groups are involved in the modification reactions, the corresponding grafting processes include the typical condensation reactions like esterification, etherification, and formation of urethanes etc. Figure 1.1 shows the basic structures of one of the most abundant polysaccharides in nature like cellulose, chitin etc. The forms of applications of these polymers have also diversified; the polymers are used not only as matrices, but also as fibers or fillers for reinforcement of other polymers. Apart from the above mentioned most abundant polymers, many other interesting polymers with good properties have also been developed in the recent years, which make the application spectrum of these materials very wide. Examples of a few of polymer materials derived from renewable resources are discussed in the next sections.

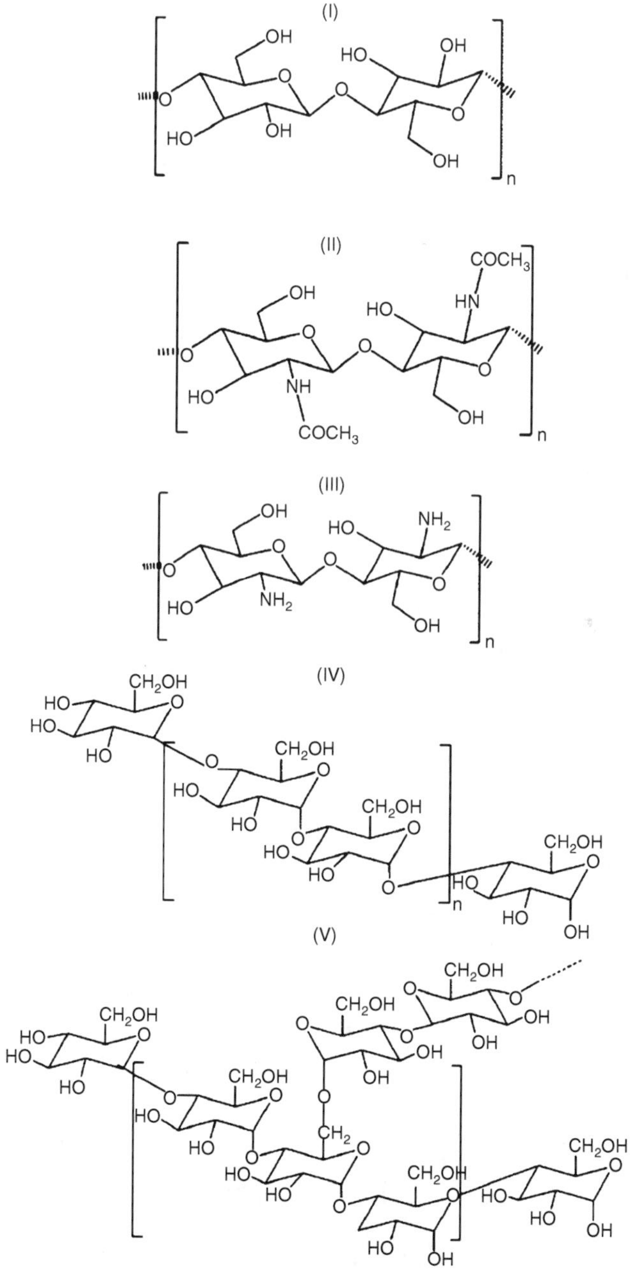

Figure 1.1 Structures of the most abundant polysaccharides in nature: (I) cellulose, (II) chitin, (III) chitosan, (IV) amylose and (V) amylopectin. Reproduced from reference 1 with permission from American Chemical Society.

1.2 Naturally Renewable Methylene Butyrolactones

Hu *et al.* [2] reported the polymerization of naturally renewable methylene butyrolactones by half-sandwich indenyl rare earth metal dialkyls. Monomers eg. α-methylene-γ-butyrolactone (MBL) and γ-methyl-α-methylene-γ-butyrolactone (MMBL) were used. Four half-sandwich dialkyl rare earth metal (REM) complexes incorporating a disilylated indenyl ligand were used for the coordination-addition polymerization of these methylene butyrolactones. Figure 1.2 shows the schematic of methylene butyrolactone monomers (M)MBL and their polymers P(M)MBL along with comparison with MMA and PMMA. Structures of half-sandwich REM dialkyl catalysts are also shown. The authors observed several differences in catalytic behavior of half-sandwich REM catalysts and well-studied sandwich REM catalysts after initial screening for the polymerization of methyl methacrylate. All four catalysts exhibited exceptional activity for polymerization of MMBL in DMF, achieving quantitative monomer conversion in <1 min with a 0.20 mol% catalyst loading.

Miyake *et al.* [3] similarly reported the living polymerization of naturally renewable butyrolactone-based vinylidene monomers

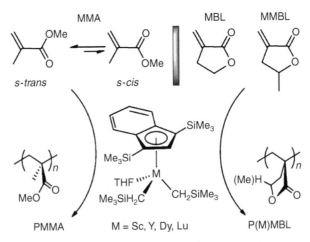

Figure 1.2 Schematic of methylene butyrolactone monomers (M)MBL and their polymers P(M)MBL along with comparison with MMA and PMMA. Structures of half-sandwich REM dialkyl catalysts are also shown. Reproduced from reference 2 with permission from American Chemical Society.

α-methylene-γ-butyrolactone (MBL) and γ-methyl-α-methylene-γ-butyrolactone (MMBL) by ambiphilic silicon propagators. Part from homopolymers, block copolymers of MBL and MMBL with MMA as well as block and statistical copolymers of MBL with MMBL with well defined characteristics could also be readily synthesized. The glass transition temperatures of the atactic homopolymers, PMBL and PMMBL, were observed to be 194 and 225°C. These values were significantly higher than the glass transition temperature of atactic PMMA (approx. 90°C for PMBL and approx 120°C for PMMBL). The polymerization of MBL in CH_2Cl_2 at ambient temperature was observed to be heterogeneous with low polymer yield and bimodal MWD. However, in the case of MMBL, a homogeneous reaction was obtained which achieved completion in 10 min even with a low catalyst loading of 0.05 mol%. Thus, polymers with controlled low to high molecular weight and narrow molecular weight distributions (1.01–1.06) were obtained. All copolymers produced also exhibited exhibit unimodal and narrow MWD's of nearly 1.03. The block copolymer PMMBL-b-PMBL displayed two glass transition peaks in the DSC signal corresponding to the PMBL and PMMBL blocks as shown in Figure 1.3a, while the statistical copolymer PMMBLco-PMBL showed only one glass transition temperature signal. As shown in Figure 1.3b, the effect of Mn on the Tg of PMMBL was also investigated. From the leveling off point of the generated curve, a critical molecular weight for PMMBL was estimated which was over 40 kg/mol.

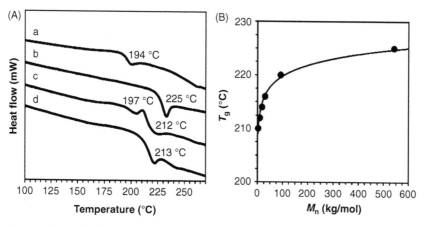

Figure 1.3 (A) DSC curves for (a) PMBL, (b) PMMBL, (c) PMMBL-b-PMBL, and (d) PMMBL-co-PMBL and (B) Plot of T_g vs M_n of atactic PMMBL. Reproduced from reference 3 with permission from American Chemical Society.

1.3   Renewable Rosin Acid-Degradable Caprolactone Block Copolymers

Wilbon *et al.* [4] reported the synthesis of block copolymers based on renewable rosin acid-degradable caprolactone by atom transfer radical polymerization and ring opening polymerization. For the two-step sequential polymerization, either poly(2-acryloyloxyethyl dehydroabietic carboxylate)-OH (PAEDA-OH) or poly(ε-caprolactone)-Br (PCL-Br) were used as macroinitiators. Two-step sequential polymerization resulted in the generation of well-defined block copolymers with low polydispersity. One-pot polymerization was also carried out with three different sequential feeds of AEDA and ε-CL monomers. The control of one-pot polymerization was observed to depend on the interactions of coexisting ATRP catalysts and ROP catalysts.

Figure 1.4 shows the DSC thermograms of the PCL-Br homopolymer and the block copolymers. PCL-Br showed a characteristic strong endothermic melting peak at approximately 55°C. The thermal behaviors of diblock copolymers depended on the length and fraction of the PCL block in the block copolymers. In the block

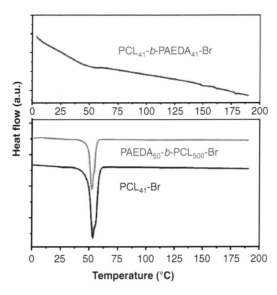

Figure 1.4 DSC thermograms of polymers PCL-Br, PAEDA-b-PCL-OH, and PCL-b-PAEDA-Br. Reproduced from reference 4 with permission from American Chemical Society.

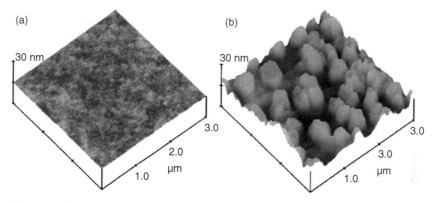

Figure 1.5 Tapping-mode AFM height images of (a) PCL_{41}-b-$PAEDA_{41}$-Br and (b) $PAEDA_{50}$-b-PCL_{500}-OH. Reproduced from reference 4 with permission from American Chemical Society.

copolymer with high fractions of PAEDA (PCL_{41}-b-$PAEDA_{41}$-Br), only the Tg of the PAEDA block at approximately 50°C was observed and suppression of the PCL crystallization resulted. For block copolymer with high fraction of the PCL block ($PAEDA_{50}$-b-PCL_{500}-OH), a strong endothermic peak at approximately 55°C corresponded to the melting of the PCL block. The AFM analysis reported in Figure 1.5 also confirmed the findings of differential scanning calorimetry. AFM images of thin films of block copolymers with a short length of the PCL block (PCL_{41}-b-$PAEDA_{41}$-Br) were observed to be very smooth, whereas the AFM height images of block copolymers with high fractions of the PCL block ($PAEDA_{50}$-b-PCL_{500}-OH) revealed the formation of small crystals.

1.4   Plant Oils as Platform Chemicals for Polymer Synthesis

Ligadas *et al.* [5] reviewed the importance of plant oils as platform chemicals for the synthesis of polyurethanes. In one such example, synthesis of vegetable oil-based polyols for polyurethane synthesis through reactions involving the ester groups was demonstrated. As shown in Figure 1.6, functionalization of oleic sunflower oil with secondary hydroxyl groups was achieved via photoperoxidation and subsequent reduction. Oleic sunflower oil was oxidized photochemically with singlet oxygen generated

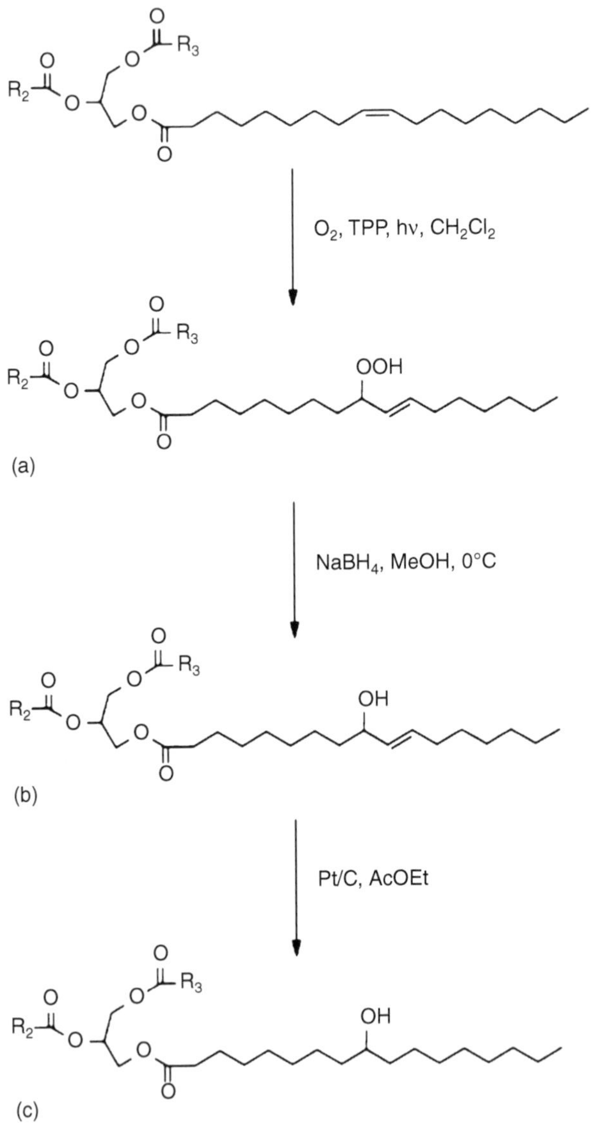

Figure 1.6 Schematic of the functionalization of oleic sunflower oil with secondary hydroxyl groups. Reproduced from reference 5 with permission from American Chemical Society.

with a high pressure sodium-vapor lamp and tetraphenylporphyrin (TPP) as sensitizer in an oxygen-saturated medium. By following this process, a mixture of isomeric allylic hydroperoxides was obtained as shown in Figure 1.6a. These were further reduced to unsaturated hydroxylated oil using sodium borohydride as shown in Figure 1.6b. The authors transformed the resulting allylic alcohol triglyceride derivative to the saturated analogues as shown in Figure 1.6c by hydrogenation at room temperature using 5% charcoal supported platinum as catalyst. The polyols were then used for the synthesis of polyurethane polymers. Similarly, synthesis of vegetable oil-based polyols through polymerization of C-C double bonds & subsequent hydrolysis of ester groups and synthesis of fatty acid-based polyols etc. were reported.

1.5   Biosourced Stereocontrolled Polytriazoles

Besset *et al.* [6] reported the synthesis of biosourced stereocontrolled polytriazoles from click chemistry step growth polymerization of diazide and dialkyne dianhydrohexitols. Figure 1.7 demonstrates the schematic of synthesis of these polymers. The obtained polymers had a good thermal degradation resistance as temperatures to 10% degradation were observed to be in the range of 325–347°C.

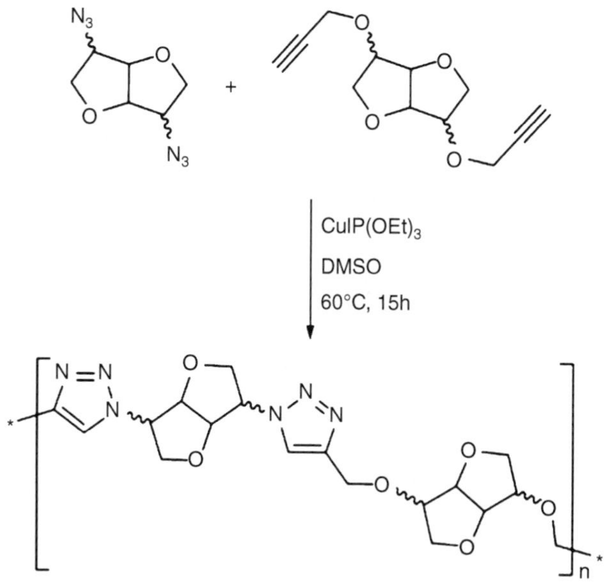

Figure 1.7 Schematic of synthesis of polytriazoles by step growth polymerization of diazides and dialkynes. Reproduced from reference 6 with permission from American Chemical Society.

The amounts of residual ashes were 25–30 wt% in the thermogravimetric analysis. The polymers displayed high glass transition temperatures in the range of 125–166°C. The molecular weights of the generated polymers were also high and lied in the range of 8–17 kg/mol. The polymers had good solubility in DMSO and DMF at room temperature. The authors reported that the monomer stereochemistry proved to be a crucial parameter aiming at generating polymers with high glass transition temperatures. Several polytriazoles were observed to exhibit high values of PDI (3.8–5.7) indicating broadening of molecular weight distribution. This may have originated from the additional presence of high molar mass species due to the formation of aggregates in hot DMSO and of low molar masses cyclic species.

1.6   Polymers from Naturally Occurring Monoterpene

Kobayashi et al. [7] reported the controlled polymerization of a cyclic diene prepared from the ring-closing metathesis of a naturally occurring monoterpene as shown in Figure 1.8. The monoterpene myrcene (1) was obtained from plants or from the pyrolysis of pinene and the cyclic diene 3-methylenecyclopentene (2) could be prepared from 1 by ring-closing metathesis reaction. Polymerization of cyclic diene was studied using radical, anionic, and cationic polymerization. Radical polymerization of cyclic diene using radical initiator AIBN exhibited low conversion in benzene solution, but bulk polymerization led to a polymeric product in 58% yield after 20 h at 80°C. Anionic polymerization was carried out using sec-butyllithium (s-BuLi) as initiator and polymerization reactions were observed to be rapid at low temperature. In the case of

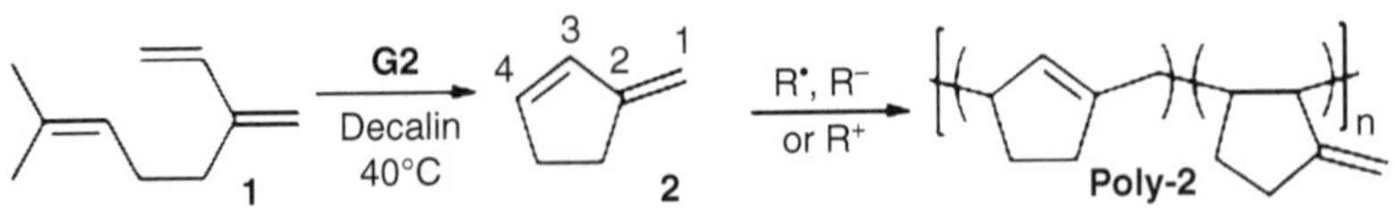

Figure 1.8 Synthesis scheme for the generation of polymer based on naturally occurring monoterpene. Reproduced from reference 7 with permission from American Chemical Society.

cationic polymerization, the polymer obtained using i-BuOCH(Cl)Me/SnCl$_4$/Et$_2$O had a broad molecular weight distribution, on the other hand, the i-BuOCH(Cl)Me/ZnCl$_2$/Et$_2$O system generated regiopure polymer with controlled molecular weight and narrow molecular weight distribution. The number average molecular weight was observed to increase wit conversion and was in good agreement with the calculated values. Apart from that, a control on the molecular weight could be achieved by changing the feed molar ratios of monomer to initiator.

1.7   Polymerization of Biosourced 2-(Methacryloyloxy)ethyl Tiglate

Kassi *et al.* [8] reported the group transfer polymerization (GTP) of tiglic acid ester after introducing it in a methacrylate monomer, 2-(methacryloyloxy)ethyl tiglate (MAET). The monomer underwent smooth polymerization to yield linear polymers of narrow molecular weight distributions. The MAET monomer was also block copolymerized with a hydrophilic methacrylate, a hydrophobic methacrylate, and a dimethacrylate to obtain, respectively, amphiphilic, double-hydrophobic, and star polymers of that tiglate ester. The authors first attempted direct polymerization of the simplest ester of tiglic acid, methyl tiglate (MT) using various controlled polymerization methods: group transfer polymerization (GTP), atom transfer radical polymerization (ATRP) and reversible addition-fragmentation chain transfer (RAFT) polymerization, but the methyl tiglate monomer could not be polymerized, thus, requiring its insertion in a methacrylate monomer.

Figure 1.9a demonstrates the schematic of the synthetic strategy followed for the preparation of MAET and its homo- and copolymerization with conventional methacrylates. Homopolymers of MAET covered a range of molecular weight values and relatively narrow distributions, corresponding to PDIs lower than 1.5 in all cases. Various copolymers were synthesized using GTP to copolymerize MAET with methyl methacrylate (MMA), 2-(dimethylamino)ethyl methacrylate (DMAEMA), or ethylene glycol dimethacrylate (EGDMA), the GPC traces of which are shown in Figure 1.9b.

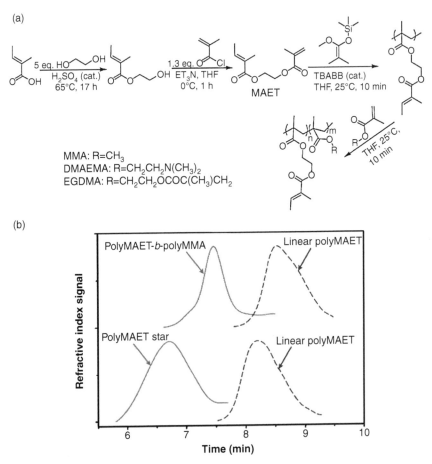

Figure 1.9 (a) Schematic of synthesis and group transfer polymerization of the tiglic acid-bearing methacrylate monomer 2-(methacryloyloxy)ethyl tiglate (MAET) and (b) GPC traces of the star homopolymer of MAET and the MAET-MMA block copolymer, as well as those of their linear poly-MAET precursors. Reproduced from reference 8 with permission from American Chemical Society.

1.8   Oxypropylation of Rapeseed Cake Residue

Serrano *et al.* [9] studied the oxypropylation of rapeseed cake residue generated in the biodiesel production process. The authors carried out the reaction by suspending the rapeseed cake residue in propylene oxide in the presence of a basic catalyst and heating the resulting mixture at 160°C in a nitrogen atmosphere which led to the synthesis of the polyol with good characteristics. Almost total conversion of the solid substrate into polol was achieved. Figure 1.10 shows the

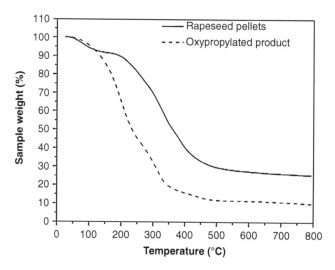

Figure 1.10 TGA thermograms of rapeseed pellets and oxypropylated product. Reproduced from reference 9 with permission from American Chemical Society.

TGA thermograms of the rapeseed pellets and oxypropylated product. Significant differences in the thermal behavior before after the oxypropylation reaction were observed. The hydrophilic character of the sample decreased after oxypropylation as the extent of the first mass loss, from room temperature to about 150°C, associated with the evaporation of water present in the samples, decreased. The introduction of propylene oxide (PO) also led to a decrease in the degradation temperature of the oxypropylated product. The differential scanning calorimetry (DSC) analysis revealed that the oxypropylated product was a branched polymer bearing numerous OH groups per macromolecule with a glass transition temperature of 33°C.

1.9 Copolymerization of Naturally Occurring Limonene

Satoh *et al.* [10] reported the living radical chain copolymerization of naturally occurring limonene with maleimide. Figure 1.11 shows the schematic of the reaction between *d*-limonene (M1) and phenylmaleimide (M2). $M_1M_2\bullet$ radical favored M_2 while $M_2M_2\bullet$ radical exclusively reacted with M_1 thus leading to 1:2 radical copolymerization.

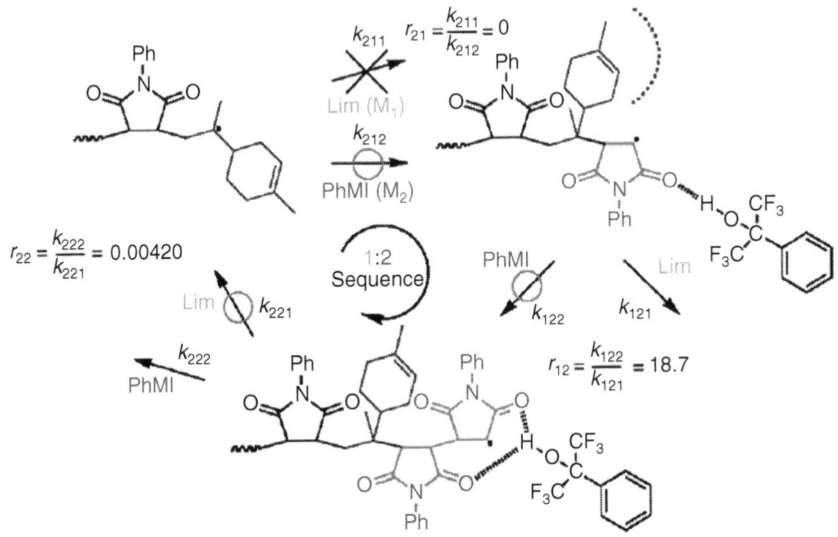

Figure 1.11 Schematics of AAB-sequence radical copolymerization of d-limonene (M1) and phenylmaleimide (M2). Reproduced from reference 10 with permission from American Chemical society.

The copolymerization reactions were examined in DMF, cumyl alcohol and fluorinated cumyl alcohol using 2,2'-azobisisobutyro-nitrile (AIBN) as a radical initiator at 60°C. The glass transition temperatures obtained for the copolymers were significantly higher (220–250°C) owing to the higher incorporation of maleimides as well as the rigid alicyclic structure of the terpenes. Apart from that, controlled/living radical polymerization was carried out with a reversible addition-fragmentation chain transfer (RAFT) agent which led to the synthesis of end-to-end sequence-regulated copolymers with controlled molecular weights.

1.10 Polymerization of Lactides

Darensbourg et al. [11] reported the ring-opening polymerization (ROP) of lactides catalyzed by natural amino-acid based zinc catalysts. All zinc complexes used in the study were observed to be very effective catalysts for the ring-opening polymerization of lactides at ambient temperature, producing polymers with controlled and narrow molecular weight distributions. The polymerization reaction using different catalysts was found to be first-order as shown in Figure 1.12.

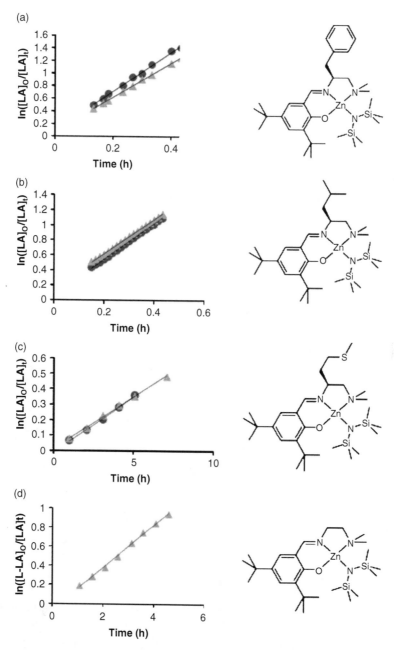

Figure 1.12 $\ln([LA]_0/[LA]_t)$ vs time plots for the ring opening polymerization of D-lactide (blue solid circles) and L-lactide (red solid triangles) catalyzed by various zinc complexes at ambient temperature. Reproduced from reference 11 with permission from American Chemical Society.

The molecular weights and polydispersity indices of the polymers after purification were determined by gel permeation chromatography (dual RI and light scattering detectors) in THF solvent using polystyrene macromolecules as a standard. The polymerization processes was observed to be similar to a living system owing to a linear relationship between molecular weight, Mn and % conversion and low polydispersity index in the range of 1.05–1.07. The differential scanning calorimetry (DSC) analysis led to the observation that the thermal properties of the generated polylactide polymers were dependant on their tacticity. Isotactically pure polylactide generated from the ring opening polymerization of L-lactide was observed to be highly crystalline with a glass transition temperature of 60°C. The polymer showed peak crystallization and melting temperatures of 90 and 178°C respectively. On the other hand, the heterotactically enriched polymers generated from the ring opening polymerization of rac-lactide exhibited glass transition temperatures which increased with increasing P_r values. The authors also reported that the reactivities of the various catalysts were greatly affected by substituents on the Schiff base ligands, with sterically bulky substituents being rate enhancing.

Jiang *et al.* [12] reported the synthesis of PEO-grafted polylactides. Novel glycolides with pendent oligo(ethylene oxide) monomethyl ether substituents were subjected to ring-opening polymerization using 4-*tert*-butylbenzyl alcohol as the initiator and Sn(2-ethylhexanote)$_2$ as the catalyst. It resulted in the synthesis of homogeneous oligo(ethylene oxide)-grafted polylactides with high molecular weights and low polydispersities. Figure 1.13 details the synthesis process. Polymers with different oligo(ethylene oxide) chains exhibited different behaviors, e.g. polymers with short oligo(ethylene oxide) chains (1 or 2 ethylene oxide repeat units) were more hydrophilic than polylactide but insoluble in water. On the other hand, polymers having 3 or 4 ethylene oxide repeat units were observed to be water-soluble.

The thermoresponsive behavior of the polymer solutions was also monitored by variable temperature dynamic light scattering measurements (DLS). When the solution of polymer poly(3,6-bis(7,10,13,16-tetraoxaheptadecyl)-1,4-dioxane-2,5-dione) was heated from 10 to 21°C, the average hydrodynamic radius of the polymer particles did not show any change. Further heating led to a sudden increase in the average hydrodynamic radius owing to polymer agglomeration. Similarly, for polymer

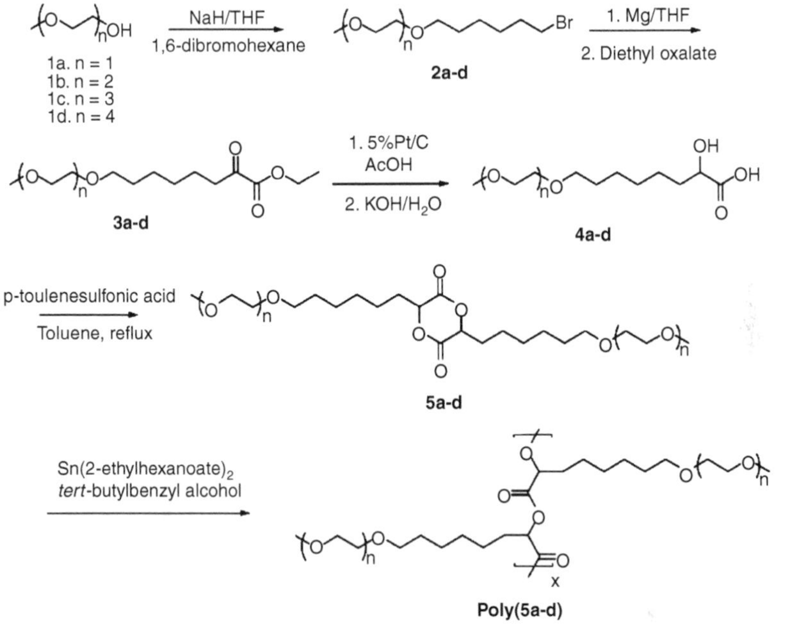

Figure 1.13 Synthesis of PEO-grafted polyglycolides. Reproduced from reference 12 with permission from American Chemical society.

poly(3,6-bis(7,10,13,16,19-pentaoxaeicosyl)-1,4-dioxane-2,5-dione) poly(5d) (3 mg/mL), between 25 and 38°C, the average hydrodynamic radius of the particles was constant, but heating the solution to 39°C induced an increase in hydrodynamic radius.

Pitet *et al.* [13] combined ring-opening metathesis polymerization and cyclic ester ring-opening polymerization to form ABA triblock copolymers from 1,5-cyclooctadiene (COD) and D,L-lactide. The authors prepared hydroxyl-functionalized telechelic polyCOD by chain transfer during ring-opening metathesis polymerization of COD using the acyclic chain transfer agent *cis*-1,4-diacetoxy-2-butene. The hydroxyl-functionalized telechelic polyCOD was used as macroinitiator for the polymerization of lactide to form a series of triblock copolymers poly(D,L-lactide)-poly(cyclooctadiene)-poly(D,L-lactide) (LCL) as shown in Figure 1.14a. DSC analysis was carried out by first heating the samples to 160°C, annealing for 5 min at 160°C, cooling to –100°C at 10°C min⁻¹ and reheating to 120°C at 10°C min⁻¹. Glass transition temperatures around 47°C for the PLA blocks and –95°C for the PCOD blocks were observed. The authors observed a decrease in both crystallinity and peak

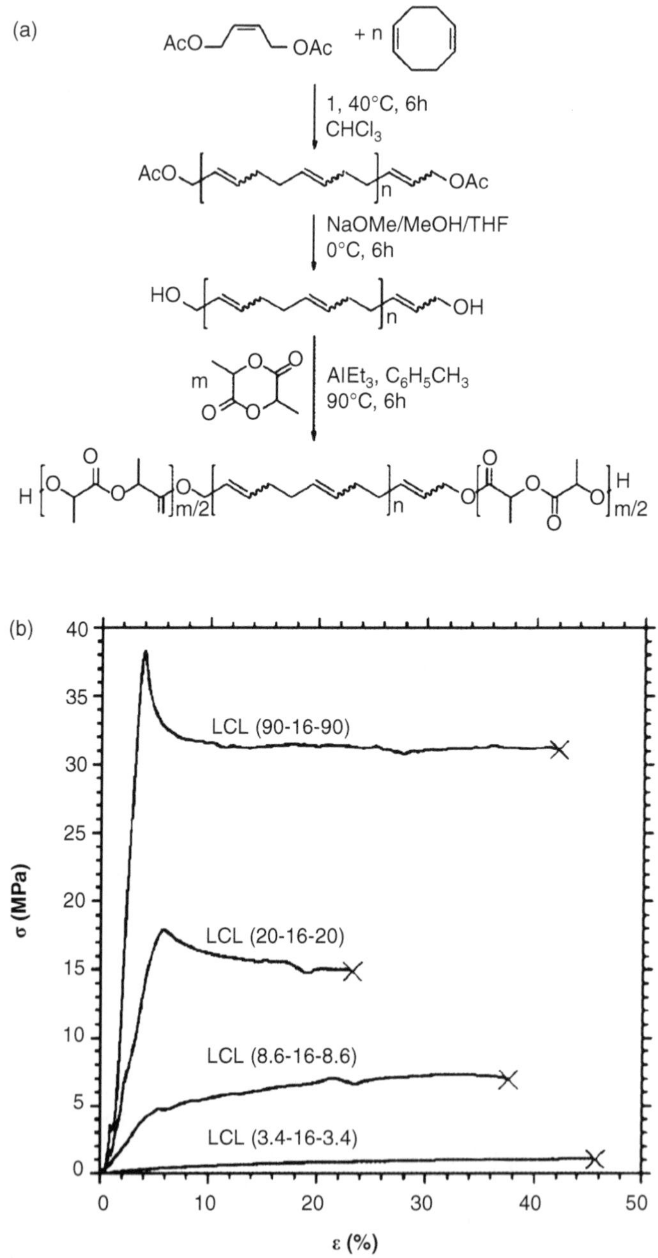

(a)

(b)

Figure 1.14 (a) Synthesis of triblock copolymer LCL and (b) tensile measurements for selected LCL triblock copolymers. The fracture point is represented by the symbol ×. Reproduced from reference 13 with permission from American Chemical Society.

melting point with increasing PLA content in the copolymers. Microphase separation was evidenced by well-defined peaks in the one-dimensional SAXS profiles. The mechanical performance of the copolymers is demonstrated in Figure 1.14b. The numbers in the parenthesis indicate the molecular weight of the blocks. The modulus, tensile strength, and yield strength of the copolymers was observed to increase with increasing molecular weight of the LCL samples. The yield strain and strain at break did not exhibit noticeable dependence on composition or molecular weight. The authors observed that the samples with the highest PLA content exhibited the largest modulus, tensile strength, and yield strength values with ultimate elongations which were an order of magnitude greater than typical values for PLA homopolymer.

1.11 Nanocomposites Using Renewable Polymers

Pranger *et al.* [14] used an in situ polymerization approach to produce polyfurfuryl alcohol (PFA) nanocomposites without the use of solvents or surfactants using either cellulose whiskers or montmorillonite clay as fillers. Figure 1.15 shows the schematic of cellulose whiskers based nanocomposites with polyfurfuryl alcohol. Figure 1.15a showed the needle-like morphology of cellulose whiskers which were produced from acid hydrolysis of microcrystalline cellulose (MCC). It was further observed that the average diameter of the whiskers was 10 nm, and the aspect ratio is in the range of 50–100. Both cellulose whiskers and montmorillonite clay first catalyzed the in-situ polymerization reaction, thereby eliminating the use of strong mineral acid catalysts, and subsequently enhancing the thermal stability of the generated nanocomposites. In the nanocomposites based on cellulose whiskers, onset of degradation (temperature at 5% weight loss) was observed to be 323°C, which was 20–30°C higher compared to the nanocomposites reinforced with montmorillonite.

1.12 Castor Oil Based Thermosets

Xia *et al.* [15] reported thermoset polymers based on castor oil which had varied crosslink densities and were prepared by ring-opening metathesis polymerization (ROMP). Two monomers based

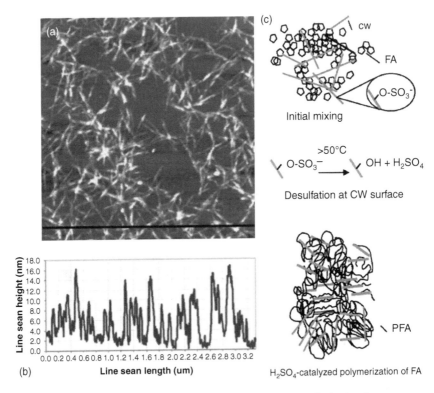

Figure 1.15 AFM image of cellulose whiskers from acid hydrolysis of microcrystalline cellulose, (b) line scan, the position of which is indicated by the bar in the AFM image and (c) schematic of the in-situ polymerization of cellulose whiskers—poly(FA) nanocomposites. Reproduced from reference 14 with permission from American Chemical Society.

on castor oil namely norbornenyl-functionalized castor oil (NCO), which had approx. 0.8 norbornene rings per fatty acid chain and norbornenyl-functionalized castor oil alcohol (NCA) with approx. 1.8 norbornene rings per fatty acid chain were prepared by the authors. Different ratios of NCO/NCA were subjected to ring-opening metathesis polymerization (ROMP) using the Grubbs catalyst, which resulted in the generation of rubbery to rigid plastics. The crosslink densities varied from 318 to 6028 mol/m³. Figure 1.16 shows the storage modulus and tan δ curves for the polymers as a function of temperature for different NCO/NCA ratios. Presence of only one peak in the tan δ versus T curves for all of the copolymers indicated the generation of homogeneous

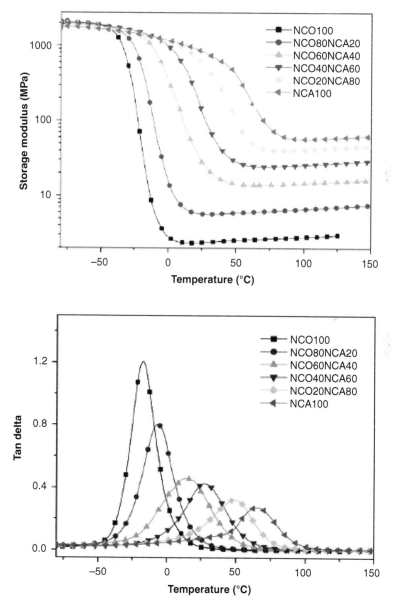

Figure 1.16 Dynamic mechanical properties of polymers. Reproduced from reference 15 with permission from Elsevier.

copolymers. The thermal properties, e.g. the glass transition temperature and room temperature storage moduli improved with an increase in the NCA content.

References

1. A. Gandini, *Macromolecules*, Vol. 41, p. 9491, 2008.
2. Y. Hu, X. Xu, Y. Zhang, Y. Chen, and E.Y.-X. Chen, *Macromolecules*, Vol. 43, p. 9328, 2010.
3. G.M. Miyake, Y. Zhang, and E.Y.-X. Chen, *Macromolecules*, Vol. 43, p. 4902, 2010.
4. P.A. Wilbon, Y. Zheng, K. Yao, and C. Tang, *Macromolecules*, Vol. 43, p. 8747, 2010.
5. G. Lligadas, J.C. Ronda, M. Galia, and V. Cadiz, *Biomacromolecules*, Vol. 11, p. 2825, 2010.
6. C. Besset, J.-P. Pascault, E. Fleury, E. Drockenmuller, and J. Bernard, *Biomacromolecules*, Vol. 11, p. 2797, 2010.
7. S. Kobayashi, C. Lu, T.R. Hoye, and M.A. Hillmyer, *Journal of the American Chemical Society*, Vol. 131, p. 7960, 2009.
8. E. Kassi, and C.S. Patrickios, *Macromolecules*, Vol. 43, p. 1411, 2010.
9. L. Serrano, M.G. Alriols, R. Briones, I. Mondragon, and J. Labidi, *Ind. Eng. Chem. Res.*, Vol. 49, p. 1526, 2010.
10. K. Satoh, M. Matsuda, K. Nagai, and M. Kamigaito, *Journal of the American Chemical Society*, Vol. 132, p. 10003, 2010.
11. D.J. Darensbourg, and O. Karroonnirun, *Inorganic Chemistry*, Vol. 49, p. 2360, 2010.
12. X. Jiang, M.R. Smith III, and G. L. Baker, *Macromolecules*, Vol. 41, p. 318, 2008.
13. L.M. Pitet, and M.A. Hillmyer, *Macromolecules*, Vol. 42, p. 3674, 2009.
14. L. Pranger, and R. Tannenbaum, *Macromolecules*, Vol. 41, p. 8682, 2008.
15. Y. Xia, and R.C. Larock, *Polymer*, Vol. 51, p. 2508, 2010.

Design, Synthesis, Property, and Application of Plant Oil Polymers

Keshar Prasain and Duy H. Hua

Department of Chemistry, 213 CBC Building, Kansas State University, Manhattan, KS, U.S.A.

Abstract

Most petroleum polymers are not biodegradable and efforts have been made to produce biodegradable and possibly biocompatible plant oil polymers. Triglyceride, a major component in natural oils, can be directly polymerized under cationic conditions or functionalized to various moieties followed by either free-radical or cross-linking polymerization reactions to generate a wide spectrum of polymers having soft, rubbery, elastic, or toughening properties. The materials are used in adhesives, plasticizers, gum-based materials, toughening of rubbers, drug carriers, flame retardants, and others. An understanding of the degree of cross linking, entanglement, and polymer network, would greatly enhance the design of polymers with desired physical properties.

Keywords: Triglyceride, free-radical, cross-linking polymerization, adhesives, flame retardants, polymer network

Abbreviations: AESO: acrylated soybean oil, AFAME: acrylated oleic methyl ester, AHMPA: 3-(acryloxy)-2-hydroxypropyl methacrylate, AIBN: 2,2'-azobis (2-methylpropionitrile), BA: bisphenol A, COGLYMA: maleic half ester of glycerolyis product of castor oil, DMA: dynamic mechanical analysis, DMF: dimethyl formamide, DMSO: dimethyl sulfoxide, DSC: differential scanning calorimetry, HEMA: 2-hydroxyethyl methacrylate, MA: maleic anhydride, MCPBA: *m*-chloroperoxybenzoic acid, MDI: 4,4'-methylen ebis(phenylisocyanate), MEKP: methyl ethyl ketone peroxide, MMA: -methyl methacrylate, MMPP: maleic anhydride grafted polypropylene, MVI: 1-methyl vinyl isocyanate, NBS: *N*-bromosuccinimide,

Vikas Mittal (ed.) Renewable Polymers, (23–68) © Scrivener Publishing LLC

NPG: neopentyl glycol, PDI: 1,4-phenylenediisocyanate, PSA: pressure-sensitive adhesives, TDI: 2,4-diisocyanatotoluene, TGA: thermogravimetric analysis, THF: tetrahydrofuran

2.1   Introduction

Polymers and polymeric composites are widely used in various industrial fields. They are lightweight and have demonstrated low assembly costs, excellent mechanical properties, high corrosion resistance, and dimensional stability. Generally, they are made of synthetic chemicals derived from petroleum. As environmentally friendly materials and applications increase, alternative sources of polymers are needed. One alternative source is the utilization of affordable composites from renewable sources, such as natural triglyceride oils, as the main component for the construction of polymers and composites [1]. Triglyceride polymers derived from oils of soybean, sunflower, cotton, and linseed, are considered to be an important class of renewable bio-based materials.

Triglycerides are major components in natural oils derived from plant and animals. They are triesters of glycerol with various fatty acids (Figure 2.1). These fatty acids differ in the carbon chain length ranging from 14 to 22 carbons and have zero to three carbon-carbon double bonds per fatty acid. Some fatty acids have additional functional groups like epoxide in vernonia oil, hydroxyl in castor and lesquerella oils, and ketone in licania oil. The type of fatty acid and the number of C=C bond determine the physical and chemical properties of the triglyceride [2].

Hydrolysis of vegetable oils provides about 15 different fatty acids. Among them the predominant five fatty acids are two saturated, palmitic and stearic, and three unsaturated, oleic, linoleic, and linolenic acid. The structure of a representative triglyceride is depicted in Figure 2.1. The fatty acid compositions of triglycerides

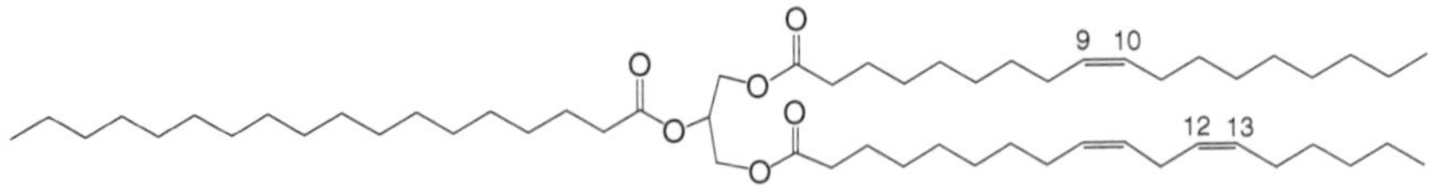

Figure 2.1 Structure of a representative triglyceride, 1-oleyl-2-stearyl-3-linoleylglycerol.

vary not only from oil to oil but also within the same oil. Fatty acid compositions in particular oil also vary on the location where the plant is grown. Fatty acids derived from nature have even number of carbon atoms, due to their biosynthesis starting from acetyl coenzyme A, a two-carbon carrier. Fatty acid distributions of some common oils are depicted in Table 2.1 [3].

Plants oils are used mainly for food purpose and as minor components in coatings, plasticizers, lubricants, agrochemicals, and inks as toughening agents or improving the fracture resistance of thermoset polymers. Polymers derived from plant oils would have low toxicity and are biodegradable and biocompatible. Presently, there is a growing interest towards the use of triglycerides as a main constituent of polymers for various applications.

2.2   Triglyceride Polymers

Plant oils have a number of reactive sites such as the double bonds, allylic carbons, ester functions, α-carbons of ester functions, and carbons attached to the oxygen of the ester functions, which may be utilized in various chemical transformations to achieve a number of synthetic intermediates for polymer syntheses. Polymers from triglycerides can be synthesized but not limited to the following strategies:

- Conversion of triglycerides to monoglycerides and diglycerides followed by copolymerization.
- Conversion of triglycerides to fatty acids followed by copolymerization.
- Functionalization of triglycerides followed by polymerization.

2.2.1   Formation and Copolymerization of Monoglycerides and Diglycerides

Alkyd resins are polyester polymers prepared from the esterification of polyalcohols, polyacids, and fatty acids [4]. They are one of the earliest polymers made from triglycerides and are used in surface coatings. Some of the polymers are described here. The materials include polyesters from monoglycerides obtained from the conversion of plant oil triglycerides via a glycerolysis followed by

Table 2.1 The percent distributions of fatty acids in various plants oils [3].

Oil	% of fatty acid (A:B, where A is the number of carbon atoms and B is the number of double bonds)							
	Palmitic 16:0	Stearic 18:0	Oleic 18:1	Linoleic 18:2	Linolenic 18:3	Gadoleic 20:1	Erucic 22:1
Canola	4.1	1.8	60.9	21	8.8	1	0.7
Corn	10.9	2	25.4	59.6	1.2	0	0
Linseed	5.5	3.5	19.1	15.3	56.6	0	0
Palm	44.4	4.1	39.3	10	0.4	0	0
Rapeseed	3	1	13.2	13.2	9	9	49.2
Soybean	11	4	23.4	53.2	7.8	0	0
High oleic*	6.4	3.1	82.6	2.3	3.7	0.4	0.1

*denoted as genetically modified soybean oil.

coupling with various polyacids, cyclic esters, and/or cyclic anhydrides. Alkyd resins have acquired wide usages because of their bioavailability and ease of preparation.

The glycerolysis reaction is a convenient reaction to convert triglycerides 2 to monoglycerides 4 (Scheme 2.1). This involves reacting triglycerides with excess of glycerol (3) and a catalytic amount of white soap at 230°C to give predominantly monoglycerides 4 as a mixture of regioisomers [4]. Likely, diglycerides 5 are also present in the soybean oil monoglycerides 4.

Monoglyceride 4 was then used to prepare monoglyceride maleate half esters by the treatment with ~2 equivalents of maleic anhydride and catalytic amounts of triphenylantimony and hydroquinone (as a free-radical scavenger) at 100°C to 800°C to give monoglyceride bis-maleate half esters 6 (Scheme 2.2) [6]. Copolymerization of compound 6 and styrene with *tert*-butyl peroxybenzoate as a free-radical initiator at 120–150°C gave rigid polymers 7 having load bearing applications. The resin has a glass-transition temperature value of 133°C (obtained from dynamic mechanical analysis; DMA) and a storage modulus value of 0.94 GPa at 35°C, and a tensile strength of 29.36 MPa and a tensile modulus of 0.84 GPa. The additive effects of rigid diols such as neopentyl glycol (NPG) or bisphenol A (BA) were studied by mixing 4 and NPG or BA separately followed by heating with maleic anhydride (MA) under similar reaction conditions as aforementioned. The resulting mixture was similarly copolymerized with styrene under free-radical

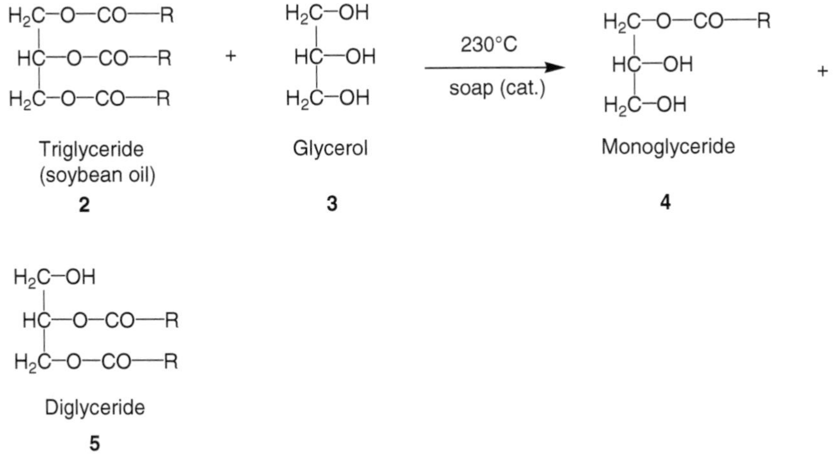

Scheme 2.1 A glycerolytic reaction of soybean triglycerides.

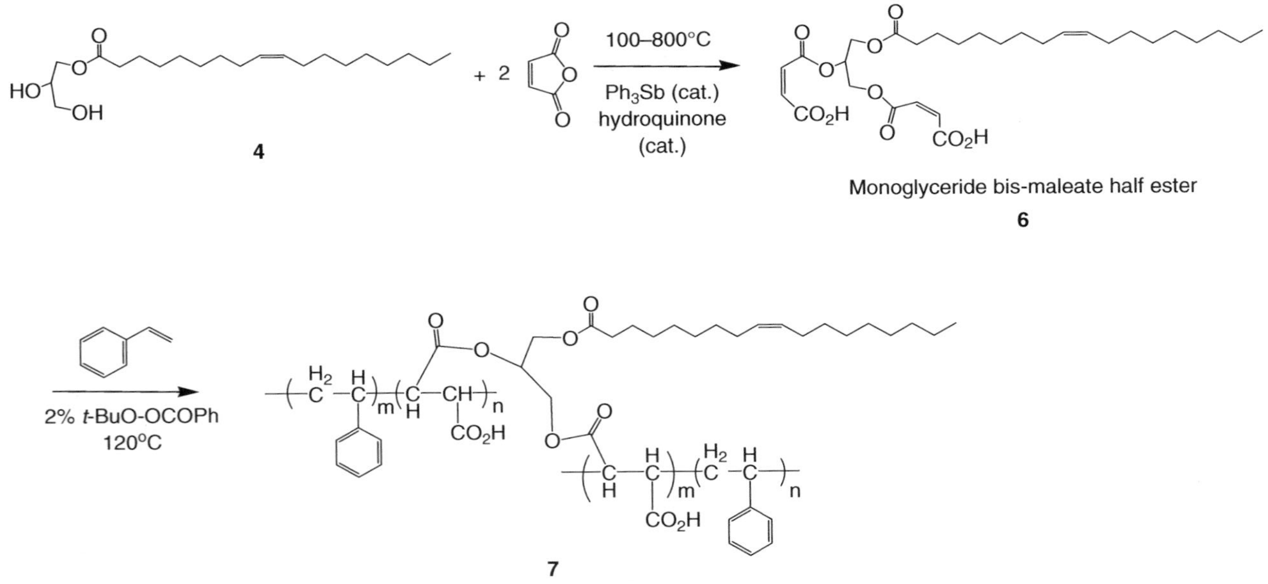

Scheme 2.2 Formation of monoglyceride bis-maleate half esters 6 and copolymerization with styrene.

conditions. The introduction of NPG/MA to 4/MA increased the glass transition temperature (Tg) and the modulus but decreased the tensile strength of the polymer formed. The introduction of BA/MA had no significant change in Tg, but a slight increase in the modulus of the polymer was found [5]. The fatty acid fragments contaminated in the above 4/MA monomers are inactive and do not participate in the polymerization, but behave as a plasticizer reducing the modulus and strength of the polymers. To decrease the plasticizing effect, castor oil, contained 87% of ricinoleic acid (possessing a C12-hydroxyl and C9,C10 double bond functions), was used in place of soybean oil, and the properties of the polymers were similarly studied. The co-polymers prepared from castor oil were found to have improved modulus, tensile strength, and glass transition temperature [6].

A mixture of partial glycerides consisted of diglycerides 9 and monoglycerides 10 was obtained from the treatment of sunflower oil and linseed oil separately with glycerol (~11% by weight) and a catalytic amount of calcium oxide at 230°C (Scheme 2.3) [7]. The acid (from fatty acids) and hydroxyl (from diglycerides and mono-glycerides) values were determined [8]. The resulting partial glyc-erides (9 & 10) were esterified with methyl methacrylate (MMA) (molar ratio of hydroxyl groups of 9 & 10 and MMA was 1:2) and a catalytic amount of calcium oxide along with hydroquinone (6% by weight; to inhibit the homopolymerization of MMA) at 200°C. Copolymerization of 11 and 12 with styrene (1:1 by weight) and benzoyl peroxide (0.5% by weight) at 100°C gave oily resins. Their flexibility, adhesion, water resistance, alkali resistance, acid resis-tance [9], and other film properties such as viscosities and hardness [7] were measured. The resins exhibited good film properties with good water, alkali, and acid resistance, and appear to be suitable for coating application.

The use of pentaerythriol (16) instead of glycerol in the alco-holysis of triglycerides was found to produce polymers with higher cross-linked densities [6]. The alcoholysis products 17 and 18 of soybean oil and castor oil upon treatment with maleic anhy-dride produced maleate half esters 19 and 20, respectively, which upon copolymerization with styrene under free-radical conditions afforded polymers with higher cross-linked densities than the poly-mers obtained from that of COGLYMA, maleate half esters derived from the glycerolysis reaction of castor oil (Scheme 2.4). This is due to the fact that alcoholysis reaction of pentaerythriol (16) with

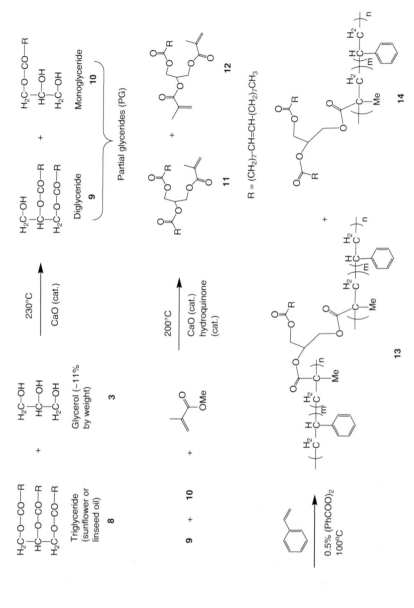

Scheme 2.3 Formation of partial glycerides/methacryate **11** and **12** and copolymerization with styrene.

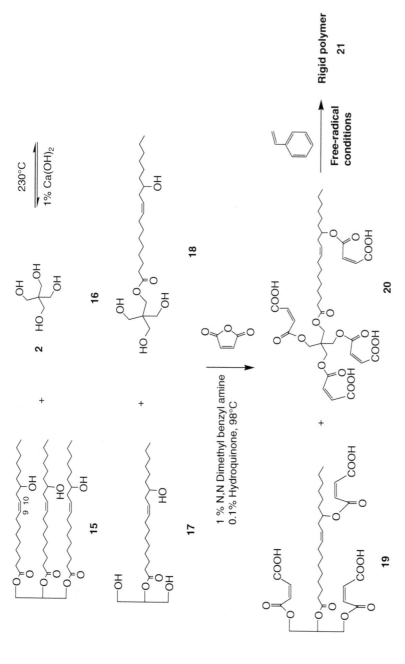

Scheme 2.4 Copolymerization of maleate half ester **20** and styrene.

triglycerides produces monoglycerides **18** with higher number of hydroxyl groups providing more reactive sites for malination, thus giving greater number of maleate half ester functions per molecule (compound **20**) of triglycerides.

It was found that values of the flexural strength, flexural modulus, storage modulus, and Tg of castor oil based rigid polymers were higher than that of soybean oil based polymers showing castor oil as a better alternative than soybean oil [10]. For castor oil based polymers, the changes in the mechanical properties like storage modulus, glass transition temperature, flexural modulus, and surface hardness by the use of pentaerythriol instead of glycerol in alcoholysis reaction were studied, and results are summarized in Table 2.2.

Results from Table 2.2 show that polymers derived from **19** and **20** have higher storage modulus, glass transition temperature, flexural modulus, flexural strength, and surface hardness than that of COGLYMA using a similar concentration of styrene.

Polymers derived from carbamoylation of partial glycerides **9** and **10** with 1-methylvinyl isocyanate were similarly investigated. Koprululu *et al.* [11] synthesized triglyceride oil-based 1-methylvinyl urethanes **22** and **23** containing vinylic moieties by starting from

Table 2.2 Mechanical properties of castor-oil-based rigid polymers prepared using 33% weight ratio of styrene.

Resin Type	Storage Modulus at 30°C (GPa)	Tg (°C)	Flexural Modulus (GPa)	Flexural Strength (MPa)	Surface Hardness (D)
Polymer **21**	2.94 (± 0.02)	149	2.17 (± 0.070)	104.6 (± 3.1)	89.3 (± 0.40)
Copolymer COGLYMA-styrene (Maleate half ester derived from glycerolysis)	2.40 (± 0.01)	124	1.76 (± 0.076)	78.89 (± 2.74)	86.1 (± 0.45)

linseed oil. Glyerolysis of linseed oil with glycerol (8.5% of the oil) at 232°C followed by the addition of 0.1% calcium hydroxide under nitrogen atmosphere for 45 minutes gave a mixture of mono- and diglycerides **9** and **10** (see Scheme 2.3). The mixture was treated with an equivalent amount of 1-methylvinyl isocyanate (MVI) in the presence of a catalytic amount (0.14 wt %) of dibutyltin dilaurate as a base to afford urethanes **22** and **23** (Scheme 2.5). The 1-methylvinyl functions of **22** and **23** were copolymerized with styrene in a weight ratio of 1:0.5, and 0.1 wt % of benzoyl peroxide (as a free-radical initiator) in toluene at 65°C to give a yellow-colored, transparent, and flexible film. The polymer film obtained has good acid resistance properties, increased thermal resistance, and low hydrophilicity. The properties likely induced from the urethane functions.

2.2.2 Copolymerization of Fatty Acids

Fatty acids are good candidates for the preparation of biodegradable polymers, as they are natural body components and hydrophobic, thus may find applications in coatings, drug carriers, and others. Teomim et al. [12] reported the synthesis of ricinoleic acid-based biopolymers which have properties desirable to be used as drug carriers. Hence, treatment of ricinoleic acid (**26**) and maleic or succinic anhydride in a molar ratio of 1:2 at 90°C for 12 hours in toluene gave ricinoleic acid maleate **28** or ricinoleic acid succinate **27**, respectively. A catalytic hydrogenation of the double bond moieties of **28** with palladium and hydrogen (80 atm.) in ethanol furnished 12-hydroxystearic acid succinate **29** (Scheme 2.6).

These diacids were used to couple with sebacic acid (decanedioic acid) to form polyanhydrides as prospective drug carriers. The synthesis involves the formations of mix anhydrides of **28**, **29**, and **27**, separately, with excess of acetic anhydride. The resulting mix anhydrides were treated with sebacic acid, separately, to form sebacic copolymers **28**-sabacic, **29**-sabasic, and **27**-sabacic, respectively by a melt condensation method under vacuum. The melt condensation method was carried out by mixing a 1:1 ratio by weight of the mix anhydride and sebacic acid at 150°C under reduced pressure at 0.3 mm Hg for 4 hours with stirring to produce the polyanhydride, a polymer. These sebacic acid copolymers, **28**-sabacic, **29**-sabacic, and **27**-sabacic, were found to have molecular weights of 31,200, 41,000 and 48,700 Daltons, respectively, obtained from GPC analyses. The polymers have low toxicity and are biodegradable due to

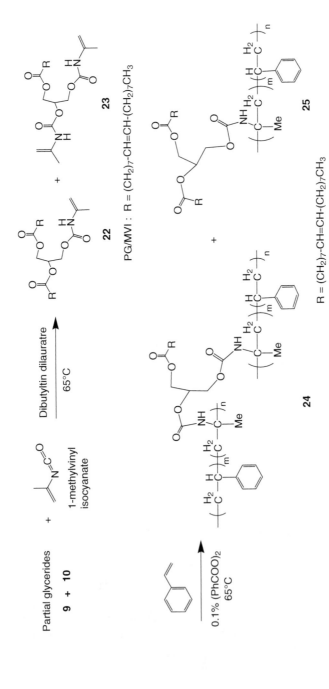

Scheme 2.5 Copolymerization of vinyl urethanes **22** and **23** with styrene.

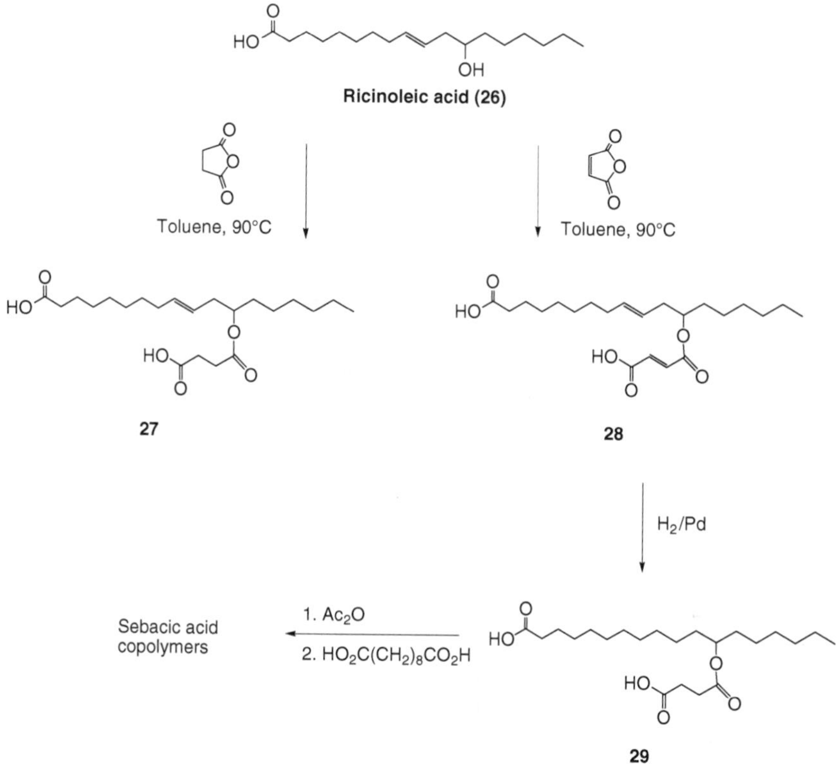

Ricinoleic acid (26)

Toluene, 90°C

Toluene, 90°C

27

28

H₂/Pd

Sebacic acid
copolymers

1. Ac₂O

2. HO₂C(CH₂)₈CO₂H

29

Scheme 2.6 Syntheses of ricinoleic acid succinate **27**, maleate **28**, and hydroxystearic acid succinate **29**, and polymerization with sebacic acid.

their rapid hydrolytic degradation via hydrolysis of the anhydride functions. The polymers are probable candidates of drug carriers.

Pressure-sensitive adhesives (PSA) are used in labels, tapes, films, postage stamps, etc. Currently, the majority of PSA are made from petroleum based acrylate monomers. Since most of the PSA are of disposable nature, it would be desirable to make these materials bio-degradable, providing an opportunity for the use of plant based fatty acids in PSA synthesis. Bunker *et al.* [13] reported the synthesis of high molecular weight polymers from methyl oleate (**30**). Oleate **30** can readily be produced from a methanolysis of triglycerides containing oleic acid in their side chain. The polymers were constructed from a polymerization reaction of an acrylate function to be installed to the double bond of methyl oleate (Scheme 2.7). Hence, epoxidation of methyl oleate with formic acid and hydrogen peroxide followed by esterification of the resulting epoxide function with acrylic acid in a ratio of 1:1.5 and a catalytic amount of hydroquinone and AMC-2 at 90°C to furnish acrylated oleic methyl ester (**32**). Self polymerization of **32** via a free-radical reaction using a catalytic amount of 2,2′-azobis(2-aminopropane) dihydrochloride as the initiator provided polymers having a molecular weight of ~10^6 g/mol, measured by size exclusion chromatography. The glass transition temperature of the polymers was measured by DSC and found to be ~ –40°C. These polymers are thought to be of considerable importance in the field of pressure-sensitive adhesives.

Copolymers utilizing a similar methodology as described above with various reactive alkenes were also synthesized. Instead of performing the aforementioned two-step process to synthesize β-hydroxy acrylate **32** from methyl oleate, a one-step synthesis of β-bromo acrylate was accomplished by Eren *et al.* [14] (Scheme 2.7) from the treatment of methyl oleate with N-bromosuccinimide (NBS) and excess of acrylic acid giving bromo acrylated methyl oleate **34**. A mixture of the regioisomers at C9, C10 was likely formed. Copolymerizations of BAMO with various alkenes, such as styrene (35% weight of **34**), vinyl acetate, and methyl methacrylate, separately, under free-radical conditions (1.5% AIBN was used) gave the corresponding high molecular weight polymers having MW of 45,000 – 120,000 Daltons. The copolymerizations of **34** and styrene, methyl methacrylate, and vinyl acetate were carried at 80°C, 65°C, and 50°C, respectively, for 24 hours. Copolymers **34**-styrene and **34**-vinyl acetate are viscous oils while **34**-methyl methacrylate is soft solid. Glass transition temperatures of **34**-styrene, **34**-vinyl acetate, and **34**-methyl methracrylate are found to be –13.5°C, –17°C, and –10°C, respectively. The presence of bromine atom in the alkyl chain adds flame retardant property to the polymers.

Cross-linkable polyesters from plant oils utilizing lipase as a catalyst were investigated. Transparent films were produced from the hardening or curing of cross-linkable polyesters with a cobalt naphthenate catalyst. This environmentally benign process was reported by Tsujimoto *et al.* [15]. The polymer synthesis involved the treatment of a mixture of divinyl sebacate, glycerol and oleic/linoleic/linolenic acids all in equivalent amounts with lipase (derived from *Candida antarctica*) at 60°C for 24 hours in a sealed tube to generate cross-linkable polyester polymers as shown in Scheme 2.8. Increases in molecular weights and yields of the polymers were found when the reactions were carried out under reduced pressure.

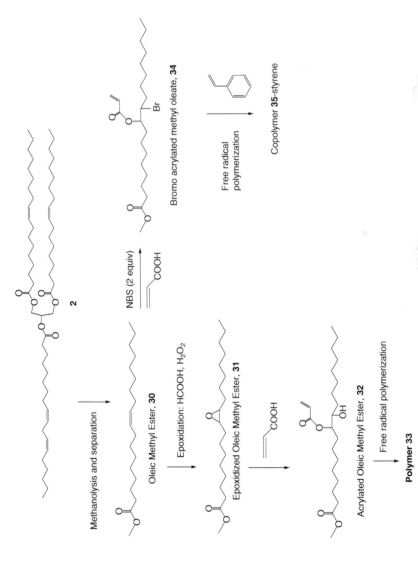

Scheme 2.7 Polymerization of acrylate oleic acid methyl ester **32** and bromo ester **34** with styrene.

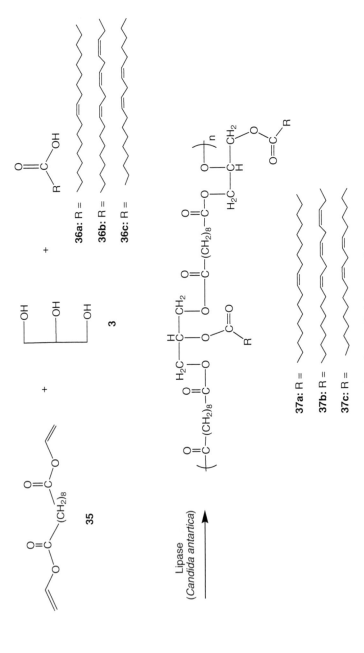

Scheme 2.8 Synthesis of polyesters using lipase-catalyzed cross linking reaction.

These cross-linakable polymers were cured by two different methods, oxidation with catalytic amounts of cobalt naphthenate (3% by weight of the polymer) under air and thermal oxidation at 150°C. In both methods, cross-linking polymers were produced after 2 hours of curing. Interestingly, the prepolymer possessing oleic acid side chain did not undergo cross-linking because of a lack of multiple double bonds in the side chain. These materials may be used in coating industries as a environmental benign and biodegradable alternatives to the petroleum based materials.

2.2.3 Polymerization of Functionalized Triglycerides

As mentioned above, triglyceride molecule contains various reactive sites. These reactive sites can be used directly or transformed into a number of polymerizable functions utilizing similar methodologies as that in the synthesis of petroleum-based polymers. The C=C bonds in triglyceride are capable of a direct polymerization through a cationic mechanism. Cationic polymerization is the mostly frequently used method towards a direct polymerization of triglycerides through carbon double bonds. Free-radical polymerization of triglycerides is rarely carried out due to the low reactivity of internal double bonds towards radicals.

A cationic copolymerization of soybean oil was reported by Larock group [16, 17] with styrene and divinylbenzene separately, and mediated by boron trifluoride diethyl etherate. Depending on the stoichiometries of soybean oil, styrene or divinylbenzene, and BF_3 used, the polymers formed range from soft rubbers to hard and tough or brittle plastic materials. Copolymers obtained from soybean oil and divinylbenzene were brittle because of their non-uniformed high cross-linking densities. The mechanical properties of the polymers, which relate to cross-linking density, apparently can be tuned by varying the amounts of comonomers such as styrene and divinylbenzene. When styrene was used as the major comonomer along with a small amount of divinylbenzene, significant improvement in mechanical properties of copolymers was achieved. These materials possess similar thermal and mechanical properties as that of industrial plastics, along with high damping and shape memory. The cationic polymerization was carried out by the addition of styrene and/or divinylbenzene to the soybean oil followed by boron trifluoride diethyl etherate, which is immiscible in the reaction medium (Scheme 2.9). The miscibility of the initiator is improved by the mixing with Norway fish oil ester resulting in a homogenous reaction. The polymerization reaction was carried out in a Teflon mold for 12 hours at 60°C. The increase of degree of cross linking enhances the glass transition temperature and affects the thermo-physical properties of the polymers.

Triglyceride itself is generally unreactive towards polymerization reaction (except for cationic polymerization involving a strong Lewis acid, boron trifluoride). Hence, it was converted into suitable polymerizable groups to affect the polymerization reactions. Epoxidation of the double bonds in the side chain of triglyceride is one of the most frequently used methods to convert triglycerides to polymerizable materials. Epoxidation reactions can be achieved by the treatment of MCPBA or hydrogen peroxide and acetic acid in the presence of Amberlite [18], and others. Epoxide ring opening reactions will afford either mono- or di-hydroxyl function utilizing sodium cyanoborohydride ($NaBH_3CN$), boron trifluoride diethyl etherate in THF [19] or in methanol (giving methoxylated polyols), 1% $HClO_4$, hydrochloric or hydrobromic acid (giving halogenated polyols), etc. Natural triglycerides possessed epoxy or hydroxyl functions are suitable for direct formation of polymers. Generally, the hydroxyl groups are condensed with a bifunctional linker to affect the polymerization. A one-step ring-opening/polymerization reaction of the epoxide moiety of triglyceride is also possible as mentioned in Scheme 2.7. Polymerizable acrylated function can also introduced into the epoxy triglycerides by the treatment with acrylic acid in the presence of a catalyst such as N,N-dimethylaniline, triethylamine, or 1,4-diazobicylco[2.2.2]octane [20].

The properties and processing ability of acrylated soybean oil can be controlled by copolymerization with styrene to achieve polymers with suitable properties for various structural applications [20]. Acrylated soybean oil **39** was generated from a partial epoxide ring opening with acrylic acid (Scheme 2.10). The properties such as moduli and glass transition temperature can be altered by changing the concentrations of styrene, the sizes of triglycerides, and functionalities of acrylated triglycerides. Hence, reactions of **39** with a linker, 1,2-cyclohexanedicarboxylic anhydride provided oligomers **40**, which upon copolymerization with styrene under free-radical reaction conditions afforded highly cross-linked rigid polymers **41**. Compound **39** also reacted with maleic anhydride forming oligomers with increasing number of double bonds. Increases of the hydroxyl or epoxy functions are desirable but can

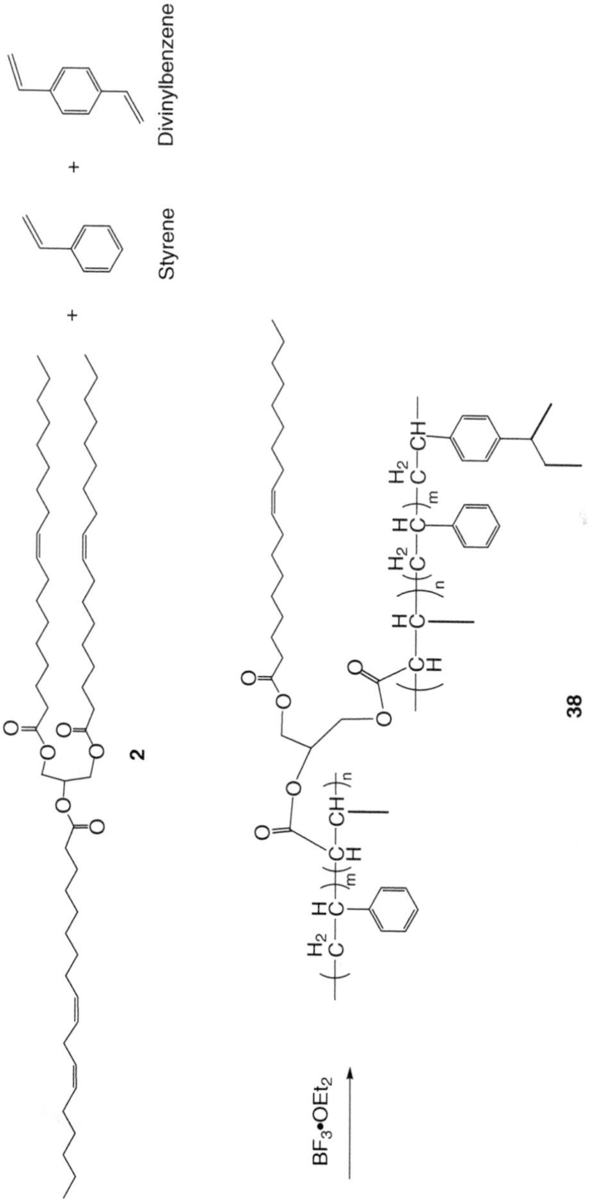

2

38

Scheme 2.9 Cationic polymerization of soybean oil with styrene and/or *p*-divinylbenzene.

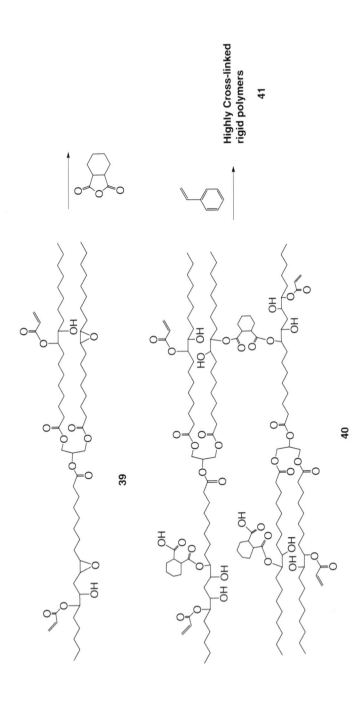

Scheme 2.10 Cross-linking of partial acrylated soybean oil with cyclohexanedicarboxylic anhydride followed by copolymerization with styrene.

also enhance the viscosity of the oligomers leading to gelation. The modified resin **40** copolymerized with styrene and cured to give highly cross-linked and rigid polymers.

A similar acrylated triglyceride, β-bromo acrylate triglycerides, derived from soybean or sunflower oil, was synthesized in one step by the treatment of the plant oil with excess of NBS (see Scheme 2.7 for a similar reaction) and acrylic acid at room temperature for one day [21]. The bromoacrylated triglycerides was copolymerized under free-radical conditions with styrene (35% by weight), 1% methyl ethyl ketone peroxide (MEKP or 2-butanone peroxide), and cobalt naphthenate (6% solution) in a Teflon mold at room temperature for two days affording flame-retardant copolymers due to the incorporation of bromine atom in the fatty acid chain. It is suggested that the bromine atom in the fatty acid chain induces halogen radical in the vapor state acting as radical scavenger and inhibits combustion. Soybean oil has an average number of 4.2 double bonds per triglyceride molecule whereas sunflower oil has only 2.9, thus soybean oil produced monomers possessing a higher number of acrylated residues per triglyceride molecule than that of the sunflower oil. Acrylated soybean oil **40** when copolymerized with styrene reaches gelation earlier than sunflower oil. The polymer obtained from soybean oil is a rigid solid whereas that of sunflower oil is a semi-rigid solid. Soybean oil based copolymer was found to be a rigid load bearing thermostat resin. Glass transition temperatures and storage modulus of polymers obtained from soybean oil and sunflower oil are summarized in Table 2.3.

Beside maleic anhydride and cyclohexanedicarboxylic anhydride, other linkers such as dicyanato agents were also incorporated into triglyceride copolymer frameworks. The resulting urethane functionality appears to enhance adhesion and toughen plastics. Eren *et al.* [22] prepared a similar bromoacrylated castor oil (**42**)

Table 2.3 Glass transition temperatures and storage modulus of copolymers from acrylated soybean and sunflower oils with styrene.

Triglyceride	Glass Transition Temperature °C	Storage Modulus Pa
Soybean oil	55–60	1.0×10^{10}
Sunflower oil	20–30	1.1×10^{8}

as described above using an equivalent amount of castor oil and acrylic acid in the presence of NBS (5.3 equivalents) (Scheme 2.11). Compound **42** was subsequently cross linked with 2,4-diisocyanatotoluene (TDI) followed by copolymerization with styrene or other active alkenes. Compound **42** was treated with TDI in a ratio of 1:2 (to maintain NCO/OH ratio of 1.54:1) at 55°C forming polyurethane prepolymer, BACOP. The acrylate function of BACOP was then allowed to copolymerize with various equal weight of alkenes, such as styrene, methyl methacrylate, 2-hydroxyethyl methacrylate (HEMA), and 3-(acryloxy)-2-hydroxypropyl methacrylate (AHPMA), separately, along with AIBN, at 50°C for 1 hour, 65°C for 12 hours, and 85–90°C for 14 hours to produce simultaneous interpenetrating networks (SINs) polymers **43** (Scheme 2.11). The resulting polymers were semi-rigid to rigid. The glass transition temperatures of polymers **42**-styrene, **42**-methyl methracrylate, **42**-HEMA, and **42**-AHPMA were 85, 126, 113.1, and 67°C, respectively, and values of storage modulus were 8.9×10^8, 8.6×10^8, 8.4×10^8 and 7.9×10^8 Pa, respectively. The SINs polymers were stable up to 200°C and lost weight rapidly around 500°C. The polymers can be used as coating and adhesive materials for steel, and as additives to reinforce rubber and toughen plastics.

To increase the molecular weight and entanglement length of polymers, a mix linkage of three components was investigated by Ozturk *et al.* [23] generating plant oil-based polymers. Polymerization of a mixture of epoxidized soybean oil-soybean oil maleic half ester (**44–46**) in the presence of different amounts of maleic anhydride grafted polypropylene (MMPP; **45**) was carried out to produce cross-linked rigid infusible polymers (Scheme 2.12). For this, **46** (11.16 mmol; anhydride groups), **45** (MW of 3900, 0.3 mmol; anhydride groups) and **44** (11.46 mmol; epoxy groups) were dissolved in a 1:1 mixture of toluene and xylene solution and heated at reflux conditions for 2 hours. After removal of the solvent, the resulting viscous oil was cured in a Teflon mold under vacuum at 180°C for 4 hours giving a yellow, transparent and flexible thermoset polymer film. Similar reactions were performed by replacing a portion of **46** with **45** in such a way that epoxy:maleate functional group ratio was kept 1:1 in each case. The effects of increasing molecular weights on the mechanical properties of the resulting polymers were investigated. DMA analyses of polymers **47a** (E43) and **47b** (PB3200) showed that the storage modulus increased with increasing **45** content, and for the same polymer content, increasing

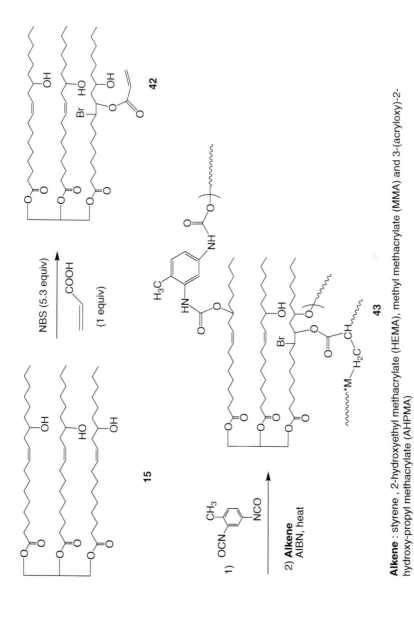

Scheme 2.11 Cross linking of bromoacrylated castor oil followed by copolymerization with alkenes.

Alkene : styrene , 2-hydroxyethyl methacrylate (HEMA), methyl methacrylate (MMA) and 3-(acryloxy)-2-hydroxy-propyl methacrylate (AHPMA)

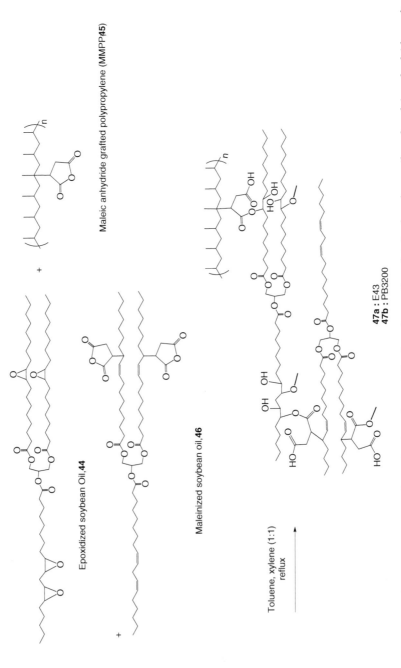

Scheme 2.12 A three-component polymerization of epoxidized soybean oil, maleinized soybean oil, and maleic anhydride grafted polypropylene.

45 molecular weight increased the storage modulus. Thermal properties of these polymers were found to be independent of the MMPP content as well as their molecular weights.

The effect of the unreactive side chains of fatty acid in properties of polymers was studied by using polyols obtained from ozonolysis and reduction reactions of triglycerides [24]. The hydroxyl functionality derived from ozonolysis-reduction process is 3, whereas the epoxidized or hydroxyl acrylated soybean oil has a functionality of 4.5 or higher. Polyurethanes synthesized from polyols obtained by the ozonloysis of soybean oil lack the dangling alkyl chains found in polymers synthesized from the epoxidized and hydroxyl acrylated soybean oil, in which the hydroxyl groups are present in the middle of the alkyl chains resulting dangling alkyl chains. These dangling chains appear to act as plasticizers and affect the properties of the polyurethanes. Ozonolyses and subsequent sodium borohydride reduction of soybean oil, canola oil and trilinolein were carried out to generate respective triols and designated as ozo-soy-polyol, ozo-canola-polyol, and ozo-model-polyol (compound **48** in Scheme 2.13). Triol **48** was allowed to react with 4,4′-methylenebis(phenylisocyanate) (MDI) at 70°C and curing at 110°C to give polyurethane **49**. Glass transition temperatures of model-, canola-, and soy-based polyurethane were found to be 53, 36 and 22°C, respectively. The azo-soy-polyurethane has Tg value of 30°C, which was lower than that of ozo-model-polyurethane. The lower functionality and the presence of saturated fatty acids in the former are suggested to induce the difference. Polyurethanes derived from ozonolysis exhibit better mechanical properties and thermal stabilities than that of polyurethanes derived from epoxidized and hydroxyl acrylated polyols.

The use of Click chemistry or a 1,3-dipolar cycloaddition reaction of alkyne and azido function was studied by Hong *et al.* [25] to generate triglyceride polymers containing 1,2,3-triazole moieties allowing an entrapment of copper sulfate catalyst used in the Click reaction (Scheme 2.14). Consequently, green to blue color materials were obtained. Epoxy triglycerides **44** was alkynated with a proprietary procedure or sodium azide to provide alkynated triglyceride **50** or azido triglyceride, respectively. A series of highly cross-linked polymers were obtained by copper-catalyzed as well as thermal cycloaddition reactions from alkynated and azidated soybean oil separately, with 1,6-diazidocyclohexane and 1,7-octadiyne, respectively. Other diazide and dialkyne molecules were also used. The polymerization

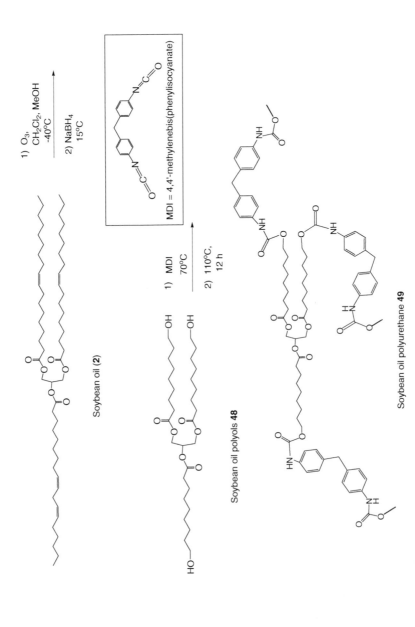

Scheme 2.13 Cleavage of the double bonds of triglyceride followed by polymerization with 4,4′-methylenebis(phenylisocyanate).

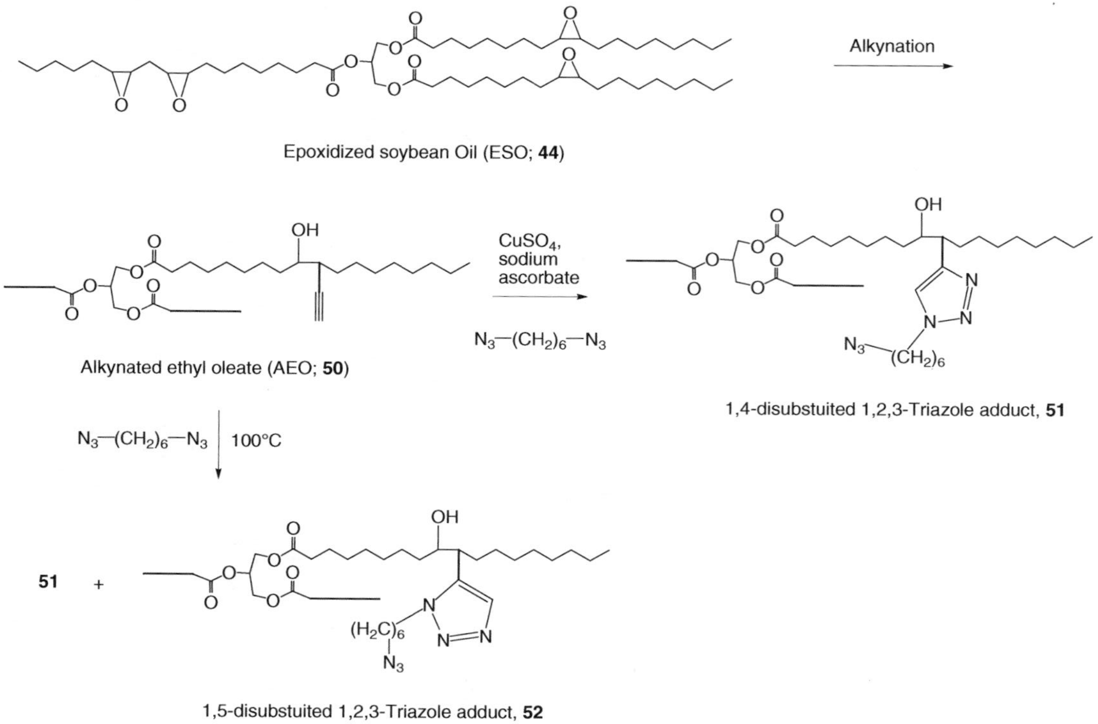

Alkynation →

Epoxidized soybean Oil (ESO; **44**)

Alkynated ethyl oleate (AEO; **50**)

CuSO₄,
sodium
ascorbate

$N_3-(CH_2)_6-N_3$

1,4-disubstituted 1,2,3-Triazole adduct, **51**

$N_3-(CH_2)_6-N_3$ | 100°C

51

+

1,5-disubstituted 1,2,3-Triazole adduct, **52**

Scheme 2.14 Polymerization of alkynated soybean oil with 1,6-diazidocyclohexane.

reactions were carried out by treating alkynated soybean oil **50**, 1,6-diazidohexane, copper sulfate, and sodium ascorbate in a 1:1 mixture of DMF and water at room temperature to give highly cross-linked polymers **52**. When an equal molar amount of alkynate soybean oil and 1,6-diazidohexane was heated at 100°C without copper catalyst for 12 to 24 hours under nitrogen atmosphere, light yellow to light brown cross-linked solid polymers were obtained. Thermal polymerization was found to provide better yields and a more homogenous cross-linking. The decomposition temperatures of these polymers were narrower by about 170°C than that of polymers obtained using the copper-catalyzed method. The glass transition temperatures of copper-catalyzed polymers range from –13 to 45°C because of the entrapped copper catalyst in the polymer networks. The thermal approach appears to provide a higher yield of polymer, which is suitable for generating triazole-vegetable-oil polymers.

Transformations of the alkenyl functions of triglyceride to various functionalities such as hydroxyl and amino moieties for subsequent polymerization were conducted by Hua group [26]. The goals were to synthesize various polymers utilizing one kind of linker, i.e. 1,4-phenylenediisocyanate (PDI), but varying the number of polymerizable functional groups on the triglyceride side chains to study their physical properties. A structure-property relationship may be drawn from the studies. To simplify the structural analysis, pure triolein (glyceryl trioleoate) was prepared from oleoic acid and glycerol (Scheme 2.15). Triolein was either triepoxidized with 3 equivalents of MCPBA to give triepoxide **54**. Triglyceride triol **55** and hexaol **56** were prepared from the reduction of triepoxide **54** with sodium cyanoborohydride and perchloric acid, respectively.

To study the effect of functionalization of one of the three side chains of glycerol trioleoate, diol in which two hydroxyl functions are on one side chain was synthesized (Scheme 2.16). For this, monoepoxidation of glyceryl trioleate with one equivalent of MCPBA affored monoepoxide (**57**), which upon treatment with 1% perchloric acid gave triglyceride diol **58**. Triamines **61** and **62** were also synthesized for the production of polymers containing urethane functions (Scheme 2.16). Bromination of triol **55** with three equivalents each of triphenylphosphine and carbon tetrabromide followed by the displacement with sodium azide afforded triazido triglyceride **60**. Reduction of the azido functions of **60** with a catalytic amount of 10% Pd/C and hydrogen gave triglyceride triamine **61** in quantitative yield. A one-pot reductive amination of azido

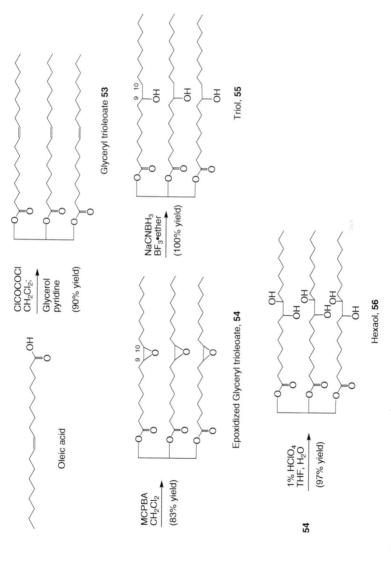

Scheme 2.15 Synthesis of triol and hexaol triglycerides from glyceryl trioleate.

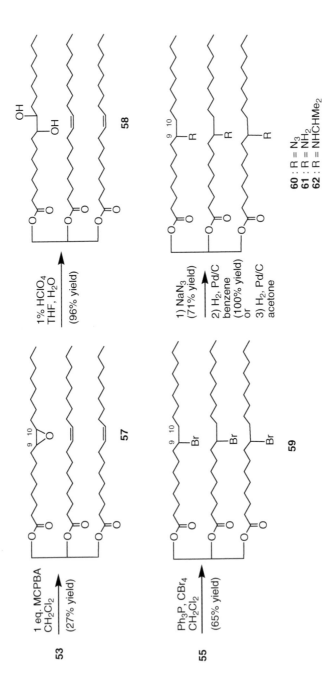

Scheme 2.16 Preparation of dihydroxyl-, triamino-, and triisopropylaminotriglycerides.

triglyceride **60** was also carried out by the treatment with 10% Pd/C and hydrogen in acetone to furnish triisopropylamine triglyceride **62**. A linker, 1,4-phenylene diisocyanate (PDI), possessing para-orientations of two isocyanato groups, was used as the crosslinking agent for the construction of rigid triglyceride polymers from the aforementioned functionalized triglycerides.

Polymerization reaction was carried out by treating diol **58** with 1 equivalent of PDI and a catalytic amount of pyridine in refluxing toluene to give polymer **63** as depicted in Scheme 2.17. Similarly, polymer **64** derived from triol **55** was prepared by the treatment with 1.5 equivalents of PDI and a catalytic amount of pyridine in refluxing toluene. Hexaol **56** was polymerized under similar reaction conditions with 3 equivalents of PDI to give polymer **65** (Scheme 2.18). Polymers containing urethane framework were similarly prepared by the treatment of triamino triglyceride **61** and triisopropylamino triglyceride **62** with PDI separately to give polymers **66** and **67**, respectively (Scheme 2.19). Polymers **63** and **64** were rubbery solids while polymer **65**, **66**, and **67** were solids. The results suggest that the lesser amounts of cross linkages from diol and triol triglycerides gave softer polymers, and greater amounts of cross linkages and urethane functionality provide rigid and toughened polymers.

The aforementioned polymers are insoluble in most organic and inorganic solvents including DMSO, DMF, chloroform, ethyl acetate, ethanol, water, etc. Solution NMR spectroscopy of these polymers cannot be obtained, however, Fourier transformed infrared (FTIR) spectra, gel contents, thermal and mechanical properties including differential scanning colorimetries (DSC), thermogravimetric analyses (TGA), and dynamic mechanical analyses (DMA) were measured [26].

DSC of these cross-linked polymers did not show melting peaks indicating that the polymers are amorphous phase (Figure 2.2). Glass transition temperatures of the polymers were obtained from their DSC thermograms and results are summarized in Table 2.4. A higher Tg value of polymer **64** implies that cross linkage framework is more regulated or uniformed than that of other polymers.

Polymer **63**, derived from diol **58**, has the lowest glass transition temperature at −2.8 °C. This might be due to the fact that only one of the three alkyl chains participates in cross-linking and the remaining two alkyl chains act internally as plasticizer. TGA measurements were carried out to obtain the thermal stability data of the polymers (Figure 2.3). Polymer **63** appears to be the most stable

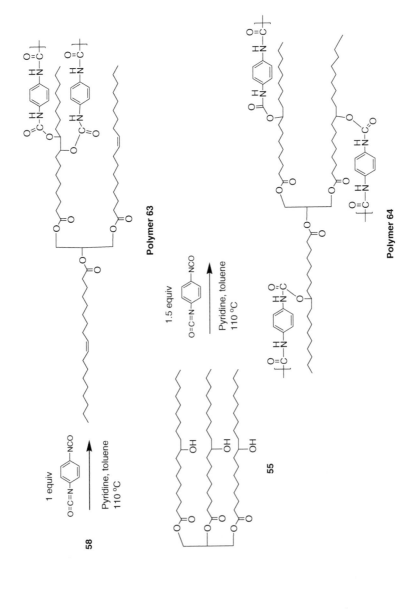

Scheme 2.17 Polymerization of diol triglyceride with 1,4-phenylene diisocyanate.

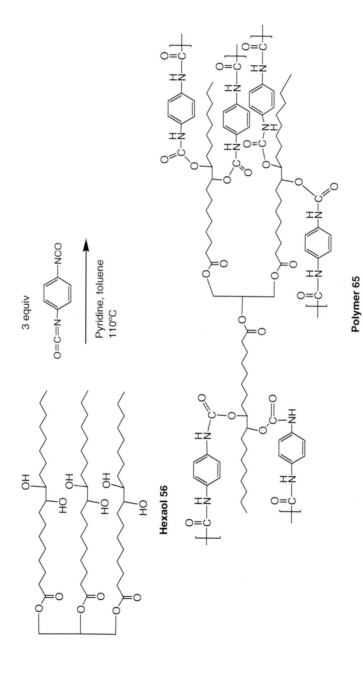

Scheme 2.18 Polymerization of hexaol triglyceride with 1,4-phenylene diisocyanate.

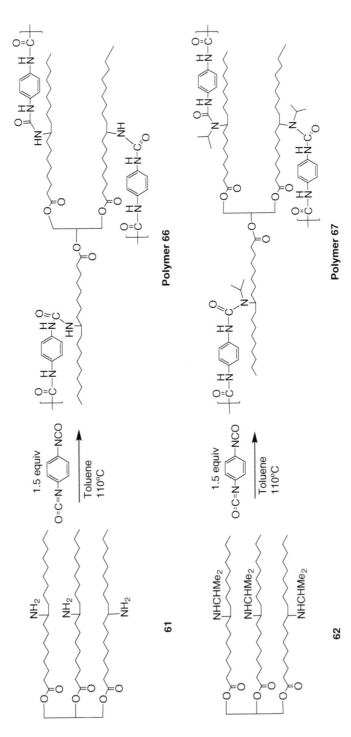

Scheme 2.19 Polymerizations of triamino and triisopropylamino triglycerides with 1,4-phenylene diisocyanate.

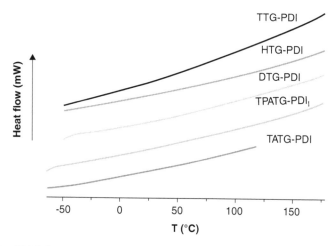

Figure 2.2 DSC thermograms of triglycerides polymers: TTG-PDI (**64**), HTG-PDI (**65**), DTG-PDI (**63**), TPATG-PDI (**67**), and TATG-PDI (**66**).

Table 2.4 Glass transition temperatures of triglycerides polymers: TTG-PDI (**64**), HTG-PDI (**65**), DTG-PDI (**63**), TPATG-PDI (**67**), and TATG-PDI (**66**).

Samples	Glass Transition Temperature (Tg, °C)
DTG-PDI (**63**)	–2.8
TTG-PDI (**64**)	10
HTG-PDI (**65**)	–1
TATG-PDI (**66**)	–20.7
TPATG-PDI (**67**)	10.2

among this set of polymers as its thermal decomposition onset temperature was at 282.6°C and 56% of the original weight was retained at 408°C.

Dynamic storage modulus of polymers **63**, **64** and **65** were measured varying frequencies at different temperatures. Polymer **63** exhibited a linear viscoelastic characteristic (Figure 2.4) at low frequency (below 10 rad/sec), which is similar to that of most thermoplastic. Hence, the polymer can likely be used as a thermoplastic.

After conducting above mentioned polymerizations of various functionalized triglycerides with one kind of linker, PDI, Hua group

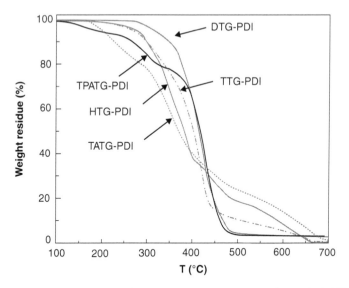

Figure 2.3 TGA thermograms of triglycerides polymers: TTG-PDI (**64**), HTG-PDI (**65**), DTG-PDI (**63**), TPATG-PDI (**67**), and TATG-PDI (**66**).

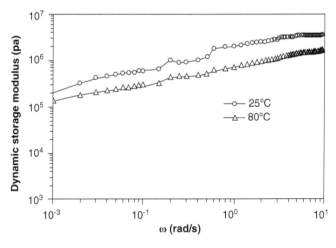

Figure 2.4 Dynamic mechanical properties of polymer **63**.

also studied triglyceride polymers from one kind of triglyceride, triol, with various cross-linking agents including adipoyl chloride (AD), azelaoyl chloride (AZ), and cyclohexane-1,4-dicarbonyl dichloride (CH) (mole ratio of **55**:linker = 1:1.5 in all cases) in the presence of pyridine at 80°C (Scheme 2.20). The polymers obtained were rubbery solids insoluble in organic and inorganic solvents

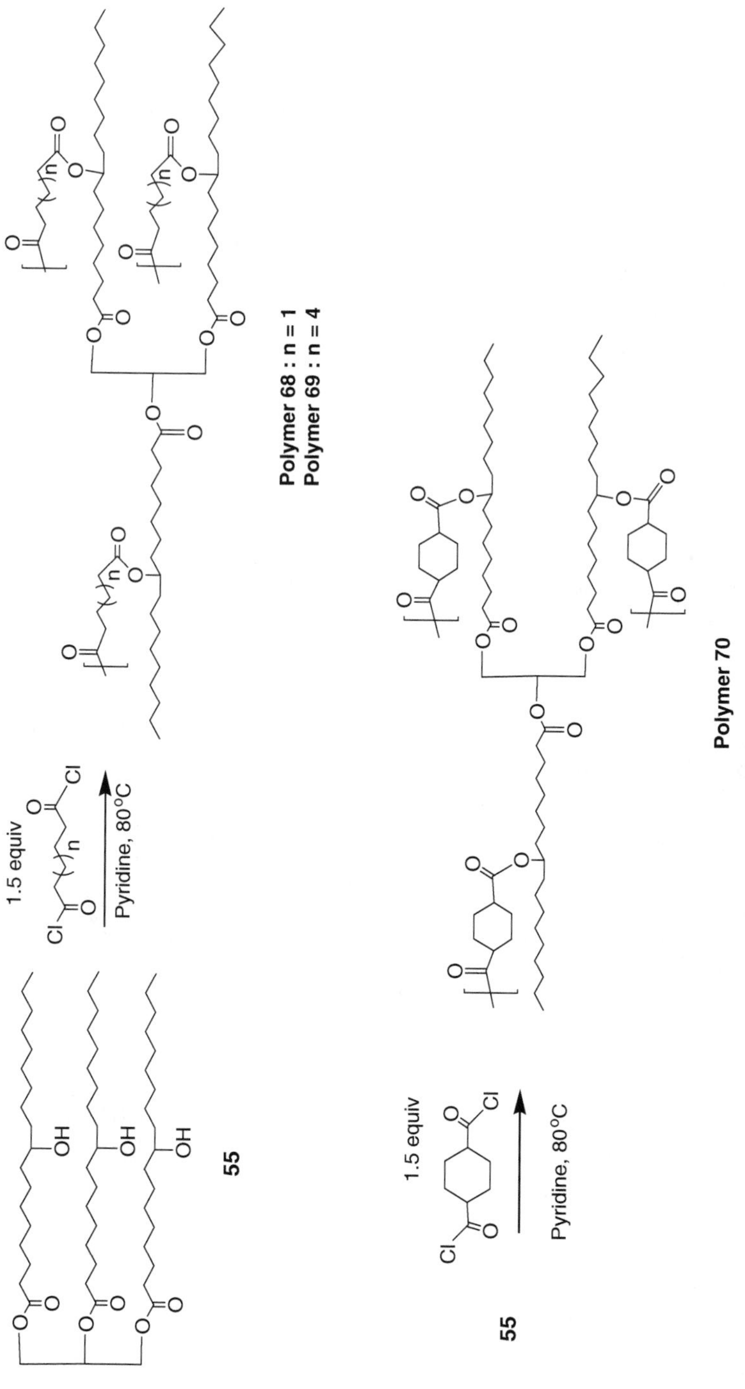

Scheme 2.20 Polymerization of triglyceride triol 55 with hexanedioyl chloride, nonanedioyl chloride, and 1,4-cyclohexanedicarbonyl dichloride.

including DMSO, methylene chloride, ethanol, ethyl acetate, and others. Thermal and mechanical properties including differential scanning colorimetry (DSC), thermogravimetric analysis (TGA), and dynamic mechanical analysis (DMA) of the synthesized polymers were measured.

The DSC graphs of TTG-AZ (**69**), TTG-AD (**68**), and TTG-CH (**70**) show glass transition temperatures at ~50°C (Figure 2.5). Prominent endothermic peaks at ~95°C were found in polymers **68** and **69** demonstrating crystalline properties of the polymers, whereas polymer **70** showed a weak peak at the same region indicative of a lesser crystalline property.

Thermo gravimetric analysis of TTG-AD (**68**), TTG-AZ (**69**), and TTG-CH (**70**) were carried and the results are presented in Figure 2.6. The result shows that polymer **69** is found to be the most stable polymers among the three having thermal decomposition onset temperature of 400°C, whereas, polymer **68** was found to be the least stable having thermal decomposition onset temperature of 200°C. On heating beyond 430°C these polymers retained only about 5–10% of their original weights.

DMA measurements of polymer **68**, **69**, and **70** were carried out at different temperatures with variable frequencies, and results are depicted in Figure 2.7, 2.8, and 2.9, respectively. Results of DMA, DSC, and TGA along with their elastic property indicate that the polymers can be applied in gum-base materials such as chewing gums and related applications. These materials are biodegradable and would be better alternatives for presently used petroleum derived gum-base materials.

It is suggested that bio-based or petroleum based resin cannot be used alone in liquid-molding process without the addition of diluents like styrene [27]. Styrene is an air pollutant that damages the

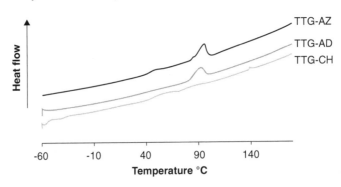

Figure 2.5 DSC of TTG-AZ (**69**), TTG-AD (**68**), and TTG-CH (**70**).

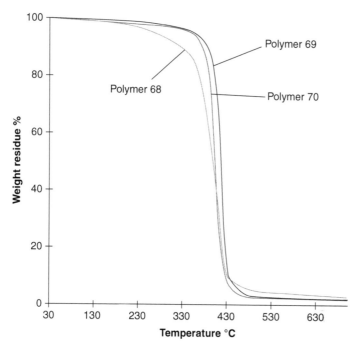

Figure 2.6 TGA of Polymer TTG-AD (**68**), TTG-AZ (**69**), and TTG-CH (**70**).

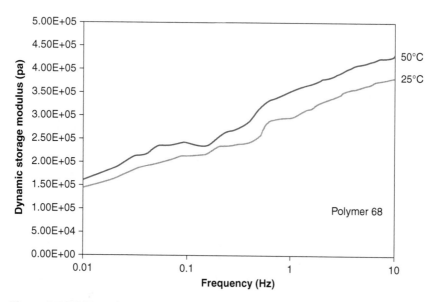

Figure 2.7 DMA analysis of TTG-AD (**68**).

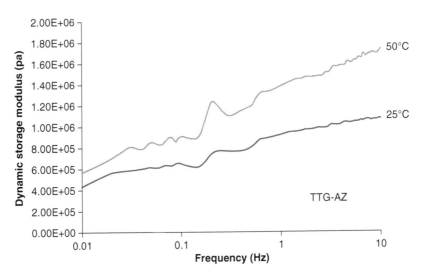

Figure 2.8 DMA analysis of TTG-AZ (**69**).

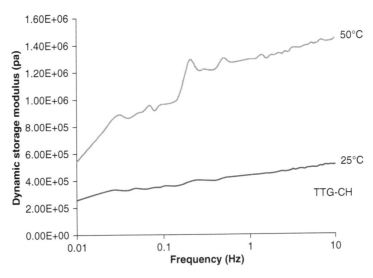

Figure 2.9 DMA analysis of TTG-CH (**70**).

environment and causes health hazards. Replacement of styrene in the synthesis of bio-based polymers was considered to be important. Campanella *et al.* [27] reported the use of acrylated epoxidized fatty acid methyl ester **32**, a diluent, as the replacement of styrene (Scheme 2.21). Bio-based cross-linkers such as acrylated epoxidized soybean oil **71** and maleinated castor oil monoglyceride **73** were

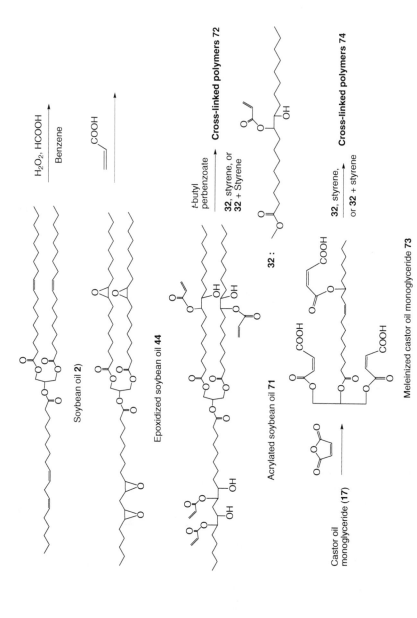

Scheme 2.21 Polymerization of hydroxylated acrylated triglycerides with monoacrylate **32**.

applied for the preparation of plant oil polymers. For the polymerization, **71** and **73** were separately copolymerized with styrene or **32** or both in the presence of *t*-butyl perbenzoate (a free-radical initiator, 1.5 wt %). The resins after being purged with nitrogen were placed in a silicon rubber mold between the aluminum plates, and cured thermally. Polymer **74** was cured at 90°C for 2 hours and post cured at 120°C for 2 hours and 160°C for 1 hour. Polymer **72** was cured at 90°C for 2 hours and 120°C for 2 hours. Polymers used **32** as a cross linker were found to be have viscosities considerably (AFAME) higher than that obtained using styrene (Table 2.5).

Table 2.5 Viscosity (η) of the bio-based triglyceride bio-based polymers copolymerized with **32**, styrene or both.

Bio-based triglyceride polymer	AFAME (wt%)	Styrene (wt%)	η (Pa s) 25°C
Polymer **72**	35	0	2.36 ± 0.03
	17.5	17.5	1.81 ± 0.06
	0	35	0.86 ± 0.03
Polymer **74**	35	0	2.08 ± 0.01
	17.5	17.5	1.77 ± 0.04
	0	35	0.39 ± 0.03

Table 2.6 Glass transition temperature and storage modulus of bio-based polymers copolymerized with **32**, styrene, and both styrene and **32**.

Bio-based triglyceride polymer	AFAME (wt%)	Styrene (wt%)	Glass transition temperature (Tg, °C)	Storage modulus (E', MPa, 30°C)
	35	0	35.0 ± 0.1	364.4 ± 0.9
Polymer **72**	17.5	17.5	50.0 ± 0.4	656.2 ± 1.4
	0	35	71.2 ± 0.2	1083.9 ± 2.2
	35	0	68.1 ± 0.3	346.2 ± 1.3
Polymer **74**	17.5	17.5	91.9 ± 0.4	1135.5 ± 1.3
	0	35	121.2 ± 0.6	1955.1 ± 4.5

High viscosities of the bio-based cross linkers using only AFAME as diluents (comonomer) makes difficult to cure these materials. Therefore, AFAME and styrene were blended together and used as diluents to achieve the viscosity requirement of composite materials. This way the amount of styrene as diluents was minimized by the use of AFAME. The thermal and mechanical properties of the polymers were also improved in the similar ways (Table 2.6).

2.3   Summary

Over the last fifteen years, various plant oil polymers were synthesized and their properties were studied. A large number of useful polymers were achieved including soft, rubbery, and rigid materials. Methodologies in generating plant oil polymers by in large involved functionalization of the double bonds of triglyceride followed by polymerization with bifunctional linkers to produce homopolymers or alkenes such as styrene to afford copolymers. A tremendous amount of data on structure-physical property relationship obtained, and careful analysis of the data to furnish a global theory on the entanglement of cross linkage and polymer net work remains. Undoubtedly, various petroleum polymers will soon be replaced by plant oil polymers, and a continuing study of plant oil polymers is warranted.

References

1. Wool, R. P. Development of affordable soy-based plastics, resins, and adhesives. *Chemtech*, **1999**, *29*, 44-48.
2. Guner, F. S.; Yagci, Y.; Erciyes, A. T. Polymers from triglyceride oils. *Prog. Polym. Sci.* 2006, 31, 633-670.
3. Liu, K. *Soybeans: Chemistry, technology, and utilization,* Chapman and Hall, New York; 1997, page 29.
4. Deligny, P.; Tuck, N. Alkyds and polyesters. In: "Resins for surface coatings", Oldring, P.K.T. edited, vol. II, Wiley, New York, 2000.
5. Can, E.; Küsefoglu, S.; Wool, R. P. Rigid thermosetting liquid molding resins from renewable resources. II. Copolymers of soybean oil monoglyceride maleates with neopentyl glycol and bisphenol A maleates. *J. Appl. Polym. Sci.* 2002, 83, 972-980.
6. Can, E.; Wool, R. P.; Küsefoglu, S. Soybean and castor oil based monomers: synthesis and copolymerization with styrene. *J. Appl. Polym. Sci.* 2006, 102, 2433-2477.

7. Akbas, T.; Beker, U. G.; Guner, F. S.; Erciyes, T.; Yagci, Y. Drying and semidrying oil macromonomers. III. Styrenation of sunflower and linseed oils. *J. Appl. Polym. Sci.* 2003, 88, 2373-2376.

8. Cocks, L. V.; van Rede, C. *Laboratory handbook for oil and fat analyses;* American Press: London, 1966.

9. Guner, F. S.; Usta, S.; Erciyes, A. T.; Yagci, Y. Styrenation of triglyceride oils by macromonomer technique. *J. Coatings Technology,* 2000, 72(907), 107-110.

10. Can, E.; Wool, R. P.; Küsefoglu, S. Soybean- and castor-oil-based thermosetting polymers: Mechanical properties. *J. Appl. Polym. Sci.* 2006, 102, 1497-1504.

11. Koprululu, A.; Onen, A.; Serhalti, I. E.; Guner, F. S. Synthesis of triglyceride-based urethane macromers and their use in copolymerization. Progress in Organic Coatings. 2008, 63, 365-371.

12. Teomim, D.; Nyska, A.; Domb, A. J. Ricinoleic acid-based biopolymers. *J. Biomed. Mater. Res.* 1999, 45, 258-267.

13. Bunker, S. P.; Wool, R. P. Synthesis and characterization of monomers and polymers for adhesives from methyl oleate. *Journal of Polymer Science: Part A: Polymer Chemistry,* 2002, 40, 451-458.

14. Eren, T.; Kusefoglu, S. H. Synthesis and characterization of copolymers of bromoacrylated methyl oleate. *J. Appl. Polym. Sci.* 2004, 94, 2475-2488.

15. Tsujimoto, T.; Uyama, H.; Kobayashi, S. Enzymatic synthesis of cross-linkables from renewable resources. *Biomacromolecules.* 2001, 2, 29-31

16. Li, F.; Larock, R. C. New soybean oil-styrene-divinylbenzene thermosetting copolymers. I. Synthesis and characterizations. *J. Appl. Polym. Sci.* 2001, 80, 658-670.

17. Li, F.; Larock, R. C. New soybean oil-styrene-divinylbenzene thermosetting copolymers. V. shape memory effect. *J. Appl. Polym. Sci.* 2002, 84, 1533-1543.

18. Goud, V. V.; Patwardhan, A. V.; Pradhan, N. C. Kinetics of in situ epoxidation of natural unsaturated triglycerides catalyzed by acidic ion exchange resin. *Eng. Chem. Res.* 2007, 46, 3078-3085.

19. Tone, H.; Nishi, T.; Oikawa, Y.; Hikota, M.; Yonemitsu, O. A stereoselective total synthesis of (9S)-9-dihydroerythronolide A from D-glucose. *Tetrahedron Letters.* 1987, 28, 4569-4572.

20. Khot, S. N.; Lascala, J. J.; Can, E.; Morye, S. S.; Williams, G. I.; Palmese, G. R.; Kusefogu, S. H.; Wool, R. P. Development and application of triglyceride-based polymers and composites. *J. Appl. Polym. Sci.* 2001, 82, 703-723.

21. Eren, T.; Kusefoglu, S. H. Synthesis and polymerization of the bromoacrylated plant oil triglycerides to rigid, flame-retardant polymers. *J. Appl. Polym. Sci.* 2004, 91, 2700-2710.

22. Eren, T.; Colak, S.; Kusefoglu, S. H. Simultaneous interpenetrating polymer networks based on bromoacrylated castor oil polyurethane. *J. Appl. Polym. Sci.* 2006, 100, 2947-2955.

23. Ozturk, C.; Kusefoglu, S. H. Polymerization of epoxidized soybean oil with maleinized soybean oil and maleic anhydride grafted polypropylene mixtures. *J. Appl. Polym. Sci.* 2010, 118, 3311-3317.

24. Petrovic, Z. S.; Zhang, W.; Javni, I. Structure and properties of polyurethanes prepared form triglyceride polyols by ozonolysis. *Biomacromolecules.* 2005, 6, 713-719.

25. Hong, J.; Luo, Q.; Shah, B. K. Catalyst- and solvent-free "Click" chemistry: A facile approach to obtain cross-linked biopolymers from soybean oil. *Biomacromolecules* 2010, 11, 2960-2965.

26. Zhao, H.; Zhang, J.; Sun, X. S.; Hua, D. H. Syntheses and properties of cross-linked polymers from functionalized triglycerides. *J. Appl. Polym. Sci.* 2008, 110, 647-656.

27. Campanella, A.; La Scala, J. J.; Wool, R. P. The use of acrylated fatty acid methyl esters as styrene replacements in triglyceride-based thermosetting polymers. *Polym. Eng. and Sci.* 2009, 49, 2384-2392.

3

Advances in Acid Mediated Polymerizations

Stewart P. Lewis[1] and Robert T. Mathers[2]

[1]*Innovative Science Corporation, Blacksburg, VA, U.S.A.*
[2]*Pennsylvania State University, New Kensington, PA, U.S.A.*

Abstract

This chapter discusses characteristics that constitute an ideal green polymerization and highlights deficiencies inherent in processes that operate by a cationic mechanism. A comprehensive review of progress made in the field of isobutene containing polymers is provided. Newer methods for the preparation of polymers synthesized from naturally derived monomers via a cationic mechanism are also discussed.

Keywords: cationic polymerization, renewable monomers polyisobutylene, butyl rubber, weakly coordinating anions

Abbreviations: BA = Brønsted acid; CCl = cumyl chloride ($C_9H_{11}Cl$); Cp = cyclopentadienyl (C_5H_5); Cp' = trimethylsilylcyclopentadienyl ($C_5H_4SiMe_3$); Cp" = tetramethylcyclopentadienyl (C_5HMe_4); Cp* = pentamethylcyclopentadienyl (C_5Me_5); CFSTR = continuous flow stirred tank reactor; COH = cumyl alcohol; CTA = chain transfer agent; CTC = charge transfer complex; DBSA = dodecylbenzenesulfonic acid; DTB = dodecyltrimethylammonium bromide; DTBMP = 2,6-di-*t*-butyl-4-methyl-pyridine; DTBP = 2,6-di-*t*-butyl-pyridine; DTFB = dodecyltrimethylammonium tetrafluoroborate; DPE = 1,1-diphenylethylene; EAS = electrophilic aromatic substitution; HFC = hydrofluorocarbon; IB = isobutene; IBVE = isobutyl vinyl ether; IP = isoprene; LA = Lewis acid; MHS = Mark-Houwink-Sakurada; MW = molecular weight; MCH = methyl cyclohexane; OTf = triflate ($CF_3SO_3^-$); PB = polybutene-1; PDI = polydispersity index; PIB = polyisobutene; PS = polystyrene; PMOS = *p*-methoxystyrene; PP = polypropylene; PFLA = perfluoroarylated Lewis acid; SDS = sodium dodecylsulfate; SHP = sterically hindered pyridine; SiCCl = silicon cumyl chloride (phenyldimethylchlorosilane);

Vikas Mittal (ed.) Renewable Polymers, (69–174) © Scrivener Publishing LLC

T = temperature; TBDCC = 5-*t*-butyl-1,3-bis(1-chloro-1-methyl)benzene;
THF = tetrahydrofuran; TMP = 2,4,4-trimethyl-1-pentene;
TMPCl = 2-chloro-2,4,4-trimethylpentane; TMSCl = trimethylsilyl chloride;
WCA = weakly coordinating anion.

3.1   Introduction

Much emphasis has been placed on refining chemical transformations to lessen their environmental impact. Green chemistry possesses one or more attributes as shown in Table 3.1 and Figure 3.1.

Table 3.1   Ideal characteristics of a green chemical process.

Ideal Green Chemistry Attributes
Less or no waste = high atom efficiency.
Reduced energy consumption.
Renewable starting materials of low or no toxicity.
Easily isolable and reusable catalysts.
Limited process steps = smaller plant footprint.

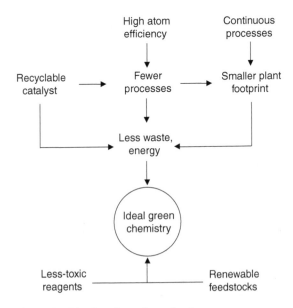

Figure 3.1 Ideal sustainable chemistry flow chart.

Many advances have been made in reducing the undesirable effects of cationic polymerizations. This chapter recaps previous work aimed at cleaner cationic olefin polymerizations in addition to newer methodologies that have vastly improved green characteristics. It is suggested that the reader who is unfamiliar with cationic olefin polymerization consult a recent overview of the topic [1].

Cationic polymerizations [2–4] present special challenges for the green chemist; for example, reduction or elimination of toxic solvents (e.g. CH_3Cl). Systems capable of operating under bulk conditions or in less benign solvents/diluents (e.g. hexane, water) are of interest but difficult. Bulk systems pose problems due to high rates (ca. $4 \pm 6 \times 10^8$ $L \cdot mol^{-1} \cdot s^{-1}$) [5, 6] and exothermicity (i.e. ~ -12 $kcal \cdot mol^{-1}$) [7] of reaction. Nonpolar media do not promote ionization essential for initiation to occur or polymerization to proceed [8] while the reactivity of most Lewis acid (LA) coinitiators and carbocations prevents widespread use of aqueous techniques [9].

The majority of cationic polymerizations rely on homogenous LA coinitiators that are difficult to reuse [9]. Even the recycling of LAs that lend themselves to removal due to their physical properties (i.e. BF_3) is problematic [10]. Heterogeneous initiator systems hold promise since they are readily isolated, which can reduce processing steps, material consumption, and waste generation.

Energy efficiency is of paramount concern due to the inverse relationship of molecular weight (MW) on temperature (T) [11, 12]. Great strides have been made in the development of initiator systems that produce higher MW materials at elevated temperatures; albeit, few have surpassed γ-radiation in terms of ultimate MW [13, 14]. In the majority of cases this is accomplished by using anions of reduced nucleophilicity that do not contribute to chain transfer (CT). The primary obstacles that must be surmounted for high temperature initiator systems are requisite use of polar solvents for some, sensitivity to trace impurities in others, and the high cost of components based on perfluoroarylated Lewis acids (PFLAs). Additional advances have been made in the area of process steps and reactor design that translate into energy savings and yet allow for use of traditional initiator components.

Finally, with the exception of terpenes [1] and questionably tetrahydrofuran (THF), most commercial polymers made by a cationic mechanism are petroleum products. Polymers with new structures made from renewable monomers that are viable replacements have yet to be synthesized. A more likely scenario is production of existing monomers from renewable starting feedstocks in place of petroleum similar to the production of butadiene and isoprene (from ethanol and limonene, respectively) during WWII [15]. Some progress [16] has been made in this area but overall, developments are lacking and this topic deserves special attention.

3.2 Problems Inherent to Cationic Olefin Polymerization

A great deal of work has been conducted with the aim of devising an ideal green cationic polymerization system for olefins; especially, in regards to isobutene (IB) based polymers. Shortcomings inherent to traditional commercial cationic polymerization processes are most evident in the $AlCl_3{}^{\dagger}$ based production of butyl rubber {i.e. poly(isobutene-co-isoprene), Chart 3.1}. Polymerization is conducted under slurry conditions in CH_3Cl, the process being detailed in a number of reviews [17–23].

As a coinitiator, $AlCl_3$ has very limited solubility in organic solvents compatible with cationic polymerization [9, 24, 25]. Even in alkyl halides, the solvent class of choice, solubility is low; as a result, it is difficult to accurately determine $[AlCl_3]$ and meter a given quantity of coinitiator [26, 27]. This complicates production and can lead to off-spec product/waste [28]. Moreover, $AlCl_3$ is not readily isolated from the reaction mixture necessitating deashing via hydrolysis with subsequent waste generation [9, 24, 25]. Halide impurities from $AlCl_3$ can also interfere with post-polymerization functionalization (e.g. halogenation) [29, 30].

$AlCl_3$ is not an efficient coinitiator in that polymerization never goes to completion [31, 32]. This may be the result of encapsulation of coinitiator and growing chain ends within butyl rubber as it

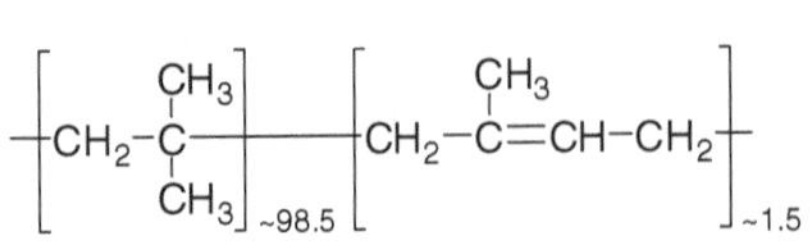

Chart 3.1 Repeating structure for butyl rubber.

---

† Throughout this work Al based LAs are represented by monomeric species for simplicity when in reality most are associated as dimers.

precipitates from slurry preventing further polymerization [33, 34]. Another plausible mode of termination involves abstraction of a β-H from the growing polymer cation by $AlCl_4^-$ with concomitant loss of HCl and formation of a stable polymer-olefin·$AlCl_3$ complex (Scheme 3.1) [34]. This might explain the observation that polymerization of IB in hydrocarbon solution (initiated by $AlCl_3$ dissolved in CH_3Cl) undergoes color changes identical to that exhibited by olefin·$AlCl_3$ complexes when aged in the presence of moisture [35]. $AlCl_3$ readily forms liquid complexes with ethylene under mild conditions [25, 36–38]. Termination with loss of chloride from anion to polymer carbocation (Scheme 3.2) is also likely and supported by findings that addition of AlR_3 or R_2AlX to dead polymer (prepared from $AlCl_3$) immediately reinitiates polymerization to high yields (Scheme 3.3) [32]. In this case the *t*-chloro endgroup functions as a *de facto* initiator and polymerization behavior is similar to organohalide carbocation synthons in conjunction with AlR_3 or R_2AlX coinitiators.

One fact that appears to be overlooked is that all $AlCl_3$/IB polymerizations involve the use of CH_3Cl and successful propagation may require coordinative complexes containing this solvent

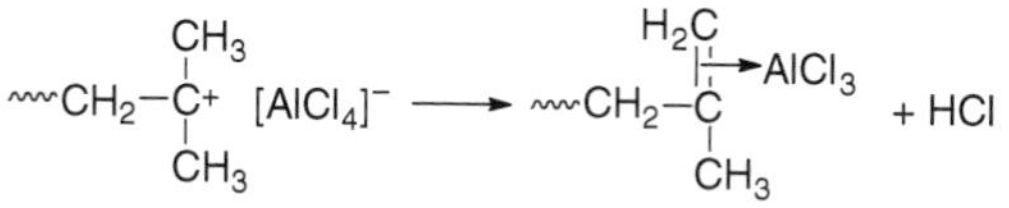

Scheme 3.1 Termination by loss of β-H from carbocationic chain end to counteranion.

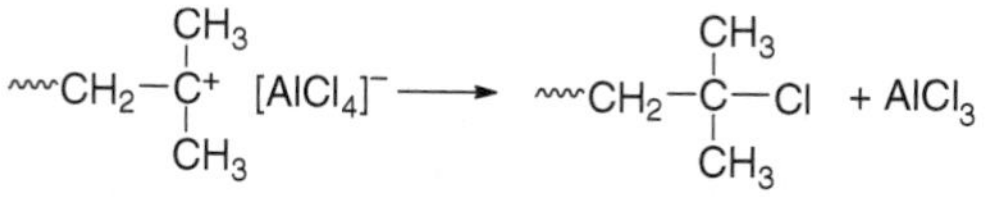

Scheme 3.2 Termination by collapse of propagating ion pair.

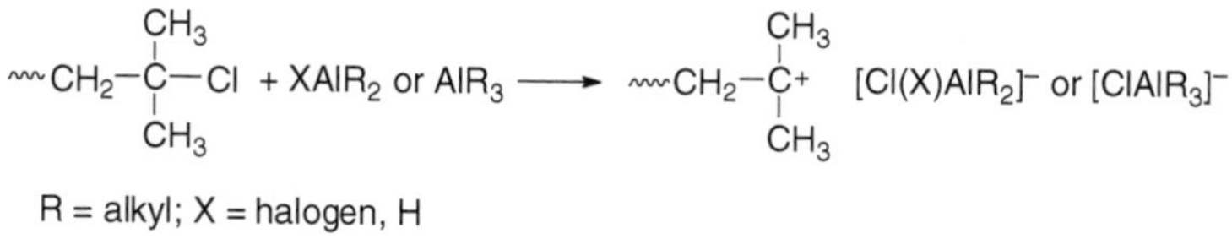

R = alkyl; X = halogen, H

Scheme 3.3 Reactivation of dead polymer by alkylaluminum coinitiators.

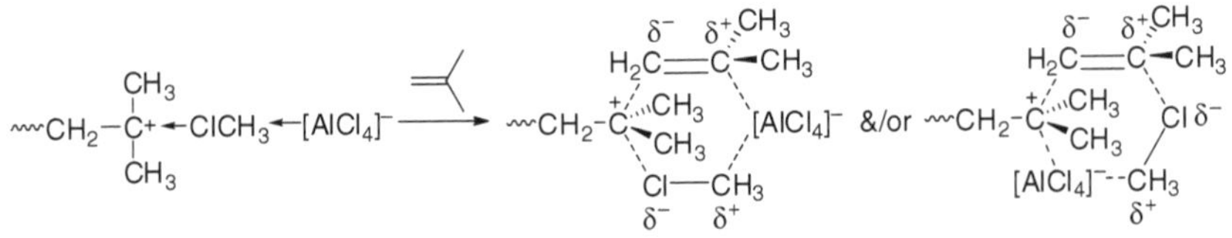

Scheme 3.4 Methyl chloride-propagating ion pair coordinative complexes.

(Scheme 3.4). The resultant ion pairs would be sufficiently separated to allow monomer to approach the propagating carbocation and any event that displaces CH_3Cl may promote ion pairing that leads to termination by transfer of chloride from $[AlCl_4]^-$ to polymer carbocation. Such behavior would explain the existence of chloro and olefinic end-groups in commercial samples of butyl rubber [32–34, 39].

The MW-T profile for the $AlCl_3$-CH_3Cl slurry system is unfavorable from an environmental standpoint in terms of energy input. Despite the use of polar CH_3Cl, the nucleophilicity of the counteranion derived from $AlCl_3$ is large enough to aid in the CT process to such an extent that high MW polymer cannot be readily made at elevated reaction temperatures. Temperatures of –103 °C are required for the synthesis of butyl rubber with desirable properties ($\overline{M}_w \sim 6 \times 10^5$ g·mol^{-1}) and polymerization is energy intensive [17–23]. Maximum MWs are obtained at T ≤ –95 °C, [17–23] most likely the result of the m.p. of CH_3Cl (–97.7 °C) [40].

The slurry process suffers from fouling; in essence, deposition of solid polymer onto heat exchangers and reactor walls [17–21, 23]. Fouling can involve polymer droplet coalescence leading to slurry instability and adherence of polymer to surfaces or by film deposition in which a polymer film directly forms on surfaces [41]. These processes not only impede flow but also limit heat transfer. This can lead to progressive slurry instability as reductions in heat transfer result in corresponding rises in temperature and viscosity resulting in additional fouling and further reduction in heat transfer. Fouling ultimately leads to periodic reactor shutdown (operational time = 10–90 hrs) and lengthy cleaning cycles (10–20 hrs) [17–21, 23]. This translates into increased energy use, materials consumption, and time input as polymer must be dissolved into an appropriate solvent (e.g. hexanes) during the cleaning process. Moreover, in order to maintain a given level of productivity a series of continuous flow stirred tank reactors (CFSTRs) that allow for alternative cycles of production and cleaning is required and results in a larger plant footprint. The slurry process is also not conducive to post-polymerization functionalization reactions (e.g. halogenation) and polymer must be dissolved in an appropriate solvent (e.g. hexane) prior to reaction, translating into additional investments in time, materials (equipment + solvent), and energy [17–21, 23, 29, 42].

From an environmental standpoint, CH_3Cl is a less than ideal polymerization diluent/coinitiator solvent given its increased toxicity [43] and higher global warming potential [44] in comparison to most hydrocarbons. Its mandatory use also presents additional complications. During workup of the slurry, CH_3Cl is subject to hydrolysis leading to the generation of materials (e.g. CH_3OH, CH_3OCH_3) that are detrimental to polymerization and must be removed [28]. Unreacted monomers must also be removed from the CH_3Cl recycle stream as on contact with fresh $AlCl_3$ (during preparation of the coinitiator solution) they oligomerize to form an insoluble polymer coating/film on its surface preventing further dissolution [45].

3.3   Progress Toward Cleaner Cationic Polymerizations

The major drawbacks inherent to cationic olefin polymerization have been implicated. Likewise, advances in the field are best illustrated in context to IB polymerization and follow below. Developments for green production of IB based polymers can be organized into two main groups, improvements to initiator systems and advances in process conditions and reactor design. Of the two, the former appears to be of greater significance compared to the latter in the context of cleaning up cationic polymerizations.

3.3.1   Improvements Resulting from Initiator System Design

By judicious choice of initiator components it is possible to achieve many of the goals of a green polymerization system. Although a great deal of research has been conducted on devising both homogeneous and heterogeneous initiator systems with improved performance characteristics, none encompass all aspects of an ideal green system. The following is a condensed overview of research conducted in both areas.

3.3.1.1 Progress in Homogeneous Initiator Systems

Homogeneous initiator systems are the most prolific and can be categorized into a number of distinctive subsets, which are not presented in chronological order.

1. Aluminum halide coinitiators of enhanced solubility.
2. Combinations of aluminum based LAs bearing different substituents.
3. Systems containing two or more LAs that do not share the same metal atom.
4. Organoaluminum coinitiators {i.e. dialkylaluminum halides (R_2AlX), trialkylaluminums (AlR_3)} with an initiator component other than adventitious moisture.
5. Initiator systems that function by electron transfer based mechanisms, often in conjunction with electromagnetic radiation.
6. Initiator systems containing components that constitute a weakly coordinating anion (WCA).
7. Organozinc halide coinitiators in combination with organohalide carbocation synthons.
8. Initiation by γ-radiation.

3.3.1.1.1 Aluminum Halide Coinitiators of Higher Solubility

In contrast to $AlCl_3$, $AlBr_3$ is soluble in a broad range of solvents including aliphatics and allows for polymerization in solution [9, 24, 25]. As such, polymerization gives rise to larger yields of butyl rubber with increased MWs and higher isoprene (IP) content at a given reaction temperature in comparison to $AlCl_3$ without the requisite use of toxic CH_3Cl (Table 3.2) [46]. Since polymerization is conducted in solution fouling is eliminated.

The mode of initiation by systems that employ aluminum halide coinitiators has been an area of debate. In the case of $AlCl_3$ in neat styrene it is evident that an initiator (e.g. adventitious moisture) is necessary for polymerization [47]. For IB in CH_2Cl_2 this becomes less clear and initiation may involve either self-ionization of the LA (Scheme 3.5), or abstraction of halide from solvent to generate a carbocation (Scheme 3.6) [48]. Support for the latter process comes from observation that when AlI_3 and n-Pr_3NI are dissolved in dichloromethane the resultant solution contains $ClAlI_3^-$, $Cl_2AlI_2^-$, $IAlCl_3^-$, and $AlCl_4^-$ in addition to AlI_4^- [25]. Polymerization of IB using $AlCl_3$ in both ^{14}C and ^{36}Cl labeled CH_3Cl results in the production of labeled polymer containing ^{14}C and ^{36}Cl radioactive chain ends, respectively [33, 39]. Due to the ambiguous nature of these experiments it is impossible to discern whether CH₃Cl is ionized directly by AlCl₃ and acts as an initiator (Scheme 3.7) or whether it reacts with propagating polymer carbocations (Scheme 3.8) and functions as a chain transfer agent (CTA). Given the higher stability of

Table 3.2 Comparison of slurry versus solution polymerization[a] of butyl rubber [46].

	Slurry Process	**Solution Process**	
Coinitiator	AlCl₃	AlBr₃	AlBr₃
Solvent/Diluent	CH₃Cl	Butanes	n-Heptane
Monomers (wt%)	$\overline{M}_v$ (kg·mol⁻¹)		
10	475	—	—
20	840	—	—
30	975	—	—
40	1,000	150	650
50	—	620	940
60	—	1,080	1,240
70	—	1,540	1,525
80	—	1,950	1,800

[a]Yields, time, and monomer feed rate were not specified. T = −95 °C, [IB] = 98.5 wt.%, [IP] = 1.5 IP wt.%.

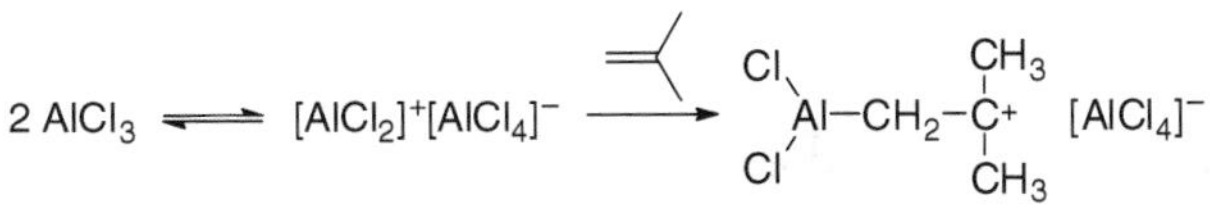

Scheme 3.5 Initiation of IB polymerization by self-ionization of aluminum chloride.

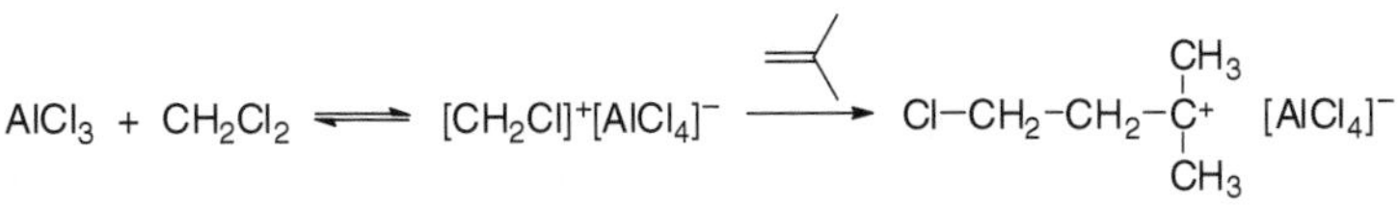

Scheme 3.6 Initiation of IB polymerization by ionization of solvent.

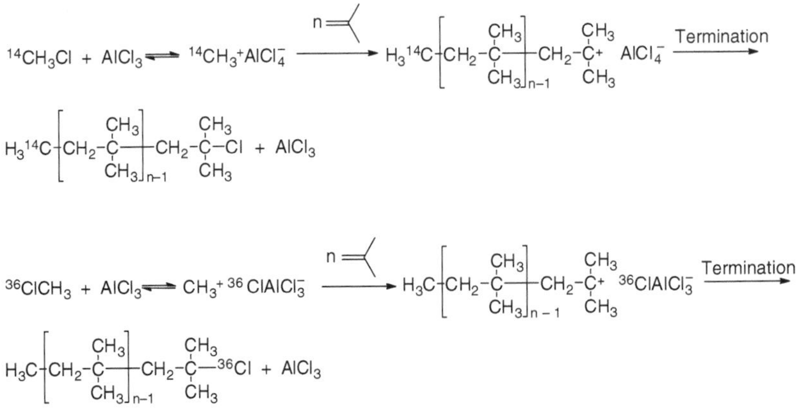

$^{14}CH_3Cl + AlCl_3 \rightleftharpoons {}^{14}CH_3{}^+AlCl_4^- \xrightarrow{n=\langle} H_3{}^{14}C\left[CH_2-\underset{CH_3}{\overset{CH_3}{C}}\right]_{n-1} CH_2-\underset{CH_3}{\overset{CH_3}{C^+}} AlCl_4^- \xrightarrow{\text{Termination}}$

$H_3{}^{14}C\left[CH_2-\underset{CH_3}{\overset{CH_3}{C}}\right]_{n-1} CH_2-\underset{CH_3}{\overset{CH_3}{C}}-Cl + AlCl_3$

$^{36}ClCH_3 + AlCl_3 \rightleftharpoons CH_3{}^+ \, {}^{36}ClAlCl_3^- \xrightarrow{n=\langle} H_3C\left[CH_2-\underset{CH_3}{\overset{CH_3}{C}}\right]_{n-1} CH_2-\underset{CH_3}{\overset{CH_3}{C^+}} \, {}^{36}ClAlCl_3^- \xrightarrow{\text{Termination}}$

$H_3C\left[CH_2-\underset{CH_3}{\overset{CH_3}{C}}\right]_{n-1} CH_2-\underset{CH_3}{\overset{CH_3}{C}}-{}^{36}Cl + AlCl_3$

Scheme 3.7 Incorporation of radioactive groups via initiation by labeled methyl chloride.

$H_3C\left[CH_2-\underset{CH_3}{\overset{CH_3}{C}}\right]_{n-1} CH_2-\underset{CH_3}{\overset{CH_3}{C^+}} AlCl_4^- \xrightarrow{{}^{14}CH_3Cl} H_3C\left[CH_2-\underset{CH_3}{\overset{CH_3}{C}}\right]_{n-1} CH_2-\underset{CH_3}{\overset{CH_3}{C}}-Cl + {}^{14}CH_3{}^+AlCl_4^-$

$\xrightarrow{n=\langle} H_3{}^{14}C\left[CH_2-\underset{CH_3}{\overset{CH_3}{C}}\right]_{n-1} CH_2-\underset{CH_3}{\overset{CH_3}{C^+}} AlCl_4^- \xrightarrow{\text{Termination}} H_3{}^{14}C\left[CH_2-\underset{CH_3}{\overset{CH_3}{C}}\right]_{n-1} CH_2-\underset{CH_3}{\overset{CH_3}{C}}-Cl + AlCl_3$

$H_3C\left[CH_2-\underset{CH_3}{\overset{CH_3}{C}}\right]_{n-1} CH_2-\underset{CH_3}{\overset{CH_3}{C^+}} AlCl_4^- \xrightarrow{{}^{36}ClCH_3} H_3C\left[CH_2-\underset{CH_3}{\overset{CH_3}{C}}\right]_{n-1} CH_2-\underset{CH_3}{\overset{CH_3}{C}}-{}^{36}Cl + CH_3{}^+ AlCl_4^-$

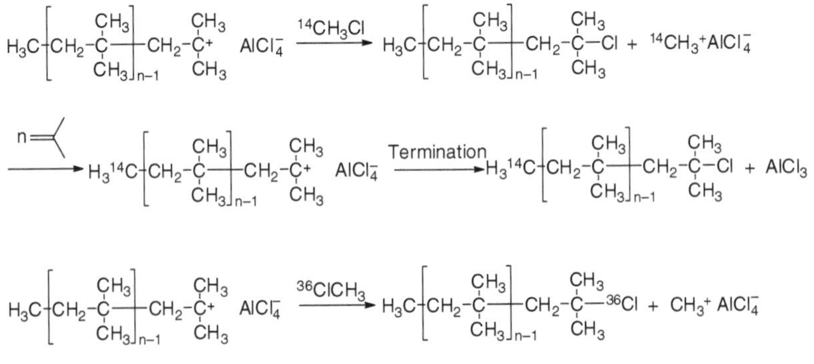

Scheme 3.8 Incorporation of radioactive groups via chain transfer to labeled methyl chloride.

the *t*-butyl carbocation versus the methyl carbocation and that ^{14}C incorporation occurs more readily at low T and low [IB], the former process seems more plausible. Ionization of solvent by LA might also explain the observation that a quiescent mixture of IB/CH₃Cl/ BF₃ prepared at −10 °C undergoes immediate polymerization once the temperature is reduced to ~ −78 °C [49]. Since such studies were conducted in the absence of sterically hindered pyridines (SHPs, Chart 3.2) the possibility that acidic impurities, especially those of low melting point (i.e. HCl), function as the initiator cannot be ruled out even at low reaction temperatures (Scheme 3.9).

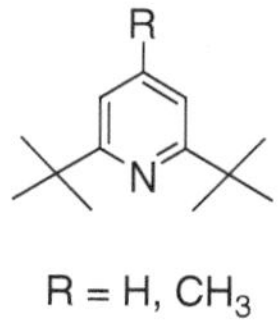

R = H, CH$_3$

Chart 3.2 Common sterically hindered pyridines.

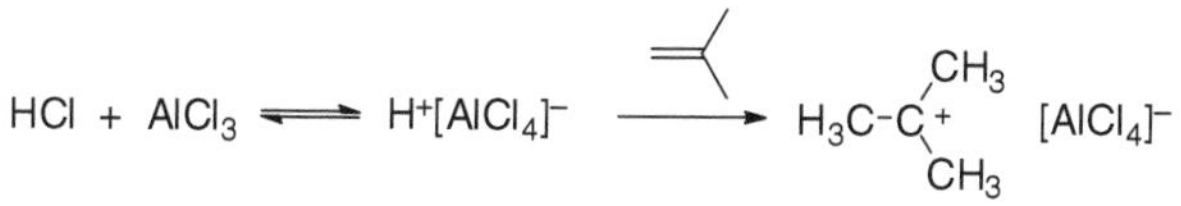

$$HCl + AlCl_3 \rightleftharpoons H^+[AlCl_4]^- \quad\quad\longrightarrow\quad\quad H_3C\text{-}\overset{CH_3}{\underset{CH_3}{C}}{}^+ \quad [AlCl_4]^-$$

Scheme 3.9 Initiation of IB polymerization by acidic impurities in aluminum chloride.

For AlBr$_3$ it would appear that initiation by self-ionization is likely. Marek and coworkers found that under high vacuum conditions in heptane, where BF$_3$ and TiCl$_4$ (coinitiators that require the presence of an initiator) are inactive, AlBr$_3$ is capable of inducing polymerization of IB [50]. This is a distinct possibility in lieu of the fact that polymerization of IB can be induced by self-ionization of the weaker LA BCl$_3$ when conducted in polar solvents [51, 52]. Under commercial reaction conditions initiation by impurities such as HBr or adventitious moisture might occur in addition to that by self-ionization and could be why polydispersity index (PDI) values for such polymerizations are higher (multiple modes of initiation) than those produced from AlCl$_3$. The enhanced MW-T profile for this system might be due to lower nucleophilicity and higher solubility of anions derived from AlBr$_3$ in comparison to those formed from AlCl$_3$.

Alkylaluminum dihalide (RAlX$_2$) coinitiators are reported to yield the highest MW polymers of the simple chemical initiator systems in polar media at temperatures above –100 °C [53]. They operate efficiently in hydrocarbons and form the basis of the only commercial solution polymerization process for butyl rubber (Table 3.3) [54, 55]. Adventitious moisture has been implicated as the initiator, present in the form of an aqua complex (i.e. **1**, Scheme 3.10) [54, 55]. Initiation most likely involves protonation of monomer by aqua complexes or haloaluminates derived from RAlX$_2$ (Schemes 3.10–3.11, respectively) [56]. Although MWs are

Table 3.3 Synthesis of butyl rubber by aqua adducts of alkyl aluminum dihalides and dialkyl aluminum halides [55].

Initiator (M)	Solvent	[IB] (vol. %)	[IP] (vol. %)	T (°C)	Time (min)	Conv. (%)	$\overline{M}_v$ (kg·mol⁻¹)
EtAlBr$_2$·H$_2$O 6.05 × 10⁻⁵	Isopentane	48.5	1.5	−85	15	18.3	70
EtAlCl$_2$·H$_2$O 1.89 × 10⁻⁴	Toluene	79.7	0.4	−80	15	12.0	56
iBuAlBr$_2$·H$_2$O 1.72 × 10⁻³	Cyclohexane	49.7	0.4	−50	18	28.4	52
Et$_2$AlCl·H$_2$O 3.94 × 10⁻⁴	Heptane	40	2.0	−85	18	16.2	68

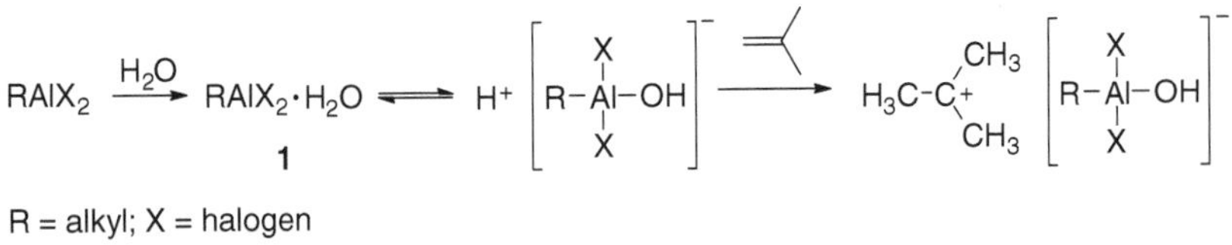

R = alkyl; X = halogen

Scheme 3.10 Initiation of IB polymerization by simple aqua complexes of alkylaluminum dihalides.

R = alkyl; X = halogen

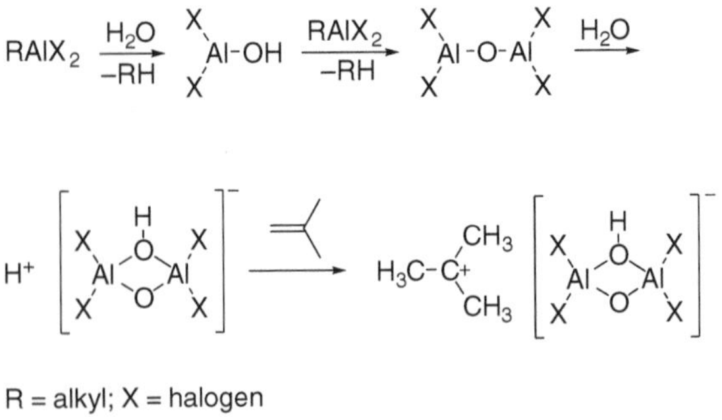

Scheme 3.11 Initiation of IB polymerization by complex haloaluminates of alkylaluminum dihalides.

depressed in aliphatic solvents in comparison to polar media, the nucleophilicity of anions derived from RAlX$_2$ coinitiators is low enough to permit production of commercially useful grades of butyl rubber in solution at temperatures of ~ −80 °C. Improvements to the MW-T profile may be attributed to anions of higher solubility that possess lower nucleophilicity compared to those produced from AlCl$_3$. These researchers also showed that aqua adducts of dialkylaluminum halides (R$_2$AlX) function in a manner similar to their RAlX$_2$ counterparts (Table 3.3) [54, 55].

3.3.1.1.2   Mixtures of Aluminum Containing LAs

An offshoot of RAlX$_2$ based systems was disclosed by Parker and coworkers (Table 3.4) [57]. This involves a mixture of R$_2$AlX and RAlX$_2$ with a 9:1 mol:mol ratio (respectively) stated as being the most active. Polymerization can be conducted in solution using nonpolar media and provides good yields of high MW butyl rubber at a reaction temperature of ~ –70 °C. A number of plausible modes of initiation can be envisioned in addition to those involving aqua complexes or haloaluminates derived from alkylaluminum halides (Schemes 3.10–3.11); such as, haloaluminates derived from dialkyl-aluminum halides (Scheme 3.12). Moreover, facile ligand exchange

Table 3.4 Preparation^a of butyl rubber in *n*-hexane using the Parker system [57].

Et$_2$AlCl (pbw)	EtAlCl$_2$ (pbw)	Time (min)	Conv. (%)	$\overline{M}_v$ (kg·mol^{-1})
1.2×10^{-3}	6.7×10^{-5}	55	24	543
1.2×10^{-3}	3.3×10^{-5}	90	23	604
—	6.7×10^{-5}	60	< 1	—

^{a}All concentrations are given on a part by weight (pbw) basis. T = –73 °C; [IB] = 98 pbw; [IP] = 2 pbw; [*n*-hexane] = 20 pbw.

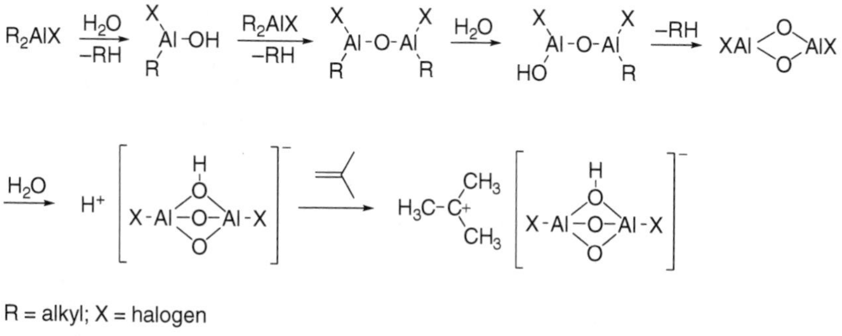

R = alkyl; X = halogen

Scheme 3.12 Initiation of IB polymerization by haloaluminates derived from dialkylaluminum halides.

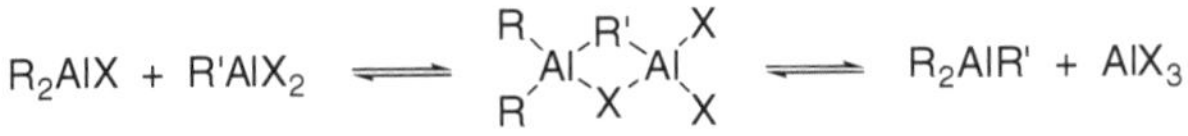

$$R_2AlX + R'AlX_2 \;\rightleftharpoons\; \underset{R \quad X \quad X}{\overset{R \quad R' \quad X}{Al \quad Al}} \;\rightleftharpoons\; R_2AlR' + AlX_3$$

R and R' = alkyl; X = halogen

Scheme 3.13 Generation of reactive coinitiators by ligand exchange.

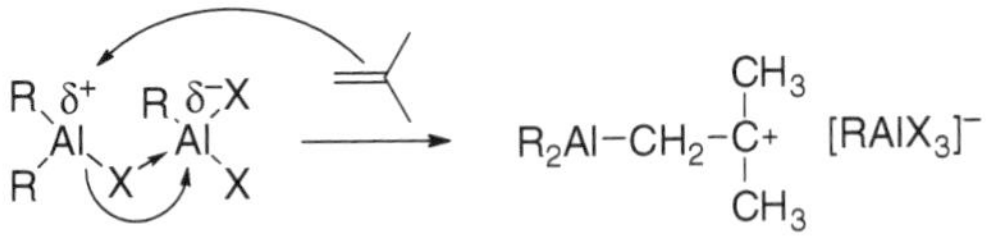

$$\underset{R \quad X \quad X}{\overset{R \,\delta^+ \quad R \,\delta^- X}{Al \quad Al}} \longrightarrow R_2Al-CH_2-\overset{CH_3}{\underset{CH_3}{\overset{|}{\underset{|}{C}}}}+ \; [RAIX_3]^-$$

R = alkyl; X = halogen

Scheme 3.14 Initiation of IB polymerization by transient aluminum cations.

between alkyl aluminum halides is known to occur readily [58, 59] and may lead to *in situ* generation of species such as alkylaluminum sesquihalides or even $AlCl_3$, which in turn ultimately coinitiate polymerization (Scheme 3.13). Such exchange reactions may also generate transitory ionic species to which monomer adds in a manner similar to self-ionization (Scheme 3.14). Moreover, as noted in Section 3.2 the LA components of this initiator system readily ionize *t*-butyl halides and would prevent irreversible termination by collapse of a propagating ion pair, thus leading to high yields.

It is interesting that an identical system [60, 61] had been described in the patent literature some years earlier by Kraus, albeit only low MW polymers were obtained. The failure of Kraus' experiments can be ascribed to the use of $CO_2(s)$ as an internal cooling agent, one which readily reacts with metal alkyls [58, 62–65]. The inability of an earlier, closely related initiator system based on $AlCl_3$ in conjunction with trialkylaluminum (AlR_3) compounds to produce high MW PIBs may result from use of ethene as an internal cooling agent [66]. Not only does this olefin form stable complexes with $AlCl_3$ [25, 36–38] but researchers [67] have shown that carbocations present in IB polymerization charges readily initiate oligomerization of ethene. Hydrocarbon suspensions of $AlCl_3$ also readily oligomerize ethene under mild conditions [36, 37].

3.3.1.1.3   Mixtures of LAs Based on Different Central Elements
The bulk of high temperature initiator systems using mixtures of LAs based on different central elements can trace their origin to two patent disclosures. In the first, Young and Kellog detail use of

LA metal alkyls, alkoxides, and oxy-alkoxides to improve the solubility of LA metal halides in hydrocarbons {e.g. Ti(OBu)$_4$ + AlCl$_3$} [68]. These initiator systems were described as being highly active for the preparation of butyl rubber in hydrocarbon solution at ~ −78 °C but no MW data was provided. Following their disclosure a number of researchers developed similar systems. For example, Strohmayer and coworkers found that alkyl aluminums and alkyl aluminum halides gave rise to relatively high MW PIBs in MeCl at increased reaction temperatures when used in conjunction with chloride bearing LAs (e.g. Et$_3$Al + SnCl$_4$; Table 3.5) [69]. It was found that unlike Ziegler catalysts which require a large excess of reducing alkyl aluminum component and do not polymerize IB, successful polymerization occurred when [alkyl Al] ≤ [chloride substituted LA]. Likewise, researchers at Mitsubishi found that VOCl$_3$, CrO$_2$Cl$_2$, and SnCl$_4$ (respectively) in conjunction with Et$_3$Al or Et$_2$AlCl form high MW PIBs at T = 0 to 10 °C in heptane [70].

The second group of mixed LA initiator systems can be traced to work by Marek *et al.* who discovered a heterogeneous initiator formed from Al(OsecBu)$_3$, BF$_3$, and TiCl$_4$ produces high MW PIBs in nonpolar media at elevated reaction temperatures [71, 72]. Following this a number of initiator systems based on BF$_3$ or TiCl$_4$ in conjunction with metal alkoxides, metal amides, and mixed metal oxide-metal alkoxides were devised. One of the earliest involved BF$_3$ in conjunction with Group 8, 9, and 10 metal alkoxides {e.g. Fe(OBu)$_3$} [73, 74]. This system is capable of producing high MW PIB/butyl rubber at elevated temperatures in hexanes (Table 3.6). A closely related system based on alkoxy aluminum and titanium halides {e.g. ClAl(OsecBu)$_2$; Cl$_3$TiOBu} in combination with BF$_3$ yields high MW polymers at high reaction temperatures in

Table 3.5 Polymerization[a] of IB by the triethylaluminum/tin (IV) chloride initiator system [69].

T (°C)	$\overline{M}_v$ (kg·mol^{-1})
−23	83.4
−35	576
−100	1,786

[a]All runs conducted in CH$_3$Cl where mol Et$_3$Al/mol SnCl$_4$ = 0.5. Details concerning the concentrations of initiator components and monomer as well as yield and reaction time not provided.

Table 3.6 Polymerization of IB[a] in *n*-hexane using the iron (III) alkoxide/boron trifluoride initiator system [73–74].

[Fe(OR)$_3$] (mM)	[IB] (vol. %)	Time (hr)	T (°C)	Conv. (%)	$\overline{M}_v$ (kg·mol^{-1})
5.0[b]	10	2	−10	72	151
10.0[b]	10	2	−60	50.5	1,303
5.0[b]	15	1	−30	62.8	352
30.0[b]	10	2	−20	100	143
10.0[b]	30	2	−20	87	150
8[c]	20	2	−50	66.1	401
3[c]	15	1	−20	56.9	121
7.5[c]	10	2	−20	96.2	143
10[c]	15	0.167	−10	64.7	88

[a]Initiator solutions made by bubbling BF$_3$(g) into 0.2 M benzene solutions of Fe(OR)$_3$ until saturated. Solvent = *n*-hexane.
[b]R = *n*-Bu.
[c]R = Et.

Table 3.7 Preparation[a] of butyl rubber in *n*-hexane using the alkoxy aluminum halide/boron trifluoride initiator system [75].

[ClAl(OsecBu)$_2$] (mM)	[BF$_3$] (mM)	Conv. (%)	$\overline{M}_v$ (kg·mol^{-1})
2.0	3.4	27.2	139
2.0	3.6	34.9	186
2.0	3.8	57.4	128
—	4.0	25.3	28

[a]Reaction conditions are as follows: [IB] = 1.2 M; [IP] = 2.3 × 10^{-2} M; solvent = *n*-hexane; T = −45 °C; time = 1 hr.

nonpolar media (Table 3.7) [75]. Mixed metal oxide-metal alkoxides {e.g. Zn[OAl(OEt)$_2$]$_2$} in combination with BF$_3$ also exhibit similar behavior (Table 3.8) [76]. By analogy, alkoxy aluminum halides

Table 3.8 Polymerization[a] of IB by the mixed metal oxide-metal alkoxide/boron trifluoride initiator system in n-heptane [76].

M[OAl(OR)$_2$]$_2$ (mmol)	[BF$_3$] (mmol)	Conv. (%)	$\overline{M}_v$ (kg·mol^{-1})
Zn[OAl(OiPr)$_2$]$_2$ 0.5	5.3	87.5	1,210
Zn[OAl(OnAm)$_2$]$_2$ 0.5	5.5	92.0	1,100
Zn[OAl(OnBu)$_2$]$_2$ 0.5	6.2	74.2	800
Zn[OAl(OEt)$_2$]$_2$ 0.5	6.5	68.8	750
—	4.0	68.5	95

[a]Reaction conditions are as follows: [IB] = 33.7 g; [n-heptane] = 150 mL; T = −65 °C; time = 1 hr.

Table 3.9 Polymerization[a] of IB by the EtAl(OEt)Cl/chloride bearing LA initiator system in methyl chloride [77].

LA (mmol)	$\overline{M}_v$ (kg·mol^{-1})
TiCl$_4$ 0.1	360
nBuOTiCl$_3$ 0.26	410
SnCl$_4$ 0.1	250
VOCl$_3$ 0.2	240

[a]Reaction conditions are as follows: [EtAl(OEt)Cl] = 2 mmol; [IB] = 28.4 g; [IP] = 0.84 g; [CH$_3$Cl] = 80 mL; T = −40 °C; time ~ 10–20 min.

(e.g. Cl$_2$AlOCH$_3$) and alkoxy alkyl aluminum halides {e.g. EtAl(OEt) Cl} were found to function in CH$_3$Cl with a variety of halogen bearing LAs (e.g. TiCl$_4$) to produce high MW butyl rubber at elevated

Table 3.10 Polymerization[a] of IB by the mixed metal amide/boron trifluoride initiator system in n-heptane [78].

Metal amide (mmol)	[BF$_3$] (mmol)	T (°C)	Conv. (%)	$\overline{M}_v$ (kg·mol^{-1})
Zn(NiBu$_2$)$_2$	8	−45	87.5	600
Ti(NEt$_2$)$_4$	4.5	−65	86.6	1,010
Al(NEt$_2$)$_{1.5}$Cl$_{1.5}$	12	−65	47.2	850
Zn(NEt$_2$)$_2$	4.5	−65	97.9	1,100
Si(NEt$_2$)$_4$	2.7	−65	47.5	320
Zn(NPh$_2$)$_2$	4.5	−65	87.5	720
Zn(NBu$_2$)$_2$	4.5	−65	90.5	910
Zr(NBu$_2$)$_2$	4.5	−65	75.1	530
—	4.5	−65	68.5	95

[a]Reaction conditions are as follows: [IB] = 33.7 g; [n-heptane] = 150 mL; time = 1 hr.

temperatures (Table 3.9) [77]. Substitution of metal (Zn, Al, Ti, Sn, Si, Zr, etc.) amide bearing LAs {e.g. Al(NEt$_2$)$_3$} for alkoxides with BF$_3$ yields high activity initiator systems capable of producing high MW PIB/butyl rubber in nonpolar solvents at high reaction temperatures (Table 3.10) [78].

The exact mode of initiation has not been determined for any of these systems. In some instances it is possible that ionization of one LA by the other occurs to generate a reactive ion pair to which monomer adds (Scheme 3.15). Ligand abstraction by a strong LA from a weaker one occurs [79–83] and direct initiation by the resulting ion pair formed from AlBr$_3$ and TiCl$_4$ has been demonstrated (Scheme 3.16) [79]. In many instances the process might actually involve ligand exchange type reactions between two different LAs resulting in momentary generation of ionic species (Scheme 3.17) or highly reactive LAs (i.e. EtAlCl$_2$, AlCl$_3$; Scheme 3.18). Monomer addition to cationic species generated in these processes would be similar to initiation by self-ionization.

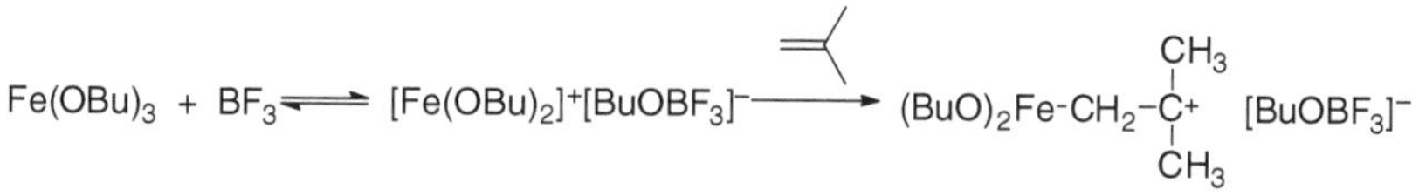

$$Fe(OBu)_3 + BF_3 \rightleftharpoons [Fe(OBu)_2]^+[BuOBF_3]^- \longrightarrow (BuO)_2Fe-CH_2-\overset{\overset{CH_3}{|}}{\underset{\underset{CH_3}{|}}{C}}^+ \ [BuOBF_3]^-$$

Scheme 3.15 Initiation of IB polymerization by ion pairs derived from mixtures of Lewis acids.

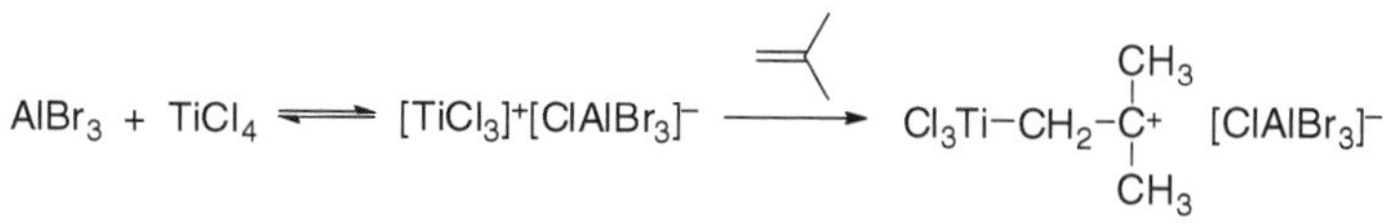

$$AlBr_3 + TiCl_4 \rightleftharpoons [TiCl_3]^+[ClAlBr_3]^- \longrightarrow Cl_3Ti-CH_2-\overset{\overset{CH_3}{|}}{\underset{\underset{CH_3}{|}}{C}}^+ \ [ClAlBr_3]^-$$

Scheme 3.16 Direct initiation of IB polymerization by the AlBr$_3$/TiCl$_4$ system.

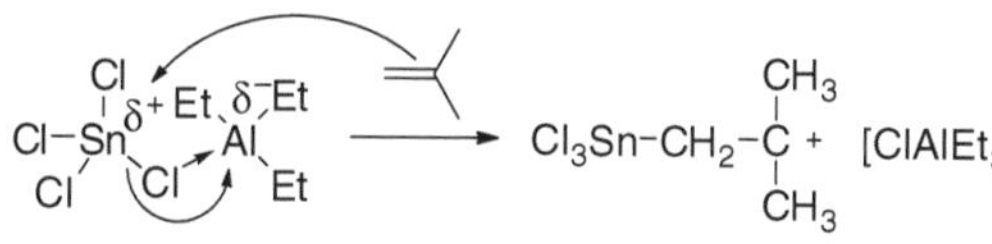

$$\longrightarrow Cl_3Sn-CH_2-\overset{\overset{CH_3}{|}}{\underset{\underset{CH_3}{|}}{C}}^+ \ [ClAlEt_3]^-$$

Scheme 3.17 Initiation of IB polymerization by transient cations generated from mixed Lewis acids.

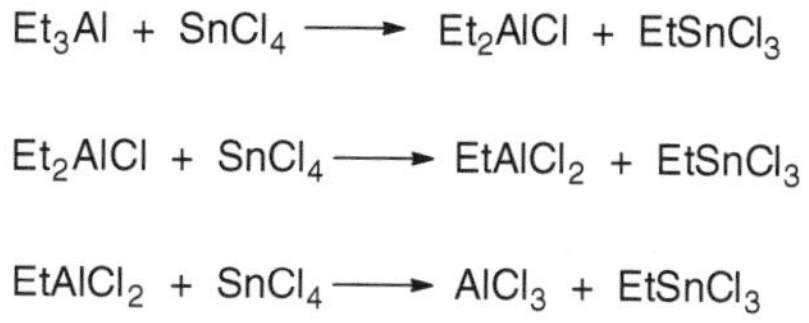

$$Et_3Al + SnCl_4 \longrightarrow Et_2AlCl + EtSnCl_3$$

$$Et_2AlCl + SnCl_4 \longrightarrow EtAlCl_2 + EtSnCl_3$$

$$EtAlCl_2 + SnCl_4 \longrightarrow AlCl_3 + EtSnCl_3$$

Scheme 3.18 Generation of highly reactive Lewis acid coinitiators by ligand exchange.

Initiation might also involve the reaction of adventitious moisture with mixed LA complexes to generate strong Brønsted acids (BAs) *in situ* that initiate polymerization by protonation of monomer (Scheme 3.19). Mixed LA complexes might form from reaction of lone pair electrons present in halide, alkoxide, or amide substituents in one of the LA components with the central atom of the other LA. This could increase the LA strength of the element bearing the donor substituent and anions derived from such mixed complexes via reaction with adventitious moisture would be bulkier and less nucleophilic in comparison to those derived from the parent LAs, thus improving the MW-T profile. Similar strong BAs

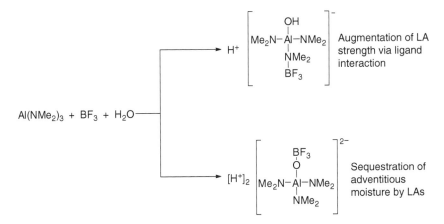

Scheme 3.19 Strong Brønsted acids derived from mixed Lewis acids and adventitious moisture.

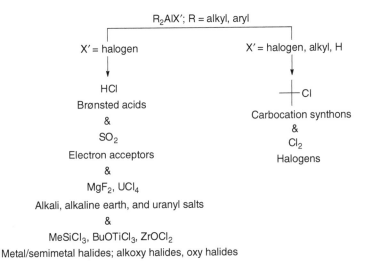

Chart 3.3 Trialkylaluminum and dialkylaluminum halide/hydride initiator systems for IB polymerization.

could be formed by concerted reaction of two dissimilar LAs with adventitious moisture.

3.3.1.1.4   Organoaluminum Coinitiators with Initiators Other Than Moisture

A vast amount of research has focused on the use of trialkylaluminum (R_3Al) and dialkylaluminum halide/hydride (R_2AlX', X'

= halogen or H) coinitiators for polymerization of IB. Both coinitiator subsets are incapable of inducing polymerization alone and are compatible with a broad range of initiators with those for the latter being the most varied (Chart 3.3). Initiator components that are useful in conjunction with Et_2AlCl include organohalide carbocation synthons (e.g. t-BuCl), hydrohalogen acids, halogens/interhalogens, electron acceptors (e.g. tetrachloro-p-benzoquinone, SO_2), alkali and alkaline earth metal salts, non-Al alkyl metal/semimetal halides, metal alkoxy halides, and metal oxy halides (e.g. $MeSiCl_3$, Cl_3TiOBu, $ZrOCl_2$). For R_3Al the number of active initiators appears to be limited primarily to organohalide carbocation synthons and halogens.

The voluminous amount of research conducted on R_2AlX based initiator systems has it origins in work conducted by Italian [84–86] and Russian [87] scientists. The former group reported that R_2AlX by itself was capable of inducing polymerization of IB to high MW at high temperatures when conducted polar (e.g. EtCl, PhCl) solvents (Table 3.11). Subsequent studies by the latter group showed that R_2AlX works effectively in conjunction with BA initiators including water complexed to salts (e.g. $LiCl \cdot nH_2O$) and carboxylic acids (e.g. CH_3CO_2H). From these results acidic impurities (i.e. HCl) are most likely the initiator in the previously mentioned R_2AlX/EtCl system.

Table 3.11 Polymerization[a] of IB by the dialkylaluminum halide/chlorinated solvent system [86].

Solvent	T (°C)	Conv. (%)	$\overline{M}_v$ (kg·mol^{-1})
EtCl	–70	41	1,600
EtCl	–50	72	1,070
EtCl	–30	100	425
EtCl	–10	100	58
ClBz	–45	27	680
ClBz	–30	26	430
ClBz	–10	31	168

[a]Reaction conditions are as follows: $[Et_2AlCl]$ = 10 mmol; [solvent] = 100 g; [IB] = 65 g; time = 1 hr.

Subsequent work involving additional BA initiators found that activity increased with increasing initiator acid strength implying a protonic mechanism (Scheme 3.20); however, conclusive proof of this was not provided (Table 3.12) [26, 27, 53]. In terms of polymer yield BAs are inefficient in conjunction with R_2AlX when compared to other initiator subsets.

Carbocation synthons [27, 53, 67, 88–91] and halogens [32, 92–100] as initiators with R_2AlX' and R_3Al were next explored. Using modeling experiments involving 2,4,4-trimethyl-1-pentene (TMP) it was shown that both LA coinitiators are capable of ionizing carbocation synthons to generate initiating carbocations (Scheme 3.21) [90, 91]. Compared to BAs, carbocation synthons are more efficient initiators and give rise to higher yields of polymers although MWs are generally lower at a given temperature (Table 3.13). TMP modeling experiments also showed that R_2AlX and R_3Al

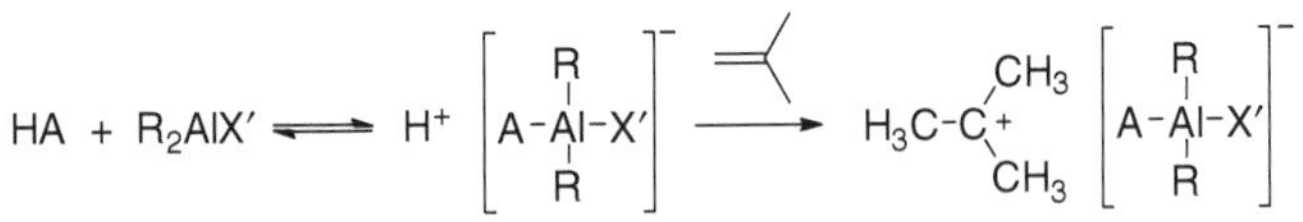

HA = Brønsted acid; R = alkyl; X' = halogen, H

Scheme 3.20 Protonic initiation of IB polymerization using Brønsted acid/dialkylaluminum halide systems.

Table 3.12 Polymerization[a] of IB by the dialkylaluminum halide/BA system in methyl chloride [26].

[HCl] (mmol)	T (°C)	Conv. (%)	$\overline{M}_v$ (kg·mol^{-1})
52	−24	20.5	143
67	−28	28.1	168
45	−55	30	397
31	−72	13.5	626
52	−77	28.1	561
660	−96	27.9	167

[a]Reaction conditions are as follows: [Et$_2$AlCl] = 5 mmol; solvent = CH$_3$Cl; [IB] = 9.69 M; time not specified.

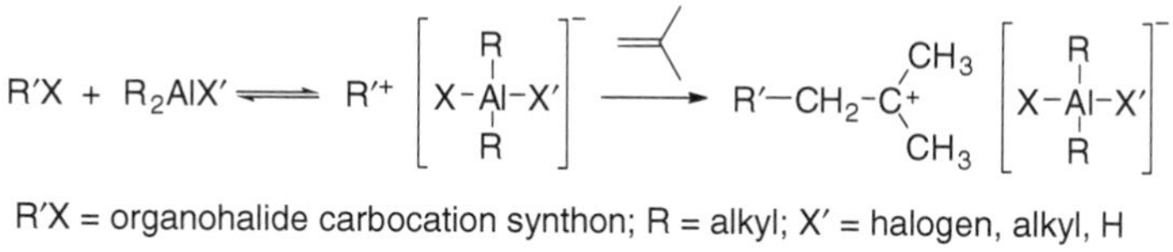

R′X = organohalide carbocation synthon; R = alkyl; X′ = halogen, alkyl, H

Scheme 3.21 Initiation of IB polymerization using carbocation synthon/dialkylaluminum halide or trialkylaluminum systems.

Table 3.13 Polymerization[a] of IB by the dialkylaluminum halide/carbocation synthon system in methyl chloride [89].

C⁺ Synthon (mmol)	Conv. (%)	$\overline{M}_v$ (kg·mol⁻¹)
tBuCl 7	2.8	667
iBuCl 58	6.8	831
secBuCl 1,200	13	701
iPrCl 1,140	86.1	253

[a]Reaction conditions are as follows: [Et₂AlCl] = 57.5 mM; [IB] = 6.27 M; solvent = CH₃Cl; T = –50 °C; time not specified.

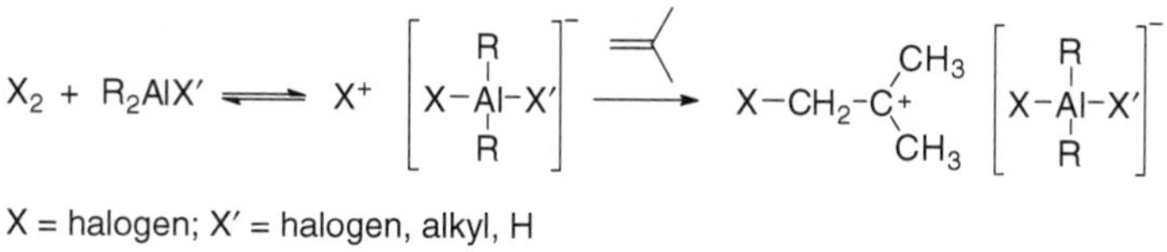

X = halogen; X′ = halogen, alkyl, H

Scheme 3.22 Initiation of IB polymerization using halogen/dialkylaluminum halide or trialkylaluminum systems.

ionize halogens to form reactive halocenium (X⁺) ions that go on to initiate polymerization (Scheme 3.22) [98, 99]. It was demonstrated that byproducts generated from the reaction of halogen with IB (via addition, Scheme 3.23) were not involved in initiation; however, the possibility that reactive LAs that can be generated from the reaction of coinitiator with halogen (via ligand exchange, Scheme 3.24) are involved in initiation cannot be completely ruled out. [98, 99].

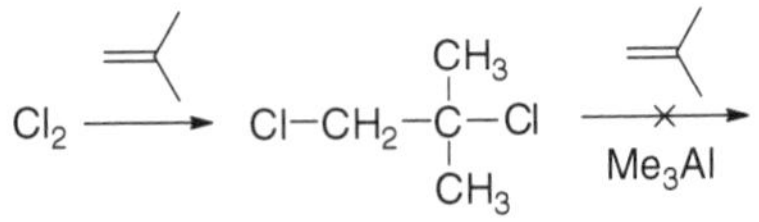

Scheme 3.23 Halogen/IB addition products do not function as initiators for IB polymerization.

$$R_nAlX_{3-n} + n\ X'_2 \longrightarrow X'_nAlX_{3-n} + n\ RX'$$

R = alkyl; X = halogen; X' = halogen

Scheme 3.24 Possible generation of highly active coinitiators via reaction of halogen with alkylaluminum compounds.

Table 3.14 Polymerization[a] of IB by the dialkylaluminum halide/halogen system in methyl chloride [100].

Halogen (mM)	T (°C)	Conv. (%)	$\overline{M}_v$ (kg·mol^{-1})
Cl$_2$ 12.6	−30	88	445
Cl$_2$ 4.2	−30	83.5	460
Cl$_2$ 12.6	−45	80.0	1,240
Cl$_2$ 4.2	−55	66.0	1,280
Br$_2$ 41.5	−30	16.9	465
Br$_2$ 41.5	−40	19.7	660
I$_2$ 18.5	−50	3.0	290

[a]Reaction conditions are as follows: [Et$_2$AlCl] = 7.7 mM; [IB] = 3.9 M; solvent = CH$_3$Cl (80 mL); time = 10 min.

Halogens were found to be highly effective initiators with MWs comparable to BA initiators and yields similar to carbocation synthons (Table 3.14).

Subsequent to work on more traditional initiators, the use of electron acceptors in conjunction with R_2AlX' was explored for homo- and copolymerization of IB [101, 102]. Polymerization is believed to perpetuate via dicarbocations derived from radical carbocations produced from reaction of IB with a semiquinone radical ultimately generated by a charge transfer complex (CTC) formed between the LA and electron acceptor (Scheme 3.25) [101]. ESR and modeling

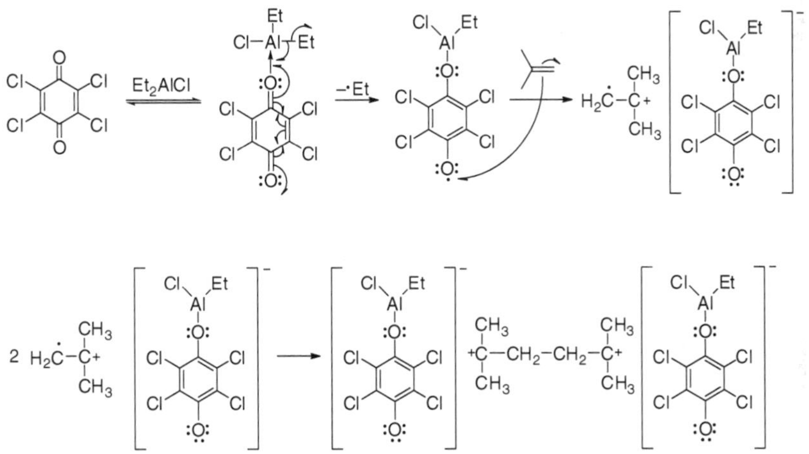

Scheme 3.25 Initiation of IB polymerization via semiquinone radicals.

Table 3.15 Synthesis of butyl rubber[a] by the dialkylaluminum halide/ electron acceptor system in methyl chloride [101].

[Et₂AlCl] (mM)	Electron Acceptor (mM)	Conv. (%)	$\overline{M}_v$ (kg·mol⁻¹)
8.3	Chloranil 0.50	57	285
8.3	Trinitrobenzene 4.16	27	375
16.7	Trinitrotoluene 3.7	18	380
16.7	Picric Acid 2.0	38	370

[a]Reaction conditions are as follows: [IB] = 3.9 M; [IP] = 97 mM; solvent = CH₃Cl (80 mL); T = −40 °C; time = 20 min.

studies with 1,1-diphenylethylene (DPE) show formation of radical carbocations whereas modeling studies provided proof of semiquinone radicals. O_2 inhibits polymerization via formation of peroxy radical carbocations providing further evidence for formation of radical carbocations [101]. From the data available (Table 3.15) it would appear that polar solvents are required for this initiator system. In comparison to other initiators, electron acceptors are highly active for polymerization of IB in conjunction with R_2AlX' where MWs decrease in the order EA > X_2 > RX.

A number of non-traditional initiators have been explored in conjunction with R_2AlX' coinitiators for IB polymerization. One

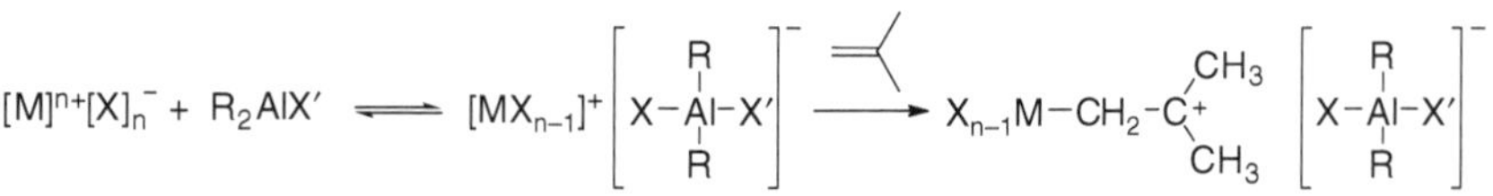

$$[M]^{n+}[X]_n^- + R_2AlX' \rightleftharpoons [MX_{n-1}]^+ \left[X-\underset{R}{\overset{R}{Al}}-X' \right]^- \longrightarrow X_{n-1}M-CH_2-\overset{CH_3}{\underset{CH_3}{C}}^+ \left[X-\underset{R}{\overset{R}{Al}}-X' \right]^-$$

M = alkali or alkaline earth metal; X = halogen, polyatomic anion; X′ = halogen, alkyl, H

Scheme 3.26 Initiation of IB polymerization by alkali/alkaline earth metal salt/dialkylaluminum halide systems.

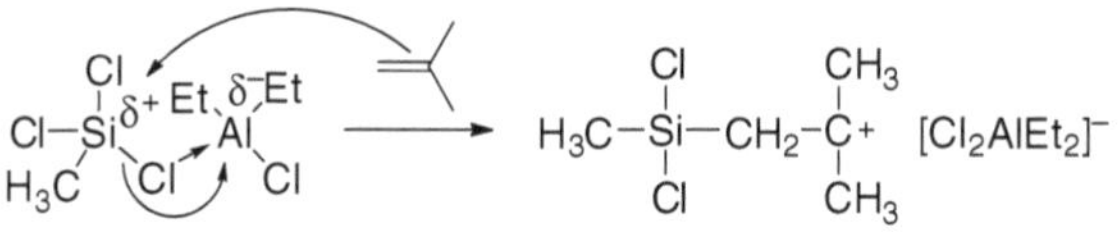

Scheme 3.27 Initiation of IB polymerization from transient cations derived from metal/semimetal halides, metal alkoxy halides, and metal oxy halides with dialkylaluminum halides.

Table 3.16 Synthesis[a] of butyl rubber by the dialkylaluminum halide/salt system in methyl chloride [103].

Salt (g)	Conv. (%)	$\overline{M}_v$ (kg·mol⁻¹)
$Mg(ClO_4)_2$ 0.490	41.2	320
MgF_2 0.522	43.7	240
UCl_4 0.071	26.4	290

[a]Reaction conditions are as follows: $[Et_2AlCl]$ = 16.7 mM; [IB] = 3.9 M; [IP] = 97 mM; solvent = CH_3Cl (80 mL); T = –35 °C; time ~ 25 min.

Table 3.17 Synthesis of butyl rubber[a] by the dialkylaluminum halide/chloride bearing LA system in methyl chloride [104].

Halide LA (mM)	Conv. (%)	$\overline{M}_v$ (kg·mol^{-1})
nBuOTiCl$_3$ 1.25	69.3	480
MeSiCl$_3$ 3.34	19.5	570
Et$_3$SnCl 1.50	69.5	360

[a]Reaction conditions are as follows: [Et$_2$AlCl] = 16.7 mM; [IB] = 3.9 M; [IP] = 97 mM; solvent = CH$_3$Cl (80 mL); T = −40 °C; time ~ 20 min.

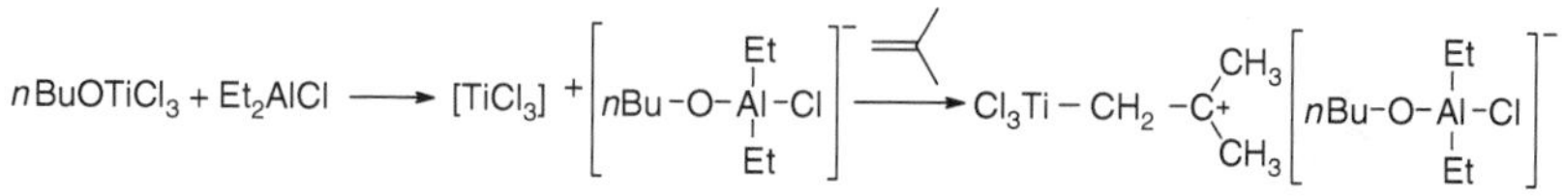

Scheme 3.28 Initiation of IB polymerization from ion pairs derived from metal/semimetal halides, metal alkoxy halides, and metal oxy halides with dialkylaluminum halides.

such class consists of alkali, alkaline earth metal, and actinide salts (Table 3.16) [103]. It is possible that these species form 1:1 addition products with R$_2$AlX′ in which metal cations are present even if transitory in nature. Initiation might involve monomer addition to such species (Scheme 3.26). Another subset contains a diverse array of metal/semimetal halides, metal alkoxy halides, and metal oxy halides (Table 3.17) [104]. Most likely these species undergo ligand type exchange reactions with R$_2$AlX′ to generate transitory cations (Scheme 3.27) or ionization to form ion pairs (Scheme 3.28) to which monomer adds, respectively. In support of the former process, studies have shown that silicon halide-containing initiators and monomers are stable towards complete ionization in conjunction with TiCl$_4$ [105, 106] and AlCl$_3$ [107]; however, studies involving Et$_2$AlCl [108] gave inconclusive results. Subsequent research involving the use of silicon halide initiators with methylaluminoxane (MAO) support the formation of ion pairs where evidence for formation of silicenium ions was detected [109].

With the exception of initiation by BAs (for Et_2AlCl) available data indicates that both R_3Al and R_2AlX' systems require MeCl to effect polymerization of IB in an efficient manner. This is especially evident for initiation by organohalide carbocation synthons. For instance, researchers at Exxon [110] note that such systems are not active for the production of butyl type rubbers in the absence of MeCl. Academic researchers have shown the R_2AlCl/carbocation synthon system only works efficiently in styrene polymerization when readily ionizable organohalides are used (i.e. $3° >> 2°$) [92] and that yields for the former ($3°$ carbocation synthons) are drastically reduced when polar solvent is omitted [53]. Similar results have been obtained for Et_2AlCl coinitiated polymerization of IB where 3 ° (e.g. t-BuCl) and resonance stabilized (e.g. $PhCH_2Cl$) carbocation synthons are the most efficient initiators [53]. Systems based on R_2AlX' give rise to higher MW polymers at a given temperature compared to those yielded by R_3Al systems for a given initiator subset.

3.3.1.1.5   Electron Transfer Based Initiator Systems

Group 5 metal halides and oxy halides (e.g. VCl_4, $VOCl_3$) are active coinitiators for the preparation of high MW IB polymers. In all cases polymerization appears to involve electron transfer processes that ultimately led to the generation of dicarbocationic species. The development of these systems can be traced to work conducted by Yamata and coworkers [111, 112]. These researchers found that $VOCl_3$ or VCl_4 in conjunction with aromatic or C=S containing "activators" allowed for the production of high MW PIBs in heptane at elevated reaction temperatures (Table 3.18). It was hypothesized that

Table 3.18 Polymerization[a] of IB by $VOCl_3$/Napthalene CTC system in n-heptane [112].

[Naphthalene] (mmol)	T (°C)	Time (hr)	Conv. (%)	$\overline{M}_v$ (kg·mol^{-1})
2.5	0	20	96	13.5
0.25	0	20	81.4	78.0
2.5	−78	40	93.5	67.7
2.5	−78	40	75	134

[a]Reaction conditions are as follows: $[VOCl_3]$ = 5 mmol; [IB] = 20 mL; n-heptane = 50 mL.

a CTC is produced from the reaction of the vanadium based coinitiator with the activator that subsequently reacts with monomer by an electron transfer process to generate radical carbocations that couple to produce dicarbocations (Scheme 3.29) [112].

Research in this area was furthered by Marek *et al.* on systems based on Group 4 and 5 metal halides (e.g. VCl_4, $TiCl_4$) [113–123]. Using ESR/EPR it was determined that polymerization involved formation of a coinitiator:monomer CTC that upon sufficient excitation undergoes electron transfer to produce an unstable radical carbocation-radical anion ion pair where the radical carbocations couple to form more stable dicarbocations (Scheme 3.30) [113, 116, 117, 120, 121, 123]. Thermal energy (even at low temperatures) is sufficient for excitation of the CTC when polymerization is run in polar solvents (i.e. CH_3Cl) or in the presence of dienes (e.g. butadiene) whereas electromagnetic radiation is required to induce polymerizations conducted under nonpolar conditions in the

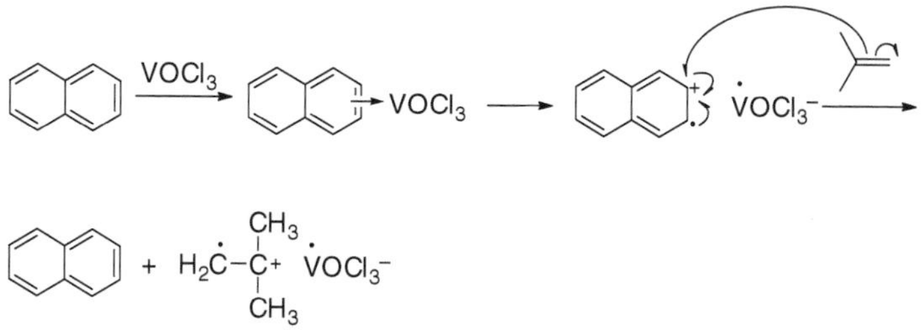

Scheme 3.29 Initiation of IB polymerization from CTCs derived from $VOCl_3$ and activators.

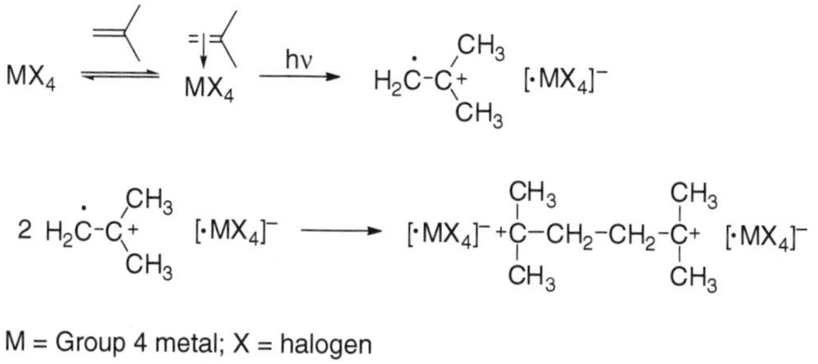

M = Group 4 metal; X = halogen

Scheme 3.30 Initiation of IB polymerization from CTCs derived from Group 4 LAs and radiation.

Table 3.19 Synthesis of butyl type rubbers[a] using VCl_4 and optional light in bulk monomer [117].

Comonomer (wt %)	Light[b] (Y/N)	T (°C)	Ind. Time[c] (min)	Pzn. Time[d] (min)	Conv. (%)	$\overline{M}_v$ (kg·mol^{-1})
—	N	−20	—	80	—	—
Isoprene 3.0	N	−40	17	10	53	Ins.[e]
Isoprene 1.0	Y	−25	0.5	5	33	300
Butadiene 3.0	N	−40	60	50	10	500
Butadiene 10.0	N	−40	45	30	46	Ins.[e]
Butadiene 10.0	Y	−40	21	20	58	470

[a]Polymerizations were performed in bulk with $[VCl_4] = 2 \times 10^{-3}$ M.
[b]Halogen bulb, 12 V, 55 W.
[c]Induction time.
[d]Polymerization time.
[e]Insoluble gel.

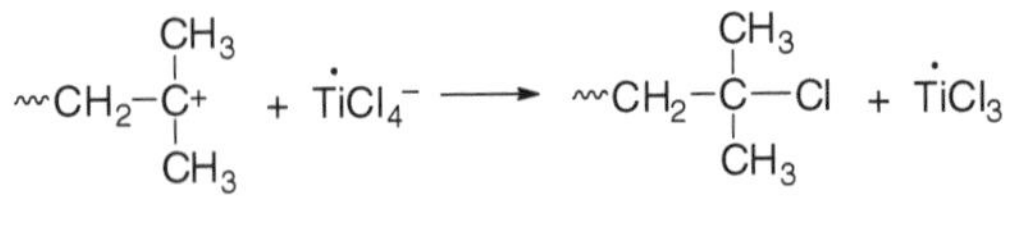

Scheme 3.31 Termination by decomposition of radical anion.

absence of diene at T < −10 °C (Table 3.19) [117]. ESR/EPR studies and inhibition of polymerization by O_2 provided proof of the existence of radical carbocations. [CTC] is directly proportional to the LA strength of the Group 4 coinitiator (as is polymerization rate) and inversely proportional to T.

Without continual irradiation, polymerizations conducted under nonpolar conditions at low T eventually cease due to instability of the radical anion (Scheme 3.31). Various additives (e.g. KOH [118], CaH_2 [119]) can be used to reduce the amount of radiation required

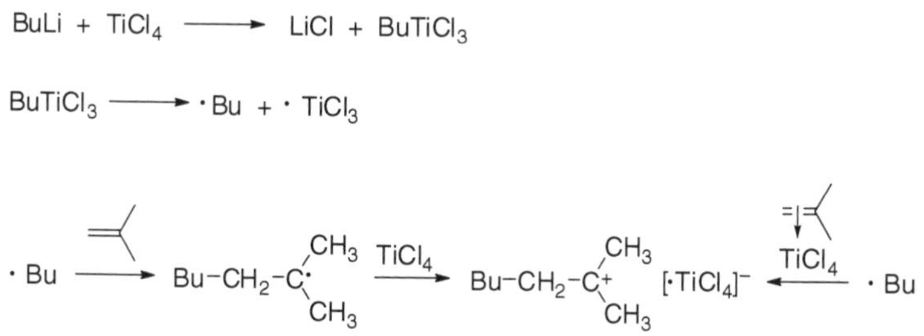
BuLi + TiCl4 → LiCl + BuTiCl3
BuTiCl3 → ·Bu + · TiCl3
·Bu
Bu–CH2–C·
CH3
CH3
TiCl4
Bu–CH2–C+
[·TiCl4]−
TiCl4
·Bu

Scheme 3.32 Initiation of IB polymerization by Group 4 LAs in conjunction with organoalkali compounds.

to induce and sustain polymerization. In a similar manner, alkyl lithium compounds can be substituted in place of electromagnetic radiation for polymerizations conducted in nonpolar media [113, 122]. Polymerization again appears to involve formation of radical carbocations but in this case they are believed to result from homolytic cleavage of unstable Group 4 alkyl compounds formed *in situ* (Scheme 3.32).

Production of gel free butyl rubber using VCl_4 systems is reported to require either polar solvents (e.g. CH_3Cl) or light (when conducted in nonpolar media) [117]. Recent work by researchers at Lanxess on the preparation of butyl rubber shows aging of the VCl_4/activator complex followed by removal of insoluble components results in a significant reduction in the gel fraction [124]. These researchers have also found inclusion of a organic nitro compounds (e.g. CH_3NO_2) in VCl_4-based initiator systems reduces the gel content of butyl rubber made in CH_3Cl but at the cost of greatly diminished yields (ca. < 6%) [125].

3.3.1.1.6 Initiator Systems Based on WCAs

Homogeneous initiator systems containing WCAs can be grouped into the following categories based on components and mode of initiation.

1. Systems based on PFLAs or their salts in conjunction with adventitious moisture.
2. Those using PFLAs or their salts in combination with carbocation synthons.
3. Systems that generate silylium tetrakis(pentafluorophenyl)borate ion pairs *in situ*.

4. Those based on metal cations paired with WCAs.

5. Systems that employ an alkylaluminoxane coinitiator.

*3.3.1.1.6.1   Adventitious Moisture in Combination with PFLAs
          and PFLA Salts*

One of the first reports of a WCA based initiator system involved application of the PFLA, tris(pentafluorophenyl)boron (**2**) with

Table 3.20 Polymerization[a] of IB by the tris(pentafluorophenyl)boron/water initiator system [128].

[H_2O] mM	Solvent	Yield (%)	$\overline{M}_n$ (kg·mol^{-1})	$\overline{M}_w/\overline{M}_n$
<0.1	CH_3Cl	17	1,330	4.2
1.0	CH_3Cl	87	411	4.2
<0.1	CH_2Cl_2/hex. 90/10 v/v	10	712	5.6
<0.1	CH_2Cl_2/hex. 80/20 v/v	< 1	—	—

[a][IB] = 3.6 M; [B(C$_6$F$_5$)$_3$] = 3.0 mM; T = –50 °C; time = 30 min.

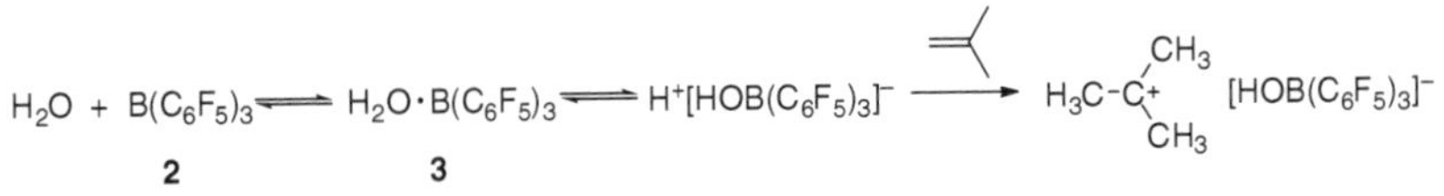

$$H_2O + B(C_6F_5)_3 \rightleftharpoons H_2O\cdot B(C_6F_5)_3 \rightleftharpoons H^+[HOB(C_6F_5)_3]^- \longrightarrow H_3C-\overset{CH_3}{\underset{CH_3}{C^+}} \quad [HOB(C_6F_5)_3]^-$$

$$\quad\quad\quad\quad \mathbf{2} \quad\quad\quad\quad\quad \mathbf{3}$$

Scheme 3.33 Initiation of IB polymerization by tris(pentafluorophenyl)boron and adventitious moisture.

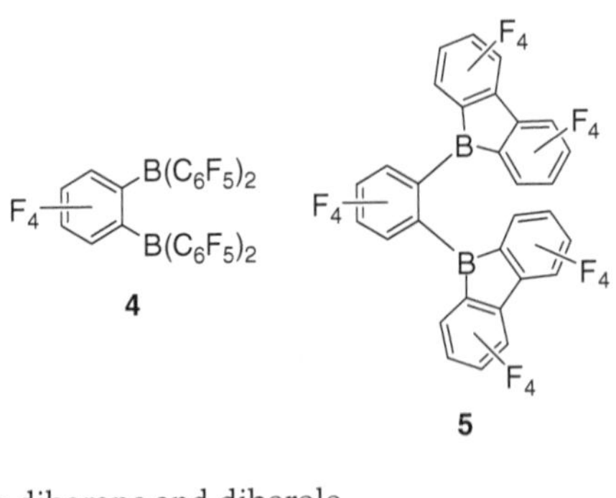

Chart 3.4 Chelating diborane and diborole.

adventitious moisture (Table 3.20) [126]. The aqua complex (3) functions as a strong BA capable of initiating polymerization under polar conditions by protonation of monomer (Scheme 3.33) and

Table 3.21 Polymerization[a] of IB using chelating diborane/water initiator system [127, 130].

Coinitiator (mM)	Temp (°C)	Yield (%)	$\overline{M}_w$ (kg·mol^{-1})	$\overline{M}_w/\overline{M}_n$
4 (2.00)[b]	−80	100	195	3.18
4 (0.480)[c]	−80	100	172	2.60
4 (2.00)[b,d]	−80	100	170	3.48
4 (0.200)[b]	−80	100	140	3.82
4 (0.0200)[b]	−80	97.7	305	1.99
4 (0.00200)[b]	−80	3.02	483	1.90
4 (2.00)[b]	−64	100	166	2.87
4 (2.00)[b]	−40	99.4	84.5	2.25
4 (2.00)[b]	−20	100	35.2	2.27
5 (0.200)[b]	−80	100	88.6	2.09
5 (0.200)[b]	−80	100	139	2.14

[a][IB] = 2.76 M; solvent = n-hexane; time = 1 hr.
[b]Coinitiator dissolved in 1 mL toluene.
[c]Coinitiator dissolved in 4.50 mL n-hexane.
[d]Solvent = CH_2Cl_2.

provides high MW PIBs at elevated temperatures. In nonpolar media this complex is not ionized sufficiently to induce polymerization [127, 128].

In contrast, the chelating diboranes (Chart 3.4) 1,2-$C_6F_4[B(C_6F_5)_2]_2$ (4) [127, 129–131] and 1,2-$C_6F_4[B(C_{12}F_8)_2]_2$ (5) [129, 130] function efficiently in nonpolar media as well as polar media being much stronger LAs than 2 (Table 3.21). Initiation is believed to result from protonation of monomer by the corresponding chelated aqua adduct (6, Scheme 3.34) [127, 132]. 4 is such an effective coinitiator that it provides high yields of PIBs even the in the presence of minute (ca. 1–2 ppm) quantities of water. MWs of PIBs produced by 4 and 5 are artificially low due to the highly exothermic nature of polymerization and inefficient heat transfer provided by magnetic stirring and external cooling used in these experiments. Support for

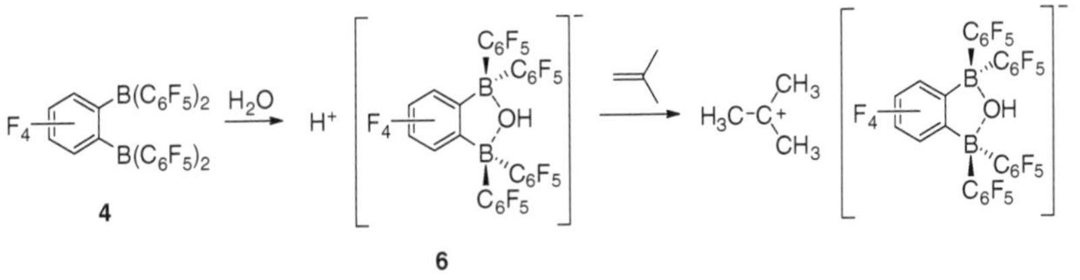

Scheme 3.34 Initiation of IB polymerization by the chelating diborane 4-aqua adduct.

Table 3.22 Chelating diborane/water system stopping[a] experiments [127].

DTBMP[b] (mM)	Yield (%)	$\overline{M}_w$ (kg·mol^{-1})	$\overline{M}_w/\overline{M}_n$
2	5.76	302	2.48
5	1.44	552	3.61
10	0.499	163	3.97
15	0.473	753	4.89
20	0	—	—

[a][IB] = 2.76 M in n-hexanes; [4] = 2.00 mM (dissolved in 1 mL toluene); T = –78 °C; time = 1 hr.
[b]DTBMP = 2,6-di-t-butyl-4-methyl-pyridine.

Table 3.23 Polymerization of IB using the PFLA salt/water initiator system [135].

[IB] (M)	$[M^+[B(C_6F_5)_4]^-]$ (mM)	$[H_2O]$ (mM)	Yield (%)	$\overline{M}_n$ (kg·mol^{-1})	$\overline{M}_w/\overline{M}_n$
0.97	1.38[a]	1.4	66	24.5	2.5
6.3	0.72[a]	3.5	30	9.90	2.1
6.3	1.3[a]	6.3	56	12.0	2.3
1.9	1.8[b]	—	—	1,610	2.0

[a]$M^+ = Ph_3C^+$; solvent = CH_2Cl_2; T = −20 °C; time = 60 min.
[b]$M^+ = Li^+$; solvent = MeCl; T = −80 °C; $[H_2O]$ and time not specified.

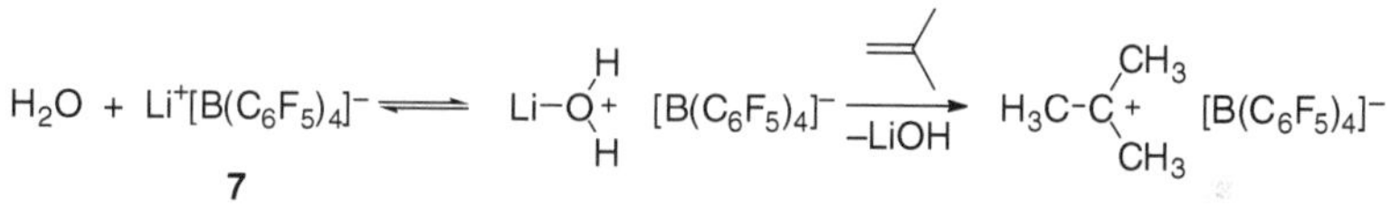

Scheme 3.35 Initiation of IB polymerization by lithium tetrakis(pentafluorophenyl) borate and adventitious moisture.

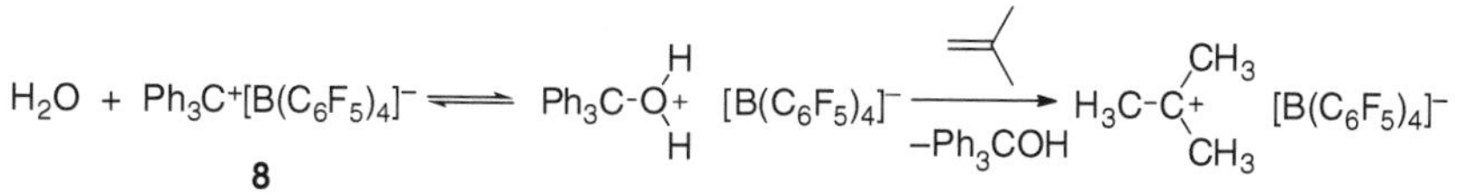

Scheme 3.36 Initiation of IB polymerization by trityl tetrakis(pentafluorophenyl) borate and adventitious moisture.

protonic initiation by these systems was provided by experiments [127, 130, 131] where purposeful addition of SHPs thwarted polymerization (Table 3.22); however, SHPs also decompose carbocations paired with WCAs and in such circumstances cannot provide conclusive evidence for proton mediated processes [127, 133].

Alkali metal {i.e. lithium tetrakis(pentafluorophenyl)borate, $Li^+[B(C_6F_5)_4]^-$, 7} and trityl {i.e. triphenylmethyl tetrakis (pentafluorophenyl)borate, $Ph_3C^+[B(C_6F_5)_4]^-$, 8} salts of PFLAs are also active coinitiators for IB polymerization [134–136]. These materials do not operate in nonpolar media and only the former coinitiator is capable of producing high MW PIBs (Table 3.23). Initiation is believed to arise from protonation of monomer by strong BAs generated *in situ* from interaction of the PFLA salt with adventitious moisture (Schemes 3.35–3.36). This was implied from experimentation using the SHP, 2,6-di-*t*-butyl-pyridine (DTBP) and the model monomer TMP in conjunction with **8**. It was reported that inclusion of the SHP prevented oligomerization of TMP with concomitant production of the corresponding SHP pyridinium ion. In the case of systems containing WCAs, SHPs are not diagnostic tools for discriminating between H$^+$ and other electrophiles since they form adducts with CpTiMe$_2^+$ [137] and decompose cumyl carbocation by β-H abstraction; [127, 133] however, despite flawed experimentation it seems likely that the proposed mechanism is operative.

3.3.1.1.6.2 Carbocation Synthons in Combination with PFLAs and PFLA Salts

Both monodentate [128, 134–136] (i.e. **2**) and bidentate [127, 129–131] (i.e. **4** and **5**) PFLAs are active coinitiators in conjunction with carbocation synthons for polymerization of IB. Similar to protonic initiation, the former (Table 3.24) is only active in polar media (e.g. CH$_2$Cl$_2$) whereas the latter function effectively regardless of solvent polarity (Table 3.25). The propensity of the carbocation synthon to undergo ionization is critical in determining overall activity, with

Table 3.24 Polymerization^a of IB by the tris(pentafluorophenyl)boron/ carbocation synthon initiator system [128].

[IB] (M)	C$^+$ Synthon (mM)	Solvent	Temp (°C)	Yield (%)	$\overline{M}_n$ (kg·mol^{-1})	$\overline{M}_w/\overline{M}_n$
3.6	COH^b 1.8	CH$_2$Cl$_2$	−30	24	159	2.2
5.1	COH^b 1.8	hexane	−30	—	—	—
5.1	TBDCC^c 2.0	CH$_2$Cl$_2$	−30	82	13.2	4.0

^{a}[B(C$_6$F$_5$)$_3$] = 3.0 mM; time = 30 min.
^{b}COH = Cumyl alcohol (2-phenylpropan-2-ol).
^{c}TBDCC = 5-*t*-butyl-1,3-bis(1-chloro-1-methylethyl)benzene.

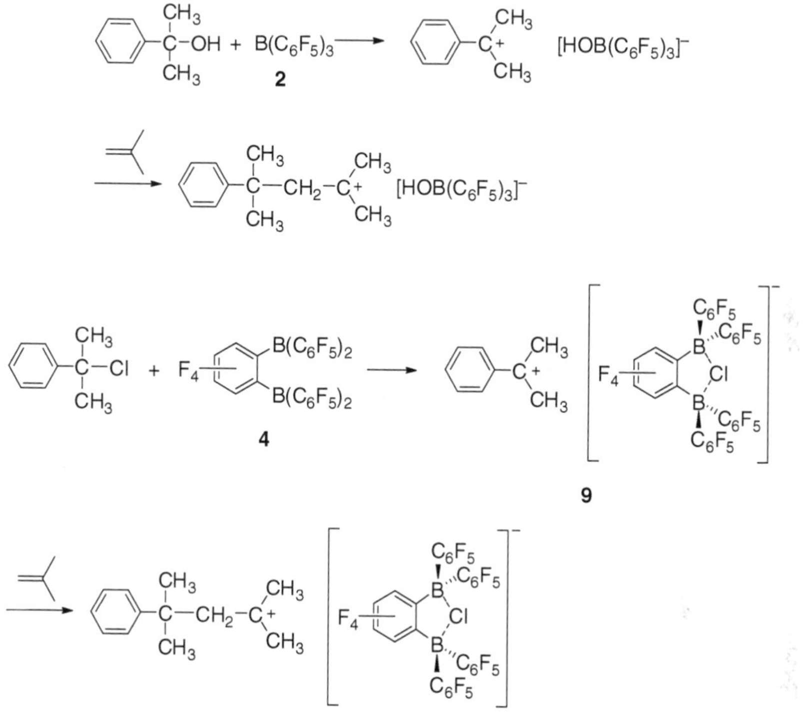

Scheme 3.37 Initiation of IB polymerization by mono and bidentate PFLAs and carbocation synthons.

3° halides being the most efficient. Initiation is believed to occur by ionization of the carbocation synthon by the PFLA to form an ion pair that then reacts with monomer (Scheme 3.37). ^{1}H and ^{19}F NMR spectroscopic analyses of modeling studies involving reaction of TMP with the 4/cumyl chloride (CCl) initiating system show the presence of 1-neopentyl-1,3,3-trimethylindane and [1,2-C$_6$F$_4${B(C$_6$F$_5$)$_2$}$_2$(µ-Cl)][127, 138]. These results support initiation involving monomer addition to the carbocationic portion of the ion pair [Cumyl][+][1,2-C$_6$F$_4${B(C$_6$F$_5$)$_2$}$_2$(µ-Cl)][(9) derived from reaction of PFLA and carbocation synthon. Additional evidence that such thermally liable ion pairs are generated from chelating diboranes 4 and 5 in conjunction with carbocation synthons was obtained using variable temperature ^{1}H and ^{19}F NMR spectroscopy [127, 131–133, 138] in addition to the synthesis of thermally stable model trityl salts [127, 131, 138].

Table 3.25 Polymerization[a] of IB by the chelating borane (**4**)/carbocation synthon initiator system [127, 130, 131].

C+ Synthon	Temp. (°C)	Yield (%)	$\overline{M}_w$ (kg·mol⁻¹)	$\overline{M}_w/\overline{M}_n$
CCl[b]	−90	50.8	512	1.68
CCl[b]	−80	42.2	361	1.72
CCl[b]	−64	36.3	267	1.61
CCl[b]	−40	27.2	155	1.61
CCl[b]	−20	15.6	99.0	1.59
CCl[c]	−90	73.0	243	1.77
CCl[c]	−80	67.4	236	1.74
CCl[c]	−64	55.6	206	1.76
CCl[c]	−40	32.6	136	2.09
CCl[c]	−20	28.9	103	1.83
CME[d]	−80	—	—	—
CME[d]	−20	—	—	—
CME[e]	−80	1.82	214	1.91

[a][diborane **4**] = 2.00 mM; [C+ Synthon] = 0.200 mM; [IB] = 2.76 M in *n*-hexane; time = 1 hr.
[b]C+ Synthon = cumyl chloride; [DTBMP] = 20.0 mM.
[c]C+ Synthon = cumyl chloride; [DTBMP] = 2.00 mM.
[d]C+ Synthon = cumyl methyl ether; [DTBMP] = 20.0 mM.
[e]C+ Synthon = cumyl methyl ether; [DTBMP] = 20.0 mM; solvent = CH_2Cl_2.

Inclusion of SHPs (e.g. 2,6-di-*t*-butyl-4-methylpyridine) were used to prevent both secondary initiation (from adventitious moisture) and CT. Indeed, SHPs do prevent these processes albeit at a price in that they readily scavenge carbocations lowering yields and preventing further polymerization (Table 3.25). Direct evidence of this behavior, which appears to be unique to WCA containing ion pairs, was provided in variable temperature ¹H and ¹⁹F NMR studies that showed SHPs readily abstract β-H from the carbocationic portion of ion pairs generated from **4** and CCl [127, 133]. Given the high activity of bidentate LAs coinitiators for polymerization

Table 3.26 Polymerization[a] of IB by the lithium tetrakis(pentafluorophenyl) borate/carbocation synthon initiator system [135].

C⁺ Synthon (mM)	[Li⁺[B(C₆F₅)₄]⁻] (mM)	Solvent	Yield (%)	$\overline{M}_n$ (kg·mol⁻¹)	$\overline{M}_w/\overline{M}_n$
TBDCC[b] 1.9	3.8	MCH/CH₂Cl₂ 60/40 v/v	100	87.1	10.9
TBDCC[b] 1.9	3.8	MCH/CH₂Cl₂ 60/40 v/v	96	53.7	4.8
TBDCC[b] 1.9	3.8	MCH	85	67.6	2.96
TMPCl[c] 1.7	1.9	MCH/CH₂Cl₂ 60/40 v/v	92	1,980	2.5

[a][IB] = 1.9 M; T = –80 °C; time not specified; MCH = methylcyclohexane.
[b]5-t-butyl-1,3-bis(1-chloro-1-methylethyl)benzene.
[c]2-chloro-2,4,4-trimethylpentane.

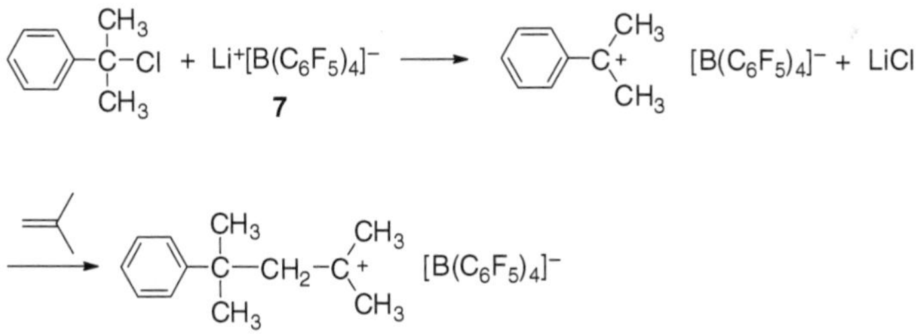

Scheme 3.38 Initiation of IB polymerization by PFLA salts and carbocation synthons.

of IB in conjunction with adventitious moisture, large quantities of SHP {[SHP] ≈ 5–10 [4]} are required to prevent protonic initiation. MWs produced from bidentate PFLA/carbocation synthon systems are lower than the theoretical value based on the ratio of moles of monomer to grams of initiator and most likely result from poor heat dissipation leading to increased CT.

Likewise, PFLA salts have been used as coinitiators in conjunction with carbocation synthons for polymerization of IB under almost completely nonpolar reaction conditions (Table 3.26) [134–136, 139]. Initiation is speculated to involve ion pair generation via metathetical chemistry (Scheme 3.38) and requires use of polar solvents, low temperatures and highly reactive carbocation synthons (i.e. 3° halides). When polymerization was conducted in polar media in the absence of an SHP the resultant polymers had high

Table 3.27 IP/IB copolymerization[a] by the bis(pentafluorophenyl) zinc·toluene/carbocation synthon initiator system [140–141].

$[Zn(C_6F_5)_2 \cdot tol]$ (mmol)	tBuCl (mmol)	Temp (°C)	Time (min)	Yield (%)	$\overline{M}_w$ (kg·mol⁻¹)	$\overline{M}_w / \overline{M}_n$
0.2	0.2	−78	30	7.0	794	1.6
0.1	0.1	−50	10	5.5	436	1.8
0.1	0.1	−35	10	7.9	310	1.7

[a][IB] = 100 mL; [IP] = 1.5 mL.

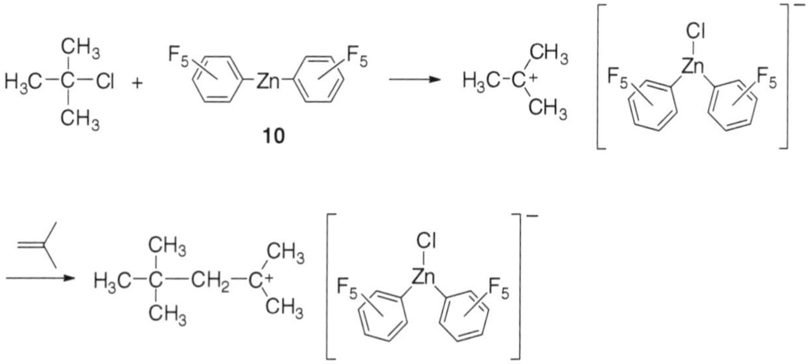

Scheme 3.39 Initiation of IB polymerization by bis(pentafluorophenyl)zinc and carbocation synthons.

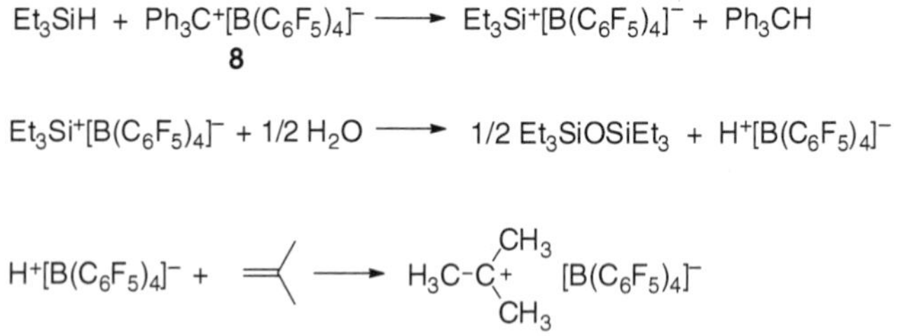

Scheme 3.40 Initiation of IB polymerization by silicenium-PFLA ion pairs generated by hydride abstraction.

PDIs which might indicate simultaneous, competitive initiation by adventitious moisture (Scheme 3.35).

More recently, the toluene adduct of bis(pentafluorophenyl) zinc (**10**) has been used as a PFLA coinitiator in combination with organohalide initiators for polymerization of IB (Table 3.27) [140,

Table 3.28 Polymerization[a] of IB by the trityl tetrakis(pentafluorophenyl) borate/triethylsilane initiator system [135].

[Et$_3$Si$^+$[B(C$_6$F$_5$)$_4$]$^-$] (mM)	Solvent	Temp (°C)	Yield (%)	$\overline{M}_n$ (kg·mol^{-1})	$\overline{M}_w/\overline{M}_n$
11.0	MCH	−20	54	0.700	2.7
9.4	MCH	−20	100	2.10	6.6
2.7	ClBz	−45	100	5.90	13.1

[a][IB] = 6.3 M; time = 10 min; MCH = methylcyclohexane.

Table 3.29 Polymerization[a] of IB by the lithium tetrakis(pentafluorophenyl) borate/trimethylsilyl chloride initiator system [144].

Time (min)	Yield (%)	$\overline{M}_n$ (kg·mol^{-1})	$\overline{M}_w/\overline{M}_n$
30	9.8	210	2.3
60	21.9	206	2.3
120	38.0	215	2.3
60	—	—	—

[a][Li$^+$[B(C$_6$F$_5$)$_4$]$^-$] = 4.0 mM; [Me$_3$SiCl] = 2.5 mM; [IB] = 9.5 mL; T = −35 °C.
[b][2,6-di-t-butyl-pyridine (DTBP)] = 4.0 mM.

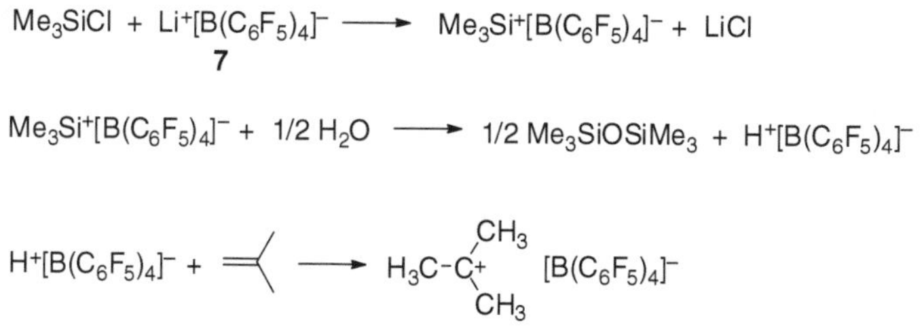

$$Me_3SiCl + Li^+[B(C_6F_5)_4]^- \longrightarrow Me_3Si^+[B(C_6F_5)_4]^- + LiCl$$
7

$$Me_3Si^+[B(C_6F_5)_4]^- + 1/2\ H_2O \longrightarrow 1/2\ Me_3SiOSiMe_3 + H^+[B(C_6F_5)_4]^-$$

$$H^+[B(C_6F_5)_4]^- + \text{>=<} \longrightarrow H_3C-\overset{CH_3}{\underset{CH_3}{C^+}}\ [B(C_6F_5)_4]^-$$

Scheme 3.41 Initiation of IB polymerization by silicenium-PFLA ion pairs generated by metathesis.

141]. Initiation is believed to involve formation of ion pairs in a manner similar to PFLAs **2**, **4** and **5** (Scheme 3.39) and evidence for this was provided by VT NMR studies as well as formation of thermally stable model trityl salts. Polymerization can be effected in nonpolar media but yields are low limiting the usefulness of this system. Reportedly, **10** is effective in combination with adventitious moisture for polymerization of IB but only to very low yields.

3.3.1.1.6.3 *Silylium Tetrakis(pentafluorophenyl)borate Initiator Systems*
The *in situ* generation of trialkylsilylium tetrakis(pentafluorophenyl) borate ion pairs by hydride abstraction [134–136] (Scheme 3.40, Table 3.28) and metathesis techniques [142–144] (Scheme 3.41, Table 3.29) for the polymerization of IB has been the focus of several research groups. The former technique has been shown to generate silylium cations [145] but no direct evidence for the formation of silicenium ion appears to have been presented for the latter approach. Initiation was envisioned to involve protonation of monomer by a strong hypothetical BA formed from reaction of the trialkylsilylium tetrakis(pentafluorophenyl)borate ion pair with adventitious moisture (Schemes 3.40–3.41). These assumptions were supported by the observation that SHPs inhibited polymerization for both routes; however, in light of more recent work where SHPs react with electrophiles other than H^+ (when paired with WCAs) such conclusions are speculative [127, 133, 137].

MWs for polymers afforded by the hydride abstraction route are low and it was rationalized that Et_3SiH functions not only as a silicenium ion precursor but also as a CTA, given the higher bond strength of C-H compared to Si-H (Scheme 3.42) [144]. Me_3SiCl (TMSCl) is unlikely to function as a CTA as the bond strength of Si-Cl exceeds that of C-Cl and may explain why MWs are higher for the metathetical route. The $Me_3SiCl/Li^+[B(C_6F_5)_4]^-$ system is unique in that it is capable of producing gel free butyl rubber in the presence of up to 12 mol% IP. Although MWs for polymers produced using the metathetical methodology [142–144] closely approximate

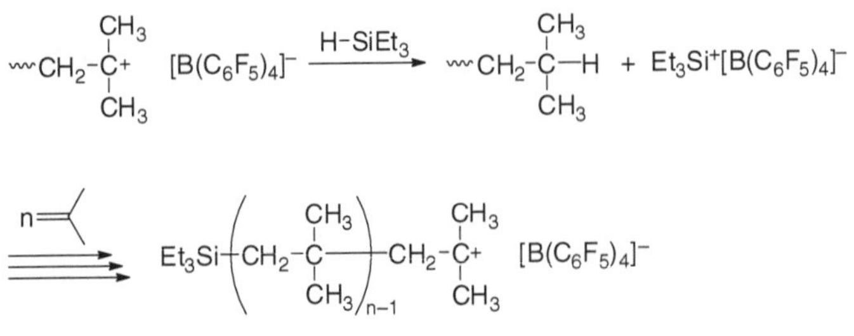

Scheme 3.42 Chain transfer by hydride abstraction from silanes.

Cp$_2$MMe$_2$; Cp'$_2$ZrMe$_2$; Cp*$_2$ZrMe$_2$; [Cp'$_2$ZrH$_2$]$_2$
(Common bis(cyclopentadienyl) complexes.)

B(C$_6$F$_5$)$_3$ (2); Ph$_3$C$^+$[B(C$_6$F$_5$)$_4$]$^-$ (8); Ph$_3$C$^+$[CN{B(C$_6$F$_5$)$_3$}$_2$]$^-$; Ph$_3$C$^+$[H$_2$N{B(C$_6$F$_5$)$_3$}$_2$]$^-$
(Common activators.)

M = Zr,Hf
Cp = cyclopentadienyl
Cp' = trimethylsilylcyclopentadienyl
Cp* = pentamethylcyclopentadienyl

Chart 3.5 Components of bis(cyclopentadienyl) WCA initiator systems for IB polymerization.

Table 3.30 Polymerization[a] of neat IB by zicronocenes in conjunction with tris(pentafluorophenyl)boron [146].

Zr Complex	Temp (°C)	[IB] (mL)	Yield %	$\overline{M}_n$ (kg·mol^{-1})	$\overline{M}_w/\overline{M}_n$
Cp*$_2$ZrMe$_2$[b]	−70	10	50	472	2.7
Cp*$_2$ZrMe$_2$[b]	−50	10	17	306	2.1
Cp*$_2$ZrMe$_2$[b]	−30	10	11	157	1.9
Cp*$_2$Zr(OH)$_2$[b]	−70	10	1	740	4.0
Cp*$_2$Zr(OH)$_2$[b]	−50	10	1	810	2.2
Cp*$_2$Zr(OH)$_2$[b]	−30	10	1	170	2.1
(Cp'$_2$ZrMe)$_2$(μ-O)[b]	−70	10	—	—	—
Cp$_2$ZrMe$_2$[c]	−78	100	8	647	3.8
Cp$_2$ZrMe$_2$[c]	−50	100	0.1	120	2.9
Cp$_2$ZrMe$_2$[c]	−30	100	0.1	72.0	2.5

[a][Zr] = 50 µmoles, [B(C$_6$F$_5$)$_3$] = 50 µmoles; initiator components dissolved in CH$_2$Cl$_2$.
[b]Time = 5 min.
[c]Time = 30 min.

those afforded by γ-radiation polymerization rates are low for a chemical initiator system and polymerization does not occur in aliphatic solvents due to poor solubility of the initiator components.

Table 3.31 Polymerization[a] of IB in chlorobenzene using metallocene/tris(pentafluorophenyl)boron [135].

[Zr Complex] (mM)	[H$_2$O] (mM)	[DTBP] (mM)	Time (min)	Yield (%)	$\overline{M}_n$ (kg·mol^{-1})	$\overline{M}_w / \overline{M}_n$
Cp*$_2$ZrMe$_2$ 1.28	ND	—	45	63	132	1.8
Cp*$_2$ZrMe$_2$ 1.28	ND	22.0	45	—	—	—
Cp*$_2$ZrMe$_2$ 9.2	3.15	44.5	45	20	162	1.8
Cp$_2$HfMe$_2$ 1.53	ND	—	90	44	150	1.8
Cp$_2$HfMe$_2$ 2.53	3.15	44.5	45	—	—	—

[a][B(C$_6$F$_5$)$_3$] = [metallocene]; [IB] = 6.3 M in ClBz; T = –20 °C.

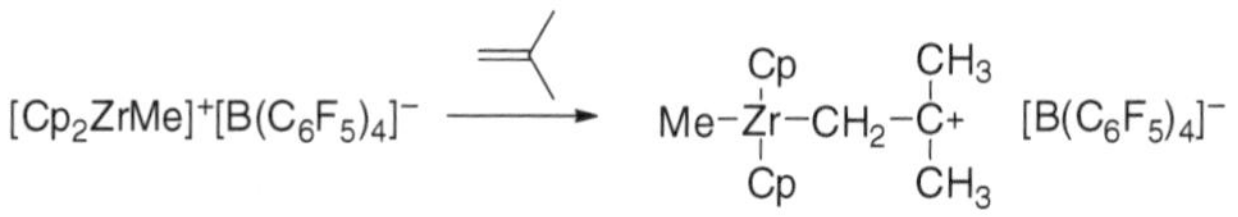

Scheme 3.43 Initiation of IB polymerization by direct addition to of bis(cyclopentadienyl) WCA ion pairs.

3.3.1.1.6.4   Metal Cation WCA Initiator Systems

Many metal cation WCA ion pair initiator systems have been explored in the polymerization of IB. These can be categorized as follows:

1. Bis(cyclopentadienyl) WCA complexes.
2. Monocyclopentadienyl WCA complexes.
3. Non-cyclopentadienyl Group 4 metal WCA complexes.
4. Transition metal nitrile WCA complexes.

Bis(cyclopentadienyl) WCA Initiator Systems

Bis(cyclopentadienyl) WCA complexes derived from dialkyl zirconocenes and PFLA activators (Chart 3.5) are active for preparation of high MW IB polymers at elevated temperatures (Tables 3.30–3.31) [134–136, 139, 146, 147]. The main initiation process is believed to

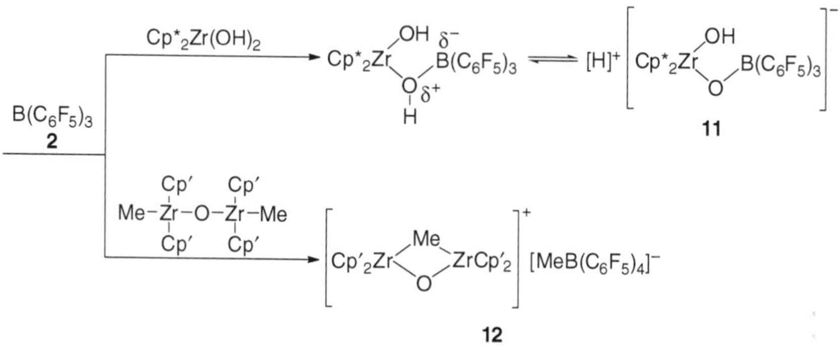

Scheme 3.44 Ion pairs generated from metallocene decomposition products.

involve direct addition of monomer to metal cation forming a carbocation (Scheme 3.43).

Another potential mode of initiation involves the formation of strong BAs derived from reaction of the metal cation with adventitious moisture and is supported by experiments involving SHPs (Table 3.31) [135]. SHPs prevent polymerization when [metallocene] ≤ [H$_2$O] but not in cases where [metallocene] ≥ [H$_2$O]. This behavior suggests that only protonic initiation occurs in the former case whereas in the latter initiation may involve both monomer addition to metal cation and protonation of monomer [135]. As mentioned previously, SHPs do not provide conclusive proof of proton mediated processes when WCAs are involved [127, 133, 137]. Subsequent work by Bochmann and coworkers demonstrated that adventitious moisture is more likely to generate metallocene decomposition products (Scheme 3.44) that either function as BAs in conjunction with PFLAs to initiate polymerization (**11**) or are completely inactive (**12**) [146]. Although polymerization using these initiator systems can be conducted in bulk, monomer conversions are typically low without chlorinated solvent. Moreover, ultra high purity reagents are required due to the sensitivity of the initiator components making large scale use impractical.

Likewise, zirconocene hydride complexes in combination with PFLA activators are highly active for polymerization of IB and its copolymerization with isoprene (Table 3.32) [148]. Initiator systems derived from cationic zirconocene hydride complexes are considerably more active in rates and yields in comparison to those based on the corresponding dimethyl zirconocene. The active initiator is a binuclear cationic trihydride WCA ion pair derived from reaction

Table 3.32 Polymerization[a] of IB by the zirconocene hydride complex/PFLA system [148].

Anion	[Init] (μM)	IP (mL)	Temp (°C)	Time (min)	Yield (%)	$\overline{M}_n$ (kg·mol⁻¹)	$\overline{M}_w/\overline{M}_n$
$[CN\{B(C_6F_5)_3\}_2]^-$	100	—	–78	4	9.2	337	2.6
$[CN\{B(C_6F_5)_3\}_2]^-$	100	—	–50	5	5.7	322	2.1
$[CN\{B(C_6F_5)_3\}_2]^-$	100	—	–35	5	7.4	269	1.9
$[CN\{B(C_6F_5)_3\}_2]^-$	70	1.5	–78	2	11	381	2.2
$[CN\{B(C_6F_5)_3\}_2]^-$	70	1.5	–50	3	7.7	284	2.0
$[CN\{B(C_6F_5)_3\}_2]^-$	70	3	–35	7	12	236	2.1
$[CN\{B(C_6F_5)_3\}_2]^-$	70	1.5	–25	3	11	148	1.9
$[H_2N\{B(C_6F_5)_3\}_2]^-$	70	1.5	–78	10	4.6	325	1.9
$[H_2N\{B(C_6F_5)_3\}_2]^-$	70	1.5	–78	3	9.8	243	2.1
$[H_2N\{B(C_6F_5)_3\}_2]^-$	70	1.5	–50	4	6.4	234	2.0
$[H_2N\{B(C_6F_5)_3\}_2]^-$	70	1.5	–35	3	15	186	1.9
$[B(C_6F_5)_4]^-$	100	—	–78	2	2.7	585	2.9
$[B(C_6F_5)_4]^-$	100	—	–50	1.5	4.5	467	1.9
$[B(C_6F_5)_4]^-$	100	—	–35	1	8.8	205	2.0
$[B(C_6F_5)_4]^-$	100	1.5	–78	30	2.7	243	2.5
$[B(C_6F_5)_4]^-$	100	1.5	–35	6	4.3	179	1.9
$[B(C_6F_5)_4]^-$	70	1.5	–35	2	6.9	198	1.9
$[B(C_6F_5)_4]^-$	70	1.5	–25	10	4.5	120	2.0

[a][IB] = 100 mL.

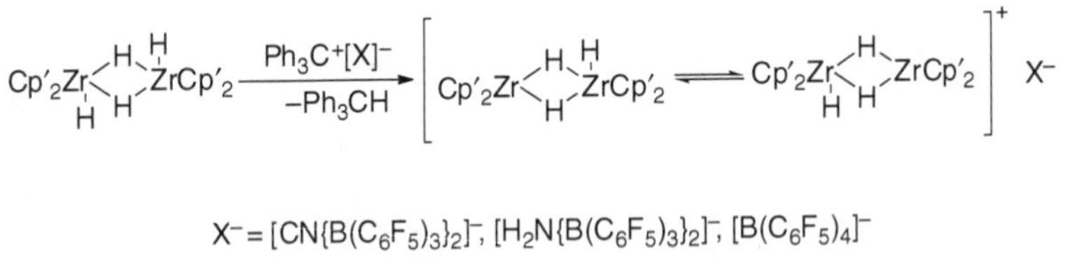

$X^- = [CN\{B(C_6F_5)_3\}_2]^-,\ [H_2N\{B(C_6F_5)_3\}_2]^-,\ [B(C_6F_5)_4]^-$

Scheme 3.45 Binuclear zirconocene hydride complexes for IB polymerization.

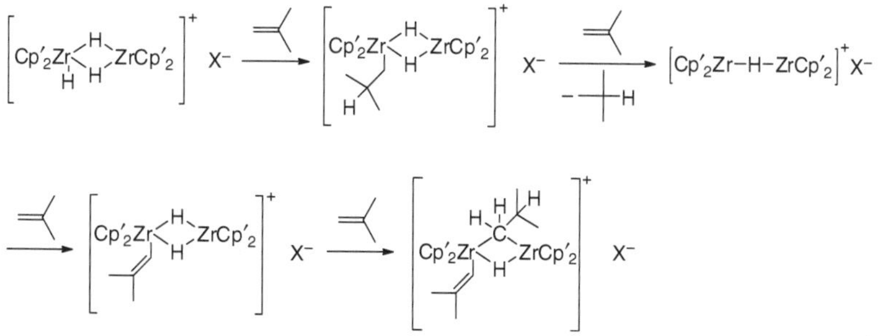

$X^- = [CN\{B(C_6F_5)_3\}_2]^-, [H_2N\{B(C_6F_5)_3\}_2]^-, [B(C_6F_5)_4]^-$

Scheme 3.46 Proposed mechanism for polymerization of IB by binuclear zirconocene hydride complexes.

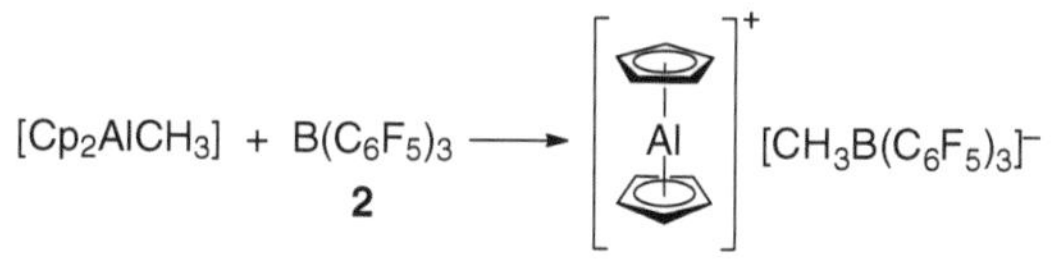

Scheme 3.47 Aluminocenium methyl-tris(pentafluorophenyl)borate ion pair.

of the corresponding binuclear zirconocene hydride complex with trityl PFLA activators (Scheme 3.45). When polymerization was conducted in a super dry system where 1 mol of IB contained <1 µmol of water, high molecular weight polymers were produced at increased reaction temperatures. Polymer MWs increase with decreasing anion nucleophilicity in the following order (highest to lowest MWs) $[CN\{B(C_6F_5)_3\}_2]^- > [H_2N\{B(C_6F_5)_3\}_2]^- \sim [B(C_6F_5)_4]^-$. Studies involving the purposeful addition of water to these systems showed that substoichiometric quantities of moisture had a dramatic, detrimental impact in reducing both yields and MWs with no polymerization occurring when $[H_2O] \geq [Cp'_2ZrH_2]_2$. The exact identity of the initiating species has yet to be determined; however, NMR and crystallographic analyses indicate that alkyl/hydride-bridged binuclear complexes formed by insertion of monomer into a Zr–H bond may be involved (Scheme 3.46) [148].

Group 3 and 13 metallocene compounds have been used as initiators for IB polymerization in combination with PFLA activators. For example, Bochmann and coworkers examined the ion pair (13) derived from reaction of Cp_2AlMe with $B(C_6F_5)_3$ (Scheme 3.47) [149, 150].

Table 3.33 Polymerization[a] of IB with bis(cyclopentadienyl)methylaluminum/tris(pentafluorophenyl)boron [149].

Temp (°C)	Yield (%)	$\overline{M}_n$ (kg·mol^{-1})	$\overline{M}_w/\overline{M}_n$
−70	4.3	1,800	2.8
−50	2.6	618	2.0
−30	1.1	318	1.8
−25	0.71	289	1.6

[a][Cp$_2$AlMe] = 50 µmol; [B(C$_6$F$_5$)$_3$] = 50 µmol; initiator components dissolved in 1.5 mL CH$_2$Cl$_2$; time = 10 min.

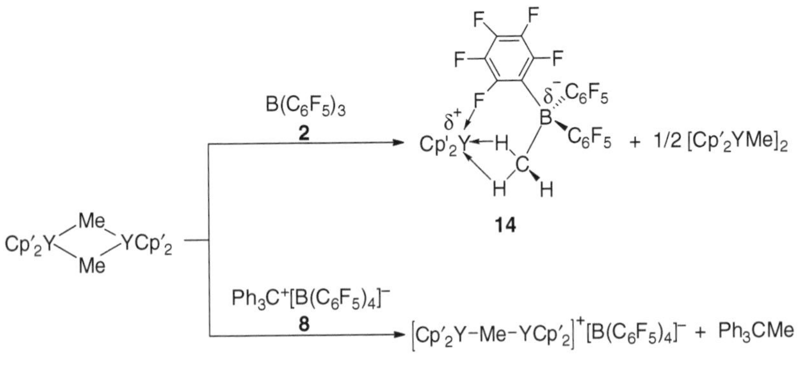

Cp' = trimethylsilylcyclopentadienyl

Scheme 3.48 Cationic lanthanide complexes used for initiation of IB polymerization.

This system gives rise to low yields of high MW PIB and butyl rubber at elevated reaction temperatures in essentially nonpolar media (Table 3.33). Preparation of **13** requires polar solvents and temperatures < −20 °C due to poor thermal stability of the aluminocenium cation. The use of bulkier cyclopentadienyl ligands leads to increased stability of the aluminocenium ion with concomitant reduction in polymerization activity where the following order (in decreasing activity) was observed: Cp$_2$Al$^+$ >> Cp''$_2$Al$^+$ > Cp*$_2$Al$^+$ (Cp'' = tetramethylcyclopentadienyl) [151, 152]. Subsequent work by Schnöckel and coworkers showed that both activity and thermal stability of the ion pair could be improved by substitution of the less coordinating anion [Al{OC(CF$_3$)$_3$}$_4$]$^-$ in place of [MeB(C$_6$F$_5$)$_3$]$^-$ [153].

Table 3.34 IB polymerization[a] by Group 4 bis(cyclopentadienyl) complex/PFLA initiator system [154].

Complex (μmol)	Temp (°C)	Time (min)	Yield (%)	$\overline{M}_n$ (kg·mol^{-1})	$\overline{M}_w/\overline{M}_n$
14[b] 50	−78	5.0	2.1	1,180	2.2
14[b] 50	−50	8.0	4.8	513	2.0
15[c] 42	−78	5.0	4.3	1,380	2.4
15[c] 42	−50	5.0	6.1	1,180	2.6

[a][IB] = 10 mL.
[b]Dissolved in 2 mL CH$_2$Cl$_2$.
[c]Dissolved in unspecified quantity toluene.

$$Cp^*TiMe_3 + B(C_6F_5)_3 \longrightarrow Cp^*TiMe_2(\mu-CH_3)B(C_6F_5)_3$$
$$\mathbf{2}$$

$$Cp^*TiMe_3 + Al(C_6F_5)_3 \longrightarrow Cp^*TiMe_2(\mu-CH_3)Al(C_6F_5)_3$$

$$Cp^*TiMe_3 + Ph_3C^+[B(C_6F_5)_4]^- \longrightarrow [Cp^*TiMe_2]^+[B(C_6F_5)_4]^- + Ph_3CCH_3$$
$$\mathbf{8}$$

$$Cp^*TiMe_3 + n\text{-}C_{18}H_{37}EH\cdot B(C_6F_5)_3 \longrightarrow [Cp^*TiMe_2]^+[\,n\text{-}C_{18}H_{37}EB(C_6F_5)_3]^- + CH_4$$

Cp* = pentamethylcyclopentadienyl; E = O, S

Scheme 3.49 Cp*TiMe$_3$ derived initiators for IB polymerization.

Likewise, ion pairs (**14** and **15**, Scheme 3.48) derived from reaction of [Cp$'_2$Y(μ-Me)]$_2$ with B(C$_6$F$_5$)$_3$ and Ph$_3$C$^+$[B(C$_6$F$_5$)$_4$]$^-$ (respectively) have been investigated in the polymerization of IB [154]. **14** is stable at room temperature in dichloromethane whereas **15** requires toluene. The activity of yttrium WCA complexes mirrors that of their aluminum counterparts in terms of low yields and MW-T profile (Table 3.34).

Monocyclopentadienyl WCA

In a manner analogous to their metallocene counterparts, 10 electron monocyclopentadienyl Group 4 WCA complexes are active for preparation of high MW IB polymers at elevated temperatures (Table 3.35) [126, 135, 137, 139, 155–157]. The majority

Table 3.35 Polymerization[a] of IB using biscyclopentadienyldimethyl-titanium/PFLA initiator systems in toluene [156].

Coinitiator	Temp (°C)	Yield (%)	$\overline{M}_n$ (kg·mol^{-1})	$\overline{M}_w/\overline{M}_n$
$B(C_6F_5)_3$	−10	6	1.65	1.6
$B(C_6F_5)_3$	−20	16	3.12	1.6
$B(C_6F_5)_3$	−30	20	4.21	1.6
$B(C_6F_5)_3$	−40	60	5.60	1.5
$B(C_6F_5)_3/n\text{-}C_{18}H_{37}OH$	−10	21	1.88	1.5
$B(C_6F_5)_3/n\text{-}C_{18}H_{37}OH$	−20	21	2.89	1.6
$B(C_6F_5)_3/n\text{-}C_{18}H_{37}OH$	−30	26	4.29	1.6
$B(C_6F_5)_3/n\text{-}C_{18}H_{37}OH$	−40	28	5.82	1.6
$Ph_3C^+[B(C_6F_5)_4]^{-b}$	−10	100	1.84	1.8
$Ph_3C^+[B(C_6F_5)_4]^{-b}$	−20	100	3.11	1.9
$Ph_3C^+[B(C_6F_5)_4]^{-b}$	−30	100	4.34	1.9
$Ph_3C^+[B(C_6F_5)_4]^{-b}$	−40	100	5.04	1.5
$Al(C_6F_5)_3$	−10	3	1.74	1.65
$Al(C_6F_5)_3$	−20	5	2.40	1.74
$Al(C_6F_5)_3$	−30	7	2.98	1.75
$Al(C_6F_5)_3$	−40	9	3.55	1.88

[a][Cp$_2$TiMe$_2$] = [Coinitiator] = 48 µmol; [IB] = 2,800 [Ti]; solvent = toluene; time = 1 hr.
[b][IB] = 11,160 [Ti].

of available polymerization data centers on use of the initiator Cp*TiMe$_3$ in combination with an assortment of activators (Scheme 3.49) including those that abstract methyl carbanion {B(C$_6$F$_5$)$_3$ and Ph$_3$C$^+$[B(C$_6$F$_5$)$_4$]$^-$} in addition to those that protolytically cleave methane {n-C$_{18}$H$_{37}$EH:B(C$_6$F$_5$)$_3$, E = O or S and n-C$_{17}$H$_{35}$CO$_2$H:[B(C$_6$F$_5$)$_3$]$_n$, n = 1–2}. Initiation is believed to involve coordination of IB to the titanium cation in an η^1 fashion followed by formation of a metal-carbon bond and generation of a carbocationic center where

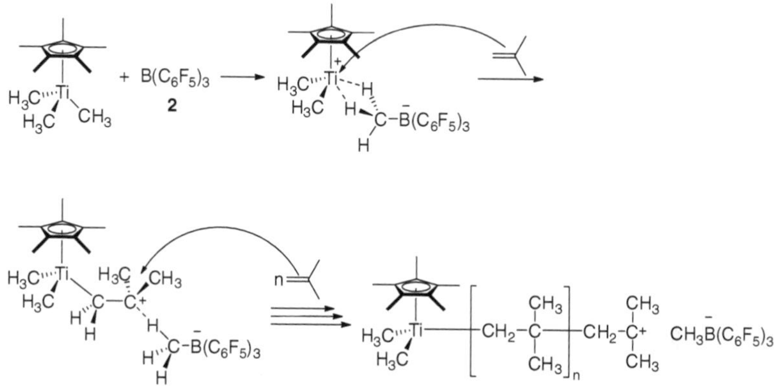

Scheme 3.50 Polymerization of IB by zwitterionic monocyclopendienyl-Ti complexes.

propagation proceeds by a classical cationic manner and polymerization perpetuates by CT (Scheme 3.50). Several key findings were used to support this mechanism [137].

1. ^{1}H and ^{13}C NMR spectroscopy studies of low MW PIBs produced from this system have structures expected to result from cationic polymerization (i.e. *t*-butyl head group and vinylidene end-group).
2. Addition of DTBP does not prevent polymerization implying protonic initiation processes are not involved.
3. [Cp*TiMe$_2$]$^+$ dimerizes 1,1-diphenylethylene to 1,1,3-triphenyl-3-methylindan.

The identity of the anion had a significant effect on MWs, which display an inverse relationship to anion nucleophilicity [156]. In particular, the low MWs produced by the system using Al(C$_6$F$_5$)$_3$ might be the result of C$_6$F$_5$ transfer from [MeAl(C$_6$F$_5$)$_3$]$^-$ to carbocationic centers resulting in termination. It is important to note that from available data polymerization does not appear to proceed in aliphatic solution and IB-IP copolymerizations require polar solvents for high conversions.

Non-Cyclopentadientyl Group 4 Metal WCA Initiators
Several non-metallocene Group 4 complexes have been explored as initiators for polymerization of IB. Bochmann and coworkers studied tris[bis(trimethysilyl)amide]methylzirconium as an initiator in conjunction with tris(pentafluorophenyl)boron for both

$$\text{MeZr}\{\text{N}(\text{SiMe}_3)_2\}_3 + \text{B}(\text{C}_6\text{F}_5)_3 \longrightarrow [\text{Zr}\{\text{N}(\text{SiMe}_3)_2\}_3]^+[\text{MeB}(\text{C}_6\text{F}_5)_3]^-$$

2   16

Scheme 3.51 Tris[bis(trimethylsilyl)amide]methylzirconium based initiators for IB polymerization.

$$\text{Ti}(\text{CH}_2\text{Ph})_4 + \text{Ph}_3\text{C}^+[\text{B}(\text{C}_6\text{F}_5)_4]^- \longrightarrow [\text{Ti}(\text{CH}_2\text{Ph})_3]^+[\text{B}(\text{C}_6\text{F}_5)_4]^- + \text{Ph}_3\text{CCH}_2\text{Ph}$$

8

Scheme 3.52 Tetrabenzyltitanium based initiators for IB polymerization.

Table 3.36 Homo and IP copolymerizations[a] of IB using the tris[bis-(trimethysilyl)amide]methylzirconium/tris(pentafluorophenyl)boron initiator system [158].

[IP] (mL)	Temp (°C)	Yield (%)	$\overline{M}_w$ (kg·mol^{-1})	$\overline{M}_w/\overline{M}_n$
—	−70	7.1	4,005	3.6
—	−50	5.7	1,670	3.5
—	−30	5.7	483	2.7
0.15	−70	13	729.5	2.2
0.15	−50	5.6	690	2.1
0.15	−30	2.8	197.5	2.9

[a][MeZr(N{SiMe$_3$}$_2$)$_3$] = 50 µmol; [B(C$_6$F$_5$)$_3$] = 50 µmol; initiator components dissolved in 1.0 mL CH$_2$Cl$_2$; [IB] = 10 mL; time = 5 min.

Table 3.37 Polymerization[a] of IB by the tetrabenzyltitanium/trityl tetrakis(pentafluorophenyl)borate initiator system [162].

Solvent	Time (min)	Temp (°C)	Yield (%)	$\overline{M}_w$ (kg·mol^{-1})	$\overline{M}_w/\overline{M}_n$
CH$_2$Cl$_2$[b]	10	−15	90	20.8	1.84
CH$_2$Cl$_2$[c]	5	−50	92	4.30	1.38
Toluene[b]	10	−15	16	203	1.51
Toluene[c]	5	−50	10	230.8	1.89

[a][Ti(CH$_2$Ph)$_4$] = 0.06 mmol; [IB] and time not specified.
[b][Ph$_3$C$^+$[B(C$_6$F$_5$)$_4$]$^-$] = 0.05 mmol.
[c][Ph$_3$C$^+$[B(C$_6$F$_5$)$_4$]$^-$] = 0.03 mmol.

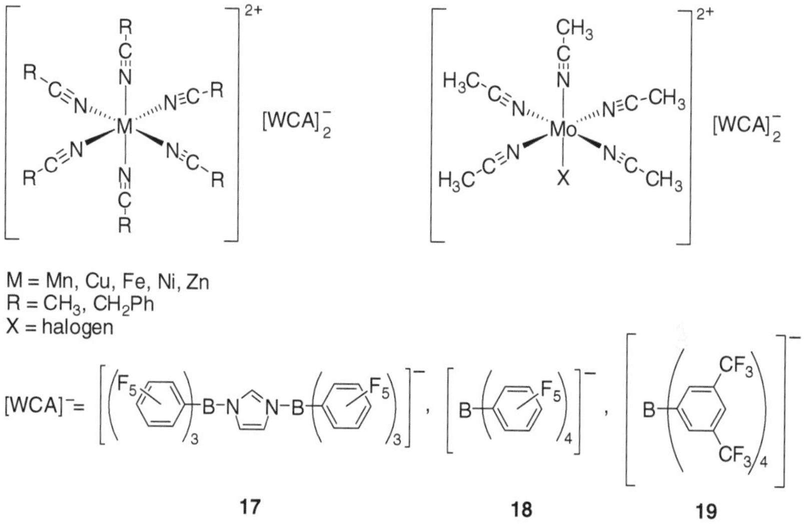

M = Mn, Cu, Fe, Ni, Zn
R = CH$_3$, CH$_2$Ph
X = halogen

17   **18**   **19**

Chart 3.6 Transition metal nitrile WCA ion pairs for IB polymerization.

IB homo- and copolymerization with IP (Scheme 3.51) [158]. The resultant ion pair (**16**) is thermally stable and produces high MW polymers at high temperatures. The activity of this initiator system (Table 3.36) is comparable to the Cp$_2$ZrMe$_2$/B(C$_6$F$_5$)$_3$ system and may stem from the fact that both share the same anion [MeB(C$_6$F$_5$)$_3$]$^-$.

Another non-metallocene initiator system derived from Ti(CH$_2$Ph)$_4$ and Ph$_3$C$^+$[B(C$_6$F$_5$)$_4$]$^-$ was investigated by the Baird research group (Scheme 3.52) [159]. The identity of the resultant ion pair was dependent on the stoichiometry of Ti:B where excess Ti(CH$_2$Ph)$_4$ leads to the production of benzyl bridged dititantium species. This system produces medium-low MW PIBs in both polar and nonpolar solvents at high temperatures where yields are higher for polymerization in polar media (Table 3.37). In addition to the low MWs the thermal instability of the initiating ion pairs and apparent inability to operate in aliphatic solvents detracts from the value of this initiator system.

Transition Metal Nitrile WCA Initiators
Kühn and coworkers have explored the use of transition metal nitrile WCA complexes (Chart 3.6) of the general formulas [M(NCR)$_6$]$^{2+}$ [WCA]$_2^-$ and [Mo(NCR)$_5$X]$^{2+}$ [WCA]$_2^-$ (M = Mn, Cu, Fe, Co, Ni, Zn; R = alkyl, aryl; X = halogen) for polymerization of IB and its copolymerization with isoprene (Table 3.38) [160–164]. These initiators only operate most efficiently within a narrow temperature range

Table 3.38 Polymerization[a] of IB in dichloromethane using Mn acetonitrile/WCA complexes [163].

[WCA][-b]	[[Mn(NCMe)$_6$]$^{2+}$ [WCA]$_2^-$] (mM)	Yield (%)	$\overline{M}_n$ (kg·mol^{-1})	$\overline{M}_w/\overline{M}_n$	Exo End Group[c] (%)
17	0.500	83	2.4	2.5	75
17	0.375	79	2.2	2.5	74
17	0.250	77	2.4	2.5	74
17	0.125	78	2.9	2.6	72
17	0.050	77	3.7	2.5	70
18	0.500	36	7.2	2.0	49
18	0.375	47	6.6	2.0	53
18	0.250	66	4.6	2.3	57
18	0.125	81	2.8	2.9	68
18	0.050	94	2.3	3.1	68
18	0.025	89	1.8	3.5	72
18	0.0125	76	1.9	3.3	70
18	0.00625	21	3.9	2.1	81
19	0.500	70	3.7	2.5	61
19	0.375	70	3.9	2.5	58
19	0.250	70	2.6	3.0	43
19	0.125	88	2.2	3.0	56
19	0.050	89	2.5	2.9	64

[a][IB] = 1.78 M; solvent = CH$_2$Cl$_2$; T = 30 °C; time = 20 hr.
[b]Refer to Chart 3.6 for structure of [WCA]$^-$.
[c]Percentage of polymer chains bearing ω-vinylidene groups.

(20–60 °C) to afford high yields of low MW PIBs. Polymerization is slow at lower temperatures whereas T > 60 °C causes decomposition of the initiator complex. None of the complexes is capable of operating efficiently in 100% aliphatic solvent although for some

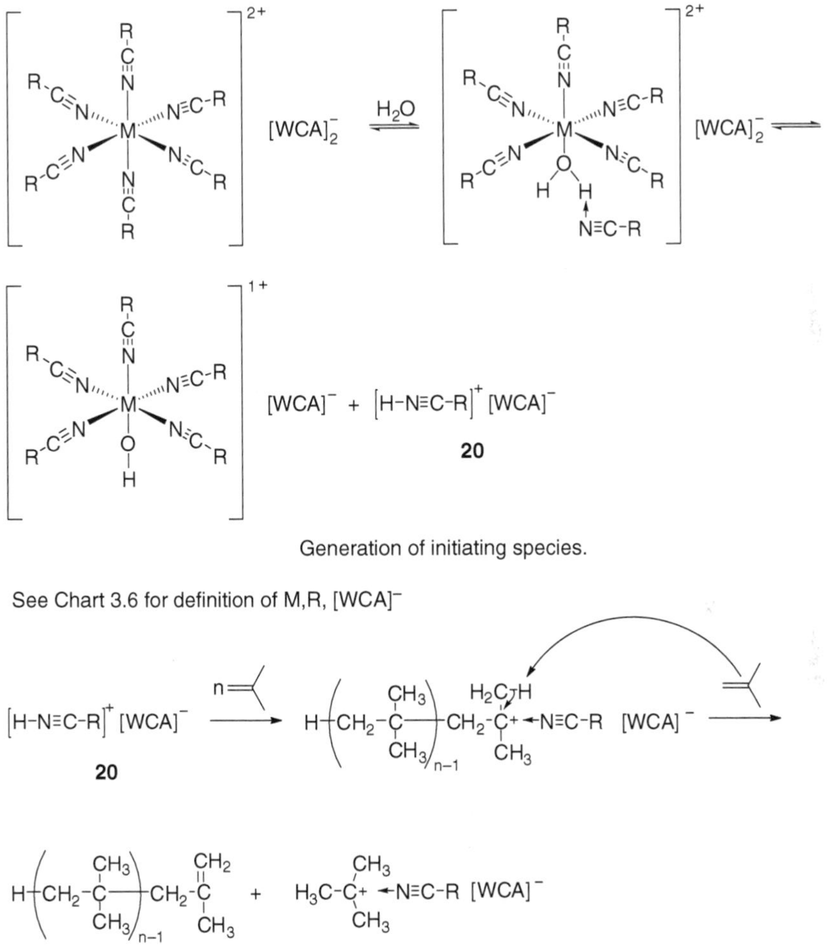

Generation of initiating species.

See Chart 3.6 for definition of M,R, [WCA]⁻

Initiation and chain transfer.

Scheme 3.53 Possible mechanism for initiation of IB polymerization by transition metal nitrile WCA ion pairs.

polymerization proceeds to high conversion in toluene. In many instances PIBs with high exo-olefinic end-group functionalities are produced. Supported versions of these initiators have been synthesized but possess little to no activity for polymerization [165–168].

The exact mode of initiation is unknown but the effects of water on polymerization behavior may signal that a strong BA (**20**) formed from the metal nitrile WCA complex and adventitious moisture is involved (Scheme 3.53). For example, MWs decrease linearly as $[H_2O]:[Mn(NCCH_3)_6]$ increases while raising the water content from 3.5 to 7 ppm results in lowering the average $\overline{M}_w$ from 3,800 to 1,400 g·mol^{-1} [162]. Moisture has no negative impact on polymerization until $[H_2O]:$[metal nitrile WCA] > 10:1 [162]. Small increases in water may lead to higher concentrations of **20** and an increase in the number of active species. Water that does not react with initiator could function as a CTA or could increase the nucleophilicity of anions in the polymerization charge, both which would reduce MWs. Once water exceeds the 10:1 threshold, inactive species are formed (Chart 3.7) and conversion drops [162]. Ab initio calculations indicate that protonation of IB by the mono H_2O-coordinated species $[Mn(H_2O)(NCMe)_5]^{2+}$ is enthalpically unfavorable; however,

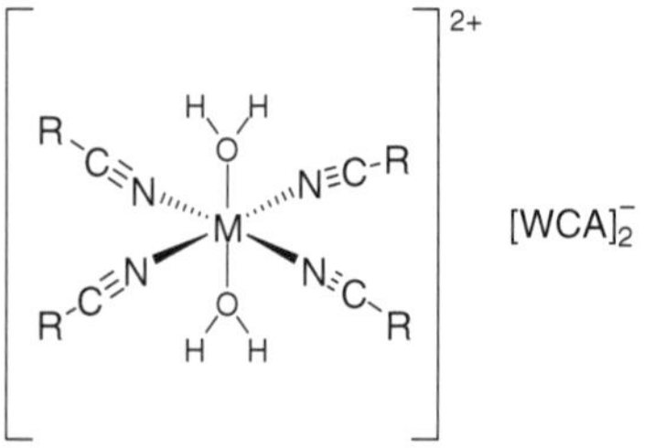

See Chart 3.6 for definition of M, R, [WCA]$^-$

Chart 3.7 Biaxially-coordinated transition metal nitrile WCA ion pairs.

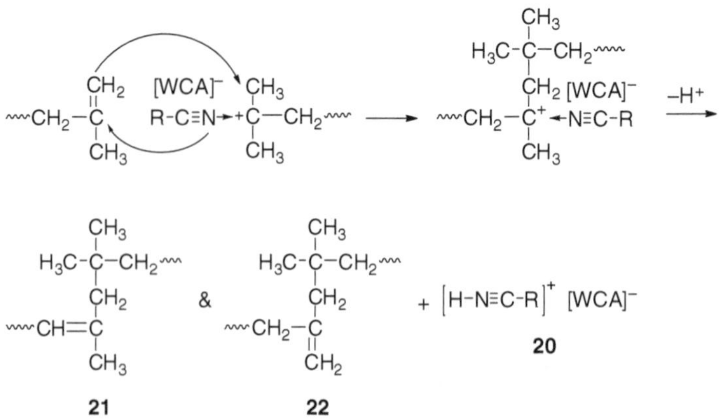

Refer to Chart 3.6 for R, [WCA]$^-$

Scheme 3.54 Plausible competitive chain transfer process for transition metal nitrile WCA ion pair IB polymerizations.

it was not possible to consider the presence of WCAs in these computations [169]. Such results may explain in part why polymerization is sluggish below 20 °C.

Metal nitrile WCA complex concentration also has a complicated effect on polymerization behavior. As its concentration is increased, $\overline{M}_w$ increases, conversion drops, and the degree of exo-olefinic functionality is reduced. This might indicate an increase in side reactions that consume chain end exo-olefinic functionalities ultimately resulting in coupling of polymer chains and production of sterically inaccessible carbocations incapable of participating in further polymerization (Scheme 3.54). Reactions of this sort might become more favorable as the concentration of BA-like species (e.g. **20**) increases while those conducive to CT (e.g. water) decrease and would favor production of **21** compared to **22** from both kinetic and thermodynamic standpoints.

In none of the aforementioned systems has conclusive evidence for polymerization of IB by a coordinative mechanism been obtained. On the contrary, polymer microstructure, the inverse dependency of MW on T, and solvent polarity effects implicate a cationic mechanism. A cationic mechanism is further supported by the fact that sterically encumbered monomers (i.e. 1,1-diphenylethylene) incapable of undergoing Ziegler type chemistry are dimerized in a fashion identical to that induced by standard cationic initiators [137].

It should also be mentioned that these systems are limited to metal cations paired with WCAs derived from PFLAs. Application of the corresponding alkylaluminoxane derived systems for cationic polymerization of IB has not been described. Such systems may not be conducive to cationic polymerization of IB due to facile transfer of alkylide from anions derived from alkyl bearing metal

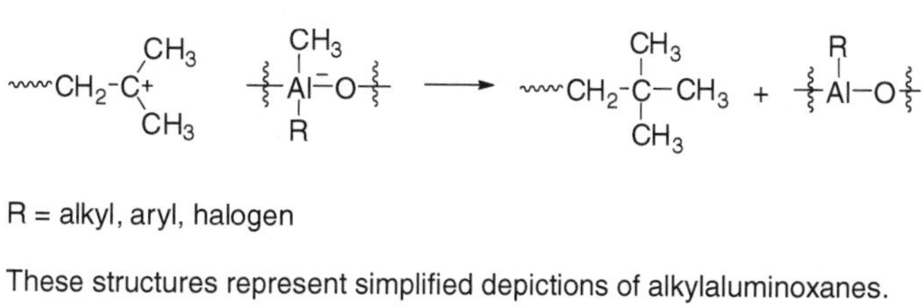

R = alkyl, aryl, halogen

These structures represent simplified depictions of alkylaluminoxanes.

Scheme 3.55 Plausible termination of IB polymerization by transfer of alkylide group from anion to propagating carbocation.

Table 3.39 IB polymerization[a] using MAO in conjunction with carbocation synthons [170].

Initiator	Solvent	Yield (%)	$\overline{M}_w$ (kg·mol^{-1})
2-Chloropropane	Toluene	90.4	351
Neopentyl chloride	Toluene	100	396
t-Butyl chloride	Toluene	87.6	30
t-Butyl chloride	Hexane	79.2	11
Chlorohexane	Hexane	45.2	715

[a][MAO] = 59.5 mM (injected as a solution in toluene); [Initiator] = 5.95 mM; [solvent] = 15 mL; T = –33 °C; time = 24 hr.

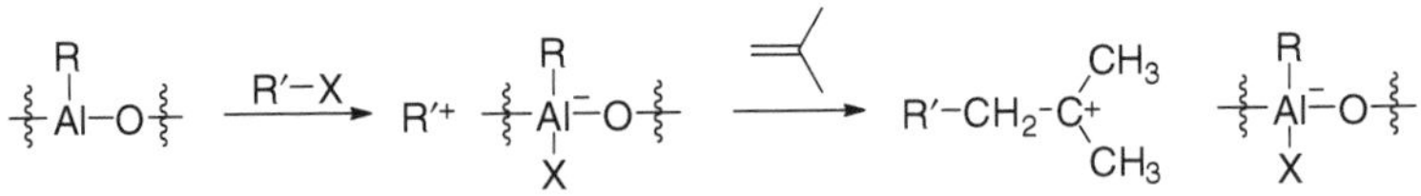

R, R' = alkyl, aryl; X = halogen

These structures represent simplified depictions of alkylaluminoxanes.

Scheme 3.56 Possible mode of initiation for alkylaluminoxane/carbocation synthon systems.

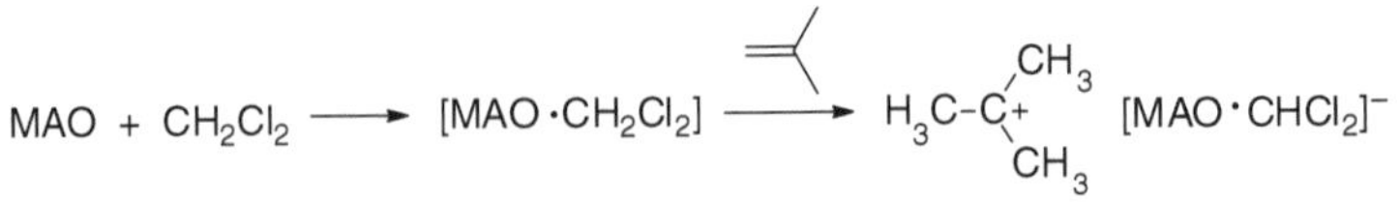

Scheme 3.57 Proposed mechanism of initiation of IB polymerization by methylaluminoxane/dichloromethane.

complexes and alkylaluminoxanes to carbocations effectively preventing/terminating polymerization (Scheme 3.55).

Despite their ability to produce high MW IB polymers at elevated temperatures the high cost of the initiator components in addition to the requisite use of high purity feedstocks (and chlorinated solvents in some instances) makes commercial practice of these systems uneconomical. Moreover, from reported data it appears that many of the metal cation WCA complex initiator systems are only useful at T ≤ −20 °C due to poor thermal stability of these ion pairs further limiting their utility.

3.3.1.1.6.5 Alkylaluminoxane Based Initiator Systems

Initiator systems based on alkylaluminoxanes (e.g. MAO) as a coinitiator with alkyl and arylhalide cation synthons have been described by industrial and academic researchers [170, 171]. This type of initiator system is capable of producing high MW PIBs in aliphatic solvents to high conversions at temperatures up to ~ −30 °C (Table 3.39). Initiation appears to involve ionization of alkyl and arylhalide cation synthons by MAO (Scheme 3.56). Even alkylhalides that are not prone to generating carbocations (e.g. CH_2Cl_2) apparently undergo ionization to form initiating carbocations. Initially it was proposed that initiation involved protonation

Table 3.40 Polymerization[a] of neat IB using MAO in conjunction with initiators other than carbocation synthons [109, 172].

Initiator[b] [mol]	Temp. (°C)	Conv. (%)	$\overline{M}_w$ (kg·mol^{-1})	$\overline{M}_w/\overline{M}_n$
TMSCl	−78	46	7,000	2.6
TMSCl	−42	19	260	2.0
TMSCl[c]	−42	25	610	2.9
TMSCl	0	4.7	120	1.3
TMSCl[c]	0	16	94	2.1
CSiCl	−78	33	31,000	22
CSiCl	−42	17	260	1.5
CSiCl	0	1.8	93	1.5
I$_2$	−78	3.2	280	2.2
I$_2$	0	6.8	53	1.9
BBr$_3$	−78	15	870	7.3
BBr$_3$	0	9.0	23	1.6

[a][MAO, based on Al] = 1.60 mmol; [IB] = 194 mmol; 1 hr.
[b][Trimethylsilylchloride, TMSCl and [I$_2$] = 40.0 μmol; [phenyldimethylchlorosilane, CSiCl] and [BBr$_3$] = 20.0 μmol. All dissolved in hexanes.
[c][MAO, based on Al] = 1.60 mmol; [IB] = 265 mmol; 1 hr.

of monomer by solvent that is activated by MAO (Scheme 3.57) as evidenced by kinetic isotopic effect (KIE = 7) where conversion for polymerization in CH_2Cl_2 was higher than in CD_2Cl_2 [171]. Such conclusions are flawed due to less stringent drying of CD_2Cl_2 in comparison to CH_2Cl_2 where the higher concentration of adventitious moisture in the former can adversely affect polymerization leading to reduced yields. The high Lewis acidity of MAO prevents use of halogenated solvents in polymerization as reaction under such conditions leads to low MW products because solvent also functions as an initiator.

At temperatures above −30 °C yields are drastically reduced limiting the usefulness of this approach. Secondly, from the reported data useful initiators were limited to alkyl and arylhalides. Indeed, subsequent investigations conducted under a joint venture between Innovative Science Corp. and Shandong Triad Engineering Ltd. showed that both ether carbocation synthons (i.e. cumyl methyl ether, CME) and acyl halide acylium ion synthetic equivalents (i.e. acetyl chloride, AcCl) were not active initiators for polymerization of neat IB in conjunction with solid MAO free of TMA [109]. Moreover, solid MAO free of TMA was also incapable of inducing polymerization of IB by itself [109].

From these investigations it was discovered that MAO is an effective coinitiator for polymerization of neat IB (or in aliphatic

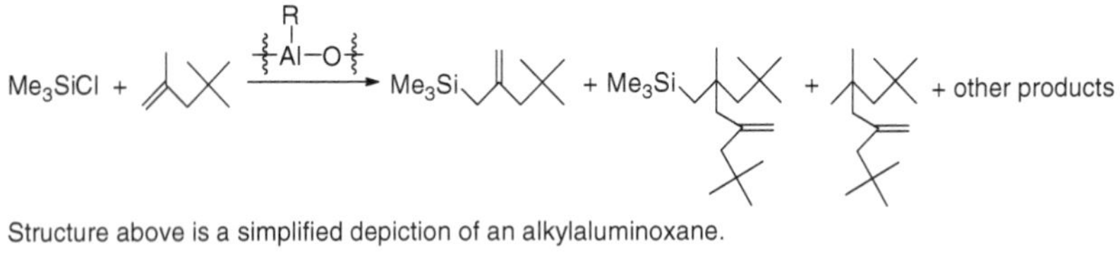

Structure above is a simplified depiction of an alkylaluminoxane.

Scheme 3.58 Modeling reactions of the silicenium synthon/MAO initiator system.

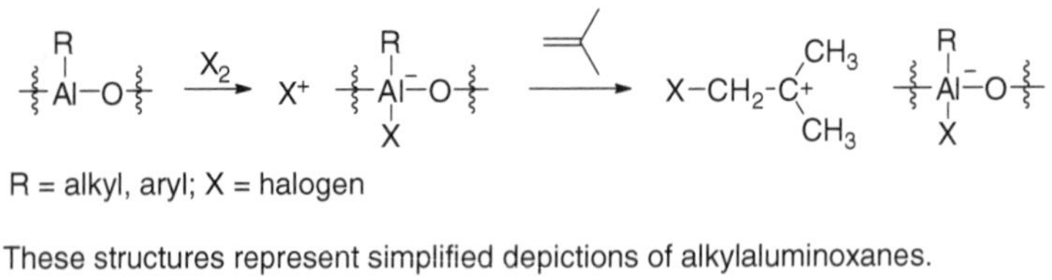

R = alkyl, aryl; X = halogen

These structures represent simplified depictions of alkylaluminoxanes.

Scheme 3.59 Proposed mechanism for initiation of IB polymerization by the halogen/MAO system.

solution) with a broad range of chemically diverse initiators including silicenium ion synthons, halogens, BAs, and boron halides (Table 3.40) [109, 172]. It should be noted that MAO possesses little to no solubility in neat IB and can be made less soluble through the use of aliphatic, especially at T<–20 °C. Since the experiments described below were run in bulk monomer, initiation may be predominantly heterogeneous in nature.

Silicon halides work efficiently as initiators in conjunction with MAO for polymerization of IB. Modeling studies using TMP in combination with TMSCl and MAO generate products consistent with initiation by direct addition of monomer to silicenium ions (Scheme 3.58). It is interesting to note that Si-X containing species

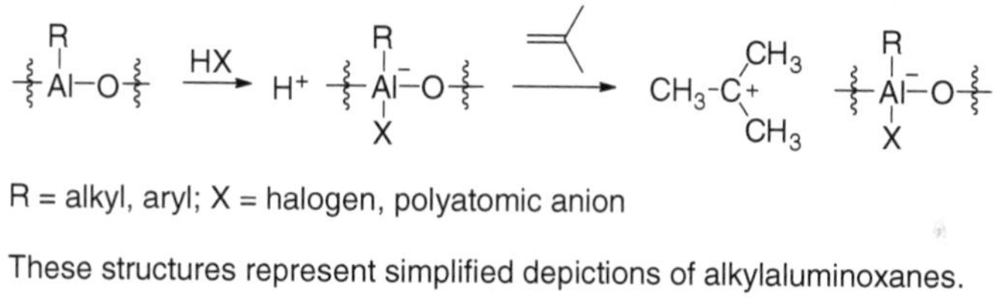
R
Al–O
HX
H+
X
CH3
CH3-C+
CH3
R = alkyl, aryl; X = halogen, polyatomic anion
These structures represent simplified depictions of alkylaluminoxanes.

Scheme 3.60 Direct protonic initiation of IB polymerization by the Brønsted acid/MAO system.

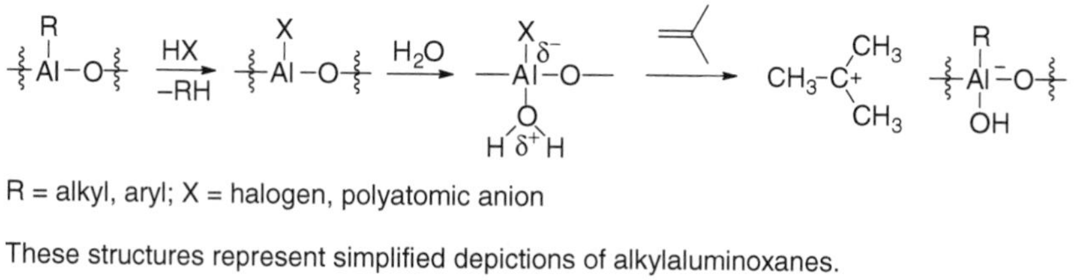
HX
–RH
H2O
δ−
H δ+ H
CH3-C+
OH
R = alkyl, aryl; X = halogen, polyatomic anion
These structures represent simplified depictions of alkylaluminoxanes.

Scheme 3.61 Indirect protonic initiation of IB polymerization by the Brønsted acid/MAO system.

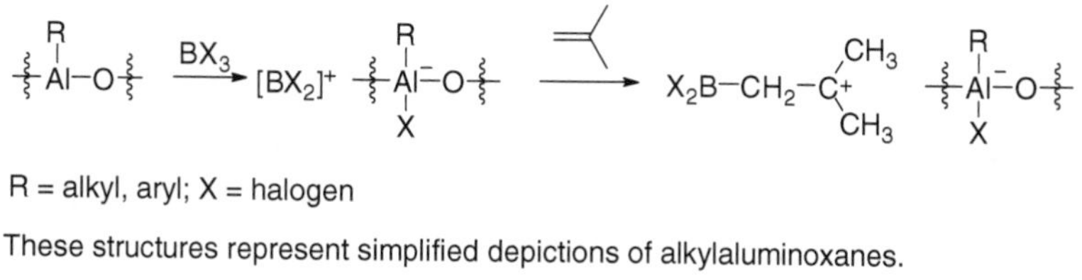
BX3
[BX2]+
X2B–CH2–C+
CH3
R = alkyl, aryl; X = halogen
These structures represent simplified depictions of alkylaluminoxanes.

Scheme 3.62 Proposed mechanism for initiation of IB polymerization by the boron halide/MAO system.

are inert for IB polymerization when used in the presence of a number of LA coinitiators including $TiCl_4$ and $AlCl_3$ [105–107]. Halogens are effective initiators with MAO. Activity is inversely proportional to the nucleophilicity of the halide ion of the parent halogen (in order of decreasing activity) $Br_2 > I_2$. It is believed that MAO ionizes the halogen to the corresponding halocenium ion to which monomer adds in a fashion reminiscent of R_2AlX'/X_2 systems (Scheme 3.59) [32, 92–100].

As is the case with R_2AlX', BAs are active initiators with MAO; albeit, less efficient in comparison to other initiator subsets. Polymerization may involve *in situ* formation of a strong BA via direct reaction of MAO with BA initiator or indirectly by reaction of adventitious moisture with a strong LA generated *in situ* (Schemes 3.60–3.61, respectively). Initiation by boron halides in conjunction with MAO is different than self-ionization processes by BCl_3 and BBr_3 alone, respectively [51, 52, 173]. The former are

Table 3.41 Polymerization of IB using MAO modified alkylaluminium based initiator systems [174–175, 177]

[MAO][a] (μL)	[Et₂AlCl] (mM)	[EtAlCl₂] (mM)	Time (min)	Yield (%)	$\overline{M}_n$ (kg·mol⁻¹)	$\overline{M}_w/\overline{M}_n$
0[b]	18.6	1.53	20	12.5	184.9	2.08
175[b]	18.6	1.53	20	19.7	203	2.71
0[c]	0	0.329	10	26.9	118.8	2.45
25[c]	0	0.329	5	40.4	150.5	1.97
150[c]	0	0.329	5	17.4	193.7	1.75

[a] 10 wt % MAO in toluene.
[b] [IB] = 9.8 M; [IP] = 235 mM; solvent = *n*-hexane; T = –60 °C.
[c] [IB] = 4.09 M; [IP] = 73.9 mM; solvent = *n*-hexane; T = –80 °C.

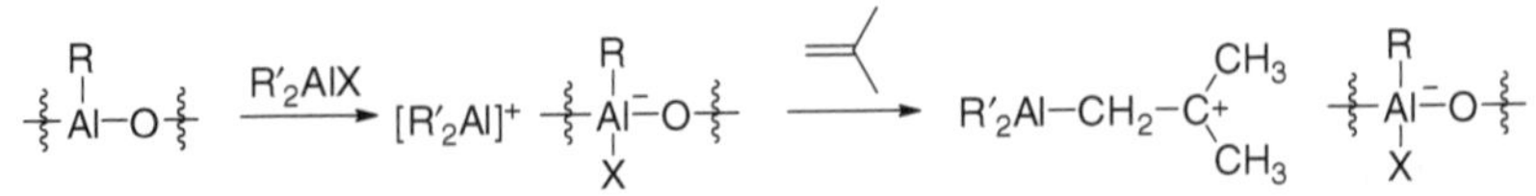

R, R' = alkyl, aryl; X = halogen

These structures represent simplified depictions of alkylaluminoxanes.

Scheme 3.63 Possible mode of IB initiation by MAO/dialkylaluminum halide systems.

extremely rapid, exothermic and operate in neat IB giving rise to moderate to low yields of high MW PIBs whereas the latter are slow, require polar solvents, and yield low MW materials. Initiation by this system may involve processes similar to those involved for mixed Lewis acid initiator systems [79–83] where the stronger LA abstracts a ligand from a weaker one (Scheme 3.62). Qualitatively the intensity of polymerization decreased with increasing B-halogen bond strength (in order of decreasing activity) $BBr_3 > BCl_3 >> BF_3$, in support of ionization of the boron halide by MAO.

Alkylaluminoxanes have also been used to modify existing initiator systems; in particular, those based on alkylaluminum dihalides. The inclusion of MAO to the Parker system (Section 3.3.1.1.2) leads to higher PDI and increased yield (Table 3.41) [174, 175]. This behavior might indicate that two forms of initiation are operating, one characteristic of the Parker system itself, the second possibly involving

Table 3.42 Polymerization[a] of IB using the organozinc halide/ carbocation synthon initiator system [180].

[EtZnCl] (mM)	[tBuCl] (mM)	Yield (%)	$\overline{M}_n$ (kg·mol⁻¹)	$\overline{M}_w/\overline{M}_n$
6.4	1.6	6.8	27	1.91
6.4	2.5	13.2	25	1.87
6.4	2.9	15.6	29	1.65
6.4	4.8	30.6	17	2.00
19.3	14.5	95.0	15	2.06
38.7	29.0	95.0	10	2.49

[a][IB] = 5.03 M; solvent = CH_2Cl_2; T = 20 °C; time = 30 min.

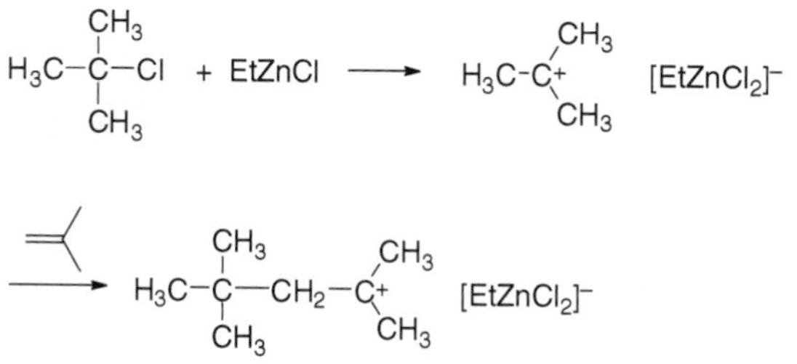

Scheme 3.64 Initiation of IB polymerization by the carbocation synthon/ organozinc halide system.

ionization of dialkylaluminum halide to dialkylaluminum cation via halide abstraction by MAO (Scheme 3.63). Such ionization is feasible given recent findings by researchers at Albemarle that show facile ligand exchange between MAO and alkylaluminum halides, a process that leads to the generation of dialkylaluminum cations [176]. An even simpler modification of the alkylaluminum dihalide based solution process with MAO has also been disclosed (Table 3.41) [177]. In this case, yields and PDI both drop and may indicate that MAO is actively scavenging the active initiator components (e.g. alkylaluminum dihalide aqua complex, 1).

In most instances initiator systems based on MAO provide a means for the preparation of high MW IB polymers to relatively high temperatures (ca. –30 °C); however, improved performance at elevated temperatures is desirable. The coinitiator is moderate in cost and possesses high activity limiting the amount required for polymerization. Moreover, polymerization can be effected in neat monomer or hydrocarbon if desired but use of chlorinated solvents prevents production of high MW polymers.

Table 3.43 Homo and IP copolymerizations[a] of IB using γ-radiation [13].

[IP] (mol %)	Temp (°C)	Irrad. Time (min)	Dose Rate (eV·g^{-1}·min^{-1})	Yield (%)	G(-m) (molecules/ 100 eV)	$\overline{M}_v$ (kg·mol^{-1})
0	0	10	7.8×10^{15}	74.8	1.1×10^7	55
0	–20	10	7.8×10^{15}	69.4	1.0×10^7	154
0	–78	20	3.1×10^{15}	6.89	1.2×10^6	5,500
3	0	115	2.3×10^{17}	8.81	3.6×10^3	159
3	–34.5	100	2.3×10^{17}	9.59	4.5×10^3	721
3	–51	206	2.3×10^{17}	10.5	2.4×10^3	1,620
3	–78	70	2.3×10^{17}	5.39	3.6×10^3	2,980
7.1	0	430	2.3×10^{17}	8.74	9.5×10^2	109
7.1	–35	405	2.3×10^{17}	6.87	7.9×10^2	181
7.1	–51	360	2.3×10^{17}	8.58	1.11×10^3	161
7.1	–78	361	2.3×10^{17}	6.98	9.0×10^2	607

[a]Details concerning the exact quantities of monomer used for each run not provided.

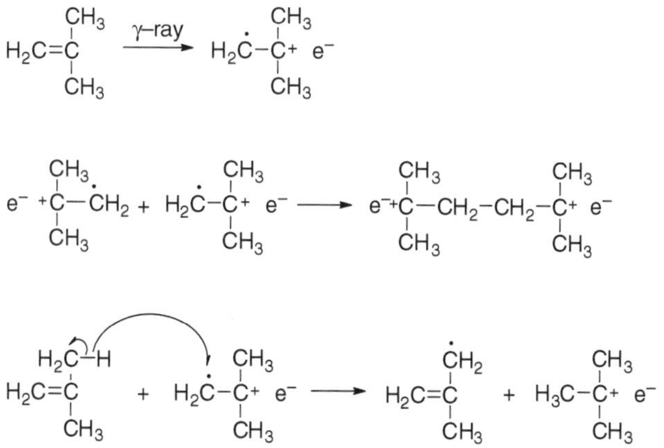

Scheme 3.65 Initiation of IB polymerization by gamma radiation.

3.3.1.1.7 Organozinc Halide/Carbocation Synthon Initiator System

Alkyl and aryl zinc halide coinitiators in conjunction with organo-halide carbocation synthons have been explored for preparation of IB polymers (Table 3.42) [178–180]. Initiation most likely involves addition of monomer to a carbocation generated by ionization of the carbocation synthon by the organozinc halide (Scheme 3.64). Due to poor solubility of the coinitiator, polymerization must be conducted in halogenated solvents at temperatures above −35 °C and generally affords low yields (ca. < 10%) of high MW polymer, severally limiting the value of this system. Increasing the size of the alkyl substituent appears to improve the efficiency of this initiator system by increasing the solubility of the coinitiator. Yields can be boosted by increasing the concentration of the initiator system to relatively high levels. One benefit to this system is that it produces PIBs with high exo-olefinic end-group functionality (ca. 92%).

3.3.1.1.8 Gamma Radiation

Initiation by γ-rays can be considered as a special type of homogeneous initiator system [13, 14]. Of the methods developed to date for cationic polymerization, it is considered to possesses the best MW-T profile (Table 3.43). Irradiation of monomer may lead to ejection of an electron to produce radical carbocations that either subsequently couple to form dicarbocations or react with monomer to form a 3° carbocation and allylic radical (Scheme 3.65) [181, 182]. Researchers speculate the high molecular weights result from unencumbered, "free" carbocations where the anion resides in a "hole" within the solvent/monomer [3, 13, 14]. Such assumptions are questionable given the low polymerization rates (10^3–10^4 times lower than chemical initiation) under bulk conditions and the fact that polymerization stops when the radiation source is removed [14]. Under such conditions propagation should proceed at high rates and polymerization would perpetuate to completion via chain transfer. It is possible that this behavior is due to a low concentration of propagating chains that undergo termination with species such as ejected electrons or impurities. Such polymerizations are highly sensitive to trace impurities. Moreover, copolymerization of IB with isoprene is 10^3 times less efficient per dose of radiation than homopolymerization [14]. Given the sensitivity to impurities, slow rates, low efficiency of copolymerization, and the dangers of radiation and associated wastes, implementation of this technology on a commercial scale does not appear feasible.

3.3.1.1.9   Conclusions on Homogeneous Initiator Systems

A number of homogeneous initiator systems with improved MW-T profiles that produce useful yields of polymer in the absence of CH_3Cl have been developed. Earlier variants comprise cheap components that lend themselves to commercialization; however, newer systems that possess the best MW-T profiles are based on costly metallocenes and PFLAs, and thus prevent large scale implementation. Furthermore, many (γ-radiation and WCA containing initiator systems) are overly sensitive to trace impurities and/or produce low yields of polymer. The major disadvantage to homogeneous systems is that workup of the reaction mixture leads to decomposition of the chemical constituents of the initiator system, thus preventing reuse.

3.3.1.2   *Developments in Heterogeneous Initiator Systems*

In an attempt to circumvent problems associated with entrainment of the initiator components in the polymer a number of systems involving heterogeneous coinitiators have been developed and can be classified as follows.

1. $AlCl_3$ supported on inorganic substrates.
2. Solid inorganic acids.
3. LAs intercalated in inorganic solids.
4. Acidic gels derived from $Al(OsecBu)_3$, BF_3 and $TiCl_4$.

5. Metal perchlorate and triflate salts and supported analogs.
6. Supported mixtures of strong and weak LAs.
7. Polymer supported Al and B containing LAs.
8. Supported coinitiators that function as solid WCAs.
9. Aqueous systems.

3.3.1.2.1   Supported AlCl$_3$

A great deal of work has been carried out on silica and alumina supported AlCl$_3$ coinitiators in context to IB polymerization [30, 183–185]. A number of approaches have been developed to produce these materials.

1. Dry mixing the support with AlCl$_3$ [186–189].
2. Reaction of the support with molten AlCl$_3$ [186–189].
3. Passing AlCl$_3$ vapors through the support material followed by optional removal of unbound AlCl$_3$ via sublimation [186–192].

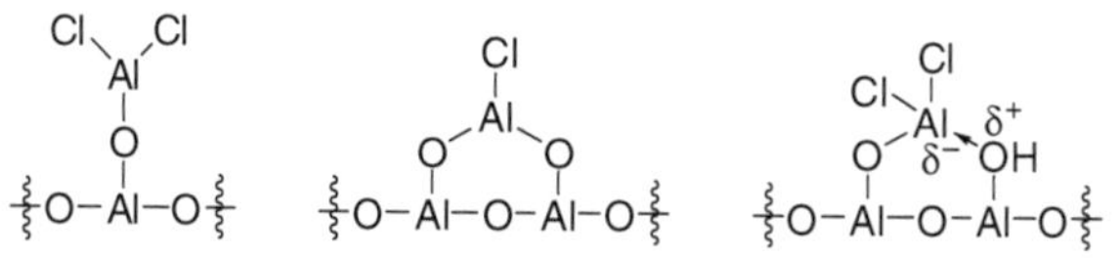

Chart 3.8 Supported AlCl$_3$ acid sites.

Table 3.44 Polymerization[a] of IB using AlCl$_3$ functionalized silica [184].

[IB] (M)	[Coinitiator][b] (g·L^{-1})	Time (min)	Yield (%)	$\overline{M}_n$ (kg·mol^{-1})	$\overline{M}_w/\overline{M}_n$
2.7	0.39	17	—	2.40	6.99
2.45	0.35	30	18.1	2.93	11.5
3[c]	0.2	10	73	1.59	10.4
2.9[c]	0.15	10	55	2.09	11.7
2.5[c]	0.12	30	38.1	2.58	12.4

[a]Solvent = heptane; T = −20 °C.
[b]TMSCl passivated silica functionalized with Et$_2$AlCl and tBuCl.
[c]NaH internal dessicant.

4. Reaction of the support with hydrocarbon, halocarbon, and hydrohalogen acid solutions of AlCl₃ [186–189, 193–198].
5. Reaction of alkylaluminum halides or alkyl aluminums with support hydroxyl groups followed by halogenation for the latter [30, 183, 184].
6. Reaction of alumina with chlorocarbons at elevated temperatures [184, 199–204].

The bulk of these materials contain a mixture (Chart 3.8) of LA (**23** and **24**) and BA (**25**) sites. Of these **24** is the least reactive and useful for polymerization. The relative contribution of **24** can be reduced through the use of supports containing predominately isolated metal hydroxyl groups which can be produced by partial passification with alkyl silyl halides or calcination. Even for supports bearing mostly **23**, none of these materials is capable of producing high MW grades of IB polymers regardless of the reaction conditions (Table 3.44).

3.3.1.2.2 Heterogeneous Inorganic Acids

Acidic clays have been used in the polymerization of styrene to yield low MW oligomers [205–208]. These materials do not appear to be active enough for the preparation of high MW grades of IB

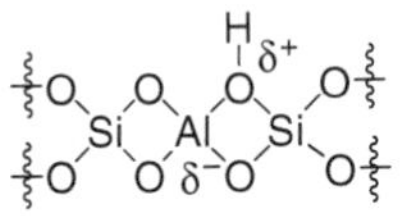

Chart 3.9 Brønsted acid sites derived from interaction of silanol groups with framework alumina.

Table 3.45 Polymerization of neat IB using 5A molecular sieves [211].

[IB] (mL)	[5A mol. sieve] (g)	T (°C)	Time (days)	Yield (%)	$\overline{M}_n$ (kg·mol⁻¹)	Exo. Funct.ᵃ
75	23	25	1.33	18	2.02	—
62	25	0	4	37	3.28	2.1
75	25	25	11	30	3.6	2.1

ᵃAverage number of exo-olefinic chain end functionalities per polymer molecule as determined by ¹H NMR spectroscopy.

Table 3.46 Polymerization[a] of IB using metal dihalides [213].

MX$_2$	Solvent	PIB:MX$_2$ (wt:wt)	$\overline{M}_v$ (kg·mol^{-1})
MgCl$_2$	—	9	45
MgCl$_2$	Heptane	76	29
CoCl$_2$	—	9	39
CoCl$_2$	MCH	2.0	24
CoCl$_2$	Heptane	2.4	46
MnCl$_2$	—	0.1	—
MnCl$_2$	Heptane	0	—

[a][Solvent:IB] = 1 vol:5 vol; T = −10 °C; time = 1 hr.

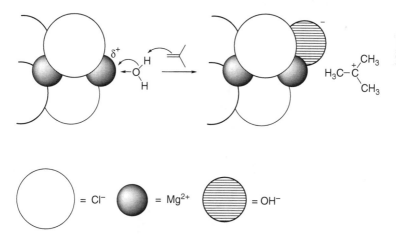

Scheme 3.66 Proposed mechanism for initiation of IB polymerization by milled magnesium chloride.

polymers. With the exception of 3Å grade, activated molecular sieves have been reported as being active coinitiators for the cationic polymerization of olefins [209]. 5-A sieves are purportedly the most active but only give rise to low yields of low molecular PIBs (Table 3.45) [210, 211]. The exo-olefinic functionality per polymer chain is relatively high at ~ 1.74. The identity of initiating species

Table 3.47 Polymerization[a] of IB in cyclohexane using salts of phosphotungstic acid [214].

Salt	Yield (%)	$\overline{M}_n$ (kg·mol^{-1})	$\overline{M}_w/\overline{M}_n$	Exo. Funct.[b] (%)
$Cs_{2.5}H_{0.5}PW_{12}O_{40}$	42	2.3	4.7	77
$Cs_{2.5}H_{0.5}PW_{12}O_{40}$	60	1.584	3.86	75
$(NH_4)_{2.5}H_{0.5}PW_{12}O_{40}$	49	1.328	11.13	24

[a][IB] = 5.25 M in cyclohexane; T = 8 °C; time = 30 min.
[b]Exo-olfenic chain end functionality determined by ^{1}H NMR.

was not identified but may be Brønsted acidic sites consisting of silanol groups activated by Al containing moieties (Chart 3.9).

$CdCl_2$ layer structure metal dihalides (e.g. $MgCl_2$) when freshly milled are active coinitiators for IB homopolymerization and its copolymerization with IP yielding high MW polymers at elevated temperatures (Table 3.46) [212, 213]. Successful polymerization requires careful manipulation of moisture level with significant reductions in both yields and MWs occurring for $MgCl_2$ containing > 1 wt % H_2O. PIBs are readily formed in aliphatic solvents using this system whereas reasonable yields of high MW butyl rubber are obtained only in the presence of polar solvents. For homopolymerization of IB, polymer yields increase while MWs decrease with increasing [$MgCl_2$]. Moreover, as the reaction temperature is lowered $MgCl_2$ becomes increasingly encapsulated in solid PIB leading to reduced activity. From these results it was believed the milling process exposes Mg^{2+} that reacts with adventitious moisture to form a strong BA that initiates polymerization (Scheme 3.66).

Heteropolyacids and their salts have been used as solid acid initiators for polymerization of IB (Table 3.47) [214–217]. These materials are only capable of affording low MW PIBs in moderate yields (ca. 60%) in aliphatic solution or bulk monomer at elevated temperatures. Cesium salts of phosphotungstic acid are the most active [214]. In some instance PIBs with relatively high exo-olefinic end-group functionality (ca. 75%) are produced. Likewise, supported transition metal oxides yield low MW PIBs and polybutenes of high exo-olefinic end-group content at elevated temperatures in aliphatic solvents [218, 219]. These materials most likely operate in a fashion similar to zeolites and acidic clays where initiation is

Table 3.48 Representative molecular weight data for PIBs prepared[a] by
Wichterle-Marek-Trekoval initiator system [71–72].

[IB]	[Initiator] (mM)	Ti/Al (mol:mol)	$\overline{M}_v$ (kg·mol^{-1})
10 wt. %	3.2	0.5	180
5 wt. %	3.2	0.5	130
10 vol. %	13.0	0.3	264
5 vol. %	13.0	0.3	238

[a]Solvent = n-hexane; T = 8 °C; time not specified.

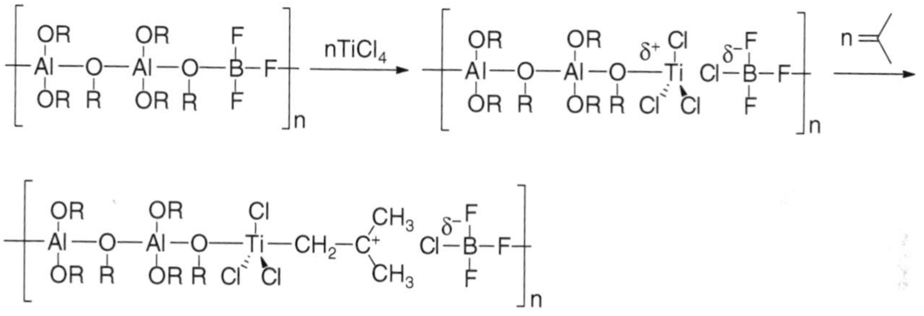

R = sec-butyl

These structures are simplified representations of products that might be yielded from
reaction of Al(OR)$_3$ with BF$_3$ assuming the former is initially present as cyclic dimers.

Scheme 3.67 Possible mechanism for IB polymerization involving ionization
of a titanium (IV) chloride by a boron fluoride/aluminum alcoholate complex.

protonic in nature and involves highly acidic groups on the surface
of these solid acids.

3.3.1.2.3   LAs Intercalated in Inorganic Matrices

Heterogeneous LAs have been synthesized by intercalation within
metal dihalides (e.g. MgCl$_2$) [220]. Intercalation is effected by mixing a hydrocarbon soluble porogen (e.g. adamantine) with the LA
and metal dihalide support in the solid state followed by solvent
extraction of the porogen. The resultant heterogeneous LAs are
active in IB polymerization in the absence of polar solvents but
yield only low MW ($\overline{M}_n \sim 600$ g·mol^{-1}) polymers at elevated temperatures. Initiation probably involves the generation of a strong BA
by reaction of adventitious moisture with LA sites that ultimately
react with monomer.

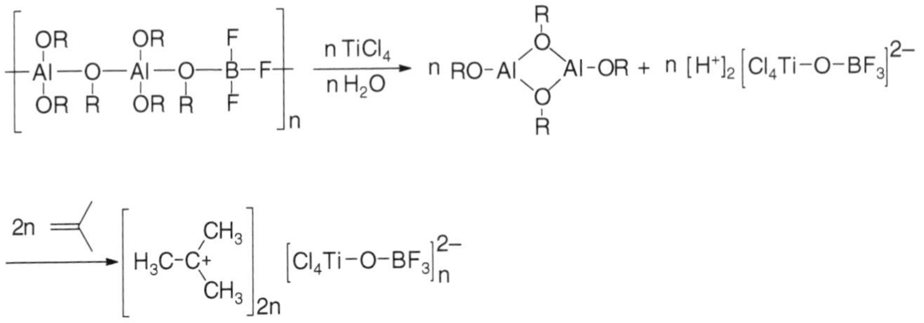

$$\left[\begin{array}{ccc} OR & OR & F \\ | & | & | \\ -Al-O-Al-O-B-F- \\ | & | & | \\ OR & R & OR \quad R \quad F \end{array}\right]_n \xrightarrow[n\,H_2O]{n\,TiCl_4} \quad n\,RO-Al \begin{array}{c} R \\ | \\ O \\ \diagdown \\ \diagup \\ O \\ | \\ R \end{array} Al-OR + n\,[H^+]_2 \left[Cl_4Ti-O-BF_3\right]^{2-}$$

$$2n \begin{array}{c} \diagup \\ \diagdown \end{array} \xrightarrow{\hspace{2cm}} \left[\begin{array}{c} CH_3 \\ | \\ H_3C-C+ \\ | \\ CH_3 \end{array}\right]_{2n} \left[Cl_4Ti-O-BF_3\right]_n^{2-}$$

R = *sec*-butyl

These structures are simplified representations of products that might be yielded from reaction of Al(OR)$_3$ with BF$_3$ assuming the former is initially present as cyclic dimers.

Scheme 3.68 Possible mechanism for IB polymerization involving Brønsted acids derived from adventitious moisture and a boron fluoride/aluminum alcoholate/titanium (IV) chloride complex.

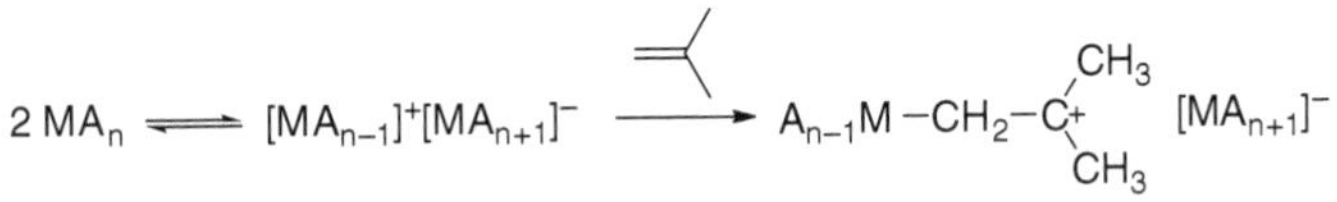

$$2\,MA_n \rightleftharpoons [MA_{n-1}]^+[MA_{n+1}]^- \xrightarrow{\qquad} A_{n-1}M-CH_2-\overset{CH_3}{\underset{CH_3}{\overset{|}{C+}}} \quad [MA_{n+1}]^-$$

M = metal with valence n
A = triflate, perchlorate

Scheme 3.69 Initiation of IB polymerization by self-ionization of metal perchlorate and triflate salts.

3.3.1.2.4   Al(OsecBu)$_3$/BF$_3$/TiCl$_4$ Acidic Network Initiator System

A complex green colored solid synthesized from Al(OsecBu)$_3$, BF$_3$, and TiCl$_4$ provides high MW PIBs at elevated temperatures in aliphatic solvents (Table 3.48) [71, 72]. Its preparation involves initial reaction of Al(OsecBu)$_3$ with BF$_3$ to yield a white precipitate that is subsequently treated with TiCl$_4$ just prior to polymerization. Both the Al(OsecBu)$_3$/BF$_3$ precipitate and its green TiCl$_4$ reaction product are thermally unstable and degrade even at temperatures below 0 °C [71]. Aging of these solids leads to reduced polymerization rates and polymer MWs limiting the usefulness of this system [71]. The exact mode of initiation has not been determined. It is plausible that the initiating species may be metal cations derived by ionization of one Lewis acid by another (Scheme 3.67) or BAs derived from adventitious moisture and mixed LA complexes (Scheme 3.68).

3.3.1.2.5 Metal Perchlorate and Triflate Based Initiator Systems
Metal perchlorate and triflate (OTf) salts have been explored for cationic polymerization under heterogeneous conditions [221, 222]. Only $Mg(ClO_4)_2$ produces high MW ($\overline{M}_n$ = 70–170 kg·mol^{-1}) PIBs at elevated temperatures (i.e. 0 °C) in neat monomer, albeit in low yields (ca. 1–4%). Numerous pieces of evidence were collected that support initiation by direct addition of monomer to exposed metal cations derived in a manner similar to self-ionization (Scheme 3.69) [221].

1. Activity increases with decreasing salt particle size.
2. Hydrolysis of polymerizations coinitiated by $Hg(ClO_4)_2$ produces elemental mercury in support of Hg-C bond formation.
3. SHPs do not affect polymerization implying non-protonic initiation.

$$[Al(SO_3CF_3)_2]^+[Al(SO_3CF_3)_4]^- \xrightarrow{[nBu_4]^+[CF_3SO_3]^-} Al(SO_3CF_3)_3 + [nBu_4]^+[Al(SO_3CF_3)_4]^-$$

Scheme 3.70 Consumption of reactive metal cations by addition of a common salt.

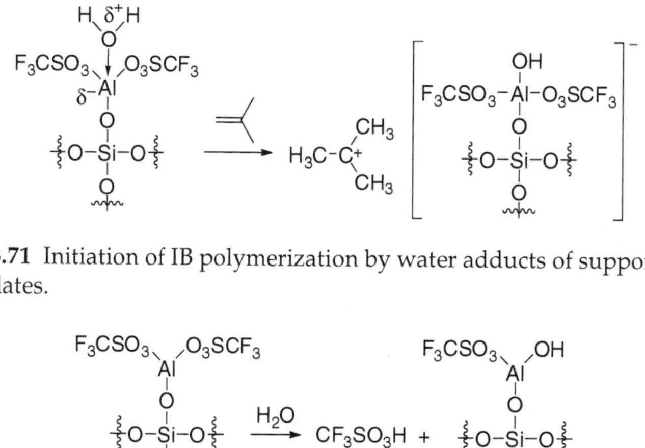

Scheme 3.71 Initiation of IB polymerization by water adducts of supported metal triflates.

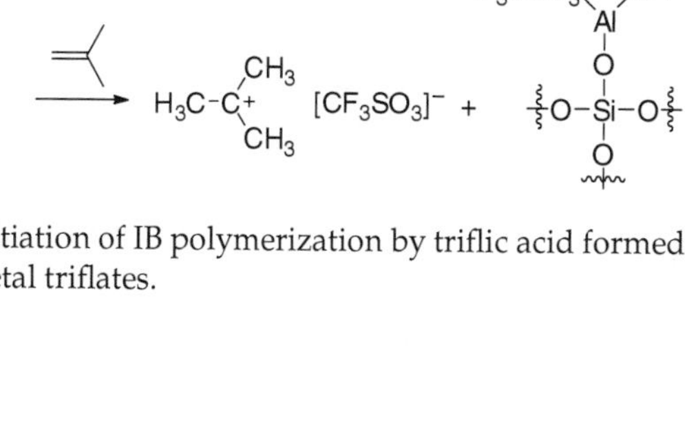

Scheme 3.72 Initiation of IB polymerization by triflic acid formed from hydrolysis of supported metal triflates.

Table 3.49 Polymerization[a] of IB using silica functionalized[b] with mixed strong/weak Lewis acids [224].

[IB] (M)	[Al] (mM)	Time (min)	Yield (%)	$\overline{M}_n$ (kg·mol^{-1})	$\overline{M}_w/\overline{M}_n$
2.5	0.35	40	56	2.90	19.7
2.8[b]	0.35	10	44	2.85	10.2
3.2[b]	0.2	40	51.5	2.58	13.4

[a]Solvent = heptane; T = −20 °C; time = 40 min.
[b]Functionalization sequence = 1) tri-*i*-butylaluminum 2) dibutyl magnesium 3) *t*BuCl.
[c]NaH internal dessicant.

4. Addition of salts that contribute common anions {e.g. [*n*Bu₄N]⁺[OTf]⁻} prevent polymerization by consuming metal cations (Scheme 3.70).
5. Al(OTf)₃ free of HOTf impurities is active for polymerization.
6. Polymerization initiated by wet salts was always slower than that using dry ones.

Supported analogs of these materials were synthesized by reacting a supported metal halide prescursor (e.g. -OAlCl₂) with an appropriate BA (e.g. CF₃SO₃H) to effect transesterficiation to form the correspond metal triflate or perchlorate {e.g. -OAl(CF₃SO₃)₂} [223, 224]. These materials are highly active for the production of low MW ($\overline{M}_n$ = 350 g·mol^{-1}) PIBs exhibiting relatively high PDIs ($\overline{M}_w/\overline{M}_n$ = 4) at elevated temperatures (30 °C) in nonpolar media. Initiation most likely arises from interaction of adventitious moisture to either form directly by complex formation (Scheme 3.71) or indirectly (via hydrolysis, Scheme 3.72) a strong BA that protonates monomer. Such multiple forms of initiation may in turn produce polymers with high PDIs.

3.3.1.2.6 Mixed Strong/Weak LA Heterogeneous Coinitiators
Heterogeneous coinitiators containing both weak and strong LA metal halide sites are active for polymerization of IB in nonpolar media to high MWs at elevated temperatures; however, polymers exhibit abnormally large PDIs potentially limiting the utility of this system (Table 3.49) [225]. These materials are prepared by first

Table 3.50 Polymerization[a] of IB using polymer supported Lewis acids [10].

Coinitiator	Solvent	T (°C)	Yield (%)	$\overline{M}_n$ (kg·mol^{-1})	$\overline{M}_w/\overline{M}_n$
PP-O-AlEtCl[a]	Hexane	−10	100	9.53	2.65
PP-O-AlEtCl[a]	Hexane	0	95	4.04	4.03
PP-O-AlEtCl[a]	Hexane	25	95	2.10	3.71
PP-O-AlEtCl[a]	CH$_2$Cl$_2$	−30	100	45.33	3.99
PP-O-AlEtCl[a]	CH$_2$Cl$_2$	0	100	24.52	3.67
PB-O-AlCl$_2$[b]	Hexane	0	70	5.45	2.63
PB-O-AlCl$_2$[b]	CH$_2$Cl$_2$	−30	100	181	4.12
PB-O-AlCl$_2$[c]	CH$_2$Cl$_2$	0	95	100	8.6
PB-OH-BF$_3$[d]	Hexane	−15	50	0.662	1.72
PB-OH-BF$_3$[d]	Hexane	0	95	0.576	1.2

[a][IB] = 0.41 M; time = 30 min.
[b][IB] = 0.79 M; time = 1 hr.
[c][IB] = 0.79 M; time = 6 hr.
[d][IB] = 0.79 M; time = 4 hr.

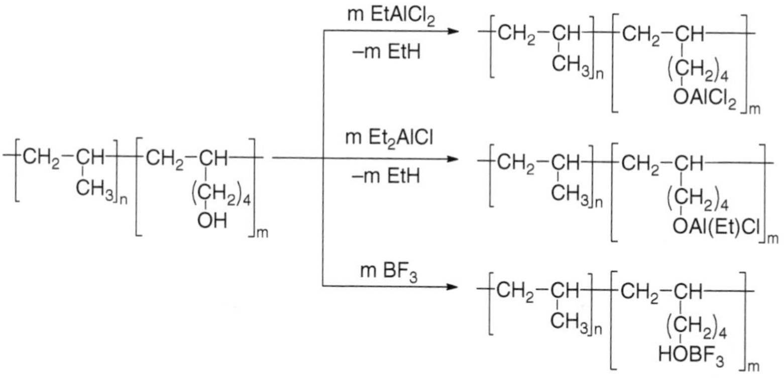

Scheme 3.73 Synthesis of supported Lewis acids by direct reaction with hydroxyl functionalized PP.

reacting hydroxyl bearing inorganic supports with a weak (e.g. Bu$_2$Mg) alkyl substituted LA and then a strong (e.g. Et$_2$AlCl) alkyl substituted LA. Residual alkyl moieties are converted to halogen

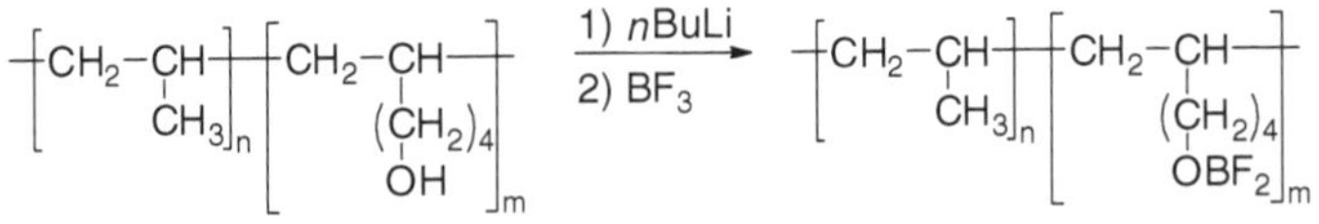

Scheme 3.74 Synthesis of supported Lewis acids by direct reaction with alkoxide functionalized PP.

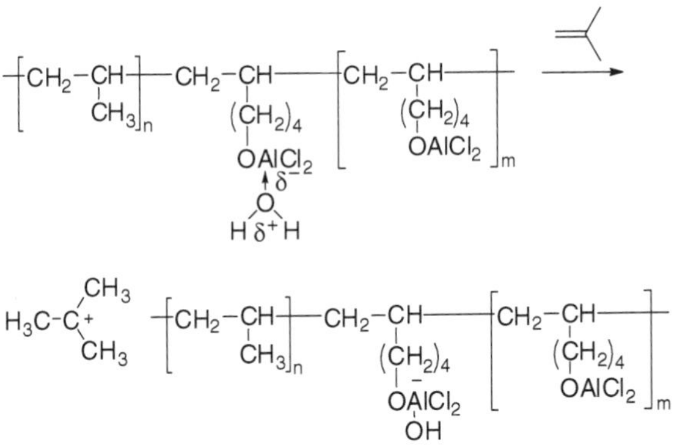

Scheme 3.75 Initiation of IB polymerization by PP supported Lewis acids.

substituents by treatment with halogens or alkyl halides to ultimately yield strong (e.g. -OAlCl$_2$) and weak (e.g. -OMgCl) acid metal halide groups. Reaction of these sites with adventitious moisture most likely forms species with high BA character that initiate polymerization by protonation of monomer. The use of weaker dialkyl LAs may actually partially passivate the support and yield isolated metal hydroxyl groups that in turn predominately form **23**.

3.3.1.2.7 Polymer Supported LA Coinitiators

Polypropylene (PP) and polybutene-1 (PB) bearing LA groups are active coinitiators for IB polymerization (Table 3.50) [10, 226–229]. PP and PB with -OAlCl$_2$, -OAlClEt, -O(H)-BF$_3$ were prepared by reacting the hydroxyl functionalized polymer with EtAlCl$_2$, Et$_2$AlCl, and BF$_3$, respectively (Scheme 3.73). PP substituted with -OBF$_2$ groups was made by converting hydroylated PP to the lithium alkoxide followed by subsequent treatment with BF$_3$ (Scheme 3.74) [226, 229]. The PP support possesses a comb-like structure in which a crystalline core of isotactic PP is covered with an amorphous surface containing hydroxyl groups allowing for production coinitiators with high surface concentrations of LA sites [226, 228].

Table 3.51 Homo[a] and IP copolymerization[b] of IB using PS supported LA coinitiators [231].

Coinitiator[c]	Time (hr)	Conv. (%)	Yield (g polymer·g Ti^{-1})	$\overline{M}_v$ (kg·mol^{-1})
PS-TiCl$_4$/Et$_2$AlCl[a]	2	74	1,935	10.26
PS-TiCl$_4$/Et$_2$AlCl[a]	6	98	2,658	21.73
PS-TiCl$_4$/Et$_2$AlCl[a]	20	99	3,702	40.10
PBDA-TiCl$_4$/Et$_2$AlCl[a]	2	14	56	3.86
PBDA-TiCl$_4$/Et$_2$AlCl[a]	6	36	143	11.21
PBDA-TiCl$_4$/Et$_2$AlCl[a]	20	97	386	37.82
PS-TiCl$_4$/Et$_2$AlCl[b]	2	32.1	782	7.45
PS-TiCl$_4$/Et$_2$AlCl[b]	8	59.4	1,448	11.87
PS-TiCl$_4$/Et$_2$AlCl[b]	24	96.7	3,367	13.18
PBDA-TiCl$_4$/Et$_2$AlCl[b]	2	10.2	90	1.22
PBDA-TiCl$_4$/Et$_2$AlCl[b]	8	34.3	303	4.18
PBDA-TiCl$_4$/Et$_2$AlCl[b]	24	87.4	772	11.67

[a][IB] = 0.1 L·min^{-1} for 70 min = 7 L total; [Coinitiator] = 0.5 g; T = 0 °C.
[b][IB] = 0.1 L·min^{-1} for 70 min = 7 L total; [IP] = 21.3 g; [Coinitiator] = 0.5 g; T = 0 °C.
[c]Please refer to Scheme 3.76 for structure of coinitiator.

The -OAlCl$_2$ and -OAlClEt derivatives produce good yields of high MW PIBs at elevated temperatures in polar (i.e. CH$_2$Cl$_2$) solvents; otherwise, in nonpolar media only low MW materials are yielded. Polymers bearing -O(H)-BF$_3$ and -OBF$_2$ groups produce only low MW polymers regardless of solvent polarity. It is likely that in all instances, adventitious moisture acts as the initiator (Scheme 3.75). Polymer microstructure is highly dependent on the identity of the LA substituent with PP-OAlCl$_2$ producing polymers with high degrees of endo-olfenic chain ends and PP-OBF$_2$ producing PIBs with predominantly exo-olfenic end-groups. An added benefit of these materials is that they can be reused repeatedly without substantial loss of activity.

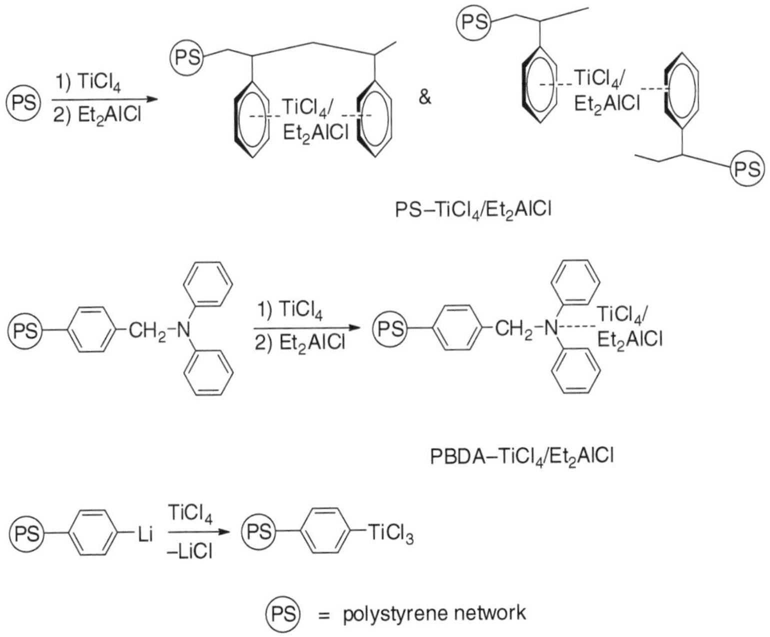

Scheme 3.76 Preparation of PS supported LA coinitiators.

In a similar manner, crosslinked polystyrene (PS) resins can be used as supports for LAs. The resultant materials are active heterogeneous coinitiators for the preparation of IB based polymers (Table 3.51) [230, 231]. Three basic groups of PS supported LAs were produced where the LA is bound by different types of bonding (Scheme 3.76). The first subset is formed by direct reaction of a LA with a standard, unfunctionalized crosslinked PS resin to form π→LA complexes. Modeling studies with 1,3-diphenylpropane suggest that for this type of coinitiator the LA molecules form a coordinative complex with phenyl rings of the PS resin where the LA is chelated between two adjacent aromatic rings. The second group possesses n→LA coordinative bonds formed by reacting a LA with a PS resin functionalized with N, O, or P n-donor atoms. The third subset is formed by reaction of aryl lithium or aryl Grignard functionalized PS resins with a LA resulting in a covalent PS-LA bond. The stability of these coinitiators in regards to leaching of LA increased in the following order: 1st group < 2nd group << 3rd group. They can be reused in many instances and regenerated as needed. TiCl$_4$/Et$_2$AlCl functionalized resins belonging to the first

Table 3.52 Polymerization[a] of neat IB using supported MAO coinitiators [109, 172].

Support	Initiator[b] [mol]	Temp. (°C)	Time (hr)	Conv. (%)	$\overline{M}_w$ (kg·mol⁻¹)	$\overline{M}_w/\overline{M}_n$
Silica	TMSCl	−78	1	20	710	8.4
Silica	TMSCl	−42	1	19	530	9.6
Silica	TMSCl	0	1	23	77	2.1
Silica	TMSCl	25	1	40	19	2.7
Silica	EtBr	0	2	92	59	4.0
Alumina	EtBr	0	2	94	54	3.6
MCM-41[c]	EtBr	0	1.25	99	35	2.6
B(OH)$_3$	EtBr	20	4	93	32	3.7
Silica	EtBr	26	1	94	19	3.8
Alumina	Cl$_2$CHCH$_3$	26	5	99	45	2.4
Alumina	CHCl$_3$	25	12	99	47	2.4

[a][MAO, based on Al] = 0.862 mmol; [IB] = 194 mmol.
[b][trimethylsilyl chloride, TMSCl] = 40.0 μmol; [EtBr], [Cl$_2$CHCH$_3$],
[CHCl$_3$] = 20.0 μmol. All dissolved in hexanes.
[c][MAO, based on Al] = 0.431 mmol; [IB] = 265 mmol.

(i.e. $\pi\rightarrow$LA) and second subsets (i.e. $n\rightarrow$LA) provided good yields of medium MW PIBs and low MW butyl at elevated temperature in neat monomer(s), albeit very slowly. Initiation may involve homolytic cleavage of unstable Group 4 alkyl compounds formed *in situ* via a process similar to that proposed for systems derived from alkyl lithium compounds in conjunction with Group 4 LAs (Scheme 3.32) or ligand exchange type reactions that generate transitory cations (Scheme 3.27) and/or ion pairs (Scheme 3.28).

Polymer supported LAs show enhanced stability in comparison to their non-supported counterparts. They be reused, in many cases without substantial loss of activity. The utility of these materials is limited because chlorinated solvents are required for high MWs, hydroxyl functionalized PP and PB supports require special/costly synthesis, and polymerization rates are low for PS supported LAs.

3.3.1.2.8   Heterogeneous WCA-Based Initiator Systems

Industrial and academic researchers have investigated supported MAO as a solid coinitiator for IB polymerization (Table 3.52) [109, 172]. Although inactive for polymerization of neat IB by themselves, supported versions of MAO were compatible with the same initiators that were useful with unsupported solid MAO (Section 3.1.1.6.5). The chemistry of initiation is believed to mirror that for unsupported MAO (Schemes 3.56–3.62). In contradistinction to unsupported MAO, supported analogs possessed superior activity with every initiator subset with few exceptions, especially at T > –30 °C. Only the when the coinitiator became encapsulated in the early stages of polymerization did supported versions fail to outperform their unsupported analogs and this can be circumvented by use of aliphatic solvents. Most polymerizations involving supported MAO with neat IB were explosively violent and rapid, leading to instantaneous reflux, high pressures (up to 160 psi), and concomitant solidification of the charge. As a result, MWs obtained are depressed by excessive chain transfer due to excessive heat buildup and are below that which can be theoretically made if proper provisions for cooling are made. Activity appeared to depend (qualitatively) on the identity of the support material and studies on the effect that the support has are underway [109, 172].

Table 3.53 Aqueous suspension polymerization[a] of PMOS via PMOS-HCl adduct/Yb(OTf)$_3$ system [234].

Time[b] (hr)	Conv. (%)	$\overline{M}_n$ (kg·mol^{-1})	$\overline{M}_w/\overline{M}_n$
2	9	0.800	1.43
88	28	1.900	1.44
200	98	2.900	1.40

[a][PMOS] = 3.0 M; [CH$_3$CH(p-C$_6$H$_4$-OCH$_3$)Cl] = 60 mM; [Yb(OTf)$_3$] = 300 mM; T = 30 °C.
[b]Estimated from graphical data.

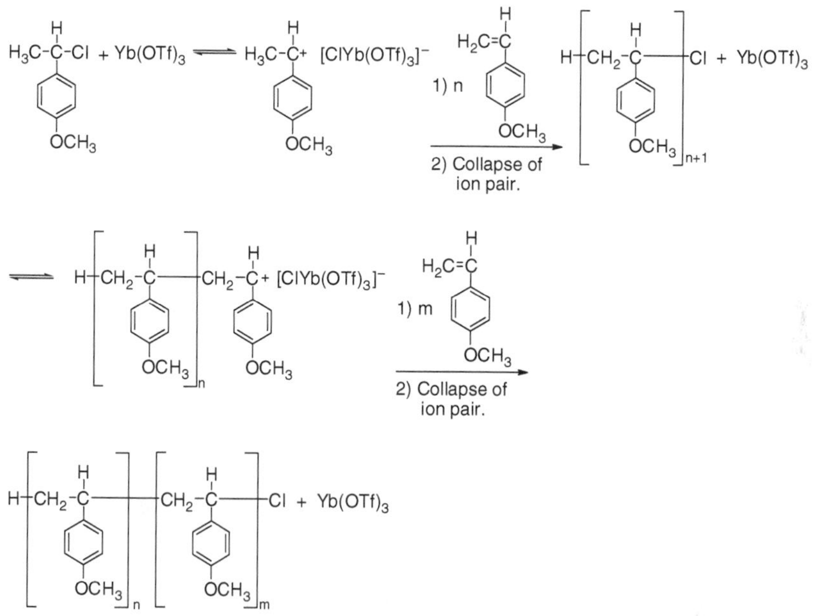

Scheme 3.77 Polymerization of PMOS by reversible ionization of ω-Cl end groups by Yb(OTf)$_3$.

3.3.1.2.9   Aqueous Polymerization Systems

Given the hydrolytic instability of most LA coinitiators and the high reactivity of carbocations, few aqueous cationic olefin polymerization systems have been developed. An early attempt at polymerization of IB in aqueous acids only produced dimer and trimer and required high temperatures (ca. 100 °C) for reasonable conversions [232]. For many years ionic aqueous techniques were limited to anionic and cationic ring opening polymerization of cyclosiloxanes [233]. Recently, two distinct approaches have been devised to effect cationic olefin polymerization in aqueous media and are summarized below.

3.3.1.2.9.1   *Aqueous Cationic Olefin Polymerization Using Hydrolytically Stable LAs*

The first significant advances in aqueous cationic olefin polymerization were made by Sawamoto and coworkers using highly reactive monomers isobutyl vinyl ether (IBVE) and *p*-methoxystyrene (PMOS) in conjunction with hydrolytically stable LA coinitiators (Table 3.53) [234–237]. Initial work focused on polymerization of

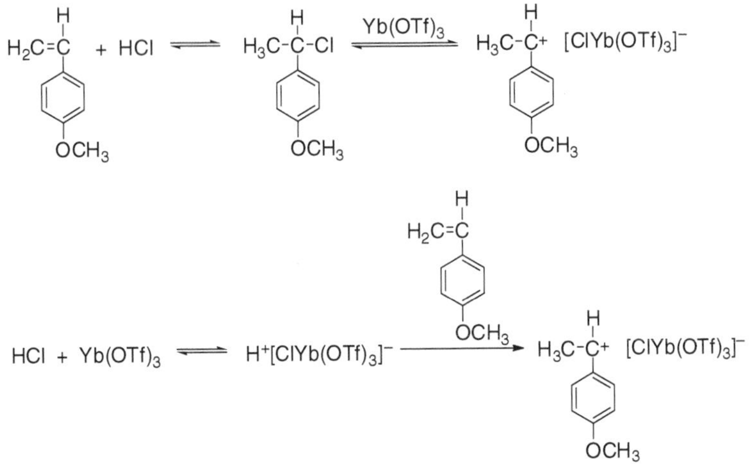

Scheme 3.78 Possible modes of initiation for PMOS by the BA/Yb(OTf)$_3$ system.

IBVE and PMOS using lanthanide triflate coinitiators {e.g. Yb(OTf)$_3$} in conjunction with the corresponding HCl·monomer adduct {i.e. CH$_3$CH(R)Cl; R = OiBu, p-C$_6$H$_4$-OCH$_3$, respectively} initiator [234]. Large amounts of water were used (organic:aqueous phase = 1:1 vol:vol) and neither coinitiator, initiator, or triflic acid (a possible hydrolysis product) alone is capable of inducing polymerization under these conditions. It was speculated that initiation involved ionization of the carbocation synthon by Yb(OTf)$_3$ entering the monomer droplet from the aqueous phase (Scheme 3.77).

For polymerization of PMOS, the $\overline{M}_n$ increased in direct proportion to monomer conversion (slowly) before leveling off late in the polymerization and agreed well with the theoretical value based on the ratio of monomer (g): initiator (moles) [234]. Moreover, PDI values for poly(p-methylstyrene) were low (ca. 1.4) and according to these researchers ^{1}H NMR spectra for these polymers contain signals for α-CH$_3$ head- and ω-PhCHCl end-groups, respectively [234]. These results support initiation by the carbocation synthon with propagation involving reversible activation of ω-PhCHCl end-groups. When similar polymerizations were effected under emulsion conditions using cationic surfactants (e.g. tetramethylammonium chloride) polymerization rates were enhanced [236]. Anionic and nonionic surfactants {e.g. sodium laurylbenzenesulfonate and poly(vinyl methyl ether)} led to a reduction in polymerization rate compared to reaction in suspension which was

Table 3.54 Miniemulsion[a] of PMOS using DBSA [238].

[PMOS] (g)	[DBSA] (g)	[H$_2$O] (g)	Temp. (°C)	$\overline{M}_n$ (kg·mol^{-1})	$\overline{M}_w/\overline{M}_n$
3	0.15	5	60	1,050	1.004
3.2	0.32	8	40	1,000	—
3	0.3	15	25	1,055	—
4	0.4	10	25	1,100	1.005

[a]All ingredients mixed at 25 °C, ultrasonified, and placed into reactor prior to final reaction at specified temperature under 350 rpm stirring. Time and yields not specified.

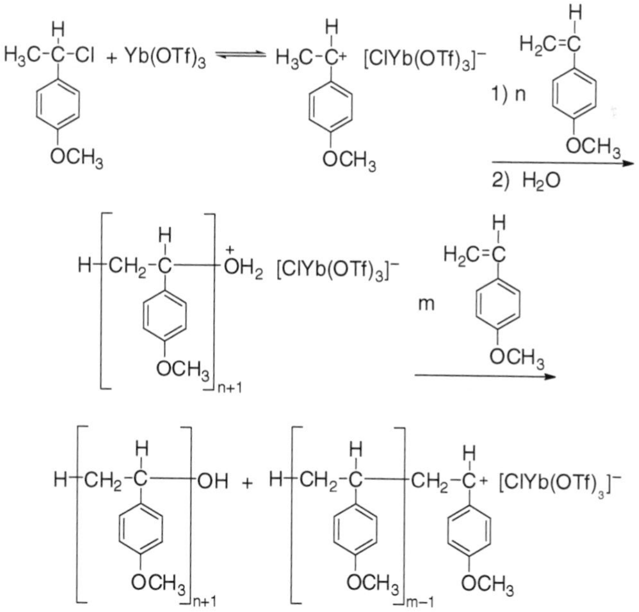

Scheme 3.79 Water mediated CT for PMOS polymerization coinitiated by Yb(OTf)$_3$.

attributed to coordination of Yb(OTf)$_3$ by these basic surfactants. Other water tolerant metal triflate and tetrafluroborate coinitiators {e.g. Sn(OTf)$_2$, Zn(OTf)$_2$, Zn(BF$_4$)$_2$} function in a manner similar to Yb(OTf)$_3$ where the polymerization rate increased with increasing Lewis acidity of the central metal [237].

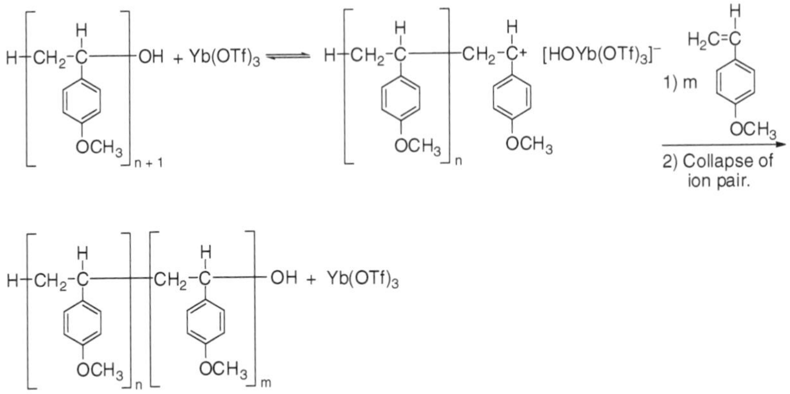

Scheme 3.80 Polymerization of PMOS by reversible ionization of ω-OH end groups by Yb(OTf)$_3$.

Table 3.55 Low temperature emulsion polymerization[a] of IB coinitiated by Yb(OTf)$_3$ [239].

[IB] (mmol)	Time (hr)	Conv. (%)	$\overline{M}_n$ (kg·mol^{-1})	$\overline{M}_w/\overline{M}_n$
89.1	2	6.6	261	1.5
178	4	14.0	502	1.5
267	6	15.0	1,100	1.4
356	8	19.0	4,200	2.6

[a][Yb(OTf)$_3$] = 8.06 mmol; [antifreeze (56.25 wt % H$_2$O, 28.12 wt % CaCl$_2$, 15.63 wt % MgCl$_2$)] = 17.0 mL; [dodecyltrimethylammonium bromide] = 0.324 mmol; [n-hexane] = 5 mL; T = –57 °C. All reactions terminated with 50 vol % THF(aq).

For polymerization of PMOS, BA initiators could be used in place of carbocation synthetic equivalents with similar results [235]. The polymerization rate was directly correlated to acid strength with the strongest acids inducing the fastest polymerization. The exact mode of initiation was not determined and could result from *in situ* formation of acid·monomer adduct followed by ionization or by protonation of monomer from the acid·Yb(OTf)$_3$ adduct (Scheme 3.78). In the case of sulfonic acid initiators, these researchers noted that ^{1}H NMR spectroscopic analysis showed the presence of α-CH$_3$ head groups in addition to ω-PhCHSO$_3$CH$_3$ and ω-PhCHOH (an expected hydrolysis product) end-groups in support of a propagation mechanism involving reversible ionization of the polymer chain end.

Polymerization by reversible termination processes as depicted in Schemes 3.77 and 3.78 is questionable in light of work conducted by Ganachaud *et al.* on emulsion polymerization of PMOS using dodecylbenezenesulfonic acid (DBSA) [238]. DBSA is unique in that it can function as both an initiator and surfactant. PMOS polymerizations using DBSA exhibit similar reaction characteristics to those coinitiated by Yb(OTf)$_3$ (Table 3.54). For example, $\overline{M}_n$ steadily increases with conversion until tapering off in the later stages of polymerization. Moreover, it was shown that Sawamoto and coworkers had misidentified ^{13}C satellites as ^{1}H NMR signals

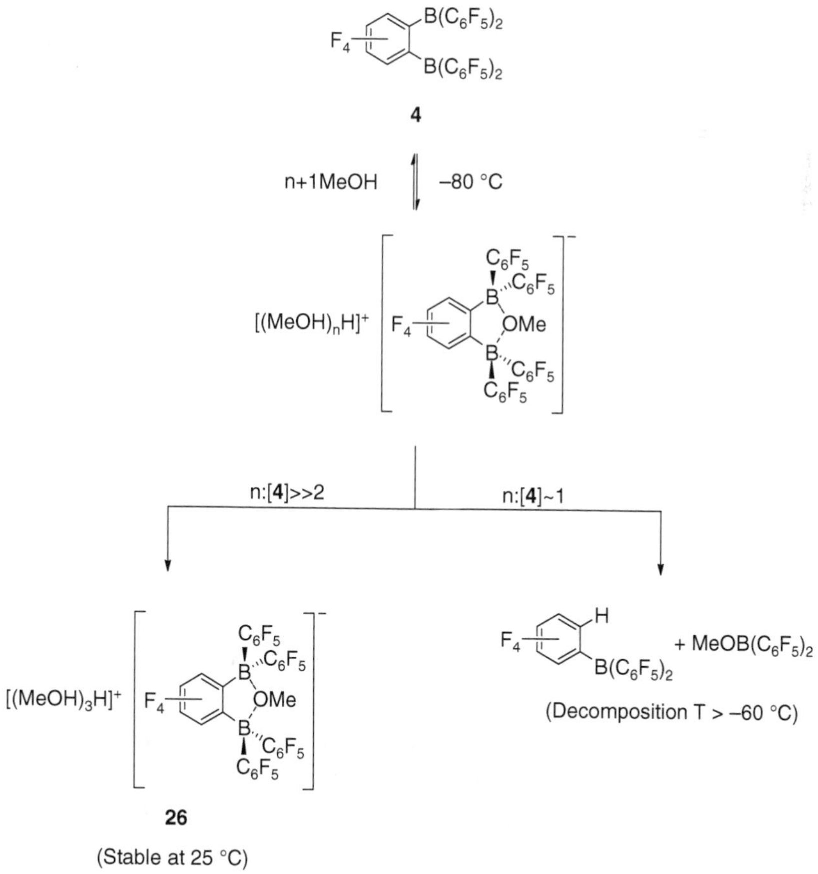

Scheme 3.81 Reaction of methanol with diborane 4.

Table 3.56 Suspension and emulsion polymerization of IB[a] using chelating diborane coinitiator [127, 130, 132].

[4] (mM)	Diluent	Soap[h]	Temp. (°C)	Conv. (%)	$\overline{M}_w$ (kg·mol^{-1})	$\overline{M}_w/\overline{M}_n$
0.426[b]	LiCl[d]	—	−60	48.1	66.1	2.55
0.426[b]	H$_2$SO$_4$[e]	—	−60	50.4	74.5	2.33
0.426[b]	HBF$_4$[f]	—	−78	85.4	138	2.16
0.628[c]	LiCl[d]	—	−60	44.5	19.8	2.30
0.628[c]	H$_2$SO$_4$[e]	—	−60	29.0	38.4	2.05
0.628[c]	HBF$_4$[f]	—	−78	57.5	50.8	2.36
0.628[c]	HCl[g]	—	−25	15.5	20.3	1.88
0.628[c]	LiCl[d]	DTB	−60	4.70	57.2	2.74
0.628[c]	LiCl[d]	SDS	−60	32.3	25.4	2.37
0.628[c]	H$_2$SO$_4$[e]	SDS	−60	17.0	45.6	2.24
0.628[c]	HBF$_4$[f]	DTFB	−78	47.0	43.7	2.22

[a]All polymerizations [IB] = 0.226 mol; [Diluent] = 18 mL.
[b]Concentration in organic phase. 4 dissolved in 10.0 mL *n*-hexane and added over 5 min.
[c]Concentration in organic phase. 4 dissolved in 1.00 mL toluene, injected rapidly.
[d][LiCl] = 23 wt%; [NaCl] = 1.2 wt%; [H$_2$O] = 75.8 wt%.
[e]38 wt% H$_2$SO$_4$ (aq).
[f]38 wt% H$_2$SO$_4$ (aq).
[g]12 N HCl (aq).
[h][Soap] = 0.100 g; DTB = dodecyltrimethylammonium bromide; SDS = sodium dodecylsulfate; DTFB = dodecyltrimethylammonium tetrafluoroborate.

belonging to ω-PhCHCl and ω-PhCHOH. Instead it was determined that PMOS made by DBSA and Yb(OTf)$_3$ based systems both contain ω-PhCHOH end-groups the signal for which is partially buried in the resonance for the *p*-methoxy group of the polymer. From this it is possible that initiation and polymerization chemistry using the Yb(OTf)$_3$/HCl·PMOS adduct system might actually involve protonic processes (and/or irreversible activation of hydroxyl functionalized polymer Schemes 3.79–3.80).

Following the work of Sawamoto and coworkers analogous research was conducted in the Kennedy group for polymerization of IB [239–240]. In contradistinction to PMOS it was reported that Yb(OTf)$_3$ was capable of inducing emulsion polymerization of IB at –57 °C by itself in the absence of purposefully added initiators [239]. Low temperature emulsions were produced using a eutectic mixture comprising 56.25 wt% H$_2$O, 28.12 wt% CaCl$_2$, and 15.63 wt% MgCl$_2$ in conjunction with the surfactant dodecyltrimethylammonium bromide. Experimentation showed polymerization to possess living characteristics (Table 3.55) [239]. Both polymer MWs and PDI values were low and polymerizations were slow. After attempts at replicating these experiments ended in failure and Yb(OTf)$_3$ was also shown to be ineffective for bulk polymerization of IB it was later discovered the actual products from previous IMA experiments were polydimethylsiloxanes and the data had been falsified [240]. Attempts at polymerizing IB in THF using Yb(OTf)$_3$ in conjunction with CCl only led to the production of poly(tetrahydrofuran) [240]. These results indicate that ROP of cyclic ethers can be accomplished using water tolerant LA coinitiators in conjunction with a suitable initiator (e.g. carbocation synthon, BA, etc.). Moreover, similar polymerization maybe feasible under aqueous conditions for cyclic ethers, which are not soluble in water.

3.3.1.2.9.2   *Aqueous Cationic Olefin Polymerization Using PFLAs*

During experimentation on solution polymerization of IB using chelating diborane coinitiator **4**, it was discovered that **4** formed a stable methanol adduct (**26**) when the reaction was conducted in excess methanol (Scheme 3.81) [127, 130, 132]. Previous attempts to generate this complex under stoichiometric conditions always led to decomposition of **4** by protolytic cleavage of B(C$_6$F$_5$)$_2$ groups [127, 241]. This finding led to the development of reaction conditions conducive for aqueous suspension and emulsion polymerization of IB (Table 3.56) [127, 130, 132, 242]. Polymerization is favored when conditions are selected such that the level of adventitious moisture in both the coinitiator stock solution as well as the monomer droplet are minimized as a strategy to reduce the effects of CT and termination [127]. This involved use of solvents of low polarity for **4** and **5** and aqueous media with high dielectric constants possessing low freezing points.

Most likely initiation is protonic in nature and involves chelated aqua adducts such as the hypothetical BA **6** (Scheme 3.34). The monodentate PFLA **2** is incapable of inducing polymerization under otherwise identical conditions. Polymerization perpetuates by chain transfer and is terminated when a growing chain migrates to the organic/aqueous interface or when it complexes an excess of water molecules within the droplet leading to inactive species. Despite the high reactivity of carbocations with water the rapidity of polymerization is such that low-moderate yields of medium-high MW PIBs and butyl rubber can be obtained under aqueous conditions. In many cases reaction is so violent as to vaporize monomer

Table 3.57 Aqueous polymerization[a] of PMOS with the $B(C_6F_5)_3$/ 1-(4-methoxyphenol)ethanol initiator system [244].

$[B(C_6F_5)_3]$ (g)	Diluent	[PMOS] (g)	Time (hr)	Yield (%)	$\overline{M}_n$ (kg·mol^{-1})	$\overline{M}_w/\overline{M}_n$
0.06	H_2O	0.48	23	99	2.88	1.75
0.14	H_2O	0.6	2	92	3.80	1.85
0.14	Buffer	0.6	20	100	4.38	1.77
0.06[b]	Buffer	0.59	21	98	3.44	2.00

[a]$[CH_3CH(p\text{-}C_6H_4\text{-}OCH_3)Cl] = 0.017$ g; [Diluent] $= 5$ mL; Buffer pH $= 7$ (NaH$_2$PO$_4$ 3.6 g·L^{-1}, Na$_2$HPO$_4$ 7.2 g·L^{-1}, NaCl $= 4.3$ g·L^{-1}, H$_2$O $= 1$L); T $= 0$ °C.
[b]No initiator added.

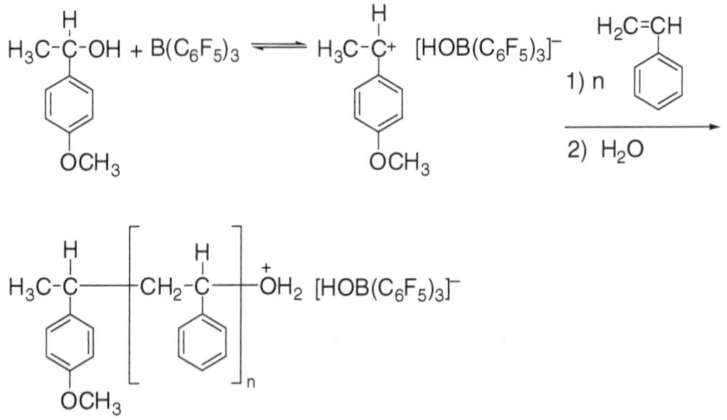

Scheme 3.82 Aqueous polymerization of styrene using the tris(pentafluorophenyl)boron/1-(4-methoxyphenyl)ethanol initiator system.

instantly even when polymerization is conducted at low temperatures. As a result, MWs are depressed not only by water mediated CT mediated but also due to the highly exothermic nature of these polymerizations.

The polymerization behavior of these systems is complex and has not been thoroughly investigated. Rational explanations can be forwarded to explain some general trends. For example, MWs increase with decreasing temperature in line with standard cationic olefin polymerization due to reduction of CT. MWs are also higher when conditions are selected to minimize the level of moisture within the monomer droplet (e.g. use of aliphatic versus aromatic solvents) and to reduce exothermicity of polymerization (e.g. gradual addition of coinitiator), both of which minimize CT. In the case of diluents that do not possess components that readily partition into the monomer droplet, the addition of surfactants that possess ionic groups of similar nucleophilicity to the diluent typically results in an increase in MW with concomitant decrease in yield {e.g. $LiCl/NaCl/MgCl_2$ eutectic + dodecyltrimethylammonium bromide (DTB); H_2SO_4 + sodium dodecylsulfate (SDS)}. This might result from displacement of water within the monomer droplet, by the hydrophobic portions of the surfactant molecules, which in turn reduces CT. Surfactants would increase the surface area of monomer droplets as well as their net surface charge and increase the likelihood of termination events involving migration of propagating polymer chains to the organic/aqueous interface resulting in lower yields.

Later research showed that $B(C_6F_5)_3$, which is hydrolytically more stable than 4 or 5, can be used to polymerize styrenic monomers [243, 244] and isoprene [245] under aqueous conditions (Table 3.57). For highly reactive monomers such as PMOS, polymerization proceeds without purposeful addition of an initiator and is believed to result from protonic processes involving the aqua adduct of 2 in a manner similar to that for IB in chlorinated solvents (Scheme 3.33). For less reactive monomers such as styrene [244] and isoprene [245] polymerization proceeded only in the presence of a carbocation synthon. ^1H NMR spectroscopy indicates that initiation occurs by ionization of the carbocation synthon (Scheme 3.82). In all instances polymerization proceeds slowly affording moderate to high yields of low MW polymers.

Despite the fact that they can be recycled, water tolerant LA coinitiators do not appear to be useful for cationic olefin polymerization under aqueous conditions. Even for the most reactive monomers polymerization is slow and MWs are low. Homopolymerization of IB and its copolymerization with IP can be accomplished in aqueous media using chelating diborane coinitiators, provided that reaction conditions are adjusted to minimize [H$_2$O] within both monomer droplet and coinitiator solution. Given the high cost of chelating diboranes and the low-moderate yields of polymers produced, these polymerizations are economically unfeasible. Aqueous polymerizations involving B(C$_6$F$_5$)$_3$ are not of value in that they cannot produce high MW polymers and polymerization rates are generally too low to be useful.

3.4   Environmental Benefits via New Process Conditions

A large volume of work has been conducted on substituting (in part or in whole) hydrofluorocarbons (HFCs) for methyl chloride in the slurry production of butyl rubber [28, 41, 246–255]. At normal polymerization temperatures (ca. –95 °C) fouling is eliminated whereas it is reduced at higher operating temperatures when compared to the methyl chloride. This in part might be attributed to a lower solubility of the diluent in monomer and polymer due to the higher dielectric constant of HFCs in comparison to methyl chloride yielding droplets of reduced tackiness [249, 250]. Critical review of the data contained in these patent filings reveals that conversion is also generally higher in HFC slurry which translates to polymer particles containing less monomer which in turn would reduce tackiness and fouling. These results are in line with the general observation that cationic polymerization generally proceeds to higher yields when conducted in media of higher dielectric strength. The addition of 3 ° halides > C4 and chalcogen containing molecules (e.g. ethers) to HFC based slurries have been reported to further reduce fouling.

Another added benefit is that unlike chlorocarbons, HFCs are not ozone depletors [256–258]. Toxicity issues are also eliminated as reflected in the use of HFCs as propellants for metered dose inhalers [258]. One major drawback to HFCs is their global warming potentials greatly exceed those of methyl chloride [44] and their discontinued use is being mandated by legislation [259]. HFCs also suffer in that they readily form azeotropes with IB and thus complicate recycling of these materials especially in light of the fact that HFC used to make the coinitiator solution must be free of monomer [28].

One point that appears to be overlooked is the upper use temperature imposed on slurry operations by the glass transition temperature (T_g) of the elastomer being produced. In theory this corresponds to approximately −70 °C for PIB and −65 °C for butyl rubber [260]. Once polymerization temperature exceeds this T_g threshold the polymer particles will no longer be in the glassy state and issues arising from fouling will be difficult if not impossible to avoid.

From a theoretical standpoint slurry processes are more prone to produce gel fractions of butyl than solution polymerization and are not conducive to the preparation of elastomers of intermediate (5–40 mol %) unsaturation. It is important to note that workers at Exxon determined that gel free butyl type rubbers of high MW were best produced by conducting polymerization in solution within a specific solvent:monomer concentration range which is dependent on solvent and comonomer identity [110, 261, 262]. This approach maximizes access of growing polymer chains ends to monomer while minimizing their interaction with polymer vinyl groups in turn maintaining high MW while preventing gellation. In essence the approach provides for high polymer solubility and high monomer concentration throughout the lifetime of polymerization. An added benefit of this approach is that it purportedly allows for azeotropic copolymerization whereas slurry processes are nonideal. Precipitation of polymer during slurry processes leads to a high localized concentration of active polymer chain ends in the vicinity polymer vinyl groups and ultimately gellation.

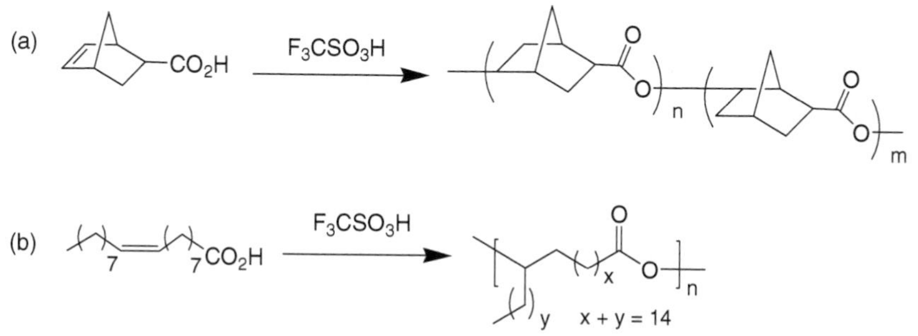

Scheme 3.83 Polymerization of monomers with alkene and carboxylic acid functional groups using trifluoromethylsulfonic acid (triflic acid).

As mentioned previously, diluent (whether CH_3Cl or HFC) must be removed and the polymer dissolved in a nonreactive solvent prior to effecting post-polymerization functionalization reactions (i.e. halogenation) [17–21, 23, 29, 42]. This is commonly done in one of two ways. The first method involves formation of a solid bale of rubber which is subsequently chopped/ground and fed into the solvent of choice. Solvent replacement involving addition of a non-reactive solvent of higher boiling point than the diluent followed by flashing of the latter and dissolution of the rubber is a second technique. Regardless of the approach, these processes are energy intensive and time consuming. As a result, solution based methods lend themselves to the production of value added products (e.g. halo-butyl rubber).

Moreover, one must consider the complexity of the slurry process itself which is illustrated by intricate reactor designs devised to reduce energy consumption and prevent fouling. Bayonet type reactors that provide a countercurrent flow of the slurry are exemplary [263]. These consist of a series of tubular heat exchangers

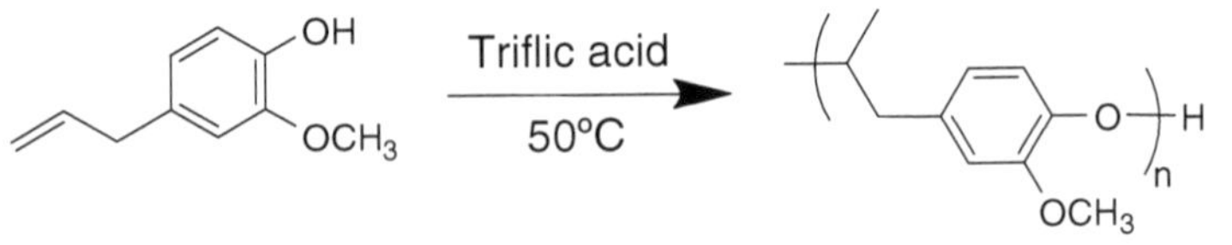

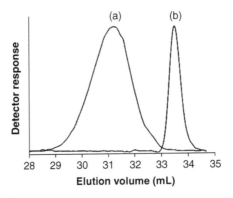

Scheme 3.84 Polymerization of neat eugenol with triflic acid. Alkene isomerization and electrophilic aromatic substitution reactions have been omitted for clarity.

Figure 3.2 Gel-permeation chromatography data (viscometer detector) for (a) the product resulting from the polymerization of neat eugenol with triflic acid (2 mol%) after 2 h at 50 °C and (b) eugenol.

grouped in bundles called bayonets where the slurry is initially pumped through a bundle contained in the central cavity of reactor and then exits the reactor in the opposite direction through bundles contained on the outside surface of the reactor. Moreover, it has been proposed that reactors with ultra-high machining tolerances be used since vessels having average surface roughness <0.3 μm reduced fouling [252]. Given the high cost of such reactors, development of alternative methods in place of the slurry process that are capable of operating continuously in lower cost equipment of simplified design are worthy targets of future research.

3.5   Cationic Polymerization of Monomers Derived from Renewable Resources

Recently, polyesters and polyethers were reported using hydrocarboxylation and hydroalkoxylation reactions, respectively [264]. These polymerizations involve the reaction of alkenes with polar functionalities, such as alcohols and carboxylic acids, in the presence of a strong Brønsted acid. For example, in Scheme 3.83a, 5-norbornene-2-carboxylic acid ($NBCO_2H$) was polymerized at 70 °C to give a polyester ($\overline{M}_w = 9,100$ g/mol) after 0.5 h. The polymerization of $NBCO_2H$ was catalyzed with triflic acid (TfOH) and conducted under ambient condition with neat monomer to eliminate the need for solvents and stringent experimental conditions. Since $NBCO_2H$ can be synthesized from cyclopentadiene and acrylic acid, the monomer is partially renewable because acrylic acid can be derived from glycerol [265]. Comparison of the $NBCO_2H$ polymerization with oleic acid (Scheme 3.83b) suggests that the strained alkene in $NBCO_2H$ is more reactive than acyclic alkenes and leads to higher conversions and molecular weight values.

Since alkenes and phenols are prevalent functional groups, many possibilities exist for polymerizing renewable molecules. Phenols are particularly interesting since they are found as components of lignin and many aromatic fragrances. We have found that eugenol, which is found in clove oil, will polymerize since it contains a phenol as well as an alkene. Scheme 3.84 illustrates the polymerization of eugenol using TfOH. The polymerization of neat eugenol at 50 °C for 2 h produced monomodal polymer with an $\overline{M}_w$ value of 2,100 g/mol (PDI = 1.3). Although the polymerization of neat eugenol solidified after 1 h at 50 °C, the polymer was readily soluble in acetone,

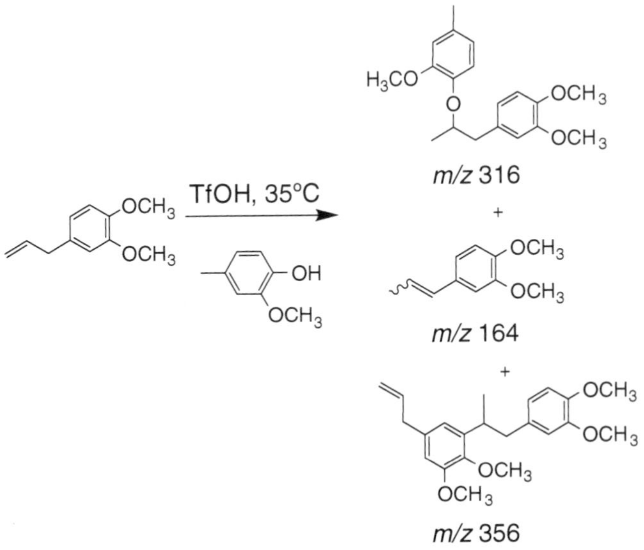

Scheme 3.85 Model reaction of a protected eugenol molecule and a substituted phenol. Although several isomers will result from electrophilic aromatic substitution, only one isomer is presented for clarity.

methanol and tetrahydrofuran. Dissolving the polymer in acetone and precipitating into petroleum ether, allowed purification of the polymer and removal of the unreacted monomer. Figure 3.2 shows the difference in elution volume for the precipitated polymer and eugenol. Similar polymerizations with triflic anhydride (Tf_2O) also resulted in monomodal polymer ($\overline{M}_w$ = 2,380 g/mol; PDI = 1.3). Tf_2O is expected to react with moisture or phenol to generate TfOH *in situ*. The polymerizations with TfOH and Tf_2O gave much higher conversions and $\overline{M}_w$ values than H_2SO_4 (~ 7% conversion).

Several control experiments were performed to determine if the polymerization was indeed acid catalyzed. First, the polymerization of eugenol was modified by adding 2 mol% of DTBP which is a hindered base that scavenges acidic protons. After 24 h, no polymer was detected by GPC. Second, when TfOH was replaced with 2 mol% of NaOTf, Na_2SO_4 or sodium dodecyl sulfate (SDS), the triflate and sulfate anions did not result in detectable amounts of polymer.

FTIR spectroscopy identified the formation of an ether linkage during the polymerization. In addition to the C-O absorbance at 1035 cm^{-1} for the methoxy group on the benzene ring, an additional ether absorbance at 1056 cm^{-1} was noted after the polymerization. The phenol O-H bending absorbance at 648 cm^{-1} and the phenol C-O stretch at 1207 cm^{-1} both decreased during the polymerization. Due to C-O-C bonds, polyethers usually have more flexibility and lower glass transition temperature (T_g) values than polyesters and polystyrene. The presence of ether bonds was reflected in DSC measurements of the glass transition temperature value ($T_g = 57$ °C; 10 °C/min) after the polymerization of eugenol with TfOH.

Based on ^{1}H NMR, the formation of methyl branches (δ 1.2 ppm) was detected during the polymerization. This would indicate that phenol undergoes nucleophilic addition on the most substituted carbon of the alkene. Since eugenol has an electron donating methoxy group, electrophilic aromatic substitution (EAS) can also occur in addition to the desired hydroalkoxylation reaction. The model reaction in Scheme 3.85 suggested that hydroalkoxylation reactions compete with alkene isomerization and EAS reactions. The existence of EAS reactions in Scheme 3.85 is expected to be more favorable than an unsubstituted phenol. Analysis of the polymerization product with gel permeation chromatography (GPC) using light scattering and viscometer detectors confirmed the polymer was branched. For example, the Mark-Houwink-Sakurada (MHS) plot for the polymerization of eugenol after 2 h at 50 °C indicated that the MHS exponent (0.45) was lower than typical values (0.6–0.8) for linear polymers in good solvents.

3.6 Sustainable Synthesis of Monomers for Cationic Polymerization

The synthesis of isoprene from natural feedstocks has been a long-standing goal of the synthetic rubber industry. Early reports that date back to 1860 involved the conversion of natural rubber to isoprene, but the yields were low [15]. Later, in 1884, conversion of turpentine in a hot iron tube to isoprene generated ~7.5% yield [15]. During World War II, turpertine and dipentene were converted to isoprene using electrically heated coils that induced pyrolysis [15]. The pyrolysis first involved the isomerization of turpentine, which contained α-pinene and β-pinene, to dl-limonene and then the subsequent pyrolysis to isoprene. More recently, a microbial fermentation process has been developed for the synthesis of isoprene [16]. This approach uses engineered strains of E. coli that contain plant derived enzymes capable of converting glucose into isoprene. Isoprene is produced as fermentation off-gas in a continuous process in yields > 60 g isoprene·L^{-1} culture broth.

Although the green production of monomers used in cationic olefin polymerization is progressing, further improvements need to be made before non-petroleum feedstocks can become viable options. In particular, use of non-food based renewable feedstocks is paramount for sustainability of green monomer production. It is also important to develop processes that operate in a more carbon neutral manner; for example, monomer synthesis via photosynthetic pathways versus fermentation processes (that also produce CO_2). Further work is required in this area before it can be successfully implemented on a large enough scale to supplant petroleum derived monomers.

Acknowledgements. RTM thanks the donors of the American Chemical Society Petroleum Research Fund for financial support. RTM also expresses gratitude to the Army Research Office for instrumentation support through the DURIP programs.

References

1. R.T. Mathers and S.P. Lewis, "Monoterpenes as Polymerization Solvents and Monomers in Polymer Chemistry," in R.T. Mathers, Meier, M.A.R eds. *Green Polymerization Methods: Renewable Starting Materials, Catalysis and Waste Reduction*, New York, Wiley-VCH, pp. 91–128, 2011.
2. J.P. Kennedy, *Cationic Polymerization of Olefins: A Critical Inventory*, New York, John Wiley and Sons, 1975.
3. J.P. Kennedy and E. Marechal, *Carbocationic Polymerization*, New York, John Wiley and Sons, 1982.
4. *Cationic Polymerizations*; K. Matyjaszewski, ed. NY, Marcel Dekker, 1996.
5. J.E. Puskas, S.W.P. Chan, K.B. Mcauley, S. Shaikh and G. Kaszas, *Journal of Polymer Science: Part A: Polymer Chemistry*, Vol. 43, p. 5394, 2005.
6. H. Schlaad, Y. Kwon, L. Sipos, R. Faust and B. Charleux, *Macromolecules*, Vol. 33, p. 8225, 2000.
7. J.P. Kennedy, *Cationic Polymerization of Olefins: A Critical Inventory*, New York, John Wiley and Sons, p. 114, 1975.
8. K. Matyjaszewski, C-H. Lin and C. Pugh, *Macromolecules*, Vol. 26, p. 2649, 1993.
9. G.A. Olah, *Friedel-Crafts Chemistry*, New York, John Wiley and Sons, 1973.
10. T.-C. Chung, F.J. Chen, J.E. Stanat and A. Kumar, US Patent 5288677, 1994.
11. P.J. Flory, *Principles of Polymer Chemistry*, Ithaca, Cornell University Press, 1971.

12. R.M. Thomas, W.J. Sparks, P.K. Frolich, M. Otto and M. Müeller-Cunradi, *Journal of the American Chemical Society*, Vol. 62, p. 276, 1940.
13. J.P. Kennedy, A. Shinkawa and F. Williams, *Journal of Polymer Science: Part A-1*, Vol. 9, p. 1551, 1971.
14. F. Williams, A. Shinkawa and J.P. Kennedy, *Journal of Polymer Science: Polymer Symposia*, Vol. 56, p. 421, 1976.
15. *Synthetic Rubber*; G.S. Whitby, C.C. Davis and R.F. Dunbrook, Eds., John Wiley and Sons: NY, NY, 1954.
16. G.M. Whited, F.J. Feher, D.A. Benko, M.A. Cervin, G.K. Chotani, J.C. McAuliffe, R.J. LaDuca, E.A. Ben-Shoshan and K.J. Sanford, *Industrial Biotechnology*, Vol. 6, p. 152, 2010.
17. A.J. Dias, "Isobutylene Copolymers," in J.C. Salamone ed., *Polymeric Materials Encyclopedia*, Boca Raton, CRC Press, pp. 3484–3492, 1996.
18. J. Duffy and G.J. Wilson, "Synthesis of Butyl Rubber by Cationic Polymerization," in B. Elvers, S. Hawkins, W. Russey and G. Schulz eds., *Ullman's Encyclopedia of Industrial Chemistry*, Weinheim, VCH Publishers, pp. 288–294, 1993.
19. J.P. Kennedy and I. Kirshenbaum, *High Polymers*, Vol. 24, p. 691, 1971.
20. E. Kresge and H.-C. Wang, "Butyl Rubber," in J.I. Kroschwitz and M. Howe-Grant eds., *Kirk-Othmer Encyclopedia of Chemical Technology*, New York, John Wiley and Sons, pp. 934–955, 1993.
21. E.N. Kresge, R.H. Schatz and H.-C. Wang, "Isobutylene Polymers," in H.F. Mark, N.M. Bikales, C.G. Overberger and G. Menges eds., *Encyclopedia of Polymer Science and Engineering*, New York, John Wiley and Sons, pp. 423–448, 1987.
22. R.M. Thomas and W.J. Sparks, US Patent 2356128, 1944.
23. R.N. Webb, T.D. Shaffer and A.H. Tsou, "Butyl Rubber," in A. Seidel ed., *Kirk-Othmer Encyclopedia of Chemical Technology*, Hoboken, N.J., John Wiley and Sons, pp. 433–458,
24. A.W. Apblett, "Aluminum: Inorganic Chemistry," in R.B. King ed., *Encyclopedia of Inorganic Chemistry*, New York, John Wiley and Sons, pp. 103–116, 1994.
25. K. Wade and A.J. Banister, "Aluminium, Gallium, Indium and Thallium," in J.C. Bailar, H.J. Emeleus, R. Nyholm and A.F. Trotman-Dickenson eds., *Comprehensive Inorganic Chemistry*, Oxford, GB, Pergamon Press, pp. 1012–1028, 1973.
26. J.P. Kennedy, US Patent 3349065, 1967.
27. J.P. Kennedy, *Journal of Polymer Science: Part A-1*, Vol. 6, p. 3139, 1968.
28. T.D. Shaffer, M.F. McDonald, D.Y.-L. Chung, R.N. Webb, D.J. Davis and P.J. Wright, US Patent Application 0234447 A1, 2008.
29. J. Duffy and G.J. Wilson, "Halobutyl Rubber " in B. Elvers, S. Hawkins, W. Russey and G. Schulz eds., *Ullman's Encyclopedia of Industrial Chemistry*, Weinheim, VCH Publishers, pp. 314–318, 1993.
30. S.C. Ho and M.M. Wu, US Patent 5326920, 1994.
31. J.P. Kennedy and R.G. Squires, *Journal of Macromolecular Science Chemistry*, Vol. A1, p. 805, 1967.
32. M.D. Maina, S. Cesca, P. Giusti, G. Ferraris and P.L. Magagnini, *Makromolekulare Chemie*, Vol. 178, p. 2223, 1977.
33. J.P. Kennedy, I. Kirshenbaum, R.M. Thomas and D.C. Murray, *Journal of Polymer Science: Part A-1*, Vol. 1, p. 331, 1963.

34. P.H. Plesch, *Journal of Polymer Science*, Vol. 12, p. 481, 1954.
35. J.P. Kennedy and R.M. Thomas, *Journal of Polymer Science*, Vol. 46, p. 233, 1960.
36. Ricard, Allenet and Cie, GB Patent 202311, 1922.
37. Ricard, Allenet and Cie, DE Patent 402990, 1924.
38. H.M. Stanley, *Transactions and Communications Journal of the Chemical Society*, Vol. 49, p. 349, 1930.
39. J.P. Kennedy and R.M. Thomas, *Journal of Polymer Science*, Vol. 45, p. 227, 1960.
40. D.R. Lide and H.P.R. Frederikse eds., *CRC Handbook of Chemistry and Physics*, Boca Raton, FL, CRC Press, 1996.
41. T.D. Shaffer, M.F. McDonald, D.Y.-L. Chung, R.N. Webb, D.J. Davis and P.J. Wright, US Patent 7402636, 2008.
42. H.-I. Paul, R. Feller, J.G.A. Lovegrove, A. Gronowski, A. Jupke, M. Hecker, J. Kirchhoff and R. Bellinghausen, WO Patent Application 006983 A1, 2010.
43. US Environmental Protection Agency, Methyl Chloride (Chloromethane), http://www.epa.gov/ttn/atw/hlthef/methylch.html, 2000.
44. P. Forster, V. Ramaswamy, P. Artaxo, T. Berntsen, R. Betts, D.W. Fahey, J. Haywood, J. Lean, D.C. Lowe, G. Myhre, J. Nganga, R. Prinn, G. Raga, M. Schulz and R.V. Dorland, "Changes in Atmospheric Constituents and in Radiative Forcing In Climate Change," in S. Solomon, D. Qin, M. Manning, Z. Chen, M. Marquis, K.B. Averyt, M.Tignor and H.L. Miller eds., *The Physical Science Basis. Contribution of Working Group I to the Fourth Assessment Report of the Intergovernmental Panel on Climate Change*, Cambridge, Cambridge University Press, pp. 129–234, 2007.
45. T. Horie, US Patent 3631013, 1971.
46. J.L. Ernst and H.J. Rose, US Patent 2772255, 1956.
47. D.O. Jordan and F.E. Treloar, *Journal of the Chemical Society*, p. 737, 1961.
48. J.H. Beard, P.H. Plesch and P.P. Rutherford, *Journal of the Chemical Society*, p. 2566, 1964.
49. J.P. Kennedy, *Cationic Polymerization of Olefins: A Critical Inventory*, New York, John Wiley and Sons, pp. 99–100, 1975.
50. M. Chmelir, M. Marek and O. Wichterle, *Journal of Polymer Science: Part C: Polymer Symposia*, Vol. 16, p. 833, 1967.
51. L. Balogh, L. Wang and R. Faust, *Macromolecules*, Vol. 27, p. 3453, 1994.
52. R. Faust and L. Balogh, US Patent 5665837, 1997.
53. J.P. Kennedy and J.K. Gillham, *Advances in Polymer Science*, Vol. 10, p. 1, 1972.
54. Scherbakova, N.V.; Petrova, V.D.; Prokofiev, Y.N.; Timofeev, E.G.; Lazariants, E.G.; Stepanov, G.A; Pautov, P.G.; Nabilkina, V.A.; Dobrovinsky, V.E.; Arkhipov, N.B.; Sobolev, V.M.; Minsker, K.S.; Emelyanova, G.V.; Vinogradova, A.V.; Lebedev, J.V.; Krapivina, K.Y.; Kolosova, E.V.; Vladykin, L.N.; Sletova, L.I.; Orlova, A.P.; Yatsyshina, T.N.; Bugrov, V.P.; Rodionova, N.N.; Tsvetkova, A.G. DE Patent 104985, 1974.
55. Scherbakova, N.V.; Petrova, V.D.; Prokofiev, Y.N.; Timofeev, E.G.; Lazariants, E.G.; Stepanov, G.A; Pautov, P.G.; Nabilkina, V.A.; Dobrovinsky, V.E.; Arkhipov, N.B.; Sobolev, V.M.; Minsker, K.S.; Emelyanova, G.V.; Vinogradova, A.V.; Lebedev, J.V.; Krapivina, K.Y.; Kolosova, E.V.; Vladykin, L.N.; Sletova, L.I.; Orlova, A.P.; Yatsyshina, T.N.; Bugrov, V.P.; Rodionova, N.N.; Tsvetkova, A.G. CA Patent 1019095, 1977.
56. Y.A. Sangalov, Y.Y. Nelkenbaum, O.A. Ponomarev and K.S. Minsker, *Vysokomol. Soedin.*, Ser. A, Vol. 21, p. 2267, 1979.

57. P.T. Parker and J.A. Hanan, US Patent 3361725, 1968.
58. T. Mole and E.A. Jeffery, *Organoaluminium Compounds*, Amsterdam, Netherlands, Elsevier, 1972.
59. G.H. Robinson, "Aluminum: Organometallic Chemistry," in R.B. King ed., *Encyclopedia of Inorganic Chemistry*, New York, John Wiley and Sons, pp. 116–139, 1994.
60. C.A. Kraus, US Patent 2220930, 1940.
61. C.A. Kraus, US Patent 2387517, 1945.
62. D.W. Marshall, US Patent 3168570, 1965.
63. S.B. Mirviss and E.J. Inchalik, US Patent 2827458, 1958.
64. R.K. Schlatzer, Jr. , US Patent 3645920, 1972.
65. G.M. Smith, S.W. Palmaka, J.S. Rogers and D.B. Malpass, US Patent 5831109, 1998.
66. D.W. Young, US Patent 2924579, 1960.
67. A. Priola, S. Cesca and G. Ferraris, *Die Makromolekulare Chemie*, Vol. 160, p. 41, 1972.
68. D.W. Young and H.B. Kellog, US Patent 2440498, 1948.
69. H.F. Strohmayer, L.S. Minckler, Jr., J.P. Simko, Jr. and E.L. Stogryn, US Patent 3066123, 1962.
70. S. Tanaka, A. Nakamura and E. Kubo, US Patent 3324094, 1967.
71. Y. Imanishi, R. Yamamoto, T. Higashimura, J.P. Kennedy and S. Okamura, *Journal of Macromolecular Science Chemistry*, Vol. A1, p. 877, 1967.
72. O. Wichterle, M. Marek and I. Trekoval, *Journal of Polymer Science*, Vol. 53, p. 281, 1961.
73. Nippon Oil Co. Ltd., GB Patent 1056730, 1967.
74. M. Miyoshi, S. Uemura, S. Tsuchiya and O. Kato, US Patent 3402164, 1968.
75. Nippon Petrochemicals Co. Ltd., G.B. Patent 1183118, 1970.
76. S. Matsushima, US Patent 3773733, 1973.
77. A. Priola, S. Cesca and G. Ferraris, US Patent 3850897, 1974.
78. S. Matsushima and K. Ueno, US Patent 3766155, 1973.
79. M. Marek and M. Chmelir, *Journal of Polymer Science: Part C*, Vol. 23, p. 223, 1968.
80. M. Marek and M. Chmelir, *Journal of Polymer Science: Part C*, Vol. 22, p. 177, 1968.
81. M. Marek, J. Pecka and V. Halaska, *Makromolekulare Chemie, Macromolecular Symposia*, Vol. 13, p. 443, 1988.
82. G. Rissoan, S. Randriamahefa and H. Cheradame, "Heterogeneous Cationic Polymerization Initiators," in R. Faust and T.D. Shaffer eds., *Cationic Polymerization*, Washington, D.C., American Chemical Society, pp. 135–150, 1997.
83. M. Webster, *Chemical Reviews*, Vol. 1, p. 87, 1966.
84. F. Gasparoni and C. Longiave, BE Patent 605351, 1961.
85. F. Gasparoni and C. Longiave, FR Patent 1292702, 1962.
86. F. Gasparoni and C. Longiave, US Patent 3123592, 1964.
87. E.I. Tinyakova, T.G. Zhuravleva, T.M. Kurengina, N.S. Kirikova and B.A. Dolgoplosk, *Doklady Akademii Nauk SSSR*, Vol. 144, p. 592, 1962.
88. J.P. Kennedy, *Journal of Macromolecular Science Chemistry*, Vol. A3, p. 885, 1969.
89. J.P. Kennedy and F.P. Baldwin, US Patent 3560458, 1971.
90. J.P. Kennedy and G.E. Milliman, *Advance in Chemistry Series*, Vol. 91, p. 287, 1969.

91. J.P. Kennedy and S. Rengachary, *Advances in Polymer Science*, Vol. 14, p. 1, 1974.
92. M. Baccaredda, P. Giusti, A. Priola and S. Cesca, GB Patent 1362295, 1974.
93. S. Cesca, P. Giusti, P.L. Magagnini and A. Priola, *Die Makromolekulare Chemie*, Vol. 176, p. 2319, 1975.
94. S. Cesca, A. Priola, M. Bruzzone, G. Ferraris and P. Giusti, *Die Makromolekulare Chemie*, Vol. 176, p. 2339, 1975.
95. P. Giusti, A. Priola, P.L. Magagnini and P. Narducci, *Die Makromolekulare Chemie*, Vol. 176, p. 2303, 1975.
96. J.P. Kennedy, US Patent 4029866, 1977.
97. J.P. Kennedy, US Patent 4081590, 1978.
98. J.P. Kennedy and S. Sivaram, *Journal of Macromolecular Science Chemistry*, Vol. A7, p. 969, 1973.
99. A. Priola, S. Cesca, G. Ferraris and M. Maina, *Die Makromolekulare Chemie*, Vol. 176, p. 2289, 1975.
100. A. Priola, G. Ferraris, M. Maina and P. Giusti, *Die Makromolekulare Chemie*, Vol. 176, p. 2271, 1975.
101. S. Cesca, A. Priola, G. Ferraris, C. Busetto and M. Bruzzone, *Journal of Polymer Science, Polymer Symposia*, Vol. 56, p. 159, 1977.
102. A. Priola, G. Ferraris and S. Cesca, US Patent 3850895, 1974.
103. A. Priola, S. Cesca and G. Ferraris, GB Patent 1407420, 1975.
104. A. Priola, S. Cesca and G. Ferraris, US Patent 3965078, 1976.
105. R. Faust, S.E. Hadjikyriacou and T. Suzuki, US Patent 6051657, 2000.
106. T.E. Hogan, D.F. Lawson, T. Yako and J.P. Kennedy, US Patent 6838539, 2005.
107. M. Licchelli, A. Greco and G. Lugli, US Patent 4829130, 1989.
108. J.P. Kennedy, V.S.C. Chang and A. Guyot, *Advances in Polymer Science*, Vol. 43, p. 1, 1982.
109. R.T. Mathers, Q. Liu, Y. Fu and S.P. Lewis, Unpublished Research Data, 2009–2011.
110. W.A. Thaler, D.J. Buckley, Sr. and J.P. Kennedy, US Patent 3856763, 1974.
111. M. Yamada, K. Shimada and T. Takemura, US Patent 3326879, 1967.
112. N. Yamada, K. Shimada and T. Hayashi, *Journal of Polymer Science, Part B, Polymer Letters*, Vol. 4, p. 477, 1966.
113. M. Marek, *Journal of Polymer Science: Polymer Symposia*, Vol. 56, p. 149, 1976.
114. M. Marek and L. Toman, *Journal of Polymer Science: Polymer Symposia*, Vol. 42, p. 339, 1973.
115. M. Marek and L. Toman, US Patent 3998713, 1976.
116. M. Marek and L. Toman, *Makromolekulare Chemie, Rapid Communications*, Vol. 1, p. 161, 1980.
117. M. Marek and L. Toman, *Journal of Macromolecular Science-Chemistry*, Vol. A15, p. 1533, 1981.
118. M. Marek, L. Toman and J. Pecka, US Patent 3897322, 1975.
119. M. Marek, L. Toman and J. Pecka, US Patent 3997417, 1976.
120. M. Marek, L. Toman and J. Pilar, *Journal of Polymer Science: Polymer Chemistry*, Vol. 43, p. 1565, 1975.
121. J. Pilar, L. Toman and M. Marek, *Journal of Polymer Science: Polymer Chemistry*, Vol. 14, p. 2399, 1976.
122. L. Toman, J. Pecka, O. Wichterle and J. Sulc, GB Patent 1349381, 1974.

123. L. Toman, J. Pilar, J. Spevacek and M. Marek, *Journal of Polymer Science: Polymer Chemistry*, Vol. 16, p. 2759, 1978.

124. G. Langstein, M. Bohnenpoll, U. Denninger, W. Obrecht and P. Plesch, US Patent 6015841, 2000.

125. G. Langstein, M. Bohnenpoll and R. Commander, US Patent Application 14726 A1, 2001.

126. M.C. Baird, US Patent 5448001, 1995.

127. S.P. Lewis, Ph.D. Dissertation, The Univ. of Akron, Diss. Abstr. Int. 2004, Vol. 65, p. 770. Cf: Chem. Abs. 2004, vol. 143, p. 173195.

128. T.D. Shaffer R. Faust and T.D. Shaffer eds., *Cationic Polymerization*, Washington, D.C., American Chemical Society, pp. 96–105, 1997.

129. S. Collins, W.E. Piers and S.P. Lewis, US Patent 7196149, 2007.

130. S.P. Lewis, C. Jianfang, S. Collins, T.J.J. Sciarone, L.D. Henderson, C. Fan, M. Parvez and W.E. Piers, *Organometallics*, Vol. 28, p. 249, 2009.

131. S.P. Lewis, W.E. Piers, N. Taylor and S. Collins, *Journal of the American Chemical Society*, Vol. 125, p. 14686, 2003.

132. S.P. Lewis, L. Henderson, M.R. Parvez, W.E. Piers and S. Collins, *Journal of the American Chemical Society*, Vol. 127, p. 46, 2005.

133. C. Jianfang, S.P. Lewis, J.P. Kennedy and S. Collins, *Macromolecules*, Vol. 40, p. 7421, 2007.

134. T.D. Shaffer, US Patent 6699938, 2004.

135. T.D. Shaffer and J.R. Ashbaugh, *Journal of Polymer Science, Part A: Polymer Chemistry*, Vol. 35, p. 329, 1997.

136. T.D. Shaffer, A.J. Dias, I.D. Finkelstein and M.B. Kurtzman, US Patent 6291389, 2001.

137. F. Barsan, A.R. Karan, M.A. Parent and M.C. Baird, *Macromolecules*, Vol. 31, p. 8439, 1998.

138. C. Jianfang, S.P. Lewis, S. Collins, T.J.J. Sciarone, L.D. Henderson, P.A. Chase, G.J. Irvine, W.E. Piers, M.R.J. Elsegood and W. Clegg, *Organometallics*, Vol. 26, p. 5667, 2007.

139. T.D. Shaffer and J.R. Ashbaugh, *Polymer Preprints American Chemical Society Division Polymer Chemistry*, Vol. 37, p. 339, 1996.

140. M. Bochmann and S. Garratt, US Patent 7041760, 2006.

141. S. Garratt, A. Guerrero, D.L. Hughes and M. Bochmann, *Angewandte Chemie International Edition*, Vol. 43, p. 2166, 2004.

142. S. Jacob, Z. Pi and J.P. Kennedy, *Polymer Bulletin*, Vol. 41, p. 503, 1998.

143. S. Jacob, Z. Pi and J.P. Kennedy, *Polymer Materials Science and Engineering*, Vol. 80, p. 495, 1999.

144. Z. Pi and J.P. Kennedy, "Cationic Polymerizations at Elevated Temperatures by Novel Initiating Systems Having Weakly Coordinating Counteranions. 1. High Molecular Weight Polyisobutylenes" in J.E. Puskas ed., *Ionic Polymerizations and Related Processes*, Dordrecht, Neth., Kluwer, pp. 1–12, 1999.

145. J.B. Lambert, S. Zhang and S.M. Ciro, *Organometallics*, Vol. 13, p. 2430, 1994.

146. A.G. Carr, D.M. Dawson and M. Bochmann, *Macromolecules*, Vol. 31, p. 2035, 1998.

147. G. Langstein, M. Bochmann, D.M. Dawson, A.G. Carr and R. Commander, DE Patent Application 19836663 A1, 2000.

148. S. Garratt, A.G. Carr, G. Langstein and M. Bochmann, *Macromolecules*, Vol. 36, p. 4276, 2003.

149. M. Bochmann and D.M. Dawson, *Angewandte Chemie International Edition English*, Vol. 35, p. 2226, 1996.
150. G. Langstein, M. Bochmann and D.M. Dawson, US Patent 5703182, 1997.
151. C.T. Burns, P.J. Shapiro, P.H.M. Budzelaar, R. Willett and A. Vij, *Organometallics*, Vol. 19, p. 3361, 2000.
152. S.-J. Lee, P.J. Shapiro and B. Twamley, *Organometallics*, Vol. 25, p. 5582, 2006.
153. M. Huber, A. Kurek, I. Krossing, R. Mulhaupt and H. Schnöckel, *Zeitschrift fur Anorganische und Allgemeine Chemie*, Vol. 635, p. 1787, 2009.
154. X. Song, M. Thornton-Pett and M. Bochmann, *Organometallics*, Vol. 17, p. 1004, 1998.
155. M.C. Baird, *Chemical Reviews*, Vol. 100, p. 1471, 2000.
156. K.R. Kumar, C. Hall, A. Penciu, M.J. Drewitt, P.J. Mcinenly and M.C. Baird, *Journal of Polymer Science, Part A: Polymer Chemistry*, Vol. 40, p. 3302, 2002.
157. K.R. Kumar, A. Penciu, M.J. Drewitt and M.C. Baird, *Journal of Organic Chemistry*, Vol. 689, p. 2900, 2004.
158. A.G. Carr, D.M. Dawson and M. Bochmann, *Macromolecular Rapid Communications*, Vol. 19, p. 205, 1998.
159. M. Lin and M.C. Baird, *Journal of Organic Chemistry*, Vol. 619, p. 62, 2001.
160. M. Bohnenpoll, J. Ismeier, O. Nuyken, M. Vierle, D.K. Schon and F. Kuhn, US Patent 7291758, 2007.
161. A.K. Hijazi, H.Y. Yeong, Y. Zhang, E. Herdtweck, O. Nuyken and F.E. Kühn, *Macromolecular Rapid Communications*, Vol. 28, p. 670, 2007.
162. Y. Li, L.T. Voon, H.Y. Yeong, A.K. Hijazi, N. Radhakrishnan, K. Köhler, B. Voit, O. Nuyken and F.E. Kühn, *Chemistry-A European Journal*, Vol. 14, p. 7997, 2008.
163. N. Radhakrishnan, A.K. Hijazi, H. Komber, B. Voit, S. Zschoche, F.E. Kühn, O. Nuyken, M. Walter and P. Hanefeld, *Journal of Polymer Science: Part A*, Vol. 45, p. 5636, 2007.
164. M. Vierle, Y. Zhang, E. Herdtweck, M. Bohnenpoll, O. Nuyken and F.E. Kuhn, *Angewandte Chemie International Edition*, Vol. 42, p. 1307, 2003.
165. S. Gago, Y.M. Zhang, A.M. Santos, K. Kohler, F.E. Kuhn, J.A. Fernandes, M. Pillinger, A.A. Valente, T.M. Santos, P.J.A. Ribeiro-Claro and I.S. Goncalves, *Microporous and Mesoporous Materials*, Vol. 76, p. 131, 2004.
166. M. Pillinger, I.S. Goncalves, A.A. Valente, P. Ferreira, D. Schon, O. Nuyken and F.E. Kuhn, *Designed Monomers and Polymers*, Vol. 4, p. 269, 2001.
167. A. Sakthivel, A.K. Hijazi, A.I. Al Hmaideen and F.E. Kuhn, *Microporous and Mesoporous Materials*, Vol. 96, p. 293, 2006.
168. S. Syukri, C.E. Fischer, A.A. Al Hmaideen, Y. Li, Y. Zheng and F.E. Kuhn, *Microporous and Mesoporous Materials*, Vol. 113, p. 171, 2008.
169. H. Chaffey-Millar and F.E. Kuhn, *Applied Catalysis A: General*, Vol. 384, p. 171, 2010.
170. G. Langstein, D. Freitag, M. Lanzendörfer and K. Weiss, US Patent 5668232, 1997.
171. A. Lisovskii, E. Nelkenbaum, V. Volkis, R. Semiat and M.S. Eisen, *Inorganica Chimica Acta*, Vol. 334, p. 243, 2002.
172. S. Lewis, US Patent Application 0273964, 2010.
173. J.P. Kennedy, S.Y. Huang and S.C. Feinberg, *Journal of Polymer Science: Polymer Chemistry*, Vol. 15, p. 2869, 1977.
174. A. Gronowski, CA Patent 2252295, 2000.

175. A. Gronowski, US Patent 6403747, 2002.
176. L. Luo, S.A. Sangokoya, X. Wu, S.P. Diefenbach and B. Kneale, US Patent Application 62492 A1, 2009.
177. A. Gronowski, US Patent 6630553, 2003.
178. M. Bochmann and K. Kulbaba, US Patent Application 0238843 A1, 2007.
179. A. Guerrero, K. Kulbaba and M. Bochmann, *Macromolecules*, Vol. 40, p. 4124, 2007.
180. A. Guerrero and K. Kulbaba, EP Patent 1834964, 2010.
181. F.W. Lampe, *Journal of Physical Chemistry*, Vol. 63, p. 1986, 1959.
182. V.T. Stannett and J. Silverman, *ACS Symposium Series*, Vol. 212, p. 435, 1983.
183. T. Cai, M. He, X. Shi, X. Wang, D. Han and L. Lu, *Catalysis Today*, Vol. 69, p. 291, 2001.
184. F.J. Chen, C.L. Deore, A. Guyot, A.L.P. Lenack and J.E. Stanat, US Patent 5789335, 1998.
185. P. Hallpap, H. Hartung, G. Heublein, D. Stadermann and G. Wolff, DE Patent Application 225423 A1, 1985.
186. J.T. Kelly, W. Schoen and C.N. Sechrist, US Patent 3248343, 1966.
187. M. Simo and F.M. McMillan, US Patent 2277512, 1942.
188. M. Simo and F.M. McMillan, US Patent 2327593, 1943.
189. S.B. Thomas and F.M. McMillan, US Patent 2311713, 1943.
190. F.M. McMillan and G.S. Parsons, US Patent 2277022, 1942.
191. R.M. Smith, US Patent 2927087, 1960.
192. S.B. Thomas, US Patent 2311712, 1943.
193. R.S. Drago and E.E. Getty, *Journal of the American Chemical Society*, Vol. 110, p. 3331, 1988.
194. R.S. Drago and E.E. Getty, US Patent 4719190, 1988.
195. R.S. Drago and E.E. Getty, US Patent 4798667, 1989.
196. R.S. Drago and E.E. Getty, US Patent 4929800, 1990.
197. E.E. Getty and R.S. Drago, *Inorganic Chemistry*, Vol. 29, p. 1186, 1990.
198. J.D. Upham, US Patent 2406869, 1946.
199. J. Estes, R.M. Suggitt and S. Kravitz, US Patent 3607959, 1971.
200. A.G. Goble and J.V. Fletcher, US Patent 3240840, 1966.
201. S. Herbstman, A.N. Webb and J.H. Estes, US Patent 4066716, 1978.
202. S. Herbstman, J.H. Estes and A.N. Webb, US Patent 4113657, 1978.
203. L.C. Mih and E.T. Child, US Patent 3523142, 1970.
204. J.W. Myers, US Patent 3449264, 1969.
205. L.M. Babcock and D.G. Morrell, WO Patent 98/30519, 1998.
206. J.A. Bittles, S.W. Benson and A.K. Chaudhuri, *Journal of Polymer Science: Part A*, Vol. 2, p. 1221, 1964.
207. J.A. Bittles, S.W. Benson and A.K. Chaudhuri, *Journal of Polymer Science: Part A*, Vol. 2, p. 1847, 1964.
208. H.M. Stanley and F.E. Salt, GB Patent 524156, 1940.
209. C.J. Norton, *Chemistry and Industry*, p. 258, 1962.
210. S.L. Mannatt, J.D. Ingham. and J.A. Miller, Jr., *Journal of Magnetic Resonance*, Vol. 10, p. 198, 1977.
211. J.A. Miller, Jr., US Patent 3634383, 1972.
212. K.S.B. Addecott, L. Mayor and C.N. Turton, *European Polymer Journal*, Vol. 3, p. 601, 1967.

213. C.N. Turton, GB Patent 1091083, 1967.
214. J.R. Johnson and J.D. Burrington, EP Patent Application 825205 A1, 1998.
215. J.W. Klinkenberg, US Patent 2982799, 1961.
216. R. Michel and L. Muller, US Patent 2301966, 1942.
217. R.B. Thompson, US Patent 2237460, 1941.
218. C. Sigwart, T. Narbeshuber, K. Eller, M. Barl, R. Fischer and E. Gehrer, US Patent 6441110, 2002.
219. C. Sigwart, T. Narbeshuber, E. Gehrer, R. Fischer, U. Steinbrenner and S. Liang, US Patent 6384154, 2002.
220. F.J. Chen, C.L. Deore, R. Spitz and A.L.P. Lenack, US Patent Application 0053835 A1, 2001.
221. J. Collomb, P. Arlaud, A. Gandini and H. Cheradame, *Proceedings IUPAC 6th International Symposium on Cationic Polymerization and Related Processes*, p. 49, 1984.
222. J. Collomb, B. Morin, A. Gandini and H. Cheradame, *European Polymer Journal*, Vol. 16, p. 1135, 1980.
223. F.J. Chen, A. Guyot, T. Hamaide and C.L. Deore, WO Patent 95/26814, 1995.
224. F.J. Chen, A. Guyot, T. Hamaide and C.L. Deore, US Patent 5607890, 1997.
225. F.J. Chen, C.L. Deore, R. Spitz and A. Guyot, US Patent 5648580, 1997.
226. T.-C. Chung, "Supported Lewis Acid Catalysts (on Polypropylene; Recoverable and Resuable)," in J.C. Salamone ed., *Polymeric Materials Encyclopedia*, Boca Raton, CRC Press, pp. 8093–8100, 1996.
227. T.-C. Chung, F.J. Chen, J.E. Stanat and A. Kumar, US Patent 5409873, 1995.
228. T.-C. Chung and A. Kumar, *Polymer Bulletin*, Vol. 28, p. 123, 1992.
229. T.-C. Chung, A. Kumar and D. Rhubright, *Polymer Bulletin*, Vol. 30, p. 385, 1993.
230. R. Ran, *Journal of Polymer Science, Polymer Chemistry Edition*, Vol. 31, p. 1561, 1993.
231. R. Ran, "Support Catalysts (Lewis Acid and Ziegler-Natta)," in J.C. Salamone ed., *Polymeric Materials Encyclopedia*, Boca Raton, FL, CRC Press, pp. 8063–8078, 1996.
232. H.W. Hedrik and J.A. van Melsen, US Patent 2133732, 1938.
233. J.F. Hyde and J.R. Wehrly, US Patent 2891920, 1959.
234. K. Satoh, K. Masami and M. Sawamoto, *Macromolecules*, Vol. 32, p. 3827, 1999.
235. K. Satoh, K. Masami and M. Sawamoto, *Journal of Polymer Science Part A. Polymer Chemistry*, Vol. 38, p. 2728, 2000.
236. K. Satoh, K. Masami and M. Sawamoto, *Macromolecules*, Vol. 33, p. 4660, 2000.
237. K. Satoh, K. Masami and M. Sawamoto, *Macromolecules*, Vol. 33, p. 5836, 2000.
238. S. Cauvin, A. Sadoun, R.D. Santos, J. Belleney, F. Ganachaud and P. Hemery, *Macromolecules*, Vol. 35, p. 7919, 2002.
239. S.A. Yankovski and J.P. Kennedy, Unpublished Research Data, Nov–Dec 2000.
240. S.P. Lewis and J.P. Kennedy, Unpublished Research Data, Nov 2000–April 2001.
241. L.D. Henderson, W.E. Piers, G.J. Irvine and R. McDonald, *Organometallics*, Vol. 21, p. 340, 2002.
242. J.P. Kennedy, S. Collins and S.P. Lewis, US Patent 7202317, 2007.
243. S.V. Kostjuk and F. Ganachaud, *Macromolecules*, Vol. 39, p. 3110, 2006.
244. S.V. Kostjuk, A.V. Radchenko and F. Ganachaud, *Macromolecules*, Vol. 40, p. 482, 2007.

245. S.V. Kostjuk, S. Ouardad, F. Peruch, A. Deffieux., C. Absalon, J.E. Puskas and F. Ganachaud, Vol. 44, p. 1372, 2011.

246. D.Y.-L. Chung, R.N. Webb, M.F. McDonald, Y.-J. Chen, R.D. Hembree and J.P. Soisson, WO Patent Application 2004/014968 A2, 2004.

247. M.F. McDonald, S.T. Milner, T.D. Shaffer and R.N. Webb, US Patent 7214750, 2007.

248. M.F. McDonald, S.T. Milner, T.D. Shaffer and R.N. Webb, US Patent Application 0262180 A1, 2008.

249. M.F. McDonald, S.T. Milner, T.D. Shaffer and R.N. Webb, US Patent 7423100, 2008.

250. S.T. Milner, M.G. Matturro, T.D. Shaffer, R.N. Webb, D.Y.-L. Chung and M.F. McDonald, US Patent 7723447, 2008.

251. T.D. Shaffer, S.T. Milner, D.Y.-L. Chung, M.F. McDonald, R.N. Webb and P.J. Wright, US Patent Application 0290049 A1, 2008.

252. T.D. Shaffer, P.J. Wright, D.J. Davis and M.F. McDonald, US Patent 7557170, 2009.

253. R.N. Webb, D.Y.-L. Chung, A.B. Donnalley, M.F. McDonald, K.W. Powers and R.H. Schatz, WO Patent Application 02/50141 A1, 2002.

254. R.N. Webb, D.Y.-L. Chung, A.B. Donnalley, M.F. McDonald, K.W. Powers and R.H. Schatz, WO Patent Application 02/50129 A1, 2002.

255. R.N. Webb, M.F. McDonald, D.Y.-L. Chung, Y.-J. Chen, R.D. Hembree and J.P. Soisson, US Patent 6858690, 2005.

256. J.S. Francisco and M.M. Maricq, *Accounts of Chemical Research*, Vol. 29, p. 391, 1996.

257. B. Saleh and M. Wendland, *International Journal of Refrigeration*, Vol. 29, p. 260, 2006.

258. W.-T. Tsai, *Chemosphere*, Vol. 61, p. 1539, 2005.

259. Directive 2006/40/EC, European Parliament, Official Journal of the European Union, 2006.

260. *Polymer Handbook*; J. Brandrup, E.H. Immergut and E.A. Grulke, Eds., Wiley-Interscience: New York, 1999.

261. W.A. Thaler and D.J. Buckley, Sr., *Rubber Chemistry and Technology*, Vol. 4, p. 960, 1976.

262. W.A. Thaler, J.P. Kennedy and D.J. Buckley, Sr., US Patent 3808177, 1974.

263. M.F. McDonald, Jr., D.J. Lawrence and D.A. Williams, US Patent 5417930, 1995.

264. R.T. Mathers, C. LeBlond, K. Damodaran, D.I. Kushner and V.A. Schram, *Macromolecules*, Vol. 41, p. 524, 2008.

265. F. Jerome and J. Barrault, "Selective conversion of glycerol into functional monomers via catalytic processes." in R.T. Mathers and M.A.R. Meier eds., *Green Polymerization Methods: Renewable Starting Materials, Catalysis and Waste Reduction.*, New York, Wiley-VCH, pp. 57–87, 2011.

4

Olive Oil Wastewater as a Renewable Resource for Production of Polyhydroxyalkanoates

Francesco Valentino[1], Marianna Villano[1], Lorenzo Bertin[2], Mario Beccari[1], and Mauro Majone[1]

[1]Department of Chemistry, Sapienza University of Rome, Rome, Italy
[2]Department of Civil, Environmental and Materials Engineering (DICAM), Alma Mater Studiorum University of Bologna, 40131 Bologna, Italy

Abstract

The use of OMEs (as well as of other agro-industrial wastes) makes it possible to easily obtain PHA copolymers of HB and HV monomers, resulting in a polymer with better physicochemical properties. The HB/HV ratio can be controlled by manipulating the relevant parameters of each stage of the process and, particularly, the acids composition from the fermentation stage. Ongoing research is mainly aimed at finding an optimal tradeoff between the high process performance (in terms of rate and yield of PHA production) and the PHA composition and properties for downstream processing. Process optimization at a higher scale and capacity is required to increase the potential for industrial application.

Keywords: Polyhydroxyalkanoates, olive oil mill wastewater, mixed microbial cultures, feast and famine, sequencing batch reactor

4.1 Polyhydroxyalkanoates (PHAs): Structure, Properties, and Applications

Global demand of plastic-based materials has increased considerably over the last years [1]. However, synthetic polymers that dominate plastics are derived from limited fossil resources and represent a serious hazard to the environment due to their high recalcitrance

Vikas Mittal (ed.) Renewable Polymers, (175–220) © Scrivener Publishing LLC

to biological degradation. The concern associated with the disposal of synthetic polymers has prompted much interest in the research of alternative biodegradable materials, such as polysaccharides, polylactides, and polyhydroxyalkanoates (PHAs); the latter being considered among the most promising candidates.

PHAs are biologically synthesized polyesters (Figure 4.1) that can be produced from renewable resources and are natural, recyclable, biocompatible, and completely biodegradable to water and carbon dioxide [2]. These biopolymers can exhibit a broad range of thermoplastic and elastomeric properties depending on the incorporated monomers. Over 150 different types of hydroxyalkanoate (HA) monomers have been identified and there is an enormous variation possible in the length and composition of the side chains. In general, PHA monomers can be classified into two main groups: (i) short chain length (SCL) monomers with 3–5 atoms of carbon and (ii) medium chain length (MCL) monomers which consist of a number of carbon atoms between 6 and 14. The average number of repeating units in PHA chain is up to around 30000. An example of SCL-PHA is the poly(3-hydroxybutyrate) (P(3HB)) that, by far, is the most widely studied polyhydroxyalkanoate. P(3HB) is a highly crystalline (55–80%), stiff, but brittle material. The elongation to break (flexibility) is about 2–10% for P(3HB), compared to 400% for polypropylene. Other mechanical properties, such as Young's modulus (3.5 GPa) and tensile strength (40 MPa) are similar to those of polypropylene [3]. The melting temperature of P(3HB) is close to 180°C and the glass transition temperature is around 4°C; but the polymer starts to be thermally degraded at 170°C [4]. This means that a reduction in molecular weight that deteriorates mechanical properties may occur during processing of pure P(3HB) in melt. However, the introduction of different HA monomers, such as

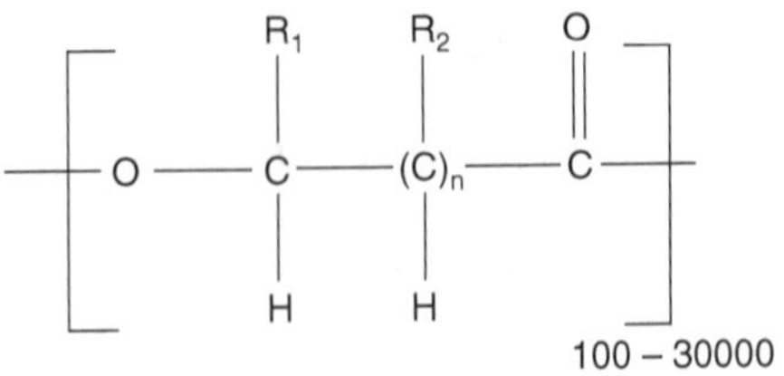

Figure 4.1 PHA general structure: n is the number of carbon atoms in the linear polyester structure, R_1 and R_2 are variable hydrocarbon side chains.

3-hydroxyvalerate (3HV) or 3-hydroxyhexanoate (3HHx), into the chain reduces polymer cristallinity and greatly improves the material's properties. The copolymers have both increased flexibility and lower melting and glass temperatures than P(3HB) [5]. On the other hand, the temperature at which degradation and decomposition of PHAs occurs is rather insensitive to composition; thus the higher distance between melting and degradation temperatures of copolymers enables their processing with reduced molecular weight losses.

The enormous possible variations in the length and composition of the side chains make the PHA polymer family suitable for an array of potential applications ranging from packaging, compost bags, agricultural films (e.g., mulch films), manufacturing disposable everyday articles (e.g., shampoo bottles and cosmetic materials), bulk commodity plastics, to medical uses [6]. Indeed, recent studies have demonstrated the use of PHAs in the production of stents and in the tissue engineering of heart valves [7].

PHAs are synthesized by many types of microorganisms, over 300 species, including both gram-negative and gram-positive bacteria (e.g., *Lampropaedia*, *Pseudomonas*, *Rhodobacter*, *Ralstonia*, *Bacillus*, *Clostridium*), aerobic photosynthetic bacteria (cyanobacteria), anaerobic photosynthetic bacteria (non-sulphur and sulphur purple bacteria), and certain Archea. They function as carbon and energy storage compounds and are present in the cell cytoplasm as insoluble granules (with diameters from 0.2 to 0.7 µm) [8, 9]. The PHA storage occurs when microorganisms are unable to grow at the same rate as they are able to take up the substrate. The restriction in growth rate can be due to limited availability of essential growth nutrients (such as nitrogen, phosphorus, sulphur or oxygen) or to limitations in anabolic enzyme levels and activity.

4.2   PHA Production Processes Employing Pure Microbial Cultures

Presently, PHA production at industrial scale is based on pure culture fermentations. The most commonly used microorganisms are *Cupriavidus necator* (formerly known as *Ralstonia eutropha* or *Alcaligenes eutrophus*), *Pseudomonas oleovorans*, *Protomonas extorquens*, *Alcaligenes latus*, or recombinant *Escherichia coli*. The culture conditions required for polymer storage are different for the different microorganisms. Most of PHA-producing bacteria require the limitation of an essential nutrient and a two-stage fed-batch process is typically employed. The culture is first supplied with a medium rich in nutrients and carbon sources and, when a high cellular density is reached, growth limiting conditions are imposed in order to induce PHA synthesis. During this limited growth stage an efficient polymer production occurs, either when the carbon source is provided under nutrient-deficient conditions (such as in *P. oleovorans* and *P. extorquens*) or when the nutrient (e.g., nitrogen source) is completely depleted in the medium (such as in *C. necator*) [10]. A more restricted group of bacteria, including *A. latus* and recombinant *E. coli* containing the PHA biosynthetic genes, does not require nutrient limitation and can accumulate PHAs during active growth [11]. For the fed-batch culture of these microorganisms the development of a nutrient feeding strategy is crucial to the success of the fermentation, since cell growth and PHA accumulation need to be balanced in order to avoid incomplete polymer accumulation. However, nutrient limitation can further enhance PHA storage also in *A. latus* [12]. Overall, even though maximum PHA contents of up to 80–90% of the cell dry weight have been reported [13, 14], well-defined substrates and aseptic process conditions are required. This typically results in high substrate costs, expensive equipments, and high energy consumption, making industrial biotechnological processes based on pure cultures unfavorable for further exploitation in PHA production. In particular, the substrate cost accounts for about 50% of the total PHA production costs, to which the carbon source contributes most significantly [15], because substrates most commonly used in the industrial processes are pure sugars, such as glucose or sucrose, or other sugar based compounds, such as corn, which have a high market price [16].

4.3   PHA Production Processes Employing Mixed Microbial Cultures

In recent years, there has been considerable interest in investigating alternative low-cost processes for PHA production and much research efforts are particularly being dedicated at exploring the use of low value substrates, such as agro-industrial wastes, and mixed microbial cultures (MMCs) [17, 18]. The combined use of waste feedstocks and MMCs is expected to decrease both investment and fermentation operating costs by simplifying equipments and saving energy (as sterile conditions are not required), as well as reducing the cost of the substrate. Indeed, mixed culture PHA production processes allow using complex substrates of undefined composition possibly contained in agro-industrial waste effluents, thereby coupling the biopolymer production to wastewater treatment.

MMCs are microbial populations operating in open biological systems, whose composition depends directly on the substrate mixture and operational conditions imposed on the open biological system. These have been so far mainly used in biological wastewater treatment plants. However, the ecological role of internal storage products such as PHAs has been recognized as an important aspect of activated sludge systems submitted to dynamic conditions, mainly caused by discontinuous feeding and variations in redox conditions (electron acceptor presence or absence) [19, 20].

PHA synthesis by mixed cultures was first observed in wastewater treatment plants designed for biological phosphorous removal (EBPR), which are operated with alternating anaerobic/aerobic cycles. The main groups of bacteria responsible for PHA accumulation in EBPR systems are polyphosphate accumulating organisms (PAOs) and glycogen accumulating organisms (GAOs). Under anaerobic conditions, PAOs and GAOs take up the carbon substrate and store it as PHAs, while glycogen, a second storage polymer, is consumed. PHAs are used in the successive aerobic phase for cell growth, maintenance, and glycogen pool replenishment. Though the carbon metabolism of PAOs and GAOs is similar, the main difference between them is related to the cycling of polyphosphate by PAOs. Under anaerobic conditions, PAOs release phosphate gaining energy for PHA accumulation, while in the presence of oxygen phosphate is taken up in excess for polyphosphate pool replenishment, and PHAs are degraded for storing energy [21]. Under anaerobic conditions, on the contrary, GAOs derive energy only by glycogen catabolism through glycolysis and no phosphate is released to the medium or accumulated inside the cells (but the one required for growth).

Significant PHA storage ability was also observed in activated sludge operating in aerobic wastewater treatment plants where selectors for bulking control were introduced [22]. This process configuration is characterized by alternating periods of excess and lack of external carbon substrates, which create internal kinetic limitations to growth and enhance PHA accumulation. Indeed, following a certain period in the absence of external carbon substrate, a decrease of the amount of intracellular components (RNA and enzymes) required for cell growth occurs. When carbon substrate becomes available again, the amount of enzymes present in the cells may not be enough to ensure that a maximum growth rate is reached and, since fewer enzymes are required for PHA storage, the latter phenomena can occur at a faster rate than cell growth and becomes dominant [23].

Therefore, an effective selective pressure to enrich activated sludge in PHA producing microorganisms is the application of a cycling feast and famine regime, that is the repeatedly alternating presence (feast phase) and absence (famine phase) of excess substrate. In this way, a competitive advantage exists for microorganisms that are quicker to store the substrate during the feast phase and reuse it as an internal carbon and energy source for cell growth and maintenance during the famine phase.

In general, the definition of feast and famine regime can be referred to both (i) the fully aerobic dynamic feeding (ADF) and (ii) the dynamic conditions of oxygen availability (anaerobic/aerobic cycles). The imposition of anaerobic/aerobic cycles is triggering for storage only if the substrate is available in the absence of the final electron acceptor limiting cell growth, whereas the presence of substrate throughout the full cycle would not enhance PHA storage. On the other hand, in anaerobic/aerobic feast and famine regime cell growth occurs only from the intracellular stored polymer, whereas under fully aerobic conditions cell growth can also occur from the external substrate simultaneously to PHA accumulation during the feast phase.

Overall, even though PHA storage from MMCs has been studied in the past mostly in relation to its relevance to wastewater treatment, current research is focusing on the use of various and ever increasing range of waste streams for their commercial production with mixed cultures.

However, some agro-industrial wastes can not be directly converted to PHAs and a preliminary treatment is usually required to convert their organic content into more direct substrates for PHA storage, such as volatile fatty acids (VFAs). As a consequence, the proposed process for mixed culture PHA production usually comprises of three stages (Figure 4.2): a first acidogenic fermentation stage, a successive culture selection stage (under feast and famine regime), and a final PHA accumulation stage after which

(a)

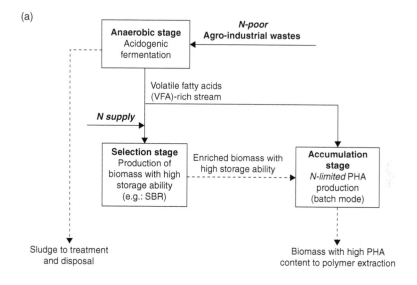

(b)

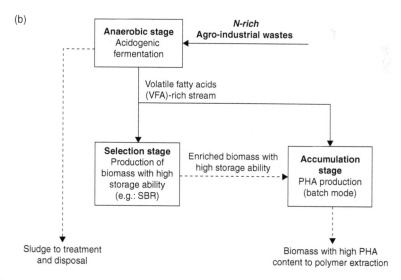

Figure 4.2 Outline of a three-stage process used for PHA production from N-poor (a) or N-rich (b) agro-industrial wastes by employing mixed microbial cultures selected under feast and famine regime (solid lines refer to liquid fluxes and dotted lines to biomass fluxes).

the polymer is extracted from the cells and purified. The performance and efficiency of the overall process largely depends on the nature of the waste used as feedstock and, in particular, its C:N ratio plays a pivotal role. As an example, the use of N-poor waste streams enhances PHA production in the accumulation stage even though it makes the supply of nitrogen sources necessary to sustain biomass growth in the selection stage (Figure 4.2A). Conversely, when N-rich wastes are used no external nitrogen supply is needed to select microorganisms with high storage ability but a lower PHA content is usually obtained in the accumulation stage. Otherwise, nutrient limiting conditions can be established in the accumulation stage by an additional step for nutrient removal from the medium (Figure 4.2B).

4.3.1   The Acidogenic Fermentation Stage: Key Aspects

In the last decade, several agro-industrial streams have been tested for mixed culture PHA production, including sugar cane and beet molasses, cheese whey effluents, plant oils, swine waste liquor, vegetable and fruit wastes, brewery wastewater, effluents of palm oil mill, olive oil mill, paper mill, pull mill, and hydrolysates of starch (e.g., corn, tapioca), cellulose and hemicellulose [24]. However, these complex materials are composed of several different organic compounds that are not all directly convertible to PHAs and the acidogenic fermentation may be required as an essential pretreatment to increase their potential as feedstock of the process. This pretreatment is particularly relevant for sugar-rich wastewaters since mixed cultures, unlike pure cultures, do not accumulate carbohydrates as PHAs [25]. Indeed, it has been observed that in the presence of both carbohydrates and acetate, the latter is stored as PHB whereas glucose or starch are stored as glycogen [26, 27]. Because acetate and other volatile fatty acids (VFAs) are more preferable as substrate for mixed culture PHA production rather than carbohydrates, acidogenic fermentation plays a pivotal role to give the best substrate for PHA synthesis.

The acidogenic fermentation is the initial phase of the anaerobic digestion of organic matter to methane and carbon dioxide (i.e., biogas). Under anaerobic conditions, soluble compounds (including carbohydrates) can be fermented into VFAs (such as acetic, propionic, butyric, and valeric acids), lactic acid, and other fermentation products, such as alcohols and hydrogen. Provided that further conversion of acidogenic fermentation intermediates into biogas is hindered, a VFA-rich stream is obtained and can be exploited for mixed culture PHA production. Methanogenic activity can be prevented by operating the fermenter at low hydraulic retention time (HRT), low temperature [28, 116], low pH [29], or by using a combination of the different approaches [30]. Clearly, the operating parameters need to be tuned in order to maximize the VFA conversion yield of the organic waste and to control the composition of the produced VFA mixture as well.

Since the physical and mechanical properties of PHAs depend on the monomer composition, which in turn is influenced by the type of substrate used, the composition of VFAs obtained during the acidogenic fermentation will influence the quality and quantity of the produced biopolymer. However, the composition of agro-industrial wastes in addition to being usually very complex and scarcely characterized with respect to specific compounds, may also contain inhibitory compounds that adversely affect both the acidogenic fermentation and the successive polymer production stages. Hence, research is needed in order to establish relations between process conditions of the acidogenic fermentation of specific wastes and the spectrum of VFAs that can be produced. In general, the relative distribution of acidogenic fermentation products depends on several factors, such as the type and concentration of feedstocks, the composition of the microbial culture, as well as the environmental and operating conditions of the fermenter (including HRT, pH, and hydrogen partial pressure). As an example, at increasing hydrogen partial pressures fermentation shifts towards the accumulation of organic acids more reduced than acetate (e.g., propionate and butyrate) [31]. With respect to pH, Horiuchi and colleagues [32] observed a change in the product spectrum of glucose fermentation due to pH shift. At acid and neutral pH, the main product was butyric acid, whereas acetic and propionic acids prevailed under basic conditions. According to the authors, the change in the main product distribution was caused by the change in the dominant microbial population.

Taking into account these considerations, the acidogenic fermentation of different agro-industrial wastewaters has been examined from the point of view of their potential application towards PHA production. Cheese whey, paper mill effluents, olive mill wastewaters, and sugar cane molasses were found to be readily and almost fully fermentable, producing mainly VFAs (acetate, propionate, butyrate, and valerate) as fermentation products [29, 30]. For cheese whey and paper mill effluents, the acidogenic fermentation was investigated both in batch and in continuous experiments and, notably, the VFA yields were higher during chemostat operation (0.75 to 0.87 gCOD gCOD^{-1}) than during batch operation (0.59 to 0.60 gCOD gCOD^{-1}) [30]. This suggested that the continuous operation allowed performing more efficient conversion through the acclimation of the anaerobic mixed culture to the feedstock. In continuous chemostat experiments with these two effluents, the HRT and pH were found to have a significant impact on outlet VFA composition, mainly with respect to the proportions of acetate, propionate, and butyrate. The concentration of propionate increased by increasing both the HRT (from 8 to 95 h) and the pH (from 3.5 to 6). The effect of pH (from 5 to 7) on organic acids composition was also examined during continuous fermentation of sugar cane molasses [29]. At pH 6 and 7, acetate and propionate were the main products, while butyrate and valerate dominated at pH 5. However, the yield of VFA production did not vary significantly with the pH, even though a slightly higher yield was achieved under more acidic conditions. This behaviour was likely due to the presence of some methanogenic activity occurring at pH 7, which was inhibited at lower pH values. Overall, both studies indicate that HRT and pH of the acidogenic fermentation are crucial parameters to act on for controlling VFA composition and, therefore, PHA monomer composition in the subsequent aerobic stages.

Reactor configuration is another operating parameter which plays an important role on the performance of the acidogenic fermentation. Biofilm systems can be used as an alternative to suspended growth reactors because the former allow reducing shock loading and/or washout problems. Typically, these factors negatively affect the productivity of suspended growth reactors operating with low growth rate biomass and high and fluctuating organic loads [33]. The acidogenic fermentation in a packed bed biofilm reactor associated with PHA synthesis has been recently investigated for olive mill effluents [28]. A partial conversion of the organic waste into VFAs and alcohols was observed, with only a small fraction (about 8%) of the substrate being converted into biogas and used for biomass growth.

In particular, up to 66% of the incoming organic matter was converted into VFAs, wich represented up to 89% of the effluent COD [116]. The reative amounts of the main VFAs was controlled by regulating the applied organic load rate (OLR).

4.3.2   The Mixed Microbial Culture (MMC) Selection Stage

In mixed culture PHA production processes, the efficiency of the selection stage is determinant on the performance of the subsequent polymer production stage. The culture selection is aimed to enrich an activated sludge in microorganisms with high and stable PHA storage ability. A main strategy to accomplish this objective is a culture exposition to alternating conditions of substrate availability and unavailability, referred to as feast and famine (FF) regime. As an example, Majone and colleagues [34] reported a 20-fold increase (405 vs. 21 mgCOD$_{PHA}$ gCOD$_{VSS}^{-1}$ h^{-1})/h) in the PHA storage rate of an activated sludge after selection under FF conditions. These conditions can be obtained in an intermittently fed reactor (e.g., sequencing batch reactor, SBR), in a plug flow reactor or in two continuous stirred reactors in series with biomass recirculation to the first reactor. The most commonly used configuration is the SBR, which operates in a succession of repeated cycles, each consisting of distinct phases, usually feed, reaction, settling, and discharge of the treated effluent. In SBR systems the substrate availability to microorganisms (feast phase) occurs during the feed phase and part of the reaction phase; whereas during the remainder of the cycle microorganisms experience the lack of external substrate (famine phase). The ratio between the length of the feast and famine phases (F/F ratio) is a crucial parameter affecting the performance of the selection stage [35, 36]. The mechanism through which the F/F ratio affects the selective pressure for PHA-storing microorganisms is related to an internal growth limitation that is necessary to induce PHA storage. If the famine phase is not long enough to ensure such an internal limitation, then the microorganisms will be better fit to grow when supplied with external carbon sources and will, therefore, accumulate less PHAs. Hence, low F/F ratios insure physiological microbial adaptation in favor of PHA synthesis in the feast phase.

Another objective of the selection stage is to produce the PHA-storing culture, to be exploited in the following accumulation stage, at the highest possible productivity (i.e., the amount of biomass produced per unit of volume of reactor and per unit of time). This allows operating the third stage at high cell densities, which is a prerequisite to obtain a high PHA productivity. In order to achieve high biomass volumetric productivities, the selection reactor needs to be operated at high OLR, which corresponds to high influent substrate concentration and/or short HRT. However, a high OLR may have a detrimental effect on the selective pressure favoring PHA-storing microorganisms by increasing the F/F ratio. The sludge retention time (SRT), which is the reciprocal of the effective specific growth rate of a culture, is also another parameter to be considered. It can be assumed that the imposition of a lower SRT will result in a higher fraction of the substrate being used for growth and, consequently, in a lower storage yield. In fact, it was found [37] that the yield of PHB storage from acetate was independent of the specific growth rate for SRT higher than 2 days and it sharply decreased at the decreasing of the SRT for values below 2 days. In other studies [38, 39], however, a culture with high storage capacity was selected even by using low SRT (1 day). Clearly an optimal tradeoff between biomass productivity and microbial selection for PHA storage needs to be carefully individuated. A recent study [35] tested the effect of the OLR on the performance of an SBR operating at 1 day SRT and fed with a mixture of organic acids (acetic, propionic, and lactic acids). The OLR was varied in the range 8.5–31.25 gCOD L^{-1} d^{-1}day by changing the influent substrate concentration. Accordingly, the F/F ratio ranged between 0.10 and 1.15. Even though, as expected, the increase in the OLR caused an increase in biomass concentration (up to about 8.7 gCOD L^{-1}), it also caused a relevant decrease of the maximal polymer production rate. The best performance, in terms of both high biomass productivity and high storage capacity, was obtained at an intermediate OLR (20 gCOD L^{-1} d^{-1}) although under this condition the process was unstable. With respect to the F/F ratio, a stable storage response was observed for values up to 0.26, a growth response for F/F ratio higher than 0.90, and an unstable response for intermediate values. Albuquerque and colleague [36] observed a similar trend by using fermented molasses at different influent concentration as feedstock of an SBR having an SRT of 10 days. At lower influent substrate concentration (up to 1.6 gCOD L^{-1}), the resulting low F/F ratio (0.21) caused the predominance of the storage response, while at higher influent concentration (2.1 gCOD L^{-1}) the F/F ratio varied from 0.53 to 1.1 and the system performance was unstable, with alternate storage and growth response.

Other operating parameters which were found to influence culture selection in SBR systems are pH and temperature. As for pH, it was found that the storage ability of culture selected at pH 7 or 8, using acetate as carbon source, was nearly the same [40]. On the other hand, Villano and colleagues [41] by using a mixture of acetate and propionate, observed that an increase of pH from 7.5 to 9.5 resulted in a decrease of storage rates and yields. In a recent study [42], long term temperature effects (in the range 15–30°C) have been investigated with main reference to PHB storage. At higher temperature (30°C) the SBR feast phase was short and the rates of acetate uptake and PHB storage were very high. At lower temperature (15°C) the feast phase was longer due to a lower rate of acetate uptake and the culture followed a strategy of direct growth on acetate rather than on PHB.

A few studies have also investigated the use of two continuous stirred reactors connected in series to select for a PHA-accumulating culture [43, 44]. In this reactor configuration mode, the first reactor mimics the feast phase while the second reactor the famine phase. The HRT ratio between the two reactors corresponds to the F/F ratio in the SBR configuration and the selection principles are the same for the two reactor operational modes. However, in the continuous FF systems, the feast reactor is continuously fed and the residual substrate concentration is higher than in SBR systems, in which feeding occurs only in a discrete short period. This implies that in two-reactor configuration a residual organic fraction may pass from the first to the second reactor, mainly depending on influent substrate concentration, HRT of the feast reactor, and substrate uptake rate. Hence, the famine reactor can receive a continuous feeding of substrate and, provided that its concentration is growth limiting, the concept of famine does not correspond to complete substrate exhaustion, as in the SBR system, but rather to a growth limitation. In the two-reactor configuration, the pressure of the FF regime is maintained only if an excessive organic load to the famine reactor is avoided and the amount of substrate available does not allow cells to grow above a certain rate.

Both in two-reactor and SBR systems, the culture selection under FF conditions is a consequence of an internal growth limitation without the need for external nutrient limitations. However, the culture selection in SBR systems has been also studied under nutrient-limiting conditions [45, 46]. The C/N ratio affects both the ratio of carbon driven toward PHA storage versus growth during the feast phase and the growth yield in the famine phase. Indeed, when nutrients are present throughout the entire SBR cycle, the selected culture can use the intracellularly stored polymer to grow during the famine phase of the cycle, obtaining a higher growth yield. Johnson and colleagues [47] investigated the effect on PHB production of the C/N ratio, moving from carbon to nitrogen limitation. They found a higher baseline PHA content in the culture selected under nitrogen limitation, but a higher maximal storage ability in the culture selected under carbon limitation (excess nutrients).

The C/N ratio adopted in the selection stage has a strong impact also on the subsequent batch accumulation stage. Indeed, for cultures selected under nutrient limitation, the storage yield decreases at increasing the nutrient concentration in the accumulation stage [46], whereas cultures selected under nutrient excess seem to be far more insensitive to the presence of nutrients, at least during the first part of the accumulation stage [48]. This aspect is particularly relevant when real complex feedstocks are used for PHA production since agro-industrial wastes may contain variable nitrogen contents and the selection of cultures that are insensitive to the presence of nutrients in the accumulation stage can be crucial for the performance of the overall process.

Although the characterization of the microbial population selected under the different FF conditions is somewhat limited, each condition was typically found to select for its own bacterial composition. Bacterial species most commonly identified belong to the genus of *Thauera, Azoarcus, Amaricoccus, Lampropedia, Zooglea, Candidatus Megamena* [35, 49, 41]. Particularly, this latter organism was found to dominate in systems fed with fermented paper mill [44] or olive mill [28] effluents.

In a recent study [50], however, it was found that the feeding pattern (slow and long *vs.* quick and short feeding) did not alter the composition of the microbial culture but rather its physiological state, with a higher PHB storage response observed when the culture was operated under short feeding regime.

4.3.3   The PHA Accumulation Stage

PHA accumulation is usually carried out in batch mode and its efficiency strongly depends on the storage capacity of the selected culture. An overview of the results of recent studies using both synthetic and real wastewaters on PHA content in the biomass and polymer composition is given in Table 4.1. The maximum attainable polymer content in the biomass is a highly relevant parameter that significantly affects the polymer extraction costs: the lower the PHA content, the higher the extraction costs. As shown in Table 4.1, an important difference is observed depending on whether the accumulation stage is performed under nutrient rich or nutrient limited conditions, with a higher PHA cell content generally achieved in the latter case with both synthetic and real feedstocks. Indeed, if the accumulation stage is performed under nutrient limitation, the driving force for storage will be kept higher until the PHA synthesis saturation level is reached. On the contrary, if the biomass is continuously exposed to nutrients and carbon substrates (nutrient-rich conditions), a growth response will progressively increase whereas the storage response will progressively decrease and, as a consequence, the maximum PHA content will be lower than the cell's maximum storage capacity.

So far, the maximum cell polymer content (PHB, 89%) with mixed cultures has been obtained by using a single synthetic carbon source (acetate) under nitrogen starvation conditions [51] and it falls within the range of the maximum values (80–90%) achieved with pure microbial cultures [14]. Notably, the highest PHA content obtained using complex feedstocks (74.6%, with fermented molasses) under nitrogen limitation [36] is close to such high levels. Lower values have been achieved with paper mill effluent both in the presence of excess nutrients (31.9%) and under N or P limitation (between 43 and 48%) [44]. The highest PHA content in the biomass presently reported under no nutrient limitation is 54% and it has been obtained by using, in the accumulation stage, fermented olive mill effluents as substrate and a mixed culture previously selected in an SBR with synthetic substrates [52]. It is worth nothing that, even though this value is lower than those obtained from complex feedstock under nitrogen limitation, the possibility to produce PHAs without nutrient limitation is interesting and promising. Indeed, since the selection stage does not require nutrient-limitation, the use of a nutrient-rich feedstock would avoid the need for nutrient supply during the selection stage while still allowing for effective polymer accumulation in the final stage.

A major advantage of using complex materials as feedstock for PHA production is the possibility to obtain as final product the copolymer P(3HB/3HV) with a variable composition, depending on the nature of the wastes. In Table 4.1, the 3HV content in the stored polymer ranges from 11% to 69% (on a molar basis), with the highest value obtained by using fermented paper mill effluents.

Table 4.1 Performance of PHA accumulation stage by mixed cultures using different substrates and selection systems.

MMCs Selection Stage (SBR)			PHA Accumulation Stage (Batch)				References
Substrate Source	Selection System	SRT (d)	Substrate Source	Presence of Nutrients	%PHA Content gPHA gVSS^{-1}	PHA Composition (mol:mol)	
Acetate	SBR-AE	10	Acetate	N-limitation	65	P(3HB)	Serafim et al., 2004 [46]
Acetate	SBR-AE	1	Acetate	N-starvation (pH 7)	89	P(3HB)	Johnson et al., 2010 [51]
	SBR-AE	1		N-limitation (pH 7)	77	P(3HB)	
	SBR-AE	1		Nutrient excess (pH 7)	69	P(3HB)	
Acetate, lactate, propionate (40%, 40%, 20%, on COD basis)	SBR-AE	1	Acetate, lactate, propionate (40%, 40%, 20%, on COD basis)	Nutrient excess	50[a]	P(3HB/3HV) 69:31	Dionisi et al., 2004 [38]

Acetate, propionate (85%, on COD basis)	SBR-AE (pH 7.5)	1	Acetate, propionate (85%, 15%, on COD basis)	Nutrient excess (pH 7.5)	31	P(3HB/3HV) 76:24	Villano et al., 2010 [41]
	SBR-AE (pH 8.5)	1		Nutrient excess (pH 8.5)	34	P(3HB/3HV) 70:30	
	SBR-AE (pH 9.5)	1		Nutrient excess (pH 9.5)	21	P(3HB/3HV) 52:48	
Fermented paper mill effluent	2 CSTR in series	7	Fermented paper mill effluent	Nutrient excess (pH 7.3)	31.9[b]	P(3HB/3HV) 31:69	Bengtsson et al., 2008 [44]
	2 CSTR in series	7		P-limitation (pH 7.3)	48.2[b]	P(3HB/3HV) 39:61	
	2 CSTR in series	7		N-limitation (pH 7.3)	44.0[b]	P(3HB/3HV) 41:59	
	2 CSTR in series	7		P and N-limitation (pH 7.3)	42.7[b]	P(3HB/3HV) 47:53	

(Continued)

Table 4.1 Performance of PHA accumulation stage by mixed cultures using different substrates and selection systems. (Continued)

MMCs Selection Stage (SBR)			PHA Accumulation Stage (Batch)				References
Substrate Source	Selection System	SRT (d)	Substrate Source	Presence of Nutrients	%PHA Content gPHA gVSS^{-1}	PHA Composition (mol:mol)	
Acetate	SBR-AE	10	Synthetic mixture simulating fermented molasses	N-limitation	NR	P(3HB/3HV) 69:31	Albuquerque et al., 2007 [29]
	SBR-AE	10	Fermented molasses	Nutrient excess (pH 7)	NR	P(3HB/3HV) 60:40	
	SBR-AE	10	Fermented molasses	Nutrient excess (pH 6)	NR	P(3HB/3HV) 69:31	
	SBR-AE	10	Fermented molasses (pH 5)	Nutrient excess (pH 5)	NR	P(3HB/3HV) 47:53	
	SBR-AE	10	Fermented molasses	N-limitation (pH 6)	NR	P(3HB/3HV) 62:38	
Fermented molasses	SBR-AE	10	Fermented molasses	N-limitation (pH 6)	NR	P(3HB/3HV) 83:17	

Fermented molasses	SBR-AE	10	Fermented molasses	N-limitation	74.6	P(3HB/3HV) 74:26	Albuquerque et al., 2010 [36]
	SBR-AE	10	Acetate, propionate, butyrate, valerate (30/20/28/22 in Cmol/100 CmolVFA)	N-limitation, pulse feeding (pH 8.2)	65	P(3HB/3HV) 70:30	Albuquerque et al., 2011 [53]
	SBR-AE	10	Acetate, propionate, butyrate, valerate (31/18/29/22 in Cmol/100 CmolVFA)	N-limitation, continuous feeding (pH 8.4)	77	P(3HB/3HV) 61:39	

(Continued)

Table 4.1 Performance of PHA accumulation stage by mixed cultures using different substrates and selection systems. (Continued)

MMCs Selection Stage (SBR)			PHA Accumulation Stage (Batch)					References
Substrate Source	Selection System	SRT (d)	Substrate Source	Presence of Nutrients	%PHA Content gPHA gVSS⁻¹	PHA Composition (mol:mol)		
Fermented molasses	SBR–AE	10	Acetate, propionate, butyrate, valerate (60/16/20/04 in Cmol/100 CmolVFA)	N-limitation, pulse feeding, (pH 8.2–9.0).	68	P(3HB/3HV) 80:20		Albuquerque et al., 2011 [53]
	SBR–AE	10	Acetate, propionate, butyrate, valerate (59/09/26/06 in Cmol/100 CmolVFA)	N-limitation, pulse feeding, (pH between 8.3–9.0).	66	P(3HB/3HV) 83:17		
	SBR–AE	10	Fermented molasses (acetate, propionate, butyrate, valerate: 60/09/25/06 in Cmol/100 CmolVFA)	N-limitation, pulse feeding, (pH between 8.4–9.1).	56	P(3HB/3HV) 85:15		

Acetate, propionate, yeast extract, peptone	SBR-AN/AE	2–12	Fermented food waste	P-limitation	60	P(3HB/3HV) 77:23	Rhu et al., 2003 [54]
Tomato cannery wastewater	SBR-AE	4	Tomato cannery wastewater	Nutrient excess	20	NR	Liu et al., 2008 [55]
Acetate, lactate, propionate (40%, 40%, 20%, on COD basis)	SBR-AE	1	Centrifuged and fermented olive mill effluents	Nutrient excess	54	P(3HB/3HV) 89:11	Dionisi et al., 2005 [52]

NR: not reported.
AN/AE: anaerobic and aerobic cycles.
AE: fully aerobic cycles.
aUnits as COD/COD.
bUnits as gPHA/gTSS.

Since, as previously stated, the physicochemical properties of the P(3HB/3HV) are strongly dependent on the 3HV fraction, a few published studies have been focused on the investigation of operating conditions affecting the final polymer composition. As an example, some authors [41, 48] found that pH has a main effect on polymer composition and, particularly, the 3HV fraction can be increased by operating the selection and the accumulation stages at different pH values. The feeding regime has also been used to control the polymer composition, with higher 3HV content obtained by using continuous feeding strategy rather than pulse feeding strategy [53]. In the latter study, according to the authors, increasing HV content resulted in a decrease of the average molecular weight, the glass transition and melting temperatures and also in a reduction in the crystallinity degree.

These results clearly indicate that the thermal and mechanical properties of PHAs produced from agro-industrial wastes by mixed cultures can be adjusted by manipulating the relevant parameters of each stage of the process.

Another important parameter to evaluate the performance of MMC processes for PHA production is the PHA volumetric productivity, which represents the amount of polymer produced per unit of volume of reactor and per day. Volumetric productivity is usually reported or can be calculated from data with reference to the accumulation stage only. Volumetric productivity of the accumulation stage depends on both the average specific rate of PHA accumulation and initial biomass concentration in the batch reactor. As for specific PHA production rates, MMCs using real complex substrates achieved values up to 0.42 $gCOD_{PHA}$ $gCOD_{VSS}^{-1}$ h^{-1} [56], whereas biomass concentration is usually around a few grams per liter. Hence, the resulting PHA volumetric productivity for the accumulation stage is in the order of magnitude of 1 $gCOD_{PHA}$ $L^{-1}h^{-1}$. On the other hand, a more accurate calculation of PHA volumetric productivity should be referred to the volume of all reactors used in the process or at least to the volume of the reactors employed for the selection and accumulation stages. In this case, data to calculate polymer productivity for the overall process involving the use of waste materials are often lacking, even though it is expected to be lower than that calculated with respect only to the accumulation stage. Some information is available on PHA volumetric productivity referred to the selection plus accumulation stages (the second and the third stage of the process, respectively) by using mixed cultures and a synthetic mixture of VFAs as substrate. As an example, Dionisi and colleagues [57] reported a value of almost 0.3 gPHA L^{-1} h^{-1}, by using a synthetic mixture of acetic, propionic, and lactic acids and performing the culture selection stage in an SBR at high OLR (20 gCOD L^{-1} d^{-1}). This value falls in the range of those reported for pure cultures by using low-cost carbon substrates (0.05–0.90 gPHA L^{-1} $h^{-1)}$ [15], but it is significantly lower than those obtained with pure culture employing well defined medium (up to 5 gPHB L^{-1} h^{-1}) [58].

In the perspective of PHA production from agro-industrial wastes, a crucial parameter to consider is the PHA yield on the substrate, which directly impacts on the amount of substrate needed. For sugar cane molasses, it was estimated that 0.26 g of PHA could be produced from 1 g of COD waste by considering the overall three-stage production process [29]. A lower value, of 0.11 g PHA for 1 g of soluble COD, has been calculated for paper mill effluents [44]. This value is similar to that obtainable from Beccari and colleagues [28] of about 0.10 g PHA (as COD) per gram of soluble COD contained in olve oil mill wastewaters.

Ongoing research is aimed at further optimizing the overall process for PHA production from agro-industrial wastes, both in terms of volumetric polymer productivity and PHA storage yield.

4.4 Olive Oil Mill Effluents (OMEs) as a Possible Feedstock for PHA Production

4.4.1 Olive Oil Production

Olive oil is produced from olive trees, each olive tree yielding between 15 and 40 kg of olives per year. Worldwide olive oil production for the year 2002 was 2 546 306 t, produced from approximately 750 million productive olive trees, the majority of which are in the Mediterranean region. Mediterranean countries alone produce 97% of the total olive oil production, while European Union (EU) countries produce 80–84%. The biggest olive oil-producing country is Spain (890 100 t in 2002), then Italy (614 950 t), Greece (402 703 t) and Turkey (168 700 t), followed by Tunisia, Portugal, Morocco and Algeria. Outside the Mediterranean basin, olives are cultivated in the Middle East, the USA, Argentina and Australia. Olive oil is produced from olives in olive mills either by the discontinuous press method or by the continuous centrifugation method (either two-phase or three phase processes). There are about 25 000 olive mills worldwide [59, 60].

Depending on the type of production process, solid and/or liquid residual streams are produced in different amounts. The residual solids contain oil, that can be recovered by means of solvent extraction [61]. Hence, the biggest problem in olive oil process is represented by the liquid wastewater to be disposed of, i.e. the Olive Oil Mill Effluents (OMEs), which is high in volume and polluting charge, with low values of BOD_5/COD ratios and pH quite low (in the range 4-6). OMEs or also called black water is a toxic effluent that results during the production of olive oil.

OMEs arising from olive processing is one of the strongest industrial effluents, with chemical oxygen demand (COD) values of up to 220 g L^{-1} and corresponding biochemical oxygen demand (BOD) values of up to 100 g L^{-1}. The OMEs arising from the milling process amounts to 0.5–1.5 m^3 per 1000 kg of olives depending on the process. The discontinuous process produces less but more concentrated wastewater (0.5–1m^3 per 1000 kg of olive) than the centrifugation process (1–1.5 m^3 per 1000 kg of olive) [62, 63].

4.4.2   Chemical and Physical Characteristic of OMEs

In general, OMEs are dark-colored and contain high amounts of many complex organic substances, some of them being not easily degradable. Deterioration of natural water bodies due to discharge of olive oil mill wastes is a serious problem as indicating by coloring, appearance of an oily shine, and increased oxygen demand. They also affect the soil quality, are toxic to plant life, and create odor nuisance when disposed into the soil. For this reasons, direct discharge of olive mill wastewaters into receiving media is not permissible and certain measures must be taken before disposal of the OMEs into the environment. In addition, it must be added the low biodegradability of this effluent that does not allow the use of technologies traditionally used for urban wastewater. In fact, the ratio between chemical oxygen demand (COD) and biochemical oxygen demand in five days (BOD_5) is rather low: in OMEs, BOD_5/COD ratio, which is an index of biodegradability, is around 0.25 to 0.35. Arpino *et al.* [64] also indicate that only 25% of the OMEs substances are completely biodegradable. Furthermore, OMEs are deficient in nutrients, especially in nitrogen content (Table 4.2).

Table 4.2 Typical OMEs composition (continuous process).

	Kornaros et al., 2010 [74]	Coskun et al., 2010 [75]	Kornaros et al., 2009 [76]	Dhouib et al., 2009 [77]	Goumenaki et al., 2007 [78]	Buobaker et al., 2007 [79]	Gonzalez et al., 2006 [80]	Sayadi et al., 2006 [81]	Beltran et al., 2005 [82]	Ammary BY, 2005 [83]	Ceccanti et al., 2003 [84]	Lyberatos et al., 2002 [85]
VSS (g L⁻¹)	34	N.A.	N.A.	38	4.6	30	N.A.	6.5	39.9	N.A.	N.A.	42
TSS (g L⁻¹)	37	N.A.	N.A.	45	5.2	35	9.0	8.9	67.1	26	N.A.	46
BOD$_5$ (g L⁻¹)	41	N.A.	41	16.3	N.A.	N.A.	96	34.4	52	31	N.A.	48
COD (g L⁻¹)	131	40.3	131	56	89.5	80	196	117	95	97	66.7	105
pH	5.0	4.6	5.0	4.95	N.A.	5	5.5	5.46	4.8	5.72	4.93	5.4
Total phenols (g L⁻¹)	6.8	N.A.	6.8	4.6	N.A.	0.1	17.5	9.2	2.2	3.6	0.98	0.01
Ammonia (g L⁻¹)	0.1	0.24	0.1	0.35	1	0.17	0.9	1.58	0.4	0.53	0.62	0.75
Phosphorous (g L⁻¹)	0.2	N.A.	0.4	N.A.	0.1	N.A.	N.A.	0.84	1.9	0.18	N.A.	0.29
COD:N	100:0.1	100:0.6	100:0.1	100:0.6	100:1.1	100:0.2	100:0.5	100:1.4	100:0.4	100:0.5:	100:0.9	100:0.7
Type of process	Three-phase	Three-phase	Three-phase	Three-phase	N.A.	Three-phase	Three-phase	Two-phase	N.A.	Three-phase	Three-phase	N.A.

N.A. not available.

The OMEs composition is not constant either qualitatively or quantitatively and varies according to cultivation soil, harvesting time, the degree of ripening, olive variety, climatic conditions, the use of pesticides and fertilisers and the duration of aging [59]. Another salient feature of this type of waste is its facility for fermenting while in storage, which gives rise to substantial changes in composition yet does not necessarily result in its complete biodegradation [65].

OMEs contain both dissolved substances (salts and simple carbohydrates) and suspended substances (pectin, oil, mucilage). OMEs are similar to a surfactant solution of high stability because of presence of a lyophilized colloidal system type complex that leads to the appearance of polarization phenomena at the interface [66]. The composition is roughly about 83–84% water, 14–15% organic matter and 1–2% mineral salts. The mineral fraction consists mainly of carbonates (21%), phosphate (14%), sodium (7%) and potassium (47%). Organic matter is composed of fats, proteins, sugars, polyalcohols, pectins, tannins and glycosides. The sugars levels is in the order of 1.6–4% (w/v), which constitute up to 60% of the total dry matter; glucose, fructose, mannose and sucrose, but also more complex polysaccharides derived from the pulp of the fruit (such as starch and hemicelluloses) are present in OMEs.

The phenolic compounds contained in OMEs can be classified roughly into two groups [67]. The first group comprises simple phenolic compounds, non-autoxidated tannins (low molecular weight) and flavonoids. The flavonoids are polyphenolic compounds possessing 15 carbon atoms.

The main flavonoids detected in OMEs are apigenin, luteolin and quercetin [68, 69]. The main phenolic acids are syringic acid, p-hydroxyphenylacetic acid, vanillic acid, veratric acid, caffeic acid, protocatechuic acid, pcoumaric acid and cinnamic acid [70]. The polyphenols of the second group, which comprise dark-coloured polymers, result from polymerisation and autoxidation of the phenolic compounds of the first group. The colour of OMEs depends on the ratio between the two groups of polyphenols. The presence of these recalcitrant organic compounds constitutes one of the major obstacles in the detoxification and purification of OMEs [71]. In addition, some of these phenols are responsible for several biological effects, including antibiosis [72] and phytotoxicity [73].

4.4.3   Wastewater Treatment and Disposal Alternatives

Main problem regarding the disposal of OMEs is to find an environmentally friendly and economically viable solution. In biochemical treatment, high capital and operating cost units with limited efficiency have to be installed due to the high organic loads and low BOD_5/COD ratios. Due to the presence of toxic organics, mostly coming from the broken seeds, these wastes are toxic to bacteria and direct biological treatment is not possible [86].

On the other hand, there are studies indicating that OMEs may also be regarded as an economic resource. Among these practices, use of OMEs as soil conditioner, biomass fuel, compost, or as starting material to obtain valuable products such as antioxidants, enzymes and biogas fuel may be counted. Recycle and reuse of process water for irrigation purposes may also be considered, but acceptable water quality must be guaranteed.

In the light of these considerations, it is clear that the problems of OMEs disposal are still remarkable because of significant technological solutions lack from an environmental point of view and economic and regulatory management. The factors that seem to have more weight to an exact size of the problem of OMEs disposal are: composition of the wastewater, in particular the presence of phytotoxic and antimicrobial substances (phenolic type compounds and long chain fatty acids); high concentration level (COD values in the range of $100–200$ g L^{-1} and BOD_5 in the range of $20–100$ g L^{-1}); waste quantity (from 40% to 110% in weight compared with olives); seasonality of the production cycle (the oil production season lasts from 2 to 6 months depending on geographical area); plant territory splitting (despite the strong upward trend in production scale, about half of the companies are small-medium-sized). To solve these problems, numerous studies aimed at developing an effective and inexpensive treatment process have been conducted in recent years.

Several treatment methods such as physicochemical, chemical, biological (aerobic and anaerobic) processes as well as evaporation (natural or forced) and land application have been proposed.

A part from biological processes, the most common techniques of OMEs disposal and/or treatment that are presently in use of investigated, are:

- Land application: some researchers have shown that the controlled spreading of raw wastewaters on agricultural land may have a positive effect on the olive plantations, as well as on other crops. However, this practice is to be considered only after a thorough evaluation of all of the environmental impacts [87].

- Chemical coagulation/flocculation followed by solid/liquid separation: different coagulants such as ferric chloride, aluminum chloride, ferrous sulfate, calcium hydroxide and their combinations sometimes added by different anionic polyelectrolytes and sulfuric acid were tested in order to remove organics prior of discharge [88].

- Evaporation—Hydrolysis—Oxidation (EHO): in this combined method, following a preconcentrating step by evaporation, hydrolysis under controlled heat input and subsequent oxidation by air take place. The EHO process is at the prototype development stage. The capital investment of a pilot plant treating olive oil wastewaters from a Greek district with an average olive oil production of 1400 tons per year was given as about 14 million Euros with an annual operating cost of nearly 1 million Euros [89].

- Membrane processes: membrane filtration such as reverse osmosis or ultrafiltration is applied after some pretreatment step. Wastewater is separated into two phases: treated wastewater (permeate) and concentrate (retentate). The concentrate can be sent to incineration or landfilled for final disposal [89].

- Electrolysis: this method is based on in situ production of strong oxidizing hydroxyl radicals, which can oxidize organics to carbon dioxide and water in an electrolysis cell using a Ti/Pt anode and stainless steel cathode. Based on reduction of COD, total organic carbon (TOC), volatile suspended solid (VSS) and phenolic compounds in a batch laboratory-scale pilot plant, OMEs electrolysis could be used as pretreatment stage followed by chemical treatment or another advanced oxidation process [90].

- Vacuum evaporation: vacuum of 5 kPa allows the OMEs to evaporate at low temperatures such as 38°C [91]. The vacuum evaporator produces two streams; a continuously produced distillate (90% of the waste input) and a discontinuous concentrate automatically discharged. The distilled liquid phase required further treatment, such as biological oxidation which preceded by pH adjustment and C:N:P correction. The concentrate can be mixed with the solid residues from olive oil mill and de-oiled, burned or used for purposes like animal fodder or fertilizer.

- Natural evaporation: this method requires extended storage time for OMEs, also depending on the climatic conditions [92]. The main disadvantages are insects and odor nuisances, as well as potential groundwater contamination if the bottom of the storage lagoon is not properly lined against infiltration. It is often necessary the separation of the solid phase that can be used as fertilizer [93].

4.4.4   Biological Wastewater Treatment

OMEs can be treated in a traditional activated sludge processes for wastewater treatment, provided that they are diluted with other less strong wastewaters; otherwise, the really high organic content and the presence of certain inhibitory compounds can negatively affect the overall biological processes. Moreover, the high yield of biomass growth during the aerobic treatment can require the addition of nutrients to balance the too high COD:N and COD:P ratios.

On the other hand, because of the high organic content of OMEs, anaerobic processes with biogas production can be the first choice for biological treatment of OMEs. Because of the quite lower yield of biomass growth under anaerobic conditions, the lack of the nutrients in OMEs has minor relevance with respect to aerobic processes. However, also under anaerobic conditions, the presence of oily and phenolic antimicrobial compounds creates a challenge for biological treatment [94]. The decrease of pH and the build up of volatile fatty acids during storage could also create inhibitory effects in anaerobic digestion [95]. If a pretreatment stage is included or OMEs are simply diluted to overcome such effects, anaerobic treatment has several advantages: low sludge generation, methane production, less energy requirement and easy restart at new seasonal production campaigns after shutdown of the previous one.

Laboratory and pilot-scale experimental studies on diluted OMEs showed that the anaerobic contact process could give high organic removal efficiencies ranging between 80–85% at 35°C and at organic load less than 4 gCOD L^{-1} d^{-1}. But for highly concentrated OMEs, the process proved unstable due to the inhibitory effects of ingredients (polyphenols and potassium). Moreover, additions of alkali substances to neutralize acidity and ammonia to furnish nitrogen for cellular biosynthesis were necessary [96].

To improve process efficiency and stability, codigestion of OMEs and sewage sludge mixture was investigated in a contact bioreactors; a laboratory-scale study showed that at an organic (COD) load rate of 4.2 g L^{-1} d^{-1} at 35°C with mixed OMEs (34%) and sewage sludge, a COD removal efficiency of 65% could be obtained [97]. Higher portions of wastewater additions caused imbalances of the process due to inhibitory effects of polyphenols. The up-flow anaerobic sludge blanket (UASB) reactor has been also proven to be suitable to treat OMEs. Experimental studies [98] showed that at 37°C and an organic load in the COD range of 12–18 g L^{-1} d^{-1}, removal efficiencies of about 70–75% were obtained by adopting a dilution ratio varying between 1:8 and 1:5. Anaerobic digestion of a mixture of OMEs (80%) plus piggery effluent (20%) was performed in an up-flow anaerobic filter [99]. This study indicated 70–80% treatment efficiency for influent COD varying between 20 and 60 g L^{-1}. The process provided the neutralization of influent, as well as the conversion of 50–70% of its phenolic content. Hence, OMEs can be degraded with high biomethanation yields (about 80%) provided that they are diluted enough, whereas remarkable inhibition of methanogenesis occurred at higher concentrations [100].

Several authors suggested that main responsible for methanogenesis inhibition in OMEs are lipids, and namely oleic acid [101], even though these compounds are easily and fully degraded under non-inhibiting conditions. Phenolic compounds are also of concern in anaerobic treatment of OMEs because they contribute to the overall inhibition effect, even though to a less extent than lipids. Moreover, this phenolic fraction is partially recalcitrant being its degradation not complete also under non-inhibiting conditions [100, 101]. These evidence suggested the opportunity of a physical-chemical pre-treatment carried out to remove lipids and phenolic compounds, as more selectively as possible, before anaerobic digestion. The addition of $Ca(OH)_2$ up to pH 6.5 and bentonite at 15 g L^{-1}, followed by centrifugation, turned out to remove lipids by 99.5% and phenolic compounds by about 45%. Such a pre-treatment allowed to achieve high bioconversion into methane of the residual OMEs, even at very low dilution ratios (1:1.5) [102]. Furthermore, the highest biomethanation yields were obtained when no solid/liquid separation before the biological process was performed, thus showing that adsorption of COD through the addition of lime and bentonite was able to decrease inhibition effects also in the presence of the biomass and that at least a fraction of the removed COD was biodegraded when slowly released back from the solid phase during the biological test [102]. Moving to continuous lab-scale experiments, further work has confirmed that the mixture OMEs–Ca(OH)$_2$–bentonite, fed to a methanogenic reactor, without providing an intermediate phase separation, gave way to high COD removal and high biogas production even at very low dilution ratios [103, 104, 105]. The best result in terms of methane production were obtained by addition of Ca(OH)$_2$ up to pH 6.5 and of 10 g L^{-1} of bentonite; in lab-scale continuous methanogenic reactor fed with the substrate without intermediate solid/liquid separation, very satisfactory performances were obtained at an organic load of 8.2 gCOD L^{-1} d^{-1} and at an dilution ratio of 1:1.5 in which the total COD removal was 91%, the biogas production was 0.80 gCOD$_{CH4}$ gCOD$_{TOT}$$^{-1}$, the removal of lipids and phenols, in term of COD, was 98% and 63% respectively.

Despite the encouraging results resumed above, the problem remained, in that the effluent from the methanogenic reactor still contained significant concentrations of residual phenolic compounds.

In order to better understand and describe the feasibility of phenolic compounds removal in the proposed treatment scheme, Beccari et al. [106] evaluated the fate of the phenolic fractions with different molecular weights during the sequence of operations (adsorption on bentonite and methanogenic digestion). The results showed that a very high percentage (above 80%) of the phenolic fraction below 500 D as molecular weight (MW) is removed by the methanogenic process whereas the phenolic fractions above 1 000 D are significantly adsorbed on bentonite.

The possibility of increasing the OMEs conversion to methane and polyphenols removal as well was investigated by employing an anaerobic OME-digesting microbial consortium passively immobilized in column reactors packed with granular activated carbon (GAC) or "Manville" silica beads (SB) [33]. Under batch conditions, both GAC- and SB-packed-bed biofilm reactors exhibited OMEs COD and phenolic compound removal efficiencies markedly higher (from 60% to 250%) than those attained in a parallel anaerobic dispersed growth reactor developed with the same inoculums. Both biofilm reactors also mediated an extensive OMEs remediation under continuous condition, where GAC-reactor was much more effective than the corresponding SB-one, and showed a higher tolerance to high and variable organic loads up to 16–17 gCOD L^{-1} d^{-1}.

The relatively low OMEs degradability has led to a deepening of the role of recalcitrant/inhibitors substances, also in relation to the interaction between acidogenesis and methanogenesis. The effect of pH, temperature and initial OMEs concentration [100] on the separate conversion to VFA or methane has been investigated in relationship to main substances (alcohols, phenols, lipids). Despite the fact that acidogenic fermentation can be less effective in terms of conversion yield under most favorable conditions, the range of acidogenic operative conditions proved wider than in the case of methanogenis, without significant lag periods; furthermore, the acidogenic fermentation was much less sensitive to inhibition than methanogenesis at high OMEs concentration. This is the starting point for an optimization of the acidogenic fermentation step of OMEs, in order to obtain high VFAs concentration, the direct precursor for the synthesis of PHAs.

4.5   OMEs as Feedstock for PHA Production

During the last years increasing attention has been paid to discovering a new viable way of handling OMEs, and a wide range of different processes have been proposed, aiming both to the reduction of the organic load and the pollutant effects and to the transformation into products of medium-high added value. An innovative strategy is to use OMEs as starting materials for the production of biodegradable polymers; that would provide high-value materials from a no cost feedstock. As already reported in section 1, bioplastics, such as polyhydroxyalkanoates (PHAs) and particularly the copolymer poly(β-hydroxybutyrate/β-hydroxyvalerate) [P(HB-HV)], are attracting much interest as alternatives to oil-based plastics, because they are biodegradable and can be formed from renewable resources.

In particular, the use of OMEs has been taken into account because of the possibility of fermenting OMEs at high rate to produce volatile fatty acids (VFAs), which are the most direct substrates for PHAs production [28]. Moreover, the acidogenic fermentation step is not inhibited to a great extent by OMEs lipidic and phenolic substances, which, on the other hand, inhibit the further step of methanogenesis [100]. Another advantage of OMEs with respect to other organic waste is that they are a liquid stream and can therefore be easily pumped and treated in a slurry-phase bioreactors [105].

As already mentioned, mixed cultures submitted to feast and famine conditions are unable to store PHAs from sugar-based compounds [26]. Thus, when the feedstock contains sugar compounds, the previous fermentation step is essential to transform carbohydrates into VFAs and other carboxylic acids, which can be further used in the selection and accumulation steps. According to the literature, the carbohydrates content of OMEs can reach up to 60% of its total dry weight, thus being 4.7–16.1 g L^{-1} [107, 108, 109]. Thus, OMEs need previous acidification; subsequently the effluent of this process can be (see section 1) treated aerobically, using selective mixed microbial cultures and appropriate conditions for the PHAs production. Thus, PHAs production from acidified OMEs has been recently investigated [28, 52, 110].

Dionisi *et al.* [52] firstly studied the feasibility of using OMEs as the substrate in the biodegradable polymer production, following their conversion to organic acid. The OMEs anaerobic fermentation step was studied both without pretreatment and with different pretreatment (i.e., centrifugation, bentonite addition, and bentonite addition followed by centrifugation) and various concentration (28.5, 36.7, 70.4 gCOD L^{-1}) in order to evaluate the final VFAs concentration and yield.

Fresh OMEs were obtained from an olive continuous centrifuge processing plant and were characterized as follow: wastewater pH was 5.2, total COD 113.8 g L^{-1}, soluble COD (filtered at 0.22 mm) 34.5 g L^{-1}, soluble polyphenols 2.2 g L^{-1}, soluble carbohydrates 3.1 g L^{-1}, Kjeldhal nitrogen 2 g L^{-1} and lipids 11.1 g L^{-1}.

OMEs centrifugation plus fermentation demonstrated to be the best pretreatment for the production of the most favorable feed for PHA storage: at initial concentration of 36.7 gCOD L^{-1}, fermentation yield increased from 25% to 35% and VFA concentration increased from 9 to 13 gCOD L^{-1} with respect to untreated OMEs. Under any conditions, the conversion yield decreased when starting from the highest concentration of 70.4 gCOD L^{-1}. Moreover, centrifugation pretreatment led to a different acid distribution; even though acids with an even number were always predominant (acetic or butyric acids), OMEs fermentation was effective in producing VFAs with an odd number of carbon atoms (propionic and valeric acids), which were absent in untreated OMEs. The latter acids allow the formation of the HV monomer, which is useful because an increasing content improves the thermal and mechanical properties of the copolymer P(HB-HV). Preliminary accumulation batch test were carried out with centrifuged and fermented OMEs, by using a MMC that had been previously acclimated on a synthetic feed containing only organic acids (VFAs and lactic acid). By comparison of PHA production from centrifuged and fermented OMEs and from the acclimation synthetic medium, good storage capacity was obtained with both media. This also indicated that centrifuged and fermented OMEs did not contain inhibiting compounds (at least at the chosen concentration), even for a non-acclimated biomass. Moreover, the initial storage yield on removed VFAs (referred only to the accumulation stage) was quite high ($\sim$1.0 mgCOD$_{PHAs}$ mgCOD$_{VFA}^{-1}$) and much more than usual with synthetic mixtures containing VFAs only. This is an indirect evidence that PHAs were also produced by readily biodegradable carbon sources other than VFAs. Even though less than maximum values were reported in the literature for pure cultures, the final PHA content in the biomass was quite high (0.54 gPHA gVSS^{-1}), also considering that a N-rich medium was used to dilute fermented OMEs.

Then, the performance of a three-stage process for polyhydroxy-alkanoates bioproduction from OMEs has been investigated (28, following the scheme of Figure 4.2): in the first anaerobic stage OMEs were fermented in a packed bed biofilm reactor into volatile fatty acids. This VFA-rich effluent was fed to the second stage, operated in an aerobic SBR, to enrich mixed cultures able to store PHAs. Finally, the storage response of the selected consortia was exploited in the third aerobic stage, operated in batch conditions.

The OMEs was taken from three phase Sant'Agata d'Oneglia olive mill, Imperia (Italy) and it was fed to the anaerobic packed bed biofilm reactor (PBBR) without previous pretreatment, addition of nutrients or dilution. The OMEs COD was about 37 g L^{-1}, partially due to VFAs (about 7 gCOD L^{-1}), its suspended and volatile solids content was 22.6 and 16.6 g L^{-1}, respectively, and the Kjendhal nitrogen amount was 4.97 g L^{-1}. The OMEs phenolic fraction was about 1.5 g L^{-1}. The wastewater pH was 5.5. PBBR technology was chosen to develop the anaerobic stage (25°C and

26.2 gCOD L^{-1} d^{-1}), since immobilized cells systems allow reduction of the risk of shock loading and/or washout problems, which can negatively affect the productivity of suspended growth reactors operating with low-growth rate biomass and high and fluctuating organic loads, as typically happens in the anaerobic digestion of OMEs [33]. The COD concentration of the fermented OMEs (33.9 g L^{-1}) was very close to the influent one, indicating that only a small influent substrate fraction (~8%) was converted into biogas and biomass growth. The VFA concentration in the effluent was 10.7 gCOD L^{-1}; this means that anaerobic fermentation stage increased the VFA percentage in the OMEs from about 18% to 32% on COD basis, with slight decrease in the pH. The acetic, butyric and propionic acid percentage in the VFA mixture produced were 62%, 22% and 12%, respectively. Fermented OMEs also contained alcohols, substrates easily convertible into PHAs, accounting for about 22% of the overall COD; namely ethanol, butanol and methanol (~80%, ~13%, ~7% of the overall alcohols respectively, on COD basis).

The SBR (second stage) was inoculated with activated sludge from a wastewater treatment plant, at an overall OLR of 8.5 gCOD L^{-1} d^{-1} (corresponding to an OLR of 2.7 gCOD L^{-1} d^{-1} in terms of VFAs only). Even if VFA were only a fraction of the overall COD, selective pressure on the enrichment of mixed cultures able to store PHAs was maintained in this stage. The maximum values of PHA storage rate and yield with fermented OMEs were about 146 mgCOD $gCOD^{-1}$ h^{-1}, and 0.36 COD COD^{-1} respectively, very similar to those obtained using a synthetic mixture of acetic and propionic acid. The SBR reactor also allowed a significant reduction of the polluting load of the fermented effluent, with COD removal efficiency of about 85%. The third stage was simulated in the SBR by means of tests at increasing loads (up to about 9 times the load of a single SBR cycle). The maximum concentration of PHAs produced and the maximum polymer content in the biomass increased almost linearly with the amount of fermented OMEs fed, indicating the possibility to operate the process at higher OLR; however, the test led to higher load exhibited lower substrate removal and storage rates suggesting that fermented OMEs exerted a slightly inhibitory effect on the activity of the enriched culture, probably due to the presence of residual lipids and phenols. The storage yield remained nearly unchanged (~0.35 COD COD^{-1}) and similar to those obtained with synthetic mixture of VFAs. These tests also revealed a significant contribution to the PHA storage of substrate other than VFAs, mostly alcohols and especially ethanol.

The feasibility of using OMEs as initial substrate for combined biohydrogen and biopolymers production, thus not only leading to an environmental friendly solution, but to a cost effective one as well, since both energy and valuable bioproducts can be generated, was investigated by Lyberatos *et al.* [110]. A two stage systems, consisting of anaerobic reactor operated at continuous mode (CSTR) and an aerobic sequential batch reactor (SBR) was developed. The anaerobic reactor was operated at different HRT with OMEs, diluted 1:4 (v/v) with tap water, as feed. The performance of CSTR was evaluated in terms of the effect of different HRTs on hydrogen production and VFAs production. It was shown that HRTs from 27 to 33 hours were favorable in terms of total acidification of the wastewater and propionate production, whereas both hydrogen production and butyrate and acetate production were favoured at HRT 14.5 hours. Subsequently, the effluent of the CSTR was collected, filtered, sterilized and used as substrate for PHAs production. The operating schedule of the SBR used in that process was determined based on the results from batch experiments carried out with acclimatized cultures of *Pseudomonas* sp (mainly *Pseudomonas putida* developed in a fill-and-draw reactor inoculated with activated sludge and fed with synthetic mixture of acetic and propionic acid) with acidified OMEs. The aerobic reactor was inoculated with the enriched PHAs producing bacteria culture mentioned above, and was operated in sequential cycles of nitrogen offer (growth phase) and nitrogen limitation (PHAs accumulation phase) by using two different feeds: in the feed used for the growth phase, nitrogen was added in the form of $(NH_4)_2SO_4$. The SBR operated with overall cycle length of 3.5 days and consequently two cycles were completed per week. During the accumulation phase, butyrate was consumed preferably, indicating that the dominant PHA produced was polyhydroxybutyrate, but the reduction of dissolved COD was always higher to VFAs uptake. This indicated that other substances apart from VFAs were also consumed, probably sugars. For the quantification of the PHAs incorporated to the cells, all the samples tested showed variations ranging from 4.02% to 8.94% PHA/dry biomass weight (procedure not optimized yet). However, the PHAs yields are promising and could be further increased according to previous studies in which is described that *P. Putida* can accumulate large amounts of PHA under different types of nutrient limitation and with different type of carbon source, such as sugar, octanoic acid and fatty acids [111, 112, 113, 114].

In the context of pure culture use, Gonzalez-Lopez *et al.* [115] described an over-producing strain of *Azotobacter chroococcum* strain H23 capable of accumulating PHAs (homo and copolymers) in chemically-defined media and in OMEs during growth without a nutrient limitation. *A. chroococcum* grown on NH_4^+ medium supplemented with OMEs formed PHAs up to 50% of the cell dry weight after 24 hours (the culture was incubated for 48 hours at 30°C with shaking at 100 rpm); the same percentages were obtained with chemically defined synthetic substrates. The results showed that OMEs support the growth of strain H23 and also that this waste could be utilized as a carbon source.

4.6   Concluding Remarks

Based on literature evidences and lab-scale proof, the valorization of OMEs as no cost feedstock for the production of polyhydroxy-alkanoates in a biotechnological three-stage process is attracting and sustainable. OMEs bring the advantage to be easily ferment-able into volatile fatty acids and other suitable substrates for PHA production. Laboratory scale studies have widely demonstrated the feasibility of each stage of the process and of the overall process itself, encouraging further development.

The use of OMEs (as well as of other agro-industrial wastes) makes it possible to easily obtain PHA copolymers of HB and HV monomers, resulting in a polymer with better physicochemi-cal properties. The HB/HV ratio can be controlled by manipu-lating the relevant parameters of each stage of the process and, particularly, the acids composition from the fermentation stage. Ongoing research is mainly aimed at finding an optimal tradeoff between the high process performance (in terms of rate and yield of PHA production) and the PHA composition and properties for downstream processing. Process optimization at a higher scale and capacity is required to increase the potential for industrial application.

References

1. S. Chanprateep, "Current trends in biodegradable polyhydroxyalkanoates", *Journal of Bioscience and Bioengineering*, Vol. 110, pp. 621–632, 2010.
2. D. Jendrossek and R. Handrick, "Microbial degradation of polyhydroxyalkanoates", *Annual Review of Microbiology*, Vol. 56, pp. 403–432, 2002.
3. S.Y. Lee, "Bacterial polyhydroxyalkanoates", *Biotechnology and Bioengineering*, Vol 49, pp. 1–14, 1996.
4. M. Kunioka and Y. Doi, "Thermal degradation of microbial copolyesters: poly(3-hydroxybutyrate-co-3-hydroxyvalerate) and poly(3-hydroxybutyrate-co-4-hydroxybutyrate)", *Macromolecules*, Vol. 23, pp. 1933–1936, 1990.
5. F. Carrasco, D. Dionisi, A. Martinelli, and M. Majone, "Thermal stability of polyhydroxyalkanoates", *Journal of Applied Polymer Science*, Vol. 100, pp. 2111–2121, 2006.
6. D.Z. Bucci, L.B.B. Tavares, and I. Sell, "PHB packaging for the storage of food products", *Polymer Testing*, Vol 24, pp. 564–571, 2005.
7. C.M. Bunger, N. Grabow, K. Sternberg, M. Goosmann, K.P. Schmitz, H.J. Kreutzer, H.S. Ince, S. Kische, C.A. Nienaber, D.P. Martin, S.F. Williams, E. Klar, and W. Schareck, "A Biodegradable Stent Based on Poly(L-Lactide) and Poly(4-Hydroxybutyrate) for Peripheral Vascular Application: Preliminary Experience in the Pig", *Journal of Endovascular Therapy*, Vol. 14, pp. 725–733, 2007.
8. A.J. Anderson and E.A. Dawes, "Occurrence, metabolism, metabolic role, and industrial uses of bacterial polyhydroxyalkanoates", *Microbiological Reviews*, Vol. 54, pp. 450–472, 1990.
9. G. Braunegg, G. Lefebvre, and K.F. Genser, "Polyhydroxyalkanoates, biopolyesters from renewable resources: Physiological and engineering aspects", *Journal of Biotechnology*, Vol. 65, pp. 127–161, 1998.
10. T. Suzuki, T. Yamane, and S. Shimizu, "Mass production of poly-β-hydroxybutyric acid by fed-batch culture with controlled carbon/nitrogen feeding", *Applied Microbiology and Biotechnology*, Vol. 24, pp. 370–374, 1986.
11. W.J. Page and A. Cornish, "Growth of Azotobacter vinelandii UWD in Fish Peptone Medium and Simplified Extraction of Poly-β-Hydroxybutyrate", *Applied Environmental Microbiology*, Vol. 59, pp. 4236–4244, 1993.
12. F. Wang and S.Y. Lee, "Poly(3-Hydroxybutyrate) Production with High Productivity and High Polymer Content by a Fed-Batch Culture of Alcaligenes latus under Nitrogen Limitation", *Applied Environmental Microbiology*, Vol. 63, pp. 3703–3706, 1997.
13. E. Grothe, M. Moo-Young, and Y. Chisti, "Fermentation optimization for the production of poly(β-hydroxybutyric acid) microbial thermoplastic", *Enzyme and Microbial Technology*, Vol. 25, pp. 132–141, 1999.
14. A. Steinbüchel and T. Lütke-Eversloh, "Metabolic engineering and pathway construction for biotechnological production of relevant polyhydroxyalkanoates in microorganisms", *Biochemical Engineering Journal*, Vol. 16, pp. 81–96, 2003.
15. B.S. Kim, "Production of poly(3-hydroxybutyrate) from inexpensive substrates", *Enzyme and Microbial Technology*, Vol. 27, pp. 774–777, 2000.

16. C.S.K. Reddy, R. Ghai, Rashmi, V.C. Kalia, "Polyhydroxyalkanoates: an overview", *Bioresource Technology*, Vol. 87, pp. 137–146, 2003.

17. M.A.M. Reis, L.S. Serafim, P.C. Lemos, A.M. Ramos, F.R. Aguiar, and M.C.M. Van Loosdrecht, "Production of polyhydroxyalkanoates by mixed microbial cultures", *Bioprocess and Biosystems Engineering*, Vol. 25, pp. 377–385, 2003.

18. H. Salehizadeh and M.C.M. Van Loosdrecht, "Production of polyhydroxyalkanoates by mixed culture: recent trends and biotechnological importance", *Biotechnology Advances*, Vol. 22, pp. 261–279, 2004.

19. M. Beccari, D. Dionisi, A. Giuliani, M. Majone, and R. Ramadori, "Effect of different carbon sources on aerobic storage by activated sludge", *Water Science and Technology*, Vol. 45, pp. 157–168, 2002.

20. H. Satoh, T. Mino, and T. Matsuo, "PHA production by activated sludge", *International Journal of Biological Macromolecules*, Vol. 25, pp. 105–109, 1999.

21. H. Pereira, P.C. Lemos, M.A.M. Reis, J.P.S.G. Crespo, M.J.T. Carrondo, and H. Santos, "Model for carbon metabolism in biological phosphorus removal processes based on in vivo13C-NMR labelling experiments", *Water Research*, Vol. 30, pp. 2128–2138, 1996.

22. M. Majone, P. Massanisso, A. Carucci, K. Lindrea, and V. Tandoi, "Influence of storage on kinetic selection to control aerobic filamentous bulking", *Water Science and Technology*, pp. 223–232, 1996.

23. G.T. Daigger and C.P.L. Grady, "The dynamics of microbial growth on soluble substrates: A unifying theory", *Water Research*, Vol. 16, pp. 365–382, 1982.

24. M.A.M. Reis, M. Albuquerque, M. Villano, and M. Majone, "Mixed culture processes for polyhydroxyalkanoate production from agro-industrial surplus/wastes as feedstocks", in M. Butler, C. Webb, A. Moreira, B. Grodzinski, Z.F. Cui, and S. Aghatos, eds., *Comprehensive Biotechnology 2nd Edition*, in press.

25. K. Dircks, J.J. Beun, M. van Loosdrecht, J.J. Heijnen, and M. Henze, "Glycogen metabolism in aerobic mixed cultures", *Biotechnology and Bioengineering*, Vol. 73, pp. 85–94, 2001.

26. F. Carta, J.J. Beun, M.C.M. van Loosdrecht, and J.J. Heijnen, "Simultaneous storage and degradation of phb and glycogen in activated sludge cultures", *Water Research*, Vol. 35, pp. 2693–2701, 2001.

27. O. Karahan, D. Orhon, M. van Loosdrecht, "Simultaneous storage and utilization of polyhydroxyalkanoates and glycogen under aerobic conditions", *Water Science & Technology*, Vol. 58, pp. 945–951, 2008.

28. M. Beccari, L. Bertin, D. Dionisi, F. Fava, S. Lampis, M. Majone, F. Valentino, G. Vallini, and M. Villano, "Exploiting olive oil mill effluents as a renewable resource for production of biodegradable polymers through a combined anaerobic-aerobic process", *Journal of Chemical Technology & Biotechnology*, Vol. 84, pp. 901–908, 2009.

29. M.G.E. Albuquerque, M. Eiroa, C. Torres, B.R. Nunes, and M.A.M. Reis, "Strategies for the development of a side stream process for polyhydroxyalkanoate (PHA) production from sugar cane molasses", *Journal of Biotechnology*, Vol. 130, pp. 411–421, 2007.

30. S. Bengtsson, J. Hallquist, A. Werker, and T. Welander, "Acidogenic fermentation of industrial wastewaters: Effects of chemostat retention time and pH on volatile fatty acids production", *Biochemical Engineering Journal*, Vol. 40, pp. 492–499, 2008.

31. M.T. Agler, B.A. Wrenn, S.H. Zinder, and L.T. Angenent, "Waste to bioproduct conversion with undefined mixed cultures: the carboxylate platform", *Trends in Biotechnology*, Vol. 29, pp. 70–78, 2011.

32. J.I. Horiuchi, T. Shimizu, K. Tada, T. Kanno, and M. Kobayashi, M., "Selective production of organic acids in anaerobic acid reactor by pH control", *Bioresource Technology*, Vol. 82, pp. 209–213, 2002.

33. L. Bertin, S. Berselli, F. Fava, M. Petrangeli-Papini, and L. Marchetti, "Anaerobic digestion of olive mill wastewaters in biofilm reactors packed with granular activated carbon and "Manville" silica beads", *Water Research*, Vol. 38, pp. 3167–3178, 2004.

34. M. Majone, M. Beccari, S. Di Gregorio, D. Dionisi, and G. Vallini, "Enrichment of activated sludge in a sequencing batch reactor for polyhydroxyalkanoate production", *Water Science and Technology*, Vol. 54, pp. 119–128, 2006.

35. D. Dionisi, M. Majone, G. Vallini, S. Di Gregorio, and M. Beccari, "Effect of the applied organic load rate on biodegradable polymer production by mixed microbial cultures in a sequencing batch reactor", *Biotechnology and Bioengineering*, Vol. 93, pp. 76–88, 2006.

36. M.G.E. Albuquerque, C.A.V. Torres, and M.A.M. Reis, "Polyhydroxyalkanoate (PHA) production by a mixed microbial culture using sugar molasses: Effect of the influent substrate concentration on culture selection", *Water Research*, Vol. 44, pp. 3419–3433, 2010.

37. J.J. Beun, K. Dircks, M.C.M. Van Loosdrecht, and J.J. Heijnen, "Poly-[beta]-hydroxybutyrate metabolism in dynamically fed mixed microbial cultures", *Water Research*, Vol. 36, pp. 1167–1180, 2002.

38. D. Dionisi, M. Majone, V. Papa, and M. Beccari, "Biodegradable polymers from organic acids by using activated sludge enriched by aerobic periodic feeding", *Biotechnology and Bioengineering*, Vol. 85, pp. 569–579, 2004.

39. K. Johnson, Y. Jiang, R. Kleerebezem, G. Muyzer, and M.C.M. van Loosdrecht, "Enrichment of a Mixed Bacterial Culture with a High Polyhydroxyalkanoate Storage Capacity", *Biomacromolecules*, Vol. 10, pp. 670–676, 2009.

40. A.S.M. Chua, H. Takabatake, H. Satoh, and T. Mino, "Production of polyhydroxyalkanoates (PHA) by activated sludge treating municipal wastewater: effect of pH, sludge retention time (SRT), and acetate concentration in influent", *Water Research*, Vol. 37, pp. 3602–3611, 2003.

41. M. Villano, M. Beccari, D. Dionisi, S. Lampis, A. Miccheli, G. Vallini, and M. Majone, "Effect of pH on the production of bacterial polyhydroxyalkanoates by mixed cultures enriched under periodic feeding", *Process Biochemistry*, Vol. 45, pp. 714–723, 2010.

42. K. Johnson, J. van Geest, R. Kleerebezem, and M.C.M. van Loosdrecht, "Short- and long-term temperature effects on aerobic polyhydroxybutyrate producing mixed cultures", *Water Research*, Vol. 44, pp. 1689–1700, 2010a.

43. M.G.E. Albuquerque, S. Concas, S. Bengtsson, and M.A.M. Reis, "Mixed culture polyhydroxyalkanoates production from sugar molasses: The use of a 2-stage CSTR system for culture selection", *Bioresource Technology*, Vol. 101, pp. 7112–7122, 2010.

44. S. Bengtsson, A. Werker, M. Christensson, and T. Welander, "Production of polyhydroxyalkanoates by activated sludge treating a paper mill wastewater", *Bioresource Technology*, Vol. 99, pp. 509–516, 2008.

45. P.C. Lemos, L.S. Serafim, and M.A.M. Reis, "Synthesis of polyhydroxyalkanoates from different short-chain fatty acids by mixed cultures submitted to aerobic dynamic feeding", *Journal of Biotechnology*, Vol. 122, pp. 226–238, 2006.

46. L.S. Serafim, P.C. Lemos, R. Oliveira, and M.A.M. Reis, "Optimization of polyhydroxybutyrate production by mixed cultures submitted to aerobic dynamic feeding conditions", *Biotechnology and Bioengineering*, Vol. 87, pp. 145–160, 2004.

47. K. Johnson, R. Kleerebezem, and M.C.M. van Loosdrecht, "Influence of the C/N ratio on the performance of polyhydroxybutyrate (PHB) producing sequencing batch reactors at short SRTs", *Water Research*, Vol. 44, pp. 2141–2152, 2010.

48. D. Dionisi, M. Beccari, S. Di Gregorio, M. Majone, M. Petrangeli Papini, and G. Vallini, "Storage of biodegradable polymers by an enriched microbial community in a sequencing batch reactor operated at high organic load rate", *Journal of Chemical Technology & Biotechnology*, Vol. 80, pp. 1306–1318, 2005.

49. L.S. Serafim, P.C. Lemos, S. Rossetti, C. Levantesi, V. Tandoi, and M.A.M. Reis, "Microbial community analysis with a high PHA storage capacity", *Water Science and Technology*, Vol. 54, pp. 183–188, 2006.

50. A.S. Ciggin, D. Orhon, S. Rossetti, and M. Majone, "Short-term and long-term effects on carbon storage of pulse feeding on acclimated or unacclimated activated sludge", *Water Research*, Vol. 45, pp. 3119–3128, 2011.

51. K. Johnson, R. Kleerebezem, M.C.M. van Loosdrecht, "Influence of ammonium on the accumulation of polyhydroxybutyrate (PHB) in aerobic open mixed cultures", *Journal of Biotechnology*, Vol. 147, pp. 73–79, 2010.

52. D. Dionisi, G. Carucci, M. Petrangeli Papini, C. Riccardi, M. Majone, and F. Carrasco, "Olive oil mill effluents as a feedstock for production of biodegradable polymers", *Water Research*, Vol. 39, pp. 2076–2084, 2005.

53. M.G.E. Albuquerque, V. Martino, E. Pollet, L. Avérous, M.A.M. Reis, "Mixed culture polyhydroxyalkanoate (PHA) production from volatile fatty acid (VFA)-rich streams: Effect of substrate composition and feeding regime on PHA productivity, composition and properties", *Journal of Biotechnology*, Vol. 151, pp. 66–76, 2011.

54. D.H. Rhu, W.H. Lee, J.Y. Kim, and E. Choi, "Polyhydroxyalkanoate (PHA) production from waste", *Water Science and Technology*, Vol. 48, pp. 221–228, 2003.

55. H.Y. Liu, P.V. Hall, J.L. Darby, E.R. Coats, P.G. Green, D.E. Thompson, and F.J. Loge, "Production of polyhydroxyalkanoate during treatment of tomato cannery wastewater", *Water Environment Research*, Vol. 80, pp. 367–372, 2008.

56. L. Serafim, P. Lemos, M. Albuquerque, and M. Reis, "Strategies for PHA production by mixed cultures and renewable waste materials", *Applied Microbiology and Biotechnology*, Vol. 81, pp. 615–628, 2008.

57. D. Dionisi, M. Majone, G. Vallini, S. Di Gregorio, and M. Beccari, "Effect of the Length of the Cycle on Biodegradable Polymer Production and Microbial Community Selection in a Sequencing Batch Reactor", *Biotechnology Progress*, Vol. 23, pp. 1064–1073, 2007.

58. S.Y. Lee, and J.-i. Choi, J.-i., "Effect of fermentation performance on the economics of poly(3-hydroxybutyrate) production by Alcaligenes latus", *Polymer Degradation and Stability*, Vol. 59, pp. 387–393, 1998.

59. M. Niaounakis, and C.P. Halvadakis, Olive-mill Waste Management: Literature Review and Patent Survey, Typothito-GeorgeDardanos Publications, Athens, Greece, 2004.

60. FAOSTAT database, http://faostat.fao.org, 2002.

61. TCD Olive, By-product resusing from olive and olive oil production, "Setting up a network of technology dissemination centres to optimize SMEs in the olive and olive oil sector", 2004.

62. M. Hamdi, "Anaerobic digestion of olive mill wastewaters", *Process Biochemistry*, Vol.31, pp. 105-110, 1996.

63. S.M. Paixao, E. Mendonca, A. Picado, and A.M. Anselmo, "Acute toxicity evaluation of olive mill wastewaters: a comparative study of three aquatic organisms". *Environmental Toxicology*, Vol. 14, pp. 263–269, 1999.

64. A. Arpino, A. Lanzani, and C. Ruffo, Atti del II convegno sui trattamenti delle acque di scarico industriali, Roma, Italia, 1978.

65. S.P. Tsonis, V.P. Tsola, and S.G. Grigoropoulos, "Systematic characterization and chemical treatment of olive oil mill wastewater", *Toxicology Environmental Chemistry*, Vol. 20/21, pp. 437–457, 1987.

66. J.A. Fiestas Ros de Ursinos, Relazione Introduttiva al tema: "Processi Microbiologici". Seminario Internazionale sul trattamento delle acque reflue degli oleifici, Lecce, Italia, 1989.

67. M. Hamdi, "Toxicity and biodegradability of olive mill wastewaters in batch anaerobic digestion, *Applied Biochemistry and Biotechnology*, Vol. 37, pp. 155–163, 1992

68. R. Maestro-Duràn, R. Borja, A. Martín, J.A. Fiestas, and J. Alba, "Biodegradaciòn de los compuestos fenòlicos presentes en el alpechìn", *Grasas Aceites*, Vol. 42, pp. 271–276, 1991.

69. M. Servili, M. Baldioli, R. Selvaggini, E. Miniati, A. Machioni, and G. Montedoro, G., "High performance liquid chromatography evaluation of phenols in olive fruit, virgin oil, vegetation waters and pomace in 1D and 2D nuclear magnetic resonance characterization", *Journal of American Oil Chemists' Society*, Vol. 76, pp. 873–882, 1999.

70. H.K. Obied, M.S. Allen, D.R. Bedgood, P.D. Prenzler, K. Robards, and R. Stockmann, "Bioactivity and analysis f biophenols recovered from olive mill waste", *Journal of Agricultural and Food Chemistry*, Vol. 53, pp. 823–837, 2005.

71. R. Borja, A. Martìn, V. Alonso, I. Garcìa, and C.J. Banks, "Influence of different aerobic pretreatments on the kinetics of anaerobic digestion of olive mill wastewater", *Water Research*, Vol. 29, pp. 489–495, 1995.

72. M.M. Rodrìguez, J. Pèrez, A. Ramos-Cormenzana, and J. Martìnez, "Effect of extracts obtained from olive oil mill waste on *Bacillus megaterium* ATCC 33085", *Journal of Applied Bacteriology*, Vol. 64, pp. 219–222, 1988.

73. R. Capasso, G. Cristinzio, A. Evidente, and F. Scognamiglio, "Isolation spectroscopy and selective phytotoxic effects of polyphenols from vegetable waste waters", *Phytochemistry*, Vol. 31, pp. 4125–4128, 1992.

74. M. Kornaros, C. Zafiri, K. Stamatelatou, S.N. Dokianakis, and M.A. Dareioti, "Exploitation of olive mill wastewater and liquid cow manure for biogas production", *Waste Management*, Vol. 30, pp. 1841–1848, 2010.

75. T. Coskun, E. Debik, and N.M. Demir, "Treatment of olive mill wastewater by nanofiltration and reverse osmosis membranes", *Desalination*, Vol. 259, pp. 65–70, 2010.

76. M. Kornaros, C. Zafiri, K. Stamatelatou, M.A. Dareioti, and S.N. Dokianakis, S.N., "Biogas production from anaerobic co-digestion of agro-industrial waste-waters under mesophilic conditions in a two-stage process", *Desalination*, Vol. 248, pp. 891–906, 2009.

77. A. Dhouib, T. Mechichi, S. Sayadi, and H. Chakroun, "High level of laccases production by Trametes trogii culture of olive mill wastewater-based media, application in textile dye decolorization", *Journal of Chemical Technology and Biotechnology*, Vol. 84, pp. 1527–1532, 2009.

78. M. Goumenaki, N. Christopoulou, I. Angelidaki, D. Georgakakis, and J. Gelegenis, "Optimitazion of biogas production from olive-oil mill waste-water, by codigesting with diluted poultry-manure", *Applied Energy*, Vol. 84, pp. 646–663, 2007.

79. F. Boubaker, and B. Cheikh Ridha, "Anaerobic co-digestion of olive mill wastewater with olive mill solid waste in a tubular digester at mesophilic temperature", *Bioresource Technology*, Vol. 98, pp. 769–774, 2007.

80. J. Gonzalez, M. Lopez, L. Sbai, and M. Mebirouk, "Olive oil mill wastewater pollution abatement by physical treatments and biodegradation with pha-nerochaetae chrysosporium", *Environmental Technology*, Vol. 27, pp. 1351–1356, 2006.

81. S. Sayadi, N. Hamad, F. Aloui, and A. Dhouib, "Pilot-plant treatment of olive mill wastewaters by Phanerochaete chrysosporium coupled to anaero-bic digestion and ultrafiltration", *Process Biochemistry*, Vol. 41, pp. 159–167, 2006.

82. J. Beltran de Heredia, and J. Garcia, "Process integration: continuous anaer-obic digestion-Ozonation treatment of olive mill wastewater", *Industrial & Engineering Chemistry Research*, Vol. 44, pp. 8750–8755, 2005.

83. B.Y. Ammary, "Treatment of olive mill wastewater using an anaerobic sequencing batch reactor", *Desalination* Vol. 177, pp. 157–165, 2005.

84. B. Ceccanti, G. Masciandaro, and S. Filidei, "Anaerobic digestion of olive oil mill effluents: evaluation of wastewater organic load and phytotoxicity reduction", *Water, Air and Soil Pollution*, Vol. 145, pp. 79–94, 2003.

85. G. Lyberatos, G.G. Aggelis, M.E. Kornaros, S.N. Dokianakis, and M.S. Fountoulakis, "Removal of phenolics in olive mill wastewater using the white-rot fungus Pleurotus ostreatus", *Water Research*, Vol. 36, pp. 4735–4744, 2002.

86. M.D. Gonzalez, E. Moreno, J. Quevedo, and A. Ramos, "Studies on antibacte-rial activity of waste waters from olive oil mill (alpechin): inhibitory activity of phenolic and fatty acids", *Chemosphere*, Vol. 20, pp. 423–432, 1990.

87. M. Demicheli, and L. Bontoux, Survey current activity on the valorization of by-products from the olive oil industry. European Commission Joint Research Centre, Final Report, www.jrc.es/projects/ff/EC/IPTS/IPTSPUBL.html, 1996.

88. A. Samsunlu, O. Tünay, Z. Öztürk, and K. Alp, Characterization and treatabil-ity of olive oil wastewaters (in Turkish), I.T.Ü. 6. Industrial Pollution Symp. Proc., 93–99, Istanbul (Turkey), 1998.

89. Improlive, http://www.fiw.rwth-aachen.de/improlive/rsanfall/abwasser/anaerob.html, 2000

90. C.J. Israilides, A.G. Vlyssides, V.N. Mourafeti, and G. Karvouni, "Olive oil wastewater treatment with the use of an electrolysis system", *Bioresource Technology*, Vol. 61, pp. 163–170, 1997.

91. Olive mill water treatment through vacuum evaporation, Technical report 7, LED Italy s.r.l., 1997.

92. E. Kasirga, "Treatment of olive oil industry wastewater by anaerobic stabilization method and development of kinetic model (in Turkish)", Unpublished PhD thesis, Dokuz Eylul University, Graduate School of Natural and Applied Sciences, Izmir, Turkey, 1988.

93. E.A. Duarte and I. Neto, "Evaporation phenomenon as a waste management technology", *Water Science & Technology*, Vol. 33, pp. 53–61, 1996.

94. N. Zouari and R. Eloouz, "Toxic effects of colored olive compounds on the anaerobic digestion of olive mill effluents in UASB-like reactors", *Journal of Chemical Technology & Biotechnology*, Vol. 66, p. 414, 1999.

95. M.C. Annesini and F. Gironi, "Olive oil mill effluent: ageing effect on evaporation behaviour", *Water Research*, Vol. 25, p.1157, 1991.

96. A. Aveni, "Biogas recovery from olive-oil mill wastewater by anaerobic digestion", *Anaerobic Digestion and Carbohydrate Hydrolysis of Waste*, pp. 489–491, 1984.

97. C. Carrieri, V. Balice, R. Rozzi, Comparison of three anaerobic treatment processes on olive mill effluents, Proc Int Conf Environmental Protection, Italy, 1988.

98. I. Ozturk, S. Sakar, G. Ubay, and V. Eroglu, "Anaerobic treatment of olive oil mill effluent", Proc. 46th Industrial Waste Conf., Purdue University, West Lafayette, IN, 1991.

99. I.P. Marques, A. Teixeira, L. Rodrigues, S. Martins Dias, J.M. Novais, "Anaerobic treatment of olive oil mill wastewater with digested piggery effluent", *Water Research*, Vol. 70, p. 1056, 1998.

100. M. Beccari, F. Bonemazzi, M. Majone, and C. Riccardi, "Interaction between acidogenesis and methanogenesis in the anaerobic treatment of olive oil mill effluents", *Water research*, Vol. 30, pp. 183–189, 1996.

101. M. Beccari, G. Carucci, M. Majone, and L. Torrisi, "Role of the lipids and phenolic compounds in the anaerobic treatment of olive oil mill effluents", *Environmental Technology*, Vol. 20, pp. 105–110, 1999.

102. M. Beccari, M. Majone, C. Riccardi, F. Savarese, and L. Torrisi, "Integrated treatment of olive oil mill effluents: effect of chemical and physical pretreatment on anaerobic treatability", *Water Science & Technology*, Vol. 40, pp. 347–355, 1999.

103. M. Beccari, M. Majone, M. Petrangeli Papini, and L. Torrisi, "Enhancement of anaerobic treatability of olive oil mill effluents by addition of $Ca(OH)_2$ and bentonite without intermediate solid/liquid separation", Proceeding of the First World Congress of the International Water Association, Paris, 2000.

104. M. Beccari, M. Petrangeli Papini, M. Majone, and L. Torrisi, "Enhancement of anaerobic treatability of olive oil mill effluents by addition of $Ca(OH)_2$ and bentonite without intermediate solid/liquid separation", *Water Science & Technology*, Vol. 43, pp. 275–282, 2001.

105. M. Beccari, G. Carucci, A.M. Lanz, M. Majone, and M. Petrangeli Papini, "Fate of the phenolic fractions in an integrated treatment of olive oil mill effluents (adsorption on bentonite, methanogenesis, activated sludge post-treatment)", in AIDIC (Ed), Proceedings of IcheaP-5, the 5th Italian Conference on Chemical and Process Engineering, Florence, Vol 1, pp. 357–362, 2001.

106. M. Beccari, G. Carucci, A.M. Lanz, M. Majone, and M. Petrangeli Papini, "Removal of molecular weight fraction of cod and phenolic compounds in an integrated treatment of olive oil mill effluents", *Biodegradation*, Vol. 13, pp. 401–410, 2002.

107. E.S. Aktas, S. Imre, and L. Ersoym, "Characterization and lime treatment of olive mill wastewater", *Water Research*, Vol. 35, pp. 2336–2340, 2001.

108. A.G. Vlyssides, M. Loizides, and P.K. Karlis, "Intergrated strategic approach for reusing olive oil extraction by-products", *J Cl Pro*, Vol. 12, pp. 603–611, 2004.

109. N. Azbar, A. Bayram, A. Filibeli, A. Muezzinoglu, F. Sengul, and A. Ozer, "A review of waste management options in olive oil production, *Critical Reviews in Environmental Science & Technology*, Vol. 34, pp. 209–247, 2004.

110. G. Lyberatos, I. Ntaikou, C. Kourmentza, E.C. Koutrouli, K. Stamatelatou, A. Zampraka, and M. Kornaros, "Exploitation of olive oil mill wastewater for combined biohydrogen and biopolymers production", *Bioresource Technology*, Vol. 100, pp. 3724–3730, 2009.

111. R.J. Sanchez, J. Schripsena, L.F. Da Silva, M.K. Taciro, J.G.C. Pradella, and J.G.C., Gomez, J.G.C., "Medium-chain-length polyhydroxyalkanoic acids (PHA) produced by Pseudomonas putida IPT 046 from renewable sources", *European Polimer Journal*, Vol. 39, pp.1385–1394, 2003.

112. S. Cardoso Diniz, M. Keico Taciro, J.G. Cabrera Gomez, and J.G. Da Cruz Pradella, "High-cell-density cultivation of Pseudomonas putida IPT 046 and mediumchain-length polyhydroxyalkanoate production from sugarcane carbohydrates", *Applied Biochemistry & Biotechnology—Part A Enzyme Engineering Biotechnology*, Vol. 119, pp. 51–69, 2004.

113. S.Y. Lee, H.H. Wong, J. Choi, S.H. Lee, S.C. Lee, and C.S. Han, "Production of medium-chain-length polyhydroxyalkanoates by high-cell-density cultivation of Pseudomonas putida under phosphorus limitation", *Biotechnology & Bioengineering*, Vol. 68, pp. 466–470, 2000.

114. M.B. Kellerhals, B. Kessler, B. Witholt, A. Tchouboukov, and H. Brandl, "Renewable long-chain fatty acids for production of biodegradable mediumchain-length polyhydroxyalkanoates (mcl-PHAs) at laboratory and pilot scales", *Macromolecules*, Vol. 33, pp. 4690–4698, 2000.

115. J. Gonzalez-Lopez, C. Pozo, M.V. Martinez-Toledo, B. Rodelas, and V. Salmeron, "Production of polyhydroxyalkanoates by *Azotobacter chroococcum* H23 in wastewater from olive oil mills (alpechin)", *International Biodeterioration & Biodegradation*, pp. 271–276, 1996.

116. L. Bertin, S. Lampis, D. Todaro, A. Scoma, G. Vallini, L. Marchetti, M. Majone, and F. Fava, "Anaerobic acidogenic digestion of olive mill wastewaters in biofilm reactors packed with ceramic filters or granular activated carbon", *Water Research*, Vol. 44, pp. 4537-4549, 2010.

5

Atom Transfer Radical Polymerization (ATRP) for Production of Polymers from Renewable Resources

Kattimuttathu I. Suresh

Organic Coatings & Polymers Division,
Indian Institute of Chemical Technology
Hyderabad, India

Abstract

The past 15 years have witnessed the development of atom transfer radical polymerization (ATRP) as a versatile living radical polymerization technique for the precision synthesis of polymers with defined molecular weight, polydispersity, topology, composition and functionality. In this chapter a brief review of atom transfer radical polymerization (ATRP) as applied to monomers from renewable resources is presented. The general considerations of ATRP along with a few case studies have been reviewed. The polymers based on renewable monomers in fact display unique properties that are not achievable in conventional petroleum based monomers, but it appears that a gap still exists, particularly in structure–property correlation and material property studies governing the design of such systems for specific applications. Within this contribution, an overview of the field by discussing the current state-of-the-art as well as selected perspectives of the utilization of renewable resources as monomer for ATRP in polymer science. This technique offer most challenges as well as opportunities in the future efforts towards a sustainable chemical industry.

Keywords: Atom transfer radical polymerization, molecular weight, polydispersity, topology, structure–property correlation, renewable monomers

5.1 Introduction

The development of many synthetic polymers that are used today may be traced to the materials demand during the Second World

Vikas Mittal (ed.) Renewable Polymers, (221–246) © Scrivener Publishing LLC

War. Eventhough, renewable resources were used in polymer production for a very long time the developments during the 1990's like the depleting and unstable fossil reserves, approaches for climate protection, secure supply of raw materials and employment, created a surge in the growth of a sustainable chemical industry based on utilization of renewable resources. The utilization of renewable raw materials is also promoted by the green chemistry initiatives. Today, the chemical industry employs a wide spectrum of renewable raw materials like plant oils, polysaccharides, sugars for preparation of fine chemicals and polymers.

The most important renewable raw materials for the chemical industry are plant oils, used for surfactants, cosmetic products, and lubricants. Unmodified vegetable oils have been used to prepare biorenewable polymers by thermal or cationic polymerization methods, taking advantage of the carbon-carbon double bonds in the fatty acid chain. Modified vegetable oils with double bonds exhibit higher reactivities and can undergo free radical polymerization to afford thermosets with good thermal and mechanical properties. Various types of polymers like polyurethanes, epoxy resins, polyesters, polyamides etc are already produced from renewable resources and these have been reviewed separately [1, 2].

The Vegetable oils extracted from various plants contain triglycerides formed between glycerol and various fatty acids possessing unsaturation in the side chain. These renewable raw feed stocks require further processing like separation and chemical treatment to obtain the desired monomers for functional polymers. The methodology to obtain functional polymers from these feedstocks could thus be summarized as in Scheme 5.1.

It means that a combination of chemistry and material science approaches are often needed to make value added functional materials from renewable resources. The gross energy requirement (GER) for these processes is still less than that of polymers from petroleum feedstocks and thus, environment friendly [1].

5.2   Atom Transfer Radical Polymerization (ATRP)

Though radical polymerization is the widely used technique to prepare polymer materials, the unwanted side reactions like irreversible chain transfer or chain termination limit the control of the radical polymerization process. The living and controlled polymerization techniques minimize the unwanted side reactions and facilitate the preparation of well defined materials. There are a number

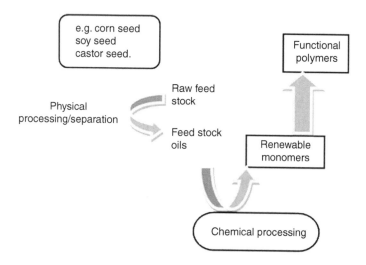

Scheme 5.1 General scheme for preparation of functional polymers from renewable resources.

of excellent reviews available on the topic [3, 4, 5]. The main feature of these techniques is that they are versatile, applicable to a large number of monomers and, tolerant to the presence of functional groups [3]. ATRP is the most frequently reported controlled polymerization technique applied to renewable monomers, whereas the other controlled radical polymerization techniques, to the best of our knowledge, have not yet been applied to such monomers.

Similar to conventional radical polymerizations the living methods are also characterized by steps of initiation and propagation for chain growth [4]. For living methods the rate of chain initiation (k_i) is fast compared to the rate of chain propagation (k_p). (Generally, a ratio of $k_i/k_p \sim 10$ is required in order to obtain polymers with a narrow molecular weight distribution). These conditions are mainly observed for the common radically polymerizable monomers like styrene or acrylates. Coming to the case of renewable monomers they have a number of structural features, which makes them different from these monomers. We will address some of these features which govern the catalyst selection when the ATRP of renewable monomers is considered. Until now very little is reported in the literature concerning living or controlled polymerization techniques particularly, atom transfer radical polymerization (ATRP) of renewable monomers in the literature. Considering the multitude of materials that can be made with ATRP [5], there exists opportunities for preparation of many functional materials with the unique performance properties using renewable resources. The fundamentals of ATRP are presented first followed by a discussion about the technique applied to renewable monomers.

5.2.1   General Considerations

Atom transfer radical polymerization (ATRP) was developed in the last decade, almost simultaneously but independently by Sawamoto [6] and Matyjaszewski [7–9]. It has evolved into one of the most successful methods for the synthesis of polymers with controlled molecular weight, molecular weight distribution and composition, without having stringent demands on water- and oxygen-removal that are necessary for other types of living polymerizations [10, 3]. The chemistry of ATRP is also tolerant to many functional groups, thus permitting the controlled synthesis of a broad range of functional polymers [3, 11]. The control results from a fast exchange reaction between propagating radicals and a dormant species [cf. Scheme 5.2], thereby the concentration of propagating radicals is kept sufficiently low, suppressing termination and transfer reactions.

The ATRP technique has been applied to a large variety of radical polymerizations of monomers, including styrene, styrene derivatives, acrylates, methacrylates [3]. The technique has been used in the controlled synthesis of cross linkable styrene-allyl methacrylate copolymers [12], and other well defined acrylic polymers and their potential applications in coatings is of high interest now a days [13, 14].

Basically, all ATRP systems are composed of monomer, initiator and the catalyst (a transition metal and a suitable ligand). The initiator is typically an alkyl halide (RX). The halide is usually bromide or chloride, although iodide based initiators have also been reported [15]. Examples of ATRP initiators include carbon tetrachloride, chloroform, benzyl halides and α-halo esters [3]. The main requirement is that the initiator must have a halogen attached to an atom containing radical stabilizing substituent. As mentioned

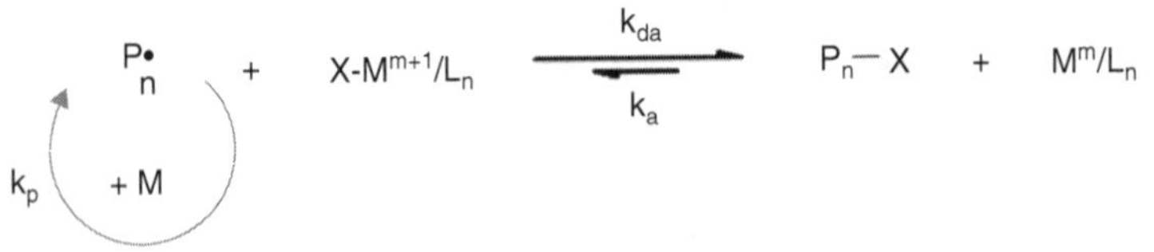

$$P_n^\bullet \; + \; X\text{-}M^{m+1}/L_n \; \underset{k_a}{\overset{k_{da}}{\rightleftharpoons}} \; P_n\text{-}X \; + \; M^m/L_n$$

$$k_p \quad + M$$

Scheme 5.2 General ATRP mechanism M; Transition metal, P: Polymer chain; L: complexing ligand; X = Br or Cl.

earlier, the initiation step must be faster than or equal to the propagation rate for a controlled polymerization [16]. Several transition metals have been used in ATRP. Copper is by far the most common metal, due to its versatility in ATRP and relatively low cost.

The metals are used in conjunction with a large variety of ligands. The ligands are an important part of the ATRP system and its role is three-fold. Firstly, the ligands solubilize the metal in the organic media. Secondly, they control selectivity by steric and electronic effects. Finally, by their electronic effects, they also affect the redox chemistry of the final metal complex [17]. Amines are the most important classes of ligands studied in ATRP [3]. Some examples are given in Figure 5.1.

The polymerization mechanism is based on a reversible halogen transfer between the dormant species (P_n-X) and the transition metal/ligand complex (M_m/L_n), resulting in the formation of propagating active species ($P_n\bullet$) and the metal complex in a higher oxidation state with an additionally coordinated halide ligand ($X-M^{m+1}/L_n$). This dynamic and rapid equilibrium minimizes the concentration of free radicals in the system and therefore the probability of radical termination reactions, and provides an equal opportunity of propagation to all polymer chains via the frequent inter conversion

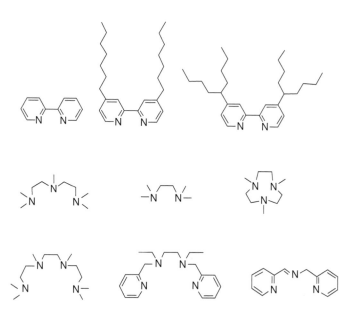

Figure 5.1 Structures of some common ligands used in copper catalyzed ATRP.

Initiation

$$R-X + Cu(I)X\,/\,ligand \overset{k_{eq}}{\leftrightarrow} R\bullet \;+\; X-Cu(II)X\,/\,ligand$$
$$(X = Cl,\,Br)$$

Propagation
$$R\bullet + monomer \overset{k_p}{\to} Pn\bullet$$

$$PnX + Cu(I)X\,/\,ligand \overset{k_{eq}}{\leftrightarrow} Pn + \; X-Cu(II)X\,/\,ligand$$

$$Pn + monomer \overset{k_p}{\to} P_{n+1}$$

Termination
$$P_n\bullet \;+ P_m\bullet \overset{k_t}{\to} P_{n+m}$$

Scheme 5.3 The mechanistic steps of ATRP [18].

between the active and the dormant species. As a result, polymers with targeted molecular weights and narrow molecular weight distributions (MWD) can be obtained by such techniques as shown in Scheme 5.3.

The polymerization takes place basically in two steps: initiation and propagation. In Figure 5.3 the mechanism is exemplified with copper(I) as catalyst. Termination reactions also occur, but no more than a few percentages of the growing chains undergo termination in ATRP [3].

5.2.2 Kinetics of ATRP

Based on the reaction scheme shown above (cf. Scheme 5.3), Matyjaszewski has described the ATRP kinetics by the following equations: [19]

$$K_{eq} = \frac{k_{act}}{k_{deact}} = \frac{[P][Cu(II)X_2]}{[Cu(I)X][PX]} \tag{1}$$

$$R_p = k_{app}[M] = k_p[P][M] = K_p K_{eq}[RX]\frac{[Cu(I)X]}{[Cu(II)X_2]}[M] \tag{2}$$

In deriving Eqn. (2) it is assumed that the termination step can be neglected and that a fast pre-equilibrium is established and hence that the propagation rate constant, k_p, is constant throughout the reaction. The polymerization rate (R_p) is first-order with respect to monomer concentration [M], initiator concentration [RX], and activator concentration [Cu(I)X] [19]. However, the reaction is not simply negative first-order with respect to deactivator concentration [Cu(II)X2]. This is due to the persistent radical effect, the irreversible formation of Cu(II)X2 in the initial stages of the polymerization [20, 21].

5.2.3   Macromolecular Architecture

The search for new, polymeric materials has traditionally focused on the chemical composition of linear and cross linked polymers. However, the architecture of a polymer will have a large influence on its final properties. Polymers with the same repeating unit can have very different physical properties depending on its structure. Therefore, the properties of a polymer can be tailored, not only by changing its chemical composition, but also by changing its architectural composition. The main feature of ATRP is that it allows synthesis of complex macromolecular architectures.

5.2.4   Choice of Reaction Medium

One of the factors that contributed to the versatility of the ATRP process is that the reaction may be carried out in homogeneous or heterogeneous aqueous medium or on solid supports. Details of such systems are summarized in several review papers [5]. To conduct ATRP in emulsion efficient transport of the monomer and catalyst from large monomer droplets to the micelles through the aqueous phase is required. There have been successful attempts to prepare the block and star polymers by ATRP in aqueous systems. The conditions have been now optimized for conducting ATRP in aqueous homogeneous systems. These techniques are advantageous when direct applications of the polymer in adhesives or coatings are desired. It was also observed that addition of small amount of water enhances the polymerization rate of hydrophobic monomers, thus advantageous to the commercial application.

5.3   Synthetic Strategies to Develop Functional Material Based on Renewable Resources - Composition, Topologies and Functionalities

Atom transfer radical polymerization (ATRP) has been effectively applied to the preparation of polymers with precisely controlled functionalities, topologies, and compositions [22, 23]. Since the technique is tolerant of functional groups. The straightforward introduction of various functionalities into the polymer structure is possible. For the synthesis of telechelic polymers with functional groups via ATRP [24] there are four major strategies.

5.3.1   Use of Functional Initiators

ATRP uses mainly alkyl halides R-X (X = Cl, Br) as initiators [25–27]. The number-average molecular weight (Mn) of polymers prepared by ATRP depends on the initial concentration ratio of monomer (M) to initiator as well as the monomer conversion through the following relation:

$$M_n = ([M]_0 / [RX]_0) \times \text{conversion} \times MW(M).$$

The alkyl halides used as initiators contain one or more halogen atoms. Depending on the exact initiator structure and the number of halogen atoms, the architecture of the prepared polymers can be varied from linear (using alkyl halides with a single halogen atom), to star-like or brush-like (multiple halogen atoms in the initiator). Star polymers can be generated using initiators with alkyl halide groups attached to a single core. Examples of some of the typical functional initiators are presented in (Figure 5.2).

There are also other methods of functionalization like 1. End group modification and 2. Polymerization of "protected" monomers, followed by post-polymerization chemical transformations. The main advantages of using functional initiators in the synthesis of polymers via ATRP are that of (a) direct functionalization, (b) no

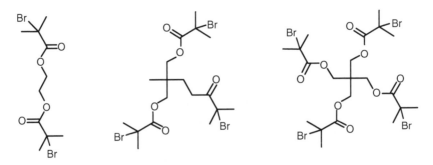

Figure 5.2 Examples initiators yielding polymeric stars of ATRP.

post-polymerization modification necessary (c)yields α-telechelic polymers and of ω-end halogen atom.

There are also alkyl halide functional initiators that yield end-functionalized polymers (See Figure 5.3). Groups such as hydroxyl in the structure of the ATRP initiator (**1**) are suitable for the synthesis of polymers that can react with molecules or surfaces with carbox-ylic acid / isocyanate groups. Polymers synthesized with initiators containing azide or alkyne groups are able to participate in click chemistry-type functionalization (**2 & 6**). Allyl group-containing initiators (**3**) yield polymers that can undergo hydrosilylation or thiolene reactions with molecules or surfaces containing Si-H or S-H bonds, respectively. Trichlorosilyl groups (**4**) react with sur-faces containing hydroxy or amine groups (including Si-OH bonds, such as those on the surface of silica particles or glass). Finally, disulfide-containing difunctional initiator (**8**) yield polymers con-taining a functional group able to react with gold surfaces, and gives the ability to degrade in reducing environments. It should be noted that polymers prepared by ATRP contain two chain ends: the α-end derived from the initiator and the ω-end, which is normally

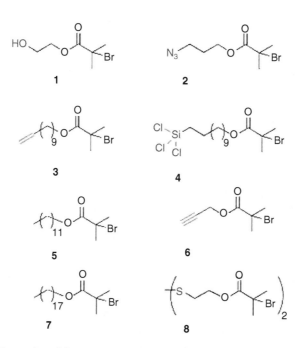

Figure 5.3 Examples of functionalized alkyl halide initiators for ATRP.

bromide or chloride. Alkyl halides (**7** and **5**) can participate in a number of nucleophilic substitution reactions, which expands the types of end-functional polymers accessible through ATRP [27].

5.3.2 Modified Processes

Classical ATRP method use large amounts of the CuX / ligand for achieving sufficient rate of polymerization and control [28, 7]. Thus, the synthesized polymers require tedious purification to remove the catalyst as it is detrimental to the properties of the synthesized polymers. Although various methods for catalyst removal have been developed, the additional purification step and the longer time needed to obtain the final product, adds to the cost and chemical wastage [29]. To overcome these difficulties different variants of the basic ATRP process have been developed like Activator Re-Generated by Electron Transfer (ARGET) [30, 31] and Continuous Activator Regeneration (ICAR) [32]. These processes decrease the amount of catalyst needed to only a few (often single-digit) ppm by using very active ligands like tris[2-(dimethylamino) ethyl]amine (Me_6TREN) and tris(2-pyridylmethyl)amine (TPMA). Both these techniques employ a reducing agent, a radical initiator such as AIBN [36] in ICAR ATRP or ascorbic acid [33], or Cu(0) [34], in ARGET ATRP.

The ARGET and ICAR ATRP processes allow chemists to reduce the amount of catalyst more than one thousand times and the polymers obtained are nearly colorless. These processes also allow preparation of well-defined block copolymers [35], polymers with high molecular weight [36], high chain-end functionality and adjustable molecular weight distribution [37]. In addition, since the ARGET and ICAR ATRP can be performed with the large excess of reducing agent, the reaction can be successfully carried out in the presence of limited amounts of air [38].

Thus, ATRP is one of the most powerful polymer synthetic methods, which allows control over molecular architecture, as evidenced by the large number of patent applications and research papers published annually. Due to developments of very active ligands and advancements in initiation processes (ARGET and ICAR ATRP), it is easy to perform any ATRP polymerization reaction and the purification of the final products is now simplified, with minimum amount of waste generation.

5.4   Sustainable Sources for Monomers with a Potential for Making Novel Renewable Polymers

Though the conventional monomers like styrene, methacrylates and acrylates, have been polymerized successfully in a controlled fashion by ATRP homopolymerization or copolymerization, in some cases like vinyl pyridine requires special conditions for the successful conduct of ATRP. When applied to renewable monomers there have been some limitations and proper reaction conditions have to be selected [39], which makes it possible even performing the reaction under ambient conditions. The major requirement for a monomer to be useful in ATRP is the presence of reactive acrylate or methacrylate groups. This limits the number of renewable monomers directly useful in ATRP. Nevertheless, it is possible that these monomers can be derivatized to acrylates or methacrylates through simple chemical transformations. Renewable monomers posses a number of unique functionality that can be modified to acrylate or methacrylate group through simple chemical transformations. The presence of multiple functionality in the renewable monomers facilitates preparation of polymers with novel properties. In the following sections we discuss briefly some of the potential renewable monomers or their derivatives polymerized by ATRP.

5.4.1   Plant Oil Derived Monomers –Fatty Acid Acrylates/ Methacrylates

Plant oils are basically triglycerides (tri-esters of glycerol with long- chain fatty acids) with varying composition of saturated and unsaturated fatty acids depending on the plant, the crop, the season, and the growing conditions [40]. The structures of the common fatty acid present in different vegetable oils are summarized in Figure 5.4 [1]. It is this structural combination that offers renewable monomers multitude of additional opportunities for chemical modifications, cross-linking or other derivatization. The direct polymerization of plant oils by radical techniques generally results in the formation of cross linked products and several methods for the (co)-polymerization of triglycerides and fatty acid derivatives have also been developed. The prominent examples for the direct

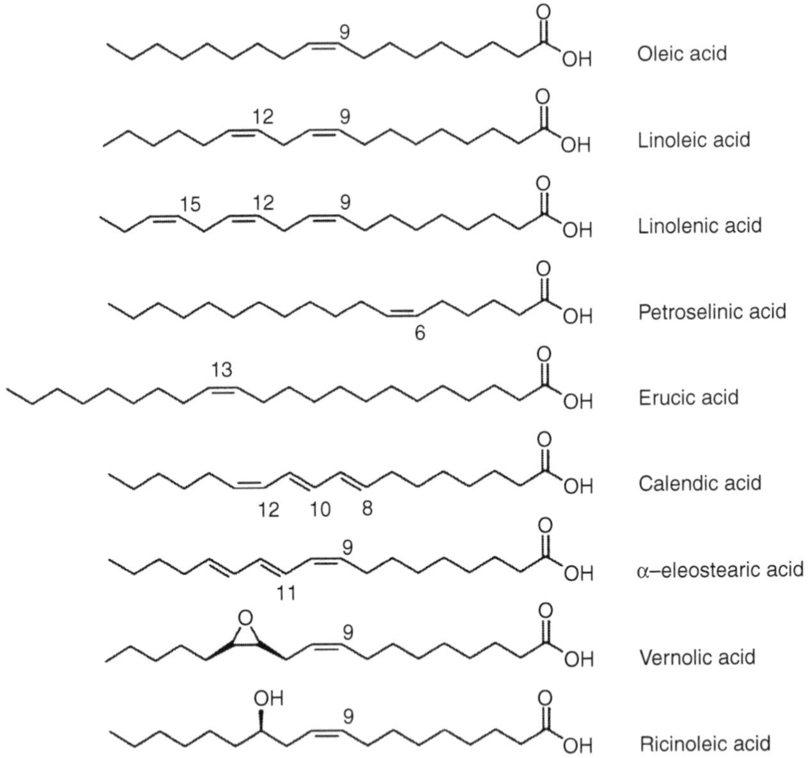

Figure 5.4 Structure of some of the common fatty acids present in vegetable oils [1]. It is worth mention that not much is reported about the ATRP of many of the fatty acid derivatives and thus potentially important for further studies.

utilization of plant oils include polyester, alkyd resins, polyure-thanes, polyamides, epoxy resins, and vinyl copolymers and has been discussed in the literature in detail [1]. Acrylates and methac-rylates of lauric acid and stearic acid have been some of the mostly studied renewable monomers in radical polymerization.

5.4.2   Monomers Prepared Through Derivatization of Vegetable Oils

Acrylates and methacrylate based on vegetable oil have been widely studied in ATRP. Mandal *et al.* developed a method for the ATRP of lauryl methacrylates under ambient conditions and obtained narrowly distributed polymers with pre- defined molecular weights. Lauryl methacrylate (LMA) is one of the most frequently studied long-chain alkyl methacrylates by various groups [39, 40, 41]. Linear poly(lauryl methacrylate) (PLMA) is an interesting material since side chains with 12 C-atoms or longer start to show side-chain crystallinity [42]. It is well known that the monomers with long alkyl chain lower the glass transition temperatures (Tg) if they are copolymerized with other monomers. These are preferred features for thermoplastic elastomers as shown for an ABA triblock copolymers of LMA and methyl methacrylate (MMA) [43].

ATRP on the other hand is, to the best of our knowledge, the only studied controlled radical polymerization technique for the polymerization of LMA, eventhough other methods like anionic polymerization has earlier been reported. For instance, the ATRP of LMA in a variety of solvents applying CuCl/N,N,N',N',N''-pentamethyl diethylenetriamine (PMDETA) as the catalyst has been investigated [44].

The preparation of high-DP (500) PLMA polymers with moderate PDI has been reported by Meier *et al.* as presented in Figure 5.5.

These examples show that it is possible to obtain materials with different physical properties and of different application possibilities from fatty alcohols by taking advantage of the synthetic potential of nature.

In the bulk ATRP of lauryl methacrylate (LMA) [41] Meier *et al.* observed that increasing the amount of ligand allows performing the polymerization of LMA with improved kinetics, leading to the synthesis of defined PLMA with targeted molecular weights and narrow polydispersity in reduced reaction times. Moreover, the polymerizations could be performed in bulk at a moderate temperature of 35°C, and high conversions above 90% were obtained with the optimized system. Subsequently, these authors have also investigated ATRP of methacrylates of different chain lengths that are

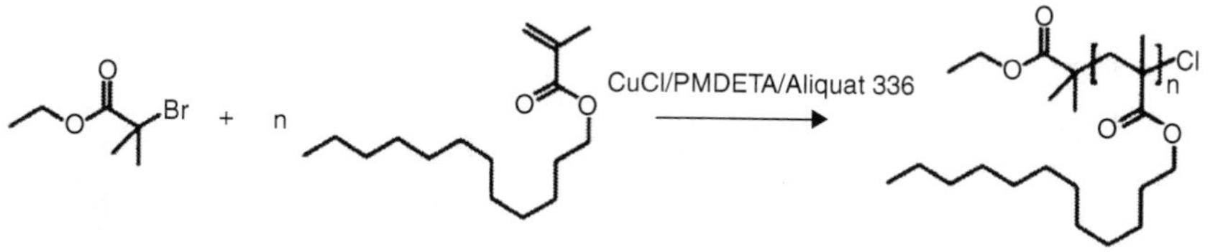

Figure 5.5 Reactions for ATRP synthesis of PLMA via ATRP [41].

derived from fatty alcohols, to establish that these monomers also can be polymerized in a controlled that fashion. It was suggested polymerization of defined polymers from fatty alcohols with different chain lengths, result in polymers with a combination of physical properties.

5.4.3 Block Co-Polymers Based on Renewable Monomers

Block copolymers are considered important materials for many nanotechnological applications. Triblock copolymers with an elastomeric center block and glassy end blocks display the properties of thermoplastic elastomers (TPE), the well-known examples being the commercially successful SBS or SIS triblock copolymers (S = styrene, B = butadiene, and I = isoprene). Such systems possess a multiphase morphology with glassy PS domains dispersed in a continuous elastomeric matrix. The glassy microdomains act as physical cross-links as well as reinforcing fillers so that attractive thermo mechanical properties are obtained without vulcanization and extraneously added fillers [45–46]. However, they have the drawback that the diene center blocks are susceptible to atmospheric oxidation and UV degradation which are overcome by the hydrogenation of the center blocks [47]. But this adds to the cost. Some attempts have been reported to high-light the unique properties of renewable block copolymers.

A facile process for the ATRP of higher alkyl methacrylates, e.g., lauryl methacrylate (LMA), at ambient temperature was developed by Mandal [48], using which the synthesis of the triblock copolymer PMMA-*b*-PLMA-*b*-PMMA (further abbreviated as MLM) was conveniently carried out. The T_g of PLMA being ca. –65°C and that of PMMA ca. 109°C. They observed that cuprous halide/linear alkylamine catalysts are solubilized in MMA at ambient temperature by catalytic amounts of quaternary ammonium halides. The chloride-based catalyst CuCl/PMDETA/AQCl (AQCl) Aliquat®336 (tricaprylyl methyl ammonium chloride) used with ethyl 2-bromoisobutyrate (EBiB) initiator provides excellent control and that too at ambient temperature which is attractive due to lower side reactions (hence better product) and cost advantage.

Mandal and Chatterjee [48] have also reported on the triblock copolymer morphology and its effect on the material properties. Saturated TPEs with poly (alkyl acrylate) center blocks and poly (methyl methacrylate) (PMMA) end blocks (MAM) attracted interest in this regard but unfortunately they were found to be inferior to the diene based TPEs as regards mechanical properties. The cause of this deficiency was traced to the high molecular weight between entanglements (M_e) of the polyacrylates compared to the polydienes as determined from dynamic mechanical analysis. These drawbacks may be overcome with suitable choice of the elastomeric block.

There are some other new monomers prepared from plant oils and polymerized by other techniques. These could be potential monomers for ATRP. For e.g. homo- and co-polymers of bromoacrylated methyl oleate (BAMO) polymers were prepared by thermal as well as photoinitiated free radical polymerization to obtain polymers with number averaged molecular weights (M_n) from 20 to 35 kDa [49]. They could be useful in the preparation of flame retardant polymers based on a renewable feed stock. Similar monomers, such as acrylated methyl oleate (AMO) as well as methyl oleate acrylamide (MOA), were obtained by acrylation of epoxidized methyl oleate [50] or the Ritter reaction of methyl oleate and acrylonitrile in the presence of $SnCl_4$ respectively, which could be successfully utilized in ATRP.

5.4.4   Cardanol – the Phenolic Lipid

Another unique monomer deserving special mention, though not a triglyceride oil is the cashew nut shell liquid obtained as a byproduct of the cashew processing industry. The main constituent of technical grade cashew nut shell Liquid is Cardanol (>97%). The major countries producing CNSL are India, Phlippines, Brazil, Indonesia etc.

Cardanol (cf. Figure 5.6), the main constituent of thermally treated cashew nut shell liquid is a valuable phenolic renewable resource

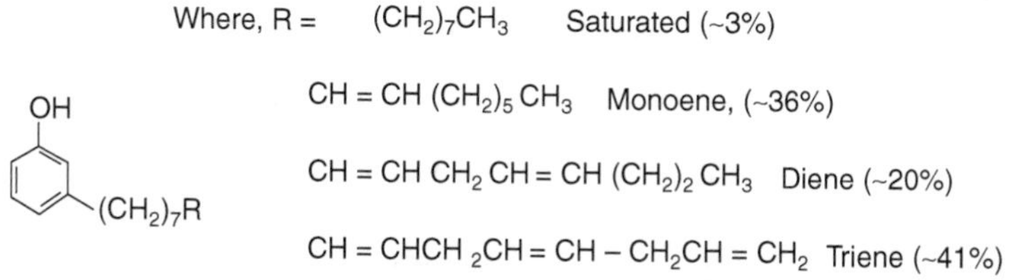

Where, R = $(CH_2)_7CH_3$ Saturated (~3%)

$CH = CH (CH_2)_5 CH_3$ Monoene, (~36%)

$CH = CH CH_2 CH = CH (CH_2)_2 CH_3$ Diene (~20%)

$CH = CHCH_2CH = CH – CH_2CH = CH_2$ Triene (~41%)

Figure 5.6 Structure of Cardanol, the main component of technical Cashew nut Shell Liquid [73].

with a 15 carbon unsaturated side chain in the m-position [51–56]. This structural combination offers a wide variety of possibilities for the synthetic polymer chemist [55, 56]. Though cardanol has long been used in various fields such as resins, friction lining materials and surface coatings development of new applications would require value added materials with defined structure. Thus, the last few decade have seen application of several novel methodologies for the preparation of value added materials starting from cardanol [57–67]. Ikeda *et al.* [61, 63] have oxidatively polymerized cashew nut shell liquid using metal complexes as catalyst to produce crosslinkable polymers for coatings with excellent hardness and gloss properties. Cardanyl acrylate or 3-pentadecenyl phenyl acrylate, obtained by the base catalysed reaction of cardanol and acryloyl chloride [68], has also received the attention of researchers world wide as a possible substitute in flexible thermoset polymer production, becasue during the polymerization in the acrylate or methacrylate the unsaturation in the side chain also undergoes polymerization. These products are light coloured in comparison to cardanol based products. It has been used in the preparation of crosslinked beads to make polymer supported reagents [69]. The gelation, free radical polymerization kinetics and crosslinking mechanism has been studied in detail, recently [70–72], implying that it could be a valuable raw material to reckon with for the production of high performance materials. In the preparation of UV curable coatings, inks as well as IPNs the primary criteria is to have prepolymers with crosslinkable groups and possessing solubility in common organic solvents. Thus, it was suggested that new methodologies for making cardanyl acrylate prepolymers of controlled molecular weight and polydispersity as well as copolymers with styrene or acrylates may give rise to resins for inks, coatings or adhesives with excellent adhesion, gloss and other properties. We found that these objectives can be combined in a classic way by the application of synthetic methodologies like ATRP to monomers obtained from renewable resources, giving rise to many new functional and value added materials. Our group first reported the ATRP of cardanyl acrylate as a new strategy to synthesize well-defined polymerizable monomers from cardanol [73]. Suresh and jayakrishna reported on random copolymers of cardanyl acrylate with MMA and observed that a thermally curable product with improved thermal decomposition by about 35°C was obtained. The procedure is outlined in Figure 5.7.

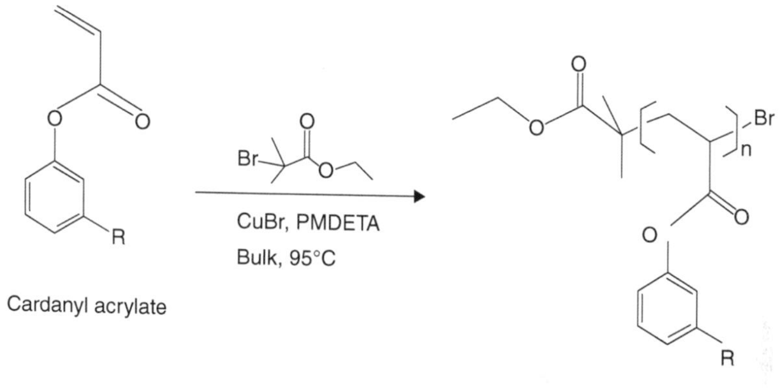

Figure 5.7 Reaction scheme for the copper catalyzed ATRP of cardanyl acrylate [73].

5.4.5   Catalyst for ATRP of Vegetable Oil Monomers

As regards the choice of ligands for the Cu-based catalyst, Matyjaszewski and co-workers tested PMDETA as a ligand for the ATRP of lauryl acrylate and found it to be unsuitable for solubility reasons even when solvents like anisole or acetone which serve well in the ATRP of MMA were used [74]. Although the systems were homogeneous to start with, the heterogeneity sets in through the precipitation of the Cu(II) complex and the polymer. However, the use of the CuBr/dNbpy catalyst eliminates the heterogeneity problem from Cu(II) complex precipitation at an elevated temperature (90°C) even without the use of a solvent. This latter catalyst was successfully used also for the ATRP of stearyl methacrylate (SMA) and stearyl acrylate (SA) in xylene at 90°C [75].

On the other hand, Xu et al. [76] claimed satisfactory control in the ATRP of lauryl methacrylate (LMA) using the CuCl/PMDETA catalyst but without any added Cu(II) complex in various solvents which were added to keep the copper complexes in solution at 110°C. However, the chain extension experiment showed bimodality and larger PDI in the chain extended polymer. Raghunadh et al. [77] also did not use any extraneously added deactivator and found that the CuBr/ PMDETA catalyst does not provide satisfactory control, while the CuBr/N(n-propyl)2-pyridylmethanimine (PPMI) catalyst does. The control improves significantly on increasing the alkyl group size in the methanimine catalyst from n-propyl to n-octyl, which increases the compatibility of the catalyst with such a highly non-polar monomer. Further improvement in control occurs on using an initiator having a large alkyl group which has been supposed to be more compatible with the monomer. Best control (PDI = 1.15) was reported when the non-polar character of the initiator was further increased by replacing the alkyl group with its perfluoro derivative, e.g. using zonyl-2-bromo-2-methyl propionate initiator.

Generally, due to their long alkyl chains the fatty acid monomers impart certain properties to the resulting polymers, such as elasticity, flexibility, high impact strength, hydrolytic stability, hydrophobicity, or lower glass transition temperatures [1]. Thus, monomers based on fatty alcohol acrylates or methacrylates are well known in industry and are applied as low Tg monomers in copolymers of various kinds. Therefore, a detailed study of the co-polymerization behavior and structure property relationship of these monomers is a prerequisite to open new avenues for material development with potential applications.

5.4.6   Rosin Gum

Rosin is obtained as the exudates of pine tree. They are already used as ingredients for ink, varnish, adhesive, chewing gum etc. but has mainly been polymerized by step growth methods. Gum rosin is a valuable resource due to its abundance, low cost, and its potential ability to be derivatized into polymerizable monomer. It is produced more than 1 million tons annually. The major component of gum rosin is abietic acid, which has a characteristic phenanthrene ring. High purity vinyl derivatives of rosin were prepared and polymerized by ATRP [78]. The authors have used dehydroabietic acid as it was more pure and derivatization gives pure monomers. The reaction scheme used for derivatization to acrylates and methacrylates with different spacers between the vinyl group and the dehydroabietic group [78]. This vary the steric effect imparted on to the vinyl group and exerted significant influence on the control of polymerization.

The synthesis of high quality rosin-derived vinyl monomers and the preparation of well-defined rosin-derived polymers using atom transfer radical polymerization (ATRP) was reported recently by Y. Zheng *et al.* [78]. The copper catalyzed ATRP of dehydroabietic

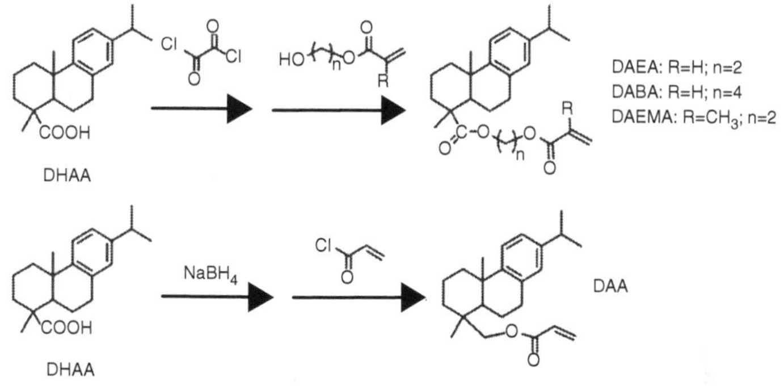

Figure 5.8 Synthesis of vinyl monomers from Gum Rosin [78].

acrylate (DAA), dehydroabietic butyl acrylate (DABA), dehydroabietic ethyl acrylate (DAEA) and dehydroabietic ethyl methacrylate (DAEMA) was reported. The general scheme for the preparation of the monomer is given in Figure 5.8. Higher polydispersity and uncontrolled polymerization with low rate was observed when the polymerization was carried out in a less polar solvent anisole. However the control of the polymerization improved and narrow disperse products obtained in THF solvent. They have used a combination of CuBr catalyst and tris [2-(dimethylamino) ethyl] amine [Me6Tren] as the ligand. The use of flexile spacer was found to improve the flexibility.

All the polymers exhibited typical thermoplastic behavior without any crystallinity. The glass transition was found to depend on the chemical structure. The highest T_g of 90°C was observed for the poly dehydroabietic ethyl methacrylate (PDAEMA). The lowest Tg of 22°C was observed for the acrylate PDAA, which was accounted for as due to the longer spacer that decreased the rotation barrier and reduced the Tg. It was shown that ATRP allows preparation of well-defined polymers with controlled molecular weight, functionality, and architectures, which are useful in developing more advanced value added materials based on renewable resources such as thermoplastic elastomers and composites.

5.4.7 Miscellaneous Monomers

There are some other renewable monomers which have gained attention of researchers recently due to their easy availability as from biomass and special properties [79, 80]. The monomers such as α-methylene -γ-butyrolactone (MBL) and γ-methyl-α-methylene-γ-butyrolactone (MMBL) are of particular interest (see Figure 5.9, below). MBL or tulipalin A is a natural substance found in tulips and the γ-methyl derivative, MMBL, is readily prepared via a two step process from the cellulosic biomass derived levulinic acid using Group 8 metals, particularly Ruthenium for this reaction [79]. MBL exhibits greater reactivity in radical polymerization as the cyclic analogue of methyl methacrylate. The cyclic ring in MBL also imparts significant enhancement of the material properties of the resulting PMBL due to the conformational rigidity of the polymer chain through incorporation of butyrolactone. For example, the Tg of PMBL is 195°C compared to the Tg of 100°C for PMMA. These polymers are potentially useful in lithographic applications.

The largest industrial economy US had initiated a programme aimed at producing these biomass derived industrial chemicals [79]. Many startup companies are formed and spent millions of dollars venture capital and Govt. funds to advance early stage technology.

The ATRP of α-methylene–γ-butyrolactone has been reported recently [80], the various copolymers and blends containing this monomer units give good optical properties, heat, weathering, scratch and solvent resistance [80]. The ATRP was carried out with CuBr/2, 2'-bipyridine catalyst and bromopropionitrile

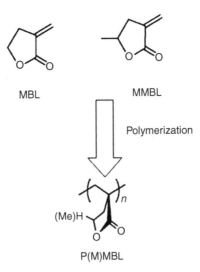

MBL   MMBL

Polymerization

(Me)H

P(M)MBL

Figure 5.9 Scheme for the Polymerization of α-methylene -γ-butyrolactone.

as initiator. Since the polymer has limited solubility in common organic solvents, DMF was used as reaction medium. More than 90% conversion was achieved in 100 minutes, with well controlled, predictable mol.wt and narrow mol.wt distribution. The technique was extended to the preparation of butyl acrylate block copolymers using the common CuBr/PMDETA catalyst system. Chain extension of MMA with MBL was also studied. The reaction was found to be fast. Difunctional PPBA macro initiator by MBL was used for preparing triblock copolymers which would be useful as thermoplastic elastomers.

Glycopolymers or glycosides are another special class of hydrophilic copolymers derived from sugar acrylates [81, 82]. The preparation of novel glycopolymers and monomers useful for their preparation has been recently disclosed in the patent application [83]. Glycopolymers comprise glycoside units and are used in cosmetic compositions. Similarly, the ATRP of sugar acrylates has been found to give hydrophilic block copolymers useful in specialty applications like drug delivery systems and for preparation of peptide conjugates. Though these are also renewable monomers in lieu of their specialty applications they are not considered in detail here.

5.5   Conclusions and Outlook

In this chapter the use of some renewable non-edible oils and gums for the synthesis of well defined polymers is reviewed. Most of these studies have been reported for the first time and not much is known about concerning the morphology and structure property correlation in these systems. From the preliminary studies reported it is inferred that ATRP allows immense opportunities to tailor the material properties as well as molecular characteristics of the obtained polymers showing that the application possibilities of these raw materials are manifold and yet to be realized for specific end user applications. The gross energy requirements as well as life-cycle assessments reveal distinct advantages for these renewable monomers. In summary, the recent developments highlighted within this contribution clearly show that there is immense potential for developing polymers of special topologies like block copolymers, graft polymers and telechelic polymers prepared by ATRP of renewable monomers. These polymers or copolymers often display special properties like capability to self assemble into multiple length scale, improved thermal stability because of their structural characteristics.

Future development of natural monomer based polymers offers very attractive advantages and assets as they descent from renewable resources. The generally good biocompatibility, the rather uncomplicated disposal after use, and in many cases unique properties making the product indispensible in meeting special end use requirements predominantly in adhesive, coatings and as thermoplastic elastomers. In short they present structural features not achievable with some of the conventional block copolymer systems. The derivatization of isolated and purified natural monomers to new classes of well defined polymers with special and useful properties could be a challenge to polymer chemists that is currently being addressed. Transferring the recent knowledge acquired in controlled synthesis of polymers, new approaches are opened up to prepare unknown compositions through derivatization and controlled radical polymerization of renewable monomer based building blocks.

References

1. M.A.R. Meier, J.O. Metzger. and U.S. Schubert, *Chem.Soc.Rev.*, Vol. 36, p. 1788, 2007.
2. G. Lligadas, J. C. Ronda, M. Galia, and V. Cadiz, *Biomacromolecules*, Vol. 11, p. 2825, 2010.
3. K. Matyjaszewski and J. Xia, *Chem. Rev.*, Vol. 101, p. 2921, 2001.
4. M. Semsarilar and S. Perrier, *Nature Chemistry*, Vol 2, p. 811, 2010.
5. N.V. Tsarevsky, and K. Matyjaszewski *Chem. Rev.*, Vol. 107, p. 2270, 2007.
6. M. Kato, M. Kamigaito, M. Sawamoto, and T. Higashimura, *Macromolecules*, Vol. 28, p. 1721, 1995.
7. J.S. Wang, and K. Matyjaszewski, *J. Am. Chem. Soc.*, Vol. 117, p. 5614, 1995.
8. J.S. Wang, and K. Matyjaszewski, *Macromolecules*, Vol. 28, p. 7901, 1995.
9. K. Matyjaszewski, S. Gaynor, and J.S. Wang, *Macromolecules*, Vol. 28, p. 2093, 1995.
10. T. Pintauer, and K. Matyjaszewski, *Coord. Chem. Rev.*, Vol. 249, p. 1155, 2005.
11. M. Kamigaito, T. Ando, M. Sawamoto, *Chem. Rev.*, Vol. 101, p. 3689, 2001.
12. R. Nagelsdiek, M. Mennicken, B. Maier, H. Keul, and H. Hoecker, *Macromolecules* Vol. 37, p. 8923, 2004.
13. H. Zhang, X. Jiang, and R. Van der Linde, *Polymer*, Vol. 45, p. 1455, 2004.
14. A.T. Slark, "Exploiting Acrylic Polymer architectures in surface coating application", A.J. Ryan (Editor) in *Emerging themes in Polymer Science*, Royal society of Chemistry, UK, p. 41, 2001.

15. Y. Kotani, M. Kamigaito, and M. Sawamoto, *Macromolecules*, Vol. 32, p. 2420, 1999.
16. K. Matyjaszewski, J.L. Wang, T. Grimaud, and D.A. Shipp, *Macromolecules*, Vol. 31, p. 1527, 1998.
17. K. Matyjaszewski, *J. Macromol. Sci.-Pure Appl. Chem.*, Vol. A34, p. 1785, 1997.
18. T.E. Patten, K. Matyjaszewski, *Adv. Mat*, Vol. 10, p. 901, 1998.
19. K. Matyjaszewski, T.E. Patten, and J. Xia, *J. Am. Chem. Soc.*, Vol. *119*, p. 674, 1997.
20. H. Fischer, *Chem. Rev.*, Vol. 101, p. 3581, 2001.
21. H. Zhang, B. Klumperman, W. Ming, H. Fischer, R. van der Linde, *Macromolecules*, Vol. 34, p. 6169, 2001.
22. K. Matyjaszewski, Y. Gnanou, and L. Leibler (Eds.), *Macromolecular Engineering. Precise Synthesis, Materials Properties, Applications*. Weinheim, Wiley-VCH, 2007.
23. K. Matyjaszewski, and T.P. Davis, *Handbook of Radical Polymerization*. Hoboken, Wiley Interscience, 2002.
24. N.V. Tsarevsky, and K. Matyjaszewski, *ACS Symp. Ser.*, Vol. 937, p. 79, 2006.
25. K. Matyjaszewski, N. V. Tsarevsky, *Nature Chem.*, Vol. 1, p. 276, **2009.**
26. M. Ouchi, T. Terashima, and M. Sawamoto, *Chem. Rev.*, Vol. 109, p. 4963, 2009.
27. V. Coessens, T. Pintauer, and K. Matyjaszewski, *Prog. Polym. Sci.*, Vol. *26*, p. 337, 2001.
28. N.V. Tsarevsky and K. Matyjaszewski, *Chem. Rev.*, Vol. 107, p. 2270, 2007.
29. N.V. Tsarevsky, and K. Matyjaszewski, *J. Polym. Sci., Part A: Polym. Chem.*, Vol. *44*, p. 5098, 2006.
30. W. Jakubowski, and K. Matyjaszewski, *Angew. Chem., Int. Ed.*, Vol. 45, p. 4482, 2006.
31. W. Jakubowski, K. Min, and K. Matyjaszewski, *Macromolecules*, Vol. 39, p. 39, 2006.
32. K. Matyjaszewski, W. Jakubowski, K. Min, W. Tang, J. Huang, W. A. Braunecker, and N. V. Tsarevsky, *Proc. Natl. Acad. Sci.*, Vol. 103, p. 15309, 2006.
33. K. Min, H. Gao, and K. Matyjaszewski, *Macromolecules*, Vol. 40, p. 1789, 2007.
34. K. Matyjaszewski, N.V. Tsarevsky, W. A. Braunecker, H. Dong, J. Huang, W. Jakubowski, Y. Kwak, R. Nicolay, W. Tang, and J.A. Yoon, *Macromolecules*, Vol. 40, p. 7795, 2007.
35. L. Mueller, W. Jakubowski, W. Tang, and K. Matyjaszewski, *Macromolecules*, Vol. 40, p. 6464, 2007.
36. W. Jakubowski, B. Kirci-Denizli, R.R. Gil, and K. Matyjaszewski, *Macromol. Chem. Phys.*, Vol. 209, p. 32, 2008.
37. J. Listak, W. Jakubowski, L. Mueller, A. Plichta, K. Matyjaszewski, and M. R. Bockstaller, *Macromolecules*, Vol. 41, p. 5919, 2008.
38. K. Matyjaszewski, H. Dong, W. Jakubowski, J. Pietrasik, and A. Kusumo, *Langmuir*, Vol. 23, p. 4528, 2007.
39. D. P. Chatterjee, and B. M. Mandal, *Polymer*, Vol. 47, p. 1812, 2006.
40. Y. Xia and R. C. Larock *Green Chem.*, Vol. 12, p. 1893, 2010.
41. G. Cayli and M. A. R. Meier, *Eur. J. Lipid Sci. Technol.*, 110, 853, 2008.
42. E. Hempel, H. Huth, M. Beiner, *Thermochim. Acta.*, Vol. 403, p. 105, 2003.
43. D. P. Chatterjee, B. M. Mandal, *Macromol Symp.*, Vol. 240, p. 224, 2006.
44. W. Xu, X. Zhu, Z. Cheng, J. Chen, *J Appl Polym Sci.*, Vol. 90, p. 1117, 2003.

45. G. Holden, T.E. Bishop, and N.R., Legge, *J. Polym. Sci. Part C*, Vol. 26, p. 37, 1969.

46. G. Holden, N.R. Legge, Thermoplastic Elastomers 2nd ed, Holden G, Legge, N.R. Quirk, R. Schroeder, H.E., Eds. *Hanser Publishers*, Munich Germany. p. 47, 1996.

47. R. Jerome, R. Fyat, Ph. Teypssie., In Thermoplastic Elastomers Legge, N.R, Holden, G., Schroeder, H.E., Eds., Hanser, Munich, p. 451, 1987.

48. D. P. Chatterjee and B. M. Mandal, *Macromolecules*, Vol. 39, p. 9192, 2006.

49. T. Eren and S.H. Kuesefoglu, *J. Appl. Polym.Sci.*, Vol. 94, p. 2475, 2004.

50. S.P. Bunker and R.P. Wool. *J. Polym. Sci. Part A-polym. Chem.*, Vol. 40, p. 451, 2002.

51. J.H.P. Tyman, *J. Chromat.*, Vol. 111, p. 277, 1975.

52. P. H. Gedam, P. S. Sampathkumaran, and M.A. Sivasamban, *Indian J. Chem.*, Vol. 10, p. 388, 1972.

53. B.G.K. Murthy, M.A. Sivasamban, and J.S. Agarwal, *J. Chromatog.*, Vol. 32, p. 519, 1968.

54. P. H. Gedam, and P. S. Sampathkumaran, *Prog. Org. Coatings.*, Vol. 14, p. 115, 1986.

55. S. Agarwal, V. Choudhary, and I. K. Varma, *Angew. Makromol. Chemie.*, Vol. 248, p. 95, 1997.

56. R. Antony, and C.K.S. Pillai, *J. Appl. Polym. Sci.*, Vol. 49, p. 2129, 1993.

57. K. Sathiyalekshmi, S. Gopalakrishnan, *Adv. in Polym. Techn.*, Vol. 23, p. 91, 2004.

58. G. Mele, R. D. Sole, G. Vasapollo, E.G. Lopez, L. Palmisano, S.E. Mazzetto, O.A. Attanasi, and P. Fillippone, *Green. Chem.*, Vol. 6, p. 604, 2004.

59. S. Suwanprasop, T. Nhujak, S. Roensumran, and A. Petsom, *Ind. Eng. Chem. Res.*, Vol. 43, p. 4973, 2004.

60. P. Das, T. Sreelatha, and A. Ganesh, *Biomass and Bioenergy.*, Vol. 27, p. 265, 2004.

61. R. Ikeda, H. Tanaka, H. Uyama, and S. Kobayashi, *Macromol. Rapid. Commun.*, Vol. 21, p. 496, 2000.

62. H.P. Bhunia, R. N. Jana, A. Basak, S. Lenka, and G. B. Nando, *J. Polym. Sci. Part A: Polym. Chem.*, Vol. 36, p. 391, 1998.

63. R. Ikeda, H. Tanaka, H. Uyama, and S. Kobayashi, *Polymer*, Vol. 43, p. 3475, 2002.

64. K.S. Alva, P. Nayak, L.J. Kumar, and S.K. Tripathy,. *J Macromol Sci: Pure and Appl. Chem.*, Vol. 34, p. 665, 1997.

65. H. P. Bhunia, A. Basak, T. K. Chaki, and G.B. Nando, *Eur. Polym. J.*, Vol. 36, p. 1157, 2000.

66. B. Philip, *Interdisciplinary Sci. Rev.*, Vol. 25, p. 220, 2000.

67. M. Saminathan, and C.K.S. Pillai, *Polymer*, Vol. 41, p. 3103, 2000.

68. G. John, and C.K.S. Pillai, *Makromol. Chem. Rapid. Commun.*, Vol. 13, p. 255, 1992.

69. G. John, K.T. Shali, and C.K.S. Pillai, *J. Appl. Polym. Sci.* Vol. 53, p. 1415, 1994.

70. L.H. Nguyen, H. Koerner, and K. Lederer, *J. Appl. Polym. Sci.*, Vol. 85, p. 2034, 2002.

71. L.H. Nguyen, H. Koerner, and K. Lederer, *J. Appl. Polym. Sci.*, Vol. 88, p. 1399, 2003.

72. L.H. Nguyen, H. Koerner, and K. Lederer, *J. Appl. Polym. Sci.*, Vol. 89, p. 2385, 2003.

73. K.I. Suresh, and M. Jaikrishna, *J. Polym. Sci. Part A: Polym Chem.*, Vol. 43, p. 5953, 2005.

74. K.L. Beers and K. Matyjaszewski, *J. Macromol. Sci. Pure Appl. Chem.*, Vol. 38, p. 731, 2001.
75. S. Qin, K. Matyjaszewski, H. Xu and S.S. Sheiko, *Macromolecules*, Vol. 36, p. 605, 2003.
76. W. Xu, X. Zhu, Z. Cheng, and J. Chen, *J. Appl. Polym. Sci.*, Vol. 90, p. 1117, 2003.
77. V. Raghunadh, D. Baskaran, and S. Sivaram, *Polymer*, Vol. 45, p. 3149, 2004.
78. Y. Zheng, K. Yao, J. Lee, D. Chandler, J. Wang, C. Wang, F. Chu, and C. Tang, *Macromolecules*, Vol. 43, p. 5922, 2010.
79. L.E. Manzer, *Topics in catalysis*, Vol. 53, p. 1193, 2010.
80. J. Mosnacek and K. Matyjaszewski, *Macromolecules,* Vol. 41, p. 5509, 2008.
81. R. Narain and S. P. Armes, *Biomacromolecules*, Vol. 4, p1746, 2003.
82. B. Voit, and D. Appelhans, *Macromol. Chem. Phys.*, Vol. 211, p. 727, 2010.
83. S. Chiron, M.P. Labeau, E. Fleury, D. Viet, S. Cottaz, H. Driguez, and S. Halila, US Patent 0281064A1, assigned to Rhodia Chimie, November 13, 2008.

6

Renewable Polymers in Transgenic Crop Plants

Tina Hausmann and Inge Broer

University of Rostock, Agricultural and Environmental Faculty,
Rostock, Germany

Abstract

Based on gene technology, great progress has been made in the production of biopolymers in plants. This is due to the careful selection of production compartments like chloroplast or endoplasmic reticulum (ER), production organs like seeds or tubers as well as the selection and combination of genes involved in the synthesis. Nevertheless, application is not in sight except for the amylopectin potato *Amflora*. This is because of several limitations that still have to be addressed. One important challenge is, the production efficiency has still to be improved, possibly by an improved supply of precursors, tissue specific expression and the selection of optimized genes. In addition, it is of major importance to analyze the causes for yield limitations, reduced seed production and other deleterious effects observed in several high production lines. The selection of the optimal production system is also of great significance.

Keywords: Cellulose, collagen, cyanophycin, elastin, lignin, PHA, rubber, silk, starch, chloroplast, endoplasmic reticulum, production efficiency

Although crop plants are mainly used to ensure the food supply of humanity, they have also served for thousands of years as origin of important renewable biomaterials like oil, fibers and wood.

Cellulose, as the main component of the plant cell wall, has traditionally been used as a source of fibers for various fabrics, wood served as building material and for generation of energy through combustion. Also, starch and rubber are natural biopolymers with a long history of application.

Vikas Mittal (ed.) Renewable Polymers, (247–304) © Scrivener Publishing LLC

The advancement of the fuel-based chemistry in the 19th century and its wide range of application largely replaced plants as production system for platform chemicals and materials like dyes or solvents [1]. Nonetheless, changes in climate and resource constraints have renewed the interest to develop plants for the sustainable synthesis of a variety of materials and chemicals required by mankind [2]. The current focus is on using plant material for the production of biofuels, like ethanol or biodiesel. But beside this application, plants are a potential source of a much wider range of useful chemicals and materials [2].

As a special ingredient, biopolymers are of great interest because they are in most cases completely biodegradable and can serve as substitutes for fossil fuel derived compounds. As their production in plants is CO_2 neutral, driven by solar energy and photosynthesis, they support a sustainable use of renewable resources.

However, most biopolymers with technical applications are often produced in agriculturally usable cultivars only in (i) low amounts, (ii) unfavorable combinations or (iii) are even absent. The development of molecular biology and plant transformation techniques opened the possibility to enhance yield and composition of traditional ingredients or to produce completely novel components of foreign origin in transgenic plants. In this chapter we discuss origin, structure and application of different polymers, with a look on transgenic attempts to improve yield, composition or handling processes. We split polymers in two groups (i) naturally in plant occurring polymers like starch, cellulose, lignin and rubber and (ii) new integrated polymers such as the fibrous proteins silk, elastin and collagen, as well as polyhydroxyalkanoates (biological polyesters) and the polypeptide cyanophycin.

6.1   Natural Plant Polymers

Plants naturally produce polymers. Polysaccharides are estimated to make up ca. 70% of all organic matter, of which 40% are cellulose, the most abundant macromolecule on earth. A second important carbohydrate starch belongs to the major component of global biomass besides lignin which comprises 15–25% of a typically woody plant [3].

A number of other natural polymers have been utilized for industrial applications; especially natural rubber is irreplaceable by synthetic equivalents because of its unique physical properties.

6.1.1   Starch

Starch is the most abundant storage carbohydrate in plants [4]. Higher plants synthesize and store starch in form of water insoluble granules in storage tissues such as seeds and tubers and in a transitory form in leaves, roots and stems. Starch belongs to the main components of tuber crops such as potato, sweet potato and cassava, as well as grain harvested from cereal crops like barley, corn, wheat, sorghum, oat and rice [5].

Starch is composed primarily of two types of glucose polymers: amylose and amylopectin.

Amylose is a mainly linear chain of α-1,4-linked glucose residues. It is infrequently branched with α-1,6-linkages (approximately one branch per 1,000 glucose residues). The degree of polymerization (dp) of amylose molecules is species dependent. It ranges from approximately 800 in corn and wheat to around 4500 in potato. In contrast, amylopectin is a much larger (dp 10^5 to 10^7) and more complex polysaccharide. It is composed of hundreds of short α-1,4 glucan chains bonded together by α-1,6-linkages (Figure 6.1). About 5% of the residues have both α-1,4 and α-1,6-linkages [6].

Amylose and amylopectin are synthesized by two different pathways having a common substrate (ADPglucose). The granule bound starch synthase (GBSS) is involved in amylose synthesis, while amylopectin is formed by the concerted action of starch synthases (SS),

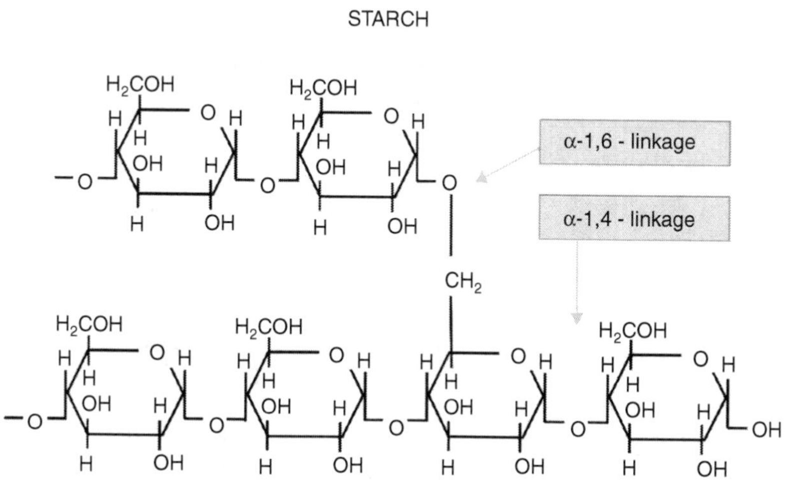

Figure 6.1 Partial chemical structure of starch.

starch branching enzymes (SBE) and starch debranching enzymes (SDE) [7] (Figure 6.2).

Amylose and amylopectin make up to 98–99% of the dry weight of natural starch granules. The remainder includes small amounts of minerals, lipids and phosphorus in the form of phosphates ester-ified to glucose hydroxyls [4].

The size and morphology of starch granules is characteristic for the species and organs in which they are produced [9]. The ratio of amylose to amylopectin contents in starch varies depending on the botanical source [10, 11].

Most wild type plant starches are composed of 15–30% amylose and 70–85% amylopectin [6].

Plants using starch as main storage compound like potato, corn, cassava and wheat provide the main sources of energy in human nutrition, but are also applied in many industrial areas (Table 6.1) like the production of adhesives, cosmetics, detergents, paper,

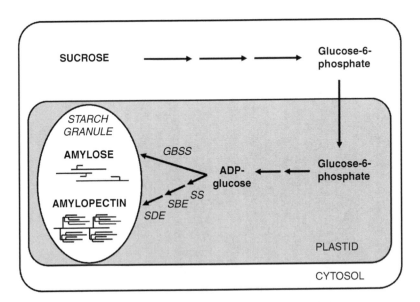

Figure 6.2 Simplified scheme of starch synthesis in storage organs (modified after [8]).
Sucrose entering the cell is converted in several steps to glucose-6-phosphate. This hexose phosphate enters the plastid, where it is metabolized to ADPglucose. ADPglucose is transformed to starch via multiple isoforms of starch synthase (SS; granule bound starch synthase (GBSS)), starch branching (SBE) and debranching (SDE) enzymes.

textiles, pharmaceuticals as well as biodegradable plastics as an substitute for petroleum-based products [6].

The different physico-chemical characteristics of different starches are utilized for various applications. For example, high amylose starches are widely used in fried snack production to create crisp, browned snacks. High amylose starches are also strong gelling agents applied in the manufacturing of jellies and confectionary. Further features of high amylose starch are transparency, flexibility, water resistance and tensile strength [12].

Starches with high amylopectin levels are, for example, used in the adhesive and paper industry, which exploits the binding and

Table 6.1 Industrial use of starch (summarized from [15, 16]).

Industry	Use of starch or modified starch
Adhesive	Adhesive production
Agrochemical	Mulches, pesticide delivery, seed coating, fertilizer
Animal feed	Pellets
Building	Mineral fibre, gypsum board, concrete
Cosmetics	Face and talcum powder
Detergent	Surfactants, builders, co-builders, bleaching agents, bleaching activators
Food	Viscosity modifier, glazing agent; contained in mayonnaise, bread, meat products, confectionery
Medical	Plasma extender/replacers, transplant organ preservation, absorbent sanitary products
Oil drilling	Viscosity modifier
Paper and board	Binding, sizing and coating
Pharmaceuticals	Diluent, binder, drug delivery, dusting powder
Plastics	Biodegradable filler, biodegradable plastic
Purification	Flocculant, for wastewater treatment
Textile	Yarn sizing, finishing and printing, fire resistance

bonding properties of amylopectin to enhance paper strength and printing properties [12]. Starch typically makes up 8% of a sheet of paper [13].

Pure amylopectin starch delivers novel textures combined with high stability, viscosity, and clarity. Applications of this improved starch are in the paper and textile industries, as well as, adhesives and construction materials [14].

However, the diverse properties of starch can also cause problems in processing due to the inconsistency of the raw material. As a consequence, chemically modified starches are used widely to overcome the variability of native starch and their lack of versatility over a wide range of processing conditions [4].

So for the most industrial uses, chemical and/or enzymatic treatments are necessary to improve the starch usability. These treatments are costly and in some cases environmentally unfriendly [6].

For many industrial applications only the thickening amylopectin part is needed, whereas the gelling amylose component is undesired since it interferes with a number of applications.

The necessary separation of amylose and amylopectin is energy intensive and not economically feasible [14].

A possible solution is the usage of modified starches with altered composition resulting in changed starch properties.

Many corn mutants exist that have been successfully used to provide modified starches [15]. A corn mutant with starch containing almost no amylose was discovered in China and characterized early in the last century [17, 18]. This and similar mutants were named waxy corn [13]. Subsequently, the granule bound starch synthase (GBSSI) has been shown to be responsible for amylose synthesis in plant storage organs, which is missing in amylose-free mutants [19].

Waxy mutants also have been identified in many other plant species like rice, barley, sorghum, and amaranth [20, 21]. Hovenkamp et al. (1987) were the first who identified an amylose-free potato mutant created by Roentgen-irradiation [20].

In corn, rice and pea, mutants with high amylose starches, (50–58%) so called *amylose extender* mutants, have been reported, which lack SBEII (reviewed by [6].)

Waxy corn has been grown to produce starch for industrial purposes since the 1940s [13].

In addition, different transgenic approaches for starch modification have been carried out, summarized in various reviews [6, 8, 22].

The approaches to obtain starch with high amylose content focus on the suppression of the starch branching enzyme (SBE) I and II isoforms. For instance, Blennow *et al.* (2005) genetically engineered potato tuber starch by SBE I and II antisense suppression. The transgenic plants produced starch with altered structural and physical properties. The lengths of amylopectin unit chains, the concentration of amylose and monoesterified phosphate were significantly increased. The most extreme line contained ca. 50% amylose as opposed to the controls having 25%. The increased phosphate concentration in the transgenic starch can give new functionality, most notably increased hydration capacity, not present in high amylose and non-phosphorylated starches prepared from other sources e.g. corn [23]. For industrial applications where only the thickening amylopectin part is needed and the gelling amylose component is undesired, the aim of transgenic approaches is to suppress the gene encoding the granule bound starch synthase (GBSSI), the key enzyme for the synthesis of amylose.

Cassava (*Manihot esculenta*) is an interesting production system for tropical areas. It is a tuberous root crop which is mainly grown as a subsistence crop in many developing countries in sub-tropical regions of Africa, Asia and Latin America. Its commercial use was first as animal feed, but has shifted since the late sixties to a source of native starch.

A cassava genotype producing amylose free starch had been created by Raemakers *et al.* (2005) by using antisense suppression of GBSSI. The modified cassava starch (<3.5% amylose) shows enhanced clarity and stability characteristics, which are important properties for the use of starch pastes in numerous applications. Since cassava can only be grown in tropical countries, this mainly offers possibilities for developing countries to export this special type of starch [24].

One development which attracted interest especially in Europe was made by Kull and colleagues. They developed in 1995 a transgenic potato which exclusively produces the amylopectin component of starch by inactivating GBSS through posttranscriptional gene silencing (PTGS). Two subgenomic fragments of the gene were expressed in antisense orientation under control of the *Cauliflower Mosaic Virus* (CaMV) 35S promoter. The resulting transgenic potato plants were effective in inhibiting amylose biosynthesis in tubers. The waxy phenotype was stable during vegetative propagation [25].

For commercial use the GBSS antisense potato variety "*Amflora*" with an increase in the branched starch component amylopectin of over 98 percent [26] was analyzed in field trials for several years to test yields, pest and disease resistance. *Amflora* maintains all the characteristics of its mother variety, the starch potato variety *Prevalent* [14]. Additionally, the allergic and toxic potentials of *Amflora* tubers were analyzed on top of possible other impacts on human health or the environment. No increased risk to humans, animals and the environment were shown in comparison to conventional potatoes [27]. In March 2010, the *Amflora* (EH92-527-1) was approved by the European Commission for cultivation and processing, which passed the proposals for ten years authorization (http://ec.europa.eu/food/dyna/gm_register/index_en.cfm).

Naturally, during the more than ten years approval procedure, further progress in potato breeding and further suitable varieties were achieved. One of these varieties has also been used for the creation of a new amylopectin potato whose approval is now being requested. The so called "*Amadea*" is based on the 2002 admitted [28] variety *Kuras* with high starch content (>20%) and high yield [29]. In contrast to *Amflora* which contains the *npt*II-gene conferring resistance against the antibiotic kanamycin, *Amadea* carries an acetohydroxyacid synthase (AHAS) marker gene conferring herbicide-resistance [29] as transformation marker.

The advances in the modification of starch to tailor-made starches can both benefit the environment by reducing chemical treatments and energy consumption and thereby supporting the industry by reducing processing costs.

6.1.2   Cellulose

Cellulose and lignin are the two most abundant biopolymers on earth [30].

All cell walls of higher plants contain besides pectin three major components

(i) cellulose, a homopolymer of β-1,4-linked glucose units (Figure 6.3), which is a flexible structural substance in the form of fibrils,

(ii) hemicellulose, a heterogeneous polysaccharide, consists of 5-carbon sugars such as xylose or arabinose along with glucose, representing a matrix in which the cellulose fibrils are embedded, and

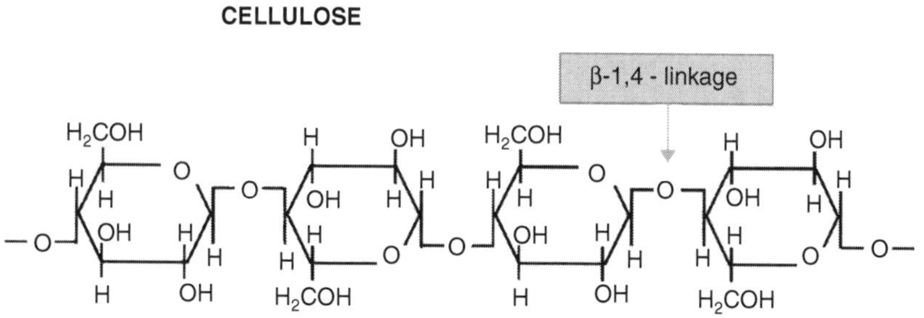

CELLULOSE

Figure 6.3 Partial chemical structure of cellulose.

(iii)   lignin, a heterogeneous hydrophobic phenolic poly-
        mer, which forms a bond between cellulose and
        hemicellulose, to give the wall water tightness and
        strength.

As lignin is largely hydrophobic and cellulose is hydrophilic, compatibility is obtained through hemicelluloses, which contains both
hydrophilic and hydrophobic sections [31].

Between 10^{10} and 10^{11} tons of cellulose are produced by nature
per annum, it is therefore the most abundantly available natural material. As a linear homopolysaccharide it is composed of
D-glucopyranose units linked by β-(1,4) linkages. Depending
strongly on the plant species, the degree of polymerization of cellulose is e.g. around 10.000 for wood and 15.000 for cotton. The
natural cellulose consist of crystalline sections held together by
strong hydrogen bonding (in-plane) and Van der Waals interactions (between planes) [31].

Despite its relative chemical simplicity, the physical and mor-
phological structure of native cellulose in higher plants is complex
and heterogeneous [32].

The quantity of cellulose in a fiber affects the properties, economics of the fiber production and the viability of the fiber for different
applications. For instance, fibers with higher cellulose content would
be suitable for textile, paper and other fibrous applications, whereas
byproducts with higher hemicellulose content would be preferable
for producing ethanol and other fermentation products [33].

Poplar wood with its high cellulose-to-lignin ratio is well-suited
for pulp and paper products [34]. Cultivated in plantations, modern hybrids exhibit short rotation time compared to those of most
other trees [34].

Since lignin prevents access of enzymes and chemicals to hemicellulose and cellulose, it reduces the degradability of the carbohydrate material [35].

Because of its recalcitrance against chemical or biochemical breakdown, lignin is the major barrier to efficient extraction of cellulose fibers for pulp and paper production and to saccharification for production of liquid biofuels [36]. In cellulosic biofuel production lignin is interfering because it inhibits the release of polysaccharides during the treatment process and absorbs the enzymes used for the saccharification [37]. Therefore, in order to obtain pure cellulose fibers e.g. for paper industry, lignins have to be eliminated from the wood chips during the pulping process to produce high-quality paper. Traditionally, natural cellulose fibers are separated from lignin using bacteria and fungi, as well as mechanical and chemical methods. By attacking the noncellulosic cell wall components, the separation of the fibers from the woody core is possible. Today, retting methods use mostly alkalis, mild acids and enzymes for fiber extraction [33].

This chemical process is very expensive and polluting along with high energy consumption [38].

Although wood is the main industrial source of cellulosic fibers, competition from diverse sectors such as furniture industries and building products, as well as the combustion of wood for energy, makes it difficult to provide all users with the amount of wood needed at a reasonable cost [32]. Therefore, fibers from fiber crops like flax, hemp and sisal, but also of agricultural by-products such as corn, wheat, rice, sorghum, barley, sugar cane, pineapple, bananas and coconut are likely to become of increasing interest (Table 6.2). These noon-woody plants generally contain less lignin than wood and thus bleaching processes are less demanding [32].

Due to the importance of lignin for cellulose isolation, most transgenic approaches to optimize the cellulose production focus on the reduction of the lignin content. This is supposed to be challenging because lignin is crucial for structural integrity of the cell wall and stiffness and strength of the stem [30].

The lignin content of angiosperms differs greatly, with grasses having 5–10% and some tropical hardwoods were reported to have 40%, an average lignin content of 20% has been estimated for modern land plants [36].

Table 6.2 Chemicals and products from cellulose isolated from alternative sources (modified after [33]).

Products and chemicals	Biomass source
Fibers for textiles and composites	Corn stover, pineapple leaf fiber, coconut fiber, banana leafs
Fibers for pulp and paper	Rice and wheat straw, corn stover
Fiber-reinforced starch foarms	Corn stalks
Lactic acid	Corn stover
Strawboard (building material)	Wheat and rice straw
Solvents	Bagasse, the fibrous material remaining after sugarcane or sorghum stalks are crushed to extract their juice to obtain sugar

Research has focused on the generation of trees with reduced lignin content or modified lignin structure that might enable easier, and less environmentally damaging, delignification procedure during chemical (Kraft) pulping [39].

The hydroxycinnamyl alcohols (or monolignols) coniferyl alcohol and sinypyl alcohol are the main building blocks of lignin, with typically minor quantities of p-coumaryl alcohol. The monolignols are synthesized from phenylalanine through the general phenylpropanoid and monolignol-specific pathways. When incorporated into the lignin polymer, the units resulting from the monolignols, are called guaiacyl (G), syringyl (S) and p-hydroxyphenyl (H) units [38]. The enzymes required for monolignol biosynthesis through phenylpropanoids are given in Figure 6.4.

First plants with reduced lignin content were identified from natural or chemically induced mutants. For example, the naturally occurring brown midrib mutations in corn (bm1 - 4), identified in the time of 1924 to 1964 [38], are associated with significant alterations in lignin composition and reduction in lignin content in the stover, associated with improved digestibility. The bm3 mutation results in reduced COMT activity, whereas bm1 mutation reduces the CAD activity [40, 41]. Inspired by the natural lignin mutants

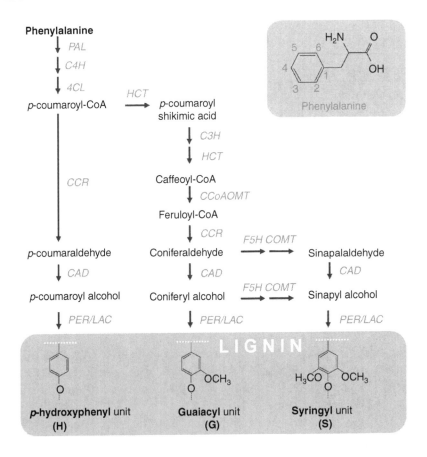

Figure 6.4 Lignin biosynthetic pathways (modified after [37]).
The enzymes involved in the pathway are: phenylalanine ammonia lyase (PAL), cinnamate 4-hydroxylase (C4H), 4-coumarate-CoA ligase (4CL), cinnamoyl-CoA reductase (CCR), hydroxycinnamoyl-CoA:shikimate hydroxycinnamoyl transferase (HCT), coumarate3-hydroxylase (C3H), caffeoyl CoA 3-O-methyltransferase (CCoAOMT), ferulate 5-hydroxylase (F5H), caffeic acid 3-O-Methyltransferase (COMT), cinnamyl alcohol dehydogenase (CAD), peroxidase (PER), and laccase (LAC).

since the beginning of 1990, many strategies were investigated to reduce lignin contents using gene technology. To knock down or silence the target gene(s) genetic transformation techniques, RNA interference (RNAi) and antisense have been deployed.

For each step of the lignin biosynthetic pathway, genes have been cloned and the impact on lignin composition and amount has been studied. Common poplars (*Populus* spp.) were investigated because they are fast growing trees with an important role in paper manufacture while they are relatively easy to modify. Other species like alfalfa (*Medicago sativa*), corn (*Zea mays*), tall fescue (*Festuca arundinacea*) a predominant cool-season gras, and tobacco (*Nicotiana tabacum*) were transformed with the aim of lignin reduction. More recently flax (*Linum usitatissimum*), the gymnosperms *Pinus radiata* and *Pinus abies* as well as rice (*Oryza sativa*) were investigated. Attempts to reduce lignin content via genetic engineering are summarized in several interesting reviews [37, 38, 42].

All in all, it can be concluded that reduction of the expression of each gene, expect F5H, reduces lignin amounts in the cell wall to varying extents [43]. Chen and colleagues (2006) down-regulated seven enzymes of the monolignol pathway independently in alfalfa and showed that the down-regulation of earlier enzymes in the pathway results in a more dramatic decrease in lignin content than did the down-regulation of later enzymes, regardless of the degree of enzyme activity reduction [44].

Wrobel-Kwiatkowska *et al.* (2007) produced low-lignin flax plants by gene silencing of the CAD. Flax (*Linum usitatissimum* L.) is a very important source of natural fibers used by the textile industry. Flax fibers contain mainly cellulose (about 70%), with hemicellulose, pectin and lignin (3–5%). Lignin presence results in reduced elastic properties of the fibers compared to non-lignocellulosic fibers, like cotton fibers which contain no lignin. The lignin content in the transgenic plants was reduced by about 40% coupled with lower content of pectin and hemicellulose but without changes in cellulose levels. Probably the changes in cell wall composition in the transgenic plants are the cause of improved stem elasticity. This might have potential commercial significance for the textile industry through improvement of the elastic properties of flax fibers. The authors also observed a decrease in resistance to the fungus *Fusarium oxysporum*. Because *Fusarium* species are used as retting organism, they assume that an increased sensitivity may lead to an improvement in flax retting [45]. Nonetheless, it has to be questioned whether the increase sensitivity to *Fusarium oxysporum* and possibly other plant pathogenic fungi might interfere with the plant production in the field.

The down-regulation of 4CL by antisense inhibition in transgenic poplars exhibited a 45% reduction of lignin in unison with 15% increase in cellulose [46]. Even the silencing of other lignin pathway enzymes led up to halving of lignin content in plants (e.g. down-regulation of CCR [47] and C3H [48] in poplar, HCT in alfalfa [49]).

Since a shift in lignin composition of poplar from G to S due to over-expression of the F5H gene without changes in lignin content increases pulping and bleaching efficiency [50], the reduction of lignin content seems not to be absolutely necessary, changes in lignin composition appear to be sufficient to enhance the pulping efficiency.

After all, lignin plays an important role in mechanical support, resistance to pest and disease, stress response (such as UV-radiation), as well as water an nutrient transport [51]. The question arises whether changes in lignin have a negative effect on plant development.

As expected, a strongly reduced lignin content is often accompanied by phenotypical changes [47, 52, 53]. Only Hu *et al.* (1999) observed higher growth rate for plants with a 45% lignin reduction compared to control plants in the greenhouse [46]. Secondary cell walls with decreased lignin contents are often characterized by collapsed vessels. This irregular xylem phenotype indicates a reduced mechanical cell wall strength [43]. Strongly reduced lignin content also results in altered plant development like reduction of biomass, height and trunk circumference or delayed flowering [47, 49, 52–55].

Field trials carried out with transgenic poplar showed that a less than 10% reduction in lignin content did not considerably change tree characteristics [56], whereas reduction by about 20% caused strongly decreased tree growth [47, 57].

The combination of lignin reducing modifications, the genetic background in which they are placed and the environment in which the resulting plant lines are grown, will subsequently determine the agricultural fitness of reduced lignin plants [58]. Therefore, field trials are necessary to analyze the performance of lignin reduced plants under natural conditions.

Beside analysis of growth indicators like height, trunk diameter or bud burst timing, also the monitoring of attacks by insects, fungi and bacteria are also of interest.

Pilate and colleagues, for instance, evaluated the agronomic performance of CAD down-regulated transgenic trees with altered lignin content and composition (4.2% decreased in lignin; increased frequency of phenolic groups [59]) in long term field trials. No significant differences between transgenic lines and wild-type trees were observed while kraft pulping experiments showed that material of transgenic poplar is more easily delignified using smaller amounts of chemicals and yield more pulp with less cellulose degradation compared with wild-type [56].

This is one of several examples that plants with moderate reductions in lignin content can provide improved digestibility or pulping efficiency [56, 59, 60] while exhibiting normal phenotype and development [53–55].

An alternative solution strategy to improve the release of polysaccharides from lignocellulosic material is the modification of lignin composition by introducing *in situ* peptide cross links into the polymer. The modified lignin should be more susceptible to protease digestion. A number of transgenic poplar lines expressing a transgene encoding a cell wall-targeted high tyrosine-content peptide showed actually enhanced susceptibility to protease digestion, resulting in a higher polysaccharide release from lignocellulose complexes [61]. Now work is ongoing to characterize the fitness of the most promising transgenic lines relative to wildtypes.

This implies that fine tuning of the lignin level will be an important effort for applications in crops, especially to enhance cellulose release in chemical processes on the one side, and to avoid phenotypical changes, on the other.

The variation of the lignin composition like the ratio between H-, G- and S-units appears to be more compatible than changes in the lignin content. Shift in G and S levels, generally have minor effects on plant development [43].

6.1.3   Rubber

Natural rubber is a polymer built-up of isoprene units $(C_5H_8)_n$ connected in a 1,4 *cis* configuration. Commercial production of rubber is almost exclusively from *Hevea brasiliensis*, the Para rubber tree, although rubber is produced in over 2500 plant species [62]. Natural rubber with its molecular structure and high molecular weight (>1 million Da), has high performance properties that cannot be achieved by synthetic rubber produced from petroleum. These unique properties include elasticity, resilience, resistance to abrasion and impact, malleability at cold temperatures and efficient heat dispersion [63]. In the last sixty years of industrial research, these properties could not be duplicated in synthetic rubber because of the unique, to some extent undefined, secondary compounds (proteins, lipids, carbohydrates, minerals) in natural rubber [3].

The annual production level of natural rubber is nearly 10^7 tones. Even if synthetic rubber copolymers, such as styrene-butadiene or acrylonitrile-butadiene, are commercially produced to similar levels, none of them would match the price-performance ratio of natural rubber [2].

Rubber is a component of latex, a milky sap defined as a colloidal suspension of a polymeric material in a liquid system and is produced in specialized cells known as laticifers [62]. The latex of *Hevea* consists of 30–50% (w/w) of natural rubber particles [64] each containing many rubber polymers suspended [62].

By tapping of the Para rubber tree (starting 5–7 years after planting) latex including rubber particles are collected. The rubber yield varies from 500 kg ha^{-1}y^{-1} in smallholder plots to more than 1500 kg ha^{-1}y^{-1} in large plantations. New *Hevea* lines in experimental plots even produce up to 3000 kg ha^{-1}y^{-1}. After 25–30 years, at the end of their productive live, the trees are used for timber production (reviewed by [65]).

Despite the economic importance of natural rubber, not all steps in its biosynthesis have been characterized in detail [66].

Natural rubber biosynthesis is located in the cytosol of the laticifers [67], where it is stored in subcellular rubber particles as an end-product [68].

The synthesis takes place on the surface of rubber particles and is catalyzed by rubber-particle bound *cis*-prenyltransferases (CPTs) by *cis*-1,4-polymerization of isoprene monomers derived from isopentenyl pyrophosphate (IPP). The initiation of this reaction also requires allylic pyrophosphates, which are produced by the action of *trans*-prenyltransferases (TPTs) [68] (Figure 6.5). The rubber molecules accumulated in homogeneous rubber particles that are surrounded by an intact monolayer membrane and particle-bound proteins [63].

Natural rubber is produced from a side branch of the isoprenoid pathway.

Natural rubber is used in over 40,000 products, including more than 400 medical devices, like surgical gloves. Also for heavy-duty tires for airplanes, trucks and busses and countless engineering and consumer products, natural rubber is irreplaceable [2, 63].

Over 90% of the annual 10^7 tones naturally produced rubber are harvested in South-East-Asia, especially in Thailand, Indonesia and Malaysia [2], with growing production in Vietnam and China [69].

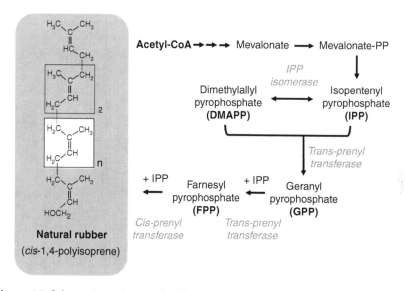

Figure 6.5 Schematic pathway of rubber synthesis (modified after [2]). The IPP isomerase converts isopentenyl pyrophosphate (IPP) to dimethylallyl pyrophosphate (DMAPP). IPP is condensed subsequently in different steps with DMAPP to generate geranyl pyrophosphate (GPP) and farnesyl pyrophosphate (FPP) through *trans*-prenyltransferases. The rubber particle bound rubber transferase (a *cis*-prenyltransferase also called rubber polymerase) binds an allylic pyrophosphate to initiate a new rubber molecule and additionally elongates the polymer by *cis*-1,4-polymerization of isopentenyl monomers derived from IPP. The structural formula shows a rubber polymer, in this representation with farnesyl pyrophosphate as initiator molecule (isoprene units are highlighted in white). Beside FPP also other allylic pyrophosphates like geranyl pyrophosphate or geranylgeranyl pyrophosphate (GGPP) could serve as initiation substrate for polymerization [63].

In contrast to this geographical distribution of successful production, the biological centre of origin of *Hevea* is the Amazon basin in South America [69]. The main reason production has switched from South America to Asia is the South American leaf blight (SALB), a fungal rubber tree disease caused by *Microcyclus ulei*, which originates in the Amazon region and is still uncontrollable either by chemical or biological control measures [3, 69]. Currently, the only barrier to prevent the spread of SALB to Asia is the strict quarantine of the rubber tree plantations [66].

These rubber plantations comprise of clonal trees. In fact, the commercial *Hevea* rubber tree is one of the most genetically restricted crops grown and therefore very susceptible to pathogen attack. The complete genetic basis for rubber tree production outside Brazil can be ascribed to only some hundred genotypes [69].

Because extensive efforts in plant breeding have thus far not yielded resistant rubber tree clones [2], the introduction of *Microcyclus ulei* into the South-East Asian areas is a continuing threat for the world market for all fine rubber products [69]. The serious nature of this threat to the world economy has led to the decision to include SALB in a list of biological weapons [69].

The urgent need to develop alternative sources of natural rubber led to research and development programs during which many plants have been investigated [70]. Although over 2500 plants species have been identified that produce rubber in their latex [62], beside the Para rubber tree only two species are well described to produce large amounts of high molecular weight rubber, an essential determinant of rubber quality: a shrub named guayule (*Parthenium argentatum* Gray) and the Russian dandelion (*Taraxacum koksaghyz*) [66]. Just before and during the World War II (WW II) these two species were successfully used for the production of rubber for army vehicle tires. Due to their relatively poor agronomic performance they were not further exploited, but the trials showed that rubber production does not have to be restricted to equatorial plant species [62]. In the year 2000 high rubber contents with high molecular weight were described in the latex of *Ficus benghalensis* by Kang *et al.*, but no information is available about its practical usage [71].

The production level of high molecular weight rubber in lettuce [67] is too low to render this plant an interesting production system [2].

Other alternative rubber-producing plants produce rubber of low-grade quality or they have not yet been studied in detail to judge their utility [66] (Table 6.3).

Guayule, a non-tropical shrub native in Mexico and parts of the U.S. South-West, is restricted to semi-arid regions [3].

It appears to be a viable alternative to *Hevea* because it produces relatively high amounts of high-quality rubber with essentially the same molecular weight as the material contained in the rubber tree [66]. It is introduced currently as a biannual or multiannual crop, also because most of the rubber is produced during winter month [66] with a peak in January [72]. Guayule needs highly specific winter temperatures for good rubber production [62]. The finding

Table 6.3 Alternative sources of poly-*cis*-isoprene (modified after [66]).

Rubber source		Content (%)	Rubber Mw (kDa)	Production T/Y (year)	Yield (kg ha⁻¹y⁻¹)
Rubber tree	*H. brasiliensis*	30–50 in latex, 2 of tree dw	1310	9 000 000 (2005)	500–3000
Guayule	*P. argentatum* Gray	3–12	1280	10 000 (1910)	300–2000
Russian dandelion	*T. koksaghyz*	Trace-30	2180	3000 (1943)	150–500
Lettuce	*Lactuca serriola*	1.6–2.2 in latex	1380	Research stage	—
Fig tree	*Ficus benghalensis*	17 in latex	1500	—	—
Rubber rabbitbrush	*Chrysothamnus nauseosus*	<7	585	—	—
Goldenrod	*Solidago virgaurea minuta*	5–12 of root dw	160–240	—	110–155
Sunflower	*Helianthus sp.*	0.1–1	69–279	Research stage	—

Abbreviation: dw, dry weight; Mw, molecular weight in kilo Daltons (kDa).

that latex from guayule does not contain the allergy causing proteins found in *Hevea* latex [73] has made guayule latex an attractive source of hypoallergenic products for the medical industry. At least 1–6% of the general world population suffers from *Hevea* latex allergies [2]. A study with workers of the guayule latex processing company Yulex Corporation investigated whether high dose occupational exposure to guayul shrub, homogenate or latex has induced a guayule-specific antibody response in the employees. No IgE antibodies specific to guayule or guayule associated allergic reactions were detected [74]. So guayule latex products would be expected to remain non-allergenic even in long- term use. The reason for this is the extremely low protein content and the general lack of allergenicity associated with only trace proteins in guayule natural rubber-containing products [74].

Unlike *Hevea*, the guayule shrub cannot be tapped for latex extraction, and the kilogram yield per hectare of rubber is only one- to two- thirds that of *Hevea* [3].

Although lots of information on the cultivation of guayule was generated by the Emergency Rubber Project (ERP) during WW II, there are still knowledge gaps in many aspects of guayule growth and development [75].

Therefore, various efforts were carried out to optimize the guayule cultivation like determination of optimum harvest time and cutting height, impact of environmental influences on rubber yield, and breeding of new more efficient guayule lines [72, 76–78]. Since crops harvested several times by cutting the branches and enabling the regrow of the plant give higher yield compared to a whole-plant harvest after 2–5 years [75], this method is now used in the commercial cultivation of guayule [66].

In contrast to the extremely labor intensive *Hevea* rubber production, guayule cultivation, harvesting and rubber production can be mechanized. To harvest the rubber, which is produced as μm-size particles in the bark parenchymal cells and not as free flowing latex, the plant material has to be disrupted thoroughly [66]. Currently, the price of the guayule rubber (costs for cultivation, harvesting, transport and processing) is clearly higher than the price of *Hevea* latex [66]. To improve the economic viability of guayule rubber, coproducts from guayule processing ranging from fuel, adhesives, coating, organic pesticides, wood preservatives and specialty chemicals (e.g. resins, leaf waxes, low molecular-weight rubber) can be used [66].

Since guayule reproduces predominantly by apomixes (asexual reproduction by seeds), it is a difficult species for traditional breeding strategies [78]. Plant transformation could be a helpful tool for overcome these difficulties.

Fortunately, guayule can be genetically modified by *Agrobacterium*-mediated transformation [79, 80].

Veatch *et al.* (2005) tried to increase the amount of allylic pyrophosphates, which are initiators of rubber biosynthesis, available to the rubber transferase by transforming guayule with one of three genes encoding various allylic pyrophosphate synthases in the rubber biosynthetic pathway. The genes, farnesyl pyrophosphate synthase, geranylgeranyl pyrophosphate synthase, and hexa-heptaprenyl pyrophosphate synthase, a mutated form of pyrophosphate synthase, all responsible for producing initiator molecules for rubber biosynthesis in the isoprenoid pathway, were separately integrated into genomes of three different guayule lines under control of a constitutive promoter. In a two-year field study with regular samplings, no effect of transgene expression on rubber production although an increase in resin content could be observed [81]. Unfortunately, none of these three genes were suitable to enhance rubber production.

Therefore, the authors now suggest the expression of a suitable rubber transferase, the enzyme responsible for the cis-1,4-polymerization of the isoprene units, in guayule [81]. It still remains to be shown whether this attempt leads to plants with a sufficient rubber production.

Russian dandelion produces a higher-molecular-mass polymer than either *Hevea* or guayule (about 2MDa), which accumulates in laticifers in the root [3]. This was discovered in Kazakhstan in the 1930s in the course of a strategic program to develop an alternative source of natural rubber in the USSR [66].

An attractive trait of the Russian dandelion is that it could be developed as an annual rubber crop for the temperate regions. For the rubber harvest, the plant roots have to be homogenized and the rubber pressed out or extracted. Yields of 150–500 kg of rubber per annum and hectare have been reported [66].

Nevertheless, converting Russian dandelion to a commercially attractive crop would require an increase in vigor and agronomic properties, for example by hybridization with the common dandelion *Taraxacum officinale* [65].

In field trials with *T. officinale* in New Zealand, 6–9 tons of dried roots per hectare were produced after six month of growth [82].

Van Beilen and Poirier speculate that by combining the large roots of the common dandelion with the improved rubber content of 20% of Russian dandelion through hybridization, theoretically a rubber yield of 1200–1800 kg rubber per hectare could be harvested [65].

Beside breeding strategies, several transgenic approaches have been carried out to transform Russian dandelion into an implementable alternative rubber source.

One example made by Wahler *et al.* (2009) focused on overcoming a disadvantage of *T. koksaghyz* polyphenol rich latex. The tendency of extruded latex to stain brown and coagulate rapidly in contact with the air made rubber extraction of *T. koksaghyz* more complex. These difficulties could be circumvented by reduction of the responsible enzyme, polyphenoloxidase (PPO).

Reduced PPO activity was reached through silencing the PPO gene by constitutive RNA interference in transgenic *T. koksaghyz* plants, which expel four to five times more latex than control plants before coagulation. Prolonged coagulation time would have the advantage that latex could be harvested by low-speed centrifugation from mechanically macerated plant material, as it is currently the case for guayule latex. The dandelion rubber could be subsequently extracted with the process currently used for *Hevea* [83].

It is supposed that the latex coagulation allows the plant to seal open wounds and to form a barrier to microorganism [83]. Hence it remains to be ascertained whether changes in the coagulation might interfere with a suitable plant protection against infections after wounding.

Basic information necessary to increase the rubber production of *T. koksaghyz* has recently been described by Schmidt and colleagues who identified, characterized and analyzed the expression of three endogenous *cis*-prenyltransferase (CTP) genes of *T. koksaghyz* (TkCTP1-3). All three genes are strongly expressed in the latex and give proteins of 32–34 kDa. The deduced amino acid sequences of TkCPT1, TkCPT2, and TkCPT3 were to 98% identical and showed high homology in five conserved regions with CPTs of *Hevea* (HRT1 and HRT2, *Hevea* rubber transferase), isolated and characterized by Asawatreratanakul *et al.* (2003) [68, 84]. Through antibody mediated detection, the researchers demonstrated for the first time in 2010 that *cis*-prenyltransferases are intrinsic parts of the rubber particle. To analyze the functionality of the TkCTPs *Nicotiana tabacum* mesophyll protoplasts were transfected with each of the tree cDNAs. Protein extracts of TkCTP expressing protoplasts showed strong IPP transferase activity, with no significant differences between the three cDNAs. The authors conclude that all three enzymes (TkCTP1-3) act cooperatively in Russian dandelion rubber biosynthesis [85].

The identification of rubber transferase genes is a crucial step and provides the opportunity for overexpression of these enzymes involved in rubber biosynthesis.

Nonetheless, before Russian dandelion can be used as rubber producing crop, the low production yields, labor-intensive cultivation, crosses and seed contamination with other dandelions, and weedy potential still have to be tackled [2].

Meanwhile, all three described plant species are transformable through *Agrobacterium*-mediated transformation [80, 83, 86]. The advantage of dandelion for transgenic modification is its short life cycle. Detectable rubber phenotypes can be obtained within 6 months while improvements in rubber production in guayule can only be assessed after 1–2 years [66]. In the case of the Para rubber tree the major challenge is not to improve rubber yield but to overcome the susceptibility to the South American leaf blight.

6.2   De Novo Synthesis of Polymers in Plants

As described above, gene technology supports breeding to increase the quality of plant components. But in contrast to all other techniques used in plant breeding, it is also able to achieve the plant based production of compounds that have never been produced in the plant kingdom before. Substances like biopolymers derived from bacteria or algae can even be produced as byproduct to traditional materials like starch, sucrose or fatty acids. The following section discusses the production of the fibrous proteins silk, collagen and elastin as well as polyhydroxyalkanoates as polyesters and cyanophycin as storage polypeptide in transgenic plants.

6.2.1   Fibrous Proteins

Natural occurring proteins such as silks from spiders and insects or collagen and elastin from mammals often exhibit interesting properties as fibers. Because of their exceptional material properties including elasticity, toughness and strength these proteins are also useful for the creation of new materials [87].

Fibrous proteins typically contain short blocks of repeated amino acids and can be regarded as elaborate block co-polymers with unique strength-to-weight, adhesive or elastic properties [2].

Protein polymers are only composed of 20 amino acids and some derivatives. Since they are encoded by genes, genetic modification allows the precise determination of both molecular mass and the sequence of amino acids. This makes it in principle possible to achieve a high degree of control of their physical properties and functionality, something which is difficult to achieve with chemical polymerization technologies [88].

An enormous combinatorial range is available by combining repeat sequences of the various natural fibrous proteins or even completely synthetic sequences [89, 90].

The development and uses of protein-based biomaterials in a variety of medical or consumer products is often limited by the difficulties in an economical feasible production of sufficient quantities and qualities of the material. Therefore, plants have been explored as a potential vector for the production of protein-based polymers [88].

6.2.1.1 Silk

Silks represent a remarkably diverse class of structural proteins in nature which are usually produced by a variety of insects and spiders. The amino acid composition, structure and properties of these fibrous proteins depend on the specific source. The most extensively characterized silks are from the domesticated silkworm, *Bombyx mori*, and from spiders (*Nephila clavipes* and *Araneus diadematus*) [91].

The silk build by the domesticated silkworm is used for the production of textiles over thousands of years [92]. The main structural protein of silkworm cocoons is fibroin, a fibrillar protein, while the material used for spider's webs is spidroin [93].

The development of novel silk-based fibers has mainly focused on the silks produced by the golden orb-weaving spider *Nephila clavipes*, which synthesizes different kinds of silks for several purposes [2].

Each of these different silks has a different amino acid composition and exhibits mechanical properties tailored to their specific functions: reproduction as cocoon capsular structures, lines for prey capture, lifeline support (dragline), web construction and adhesion [91].

Over the last decade, interest has focused on dragline silk, mainly because it is thought that such silks have the most desirable properties to be copied and used in a variety of commercial applications [94]. Dragline silks are stronger than steel, when compared on a weight basis [2, 95], have a tensile strength that is comparable to the para-aramid synthetic fiber Kevlar® and are additionally more elastic [96].

The dragline silk is primarily composed of two proteins, the major ampullate spidroins 1 (MaSp1) and 2 (MaSp2). These proteins are highly modular, each with a long repetitive sequence that is flanked on both sides by non-repetitive amino- and carboxy-termini of approximately 100 amino acids [97].

More than 90% of silk sequences consist of repeated oligopeptide motifs, rich in the amino acids alanine, glycine, glutamine and proline. In each silk protein, the composition and number of these oligopeptide motifs are important for the final silk structure [98].

The secondary structure of spider silk consists of up to 3 different structural motifs [92] (Figure 6.6):

(i) a β-turn spiral composed of multiple varying GPGXX motifs (single-letter amino acid code, were X is any amino acid and varies between or within proteins);

(ii) a helix composed of 3 amino acid repeats of GGX (3_{10}-helix); and

(iii) crystalline β-sheet domains rich in alanines, build up by $(GA)_n$ and poly(A)

The two hydrophilic domains (i and ii) are thought to confer reversible extensibility to the silk and the hydrophobic, crystalline domains (iii) are responsible for the high tensile strength of the fiber [2, 92]. Spacers of variable lengths connect these different motifs.

Biomedical applications offer the highest potential for spider silk, due to the combination of excellent mechanical properties, biocompatibility, and slow biodegradability. Silk based biomaterials also have promising scaffolds for engineering skeletal tissue such as bone, cartilage and ligaments, as well as nerve or connective tissue like skin [98]. Additional applications of silk are cosmetic products, such as soaps, shampoos, creams and nail polish. Therein silk is inserted to enhance the brightness, softness and/or toughness

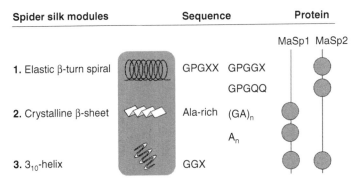

Figure 6.6 Structural modules occurring within spider silks are shown exemplarily for MaSp1 and MaSp2. Modules are similar amino acid motifs. Residues which may vary between or within proteins are presented by an X (modified after [99]).

of the product. Also, an employment of spider silk in technical textiles (used for example in parachutes and bullet-proof vests) which demand high toughness in combination with sleaziness and in technical applications such as micromechanical and electric set-ups, is possible [98].

The requirement for such an extensive use is the cost effective production in sufficient amounts. Unfortunately, the material cannot be obtained in large quantities from spiders [100]. An approach to a solution is the production of artificial spider silk fibers in diverse expression hosts (summarized by [98]).

The underlying silk genes are large (up to 15 Mbp) [98] and the silk proteins have highly repetitive amino acid sequences which retarded their expression in heterologous hosts like bacteria or yeast. Low yields or bad protein purification efficiency based on size limitations, clone instability due to homologous recombination of repetitive sequences, the formation of inclusion bodies and distinct codon usage [1, 2, 98].

In contrast, several examples exhibit the feasibility of spider silk protein production in transgenic plants. The application of plants as expression system could offer an efficient and cheap production system (roughly calculated 10–50% of production costs in bacterial fermenter) that can be scaled up very easily [100].

Previously Scheller *et al.* (2001) reported a plant production of spider silk in transgenic tobacco and potato. Expression of several artificial spider-silk genes under control of the CaMV 35S promoter where the protein is targeted to the endoplasmic reticulum (ER) resulted in an accumulation of up to 2% spider silk protein of total soluble protein (TSP) in tobacco and potato leaves and potato tubers. The genes were assembled so as to achieve very high homology to the native MaSp1 gene from *Nephila clavipes* (more than 90%). The extreme heat stability of the plant-produced spider silk proteins is used to purify the spidroins by a simple and effective procedure [100].

Menassa *et al.* (2004) reported the production of native MaSp1 and MaSp2 in transgenic tobacco plants in the greenhouse and in a field trial. Two different promoters, the enhanced CaMV 35S and the tobacco cryptic constitutive promoter (tCUP) were used, in conjugation with a plant secretory signal and an endoplasmic reticulum retention signal (KDEL). Despite the usage of genes optimized for plant expression and fusion to a translational enhancer (alfalfa mosaic virus, AMV) the expression levels achieved were relatively low (69g per hectare resulting from 0.1% TSP) in comparison to results of Scheller *et al.* (2001) [101]. It is not clear whether the small sequence differences between artificial spider silk proteins and 'native' MaSp1 and MaSp2 are responsible for these differences in expression levels [87].

CaMV 35S promoter driven expression of DP1B, a synthetic analogue of spider dragline silk protein, in *Arabidopsis thaliana* led to an accumulation of up to 6.7% of TSP, when protein accumulation was targeted to the ER of leave tissue. This level could be increased to 8.5% in the leaf apoplast and as high as 18% in seed-ER-targeted expression [102].

Though Arabidopsis is not a good candidate for commercial production, the observed findings could be used to produce a transgenic industrial crop. Seeds, especially those of legumes like soybean are cost effective protein synthesis machines (ca. 38% of dry weight is protein) and seem to be suitable for spider silk protein production [103].

In addition to the optimized production of fibrous proteins, the resulting fiber should have similar characteristics compared to the natural silk proteins from spiders. But the properties of silk fibers depend on the correct assembly of the different types of proteins by spinning. Recombinant spider silk protein (60kDa), obtained from mammalian cells could be successfully spun into filaments showing similar toughness to dragline silk but with lower tenacity [104]. Xia *et al.* (2010) succeeded in spinning of a native sized

284.9 kDa recombinant spider silk protein, expressed in *E. coli,* into fibers, which displayed mechanical properties comparable to those of native silk [97]. Genetic engineering enables the design and creation of a library of synthetic repetitive silk-like genes with given protein sequences that target specific mechanical properties in the resulting fibers [105].

Despite these promising objectives, plant made spider silk is still far away from the application and only few new results were published in the last years. The factors that limit the usage of plant made silk withal the progress in purification processes and advances in spinning remain to be ascertained.

6.2.1.2   *Collagen*

Collagens form a family of extracellular matrix proteins which represent the main structural proteins in vertebrates and many other multi-cellular organisms. The collagen molecules assemble to fibrils that are organized to fibers, providing unmatched structural integrity for the extracellular matrix [106].

Collagen molecules comprise three polypeptides, referred to as α chains, which assemble to form triple helical domains [107]. The α chains are composed of repeating Glycine-X-Y triplets in which X and Y can be any residue but are usually proline and hydroxyproline, respectively [108]. This sequence is necessary for the correct formation of the triple helix [87].

Collagens are synthesized in a soluble precursor form, procollagen. The *in vivo* association starts with three polypeptide chains, which form a triple helix. Importantly, non-triple helical extensions exist at both ends. Extracellularly, before fiber assembly, procollagens undergo specific enzymatic cleavage of the terminal propeptides (Figure 6.7). After assembly, the fibers are stabilized by covalent cross-linking which is initiated by the oxidative deamination of specific lysine and hydroxylysine residues in collagen by lysyl oxidase [109].

Type I collagen represents the prototype of fibrillar collagens, forms the predominant collagen in bone and tendon, and is found in significant quantities in skin, aorta and lung [106].

Its heterotrimer consists of two α1(I) and one α2(I) chains, nevertheless even the homotrimer $(\alpha1[I])_3$ can form a stable triple helix. Collagen's biocompatibility, biodegradability and low immunogenicity make it attractive for a number of medical applications [1].

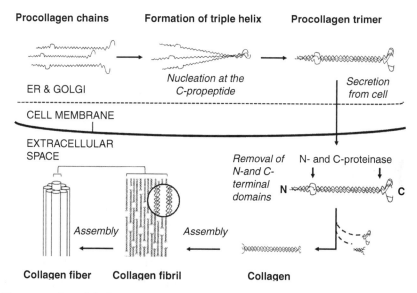

Figure 6.7 Simplified presentation of the formation and subcellular localization of natural collagen fibers (modified after [108]).

The synthesis of procollagen chains occurs in the endoplasmic reticulum (ER). Chains are put together by interactions between the C-propeptides and fold to form a triple-helical domain with globular N- and C-propeptides at the ends. Post-translational modifications taking place in the ER are not shown here. Extracellularly, after transport of procollagen across the Golgi stacks, the N- and C-propeptides are removed by proteinases from fully folded procollagen and the resulting collagen molecules are than able to assemble fibrils via covalent cross links followed by assembling to collagen fibers.

Over the past three decades, several collagen-based medical devices have been approved. Current applications include haemostasis and soft tissue repair to artificial skin, bone repair and drug delivery [110]. Collagen is applicable in tissue engineering of heart valves, ligaments and tendons, nerves, cartilage, menisci, and blood vessels.

However, to date collagen products used in pharmaceutical or biotechnological applications are extracted from animal sources. The use of such materials involves several risks including allergenicity and potential contamination with pathogens [106]. Thus, the development of alternative expression systems for collagen appears attractive.

Ruggerio *et al.* (2000) showed that tobacco plants transformed with a cDNA that encodes the human proα1(I) chain of collagen are able to produce fully processed triple helical collagen. As a consequence of the lack of pyroyl-hydroxylation in plants, the thermal stability of recombinant product was decreased in comparison of the homotrimer purified from animal tissue. Hydroxyprolin residues are necessary to stabilize the structure of the trimeric molecule [107].

To overcome this problem Merle *et al.* (2002) established transgenic tobacco plants that carry a human type-I collagen gene and ancillary one for the chimeric proline-4-hydroxylase (P4H). Thereby, the thermal stability of the plant produced collagen I was significantly improved, although collagen yield was rather low (0.14–20 mg/kg leaf material) [111].

Recently Stein *et al.* (2009) considerably improved collagen expression in tobacco by combining several transgenes in one transgenic plant. In addition to the human procollagen α1(I) and α2(I) chains, this plant expressed both subunits of the human P4H. Furthermore a human lysyl hydroxylase 3 (LH3), an enzyme that sequentially modifies lysyl residues within the collagen polypeptide to hydroxylysyl, galactosylhydroxylysyl and glucosylgalactosylhydroxylysyl, was expressed on the top of the four other genes. These lysyl carbohydrate structures are unique to collagen and have been implicated in the control of fiber diameter. Plants coexpressing all five vacuole-targeted proteins generated intact collagen yields of up to 2% of the extracted total soluble protein (200mg/kg fresh weight). Plant extracted collagen formed thermally stable triple helical structures and demonstrated biofunctionality similar to human-tissue derived collagen supporting binding and proliferation of adult peripheral blood-derived endothelial progenitor-like cells [106].

6.2.1.3   Elastin

Elastin is a structural extracellular matrix protein that is present in all vertebrate connective tissue. Its function includes the provision of elasticity and resilience to tissues, such as large elastic blood vessels (aorta), elastic ligaments, lung and skin, which are subjected to repetitive and reversible deformation [112].

Its ability to resist severe deformation without rupture makes it a particularly interesting target for tissue repair and cell culture application [112].

It is a highly insoluble, cross-linked polymer synthesized *in vivo* as soluble tropoelastin monomers (ca. 66 kDa), which become extensively cross-linked through lysine residues in the extracellular matrix, forming large complex arrays [110]. The reaction is catalyzed by lysyl oxidase, which oxidatively deaminates lysine [109].

The elastin sequence is characterized by two major domains: the first is hydrophilic and contains many cross-linked lysine and alanine residues. The second domain is hydrophobic and rich in nonpolar residues, in particular valine (V), proline (P), alanine (A) and glycine (G), characteristically present as tetra-, penta- and hexa-repeats, like V-P-G-G, V-P-G-V-G and A-P-G-V-G-V. These repeats are responsible principally for the elasticity of the protein [112].

Under suitable *in vitro* conditions (pH, temperature and ionic strength), elastin undergoes a process of ordered self-aggregation, the so called coaservation caused by multiple and specific interactions of individual hydrophobic domains. This is a reversible process by which, for example, an increase in temperature leads to soluble protein partitioning from the solvent to an aggregated phase [110]. The elastin aggregates formed through coaservation appear as ordered fibrillar structures resembling the elastic fiber core, indicating that the protein has an intrinsic ability to organize into polymeric structures [109].

Synthetic proteins made from multiple repeats of the sequence G-V-G-V-P, based on the repeating pentamer V-P-G-V-G of elastin, also display elastic properties. They are biocompatible as well as biodegradable polymers, being non-toxic and naturally resorbed by animal tissue [88]. Medical applications could be tissue engineering, wound recovering and programmed drug delivery. Non-medical applications contain super-absorbents, transducers and biodegradable plastics [88].

(G-V-G-V-P)$_{121}$ was produced in cytoplasm of transgenic tobacco plants. But only poor yields could be observed (0.01% to 0.05% of TSP) [113]. Interestingly, the expression of the transgene in the chloroplast instead of the nucleus, followed by a 100-fold higher level of polymer mRNA, did not increase the accumulation of the recombinant protein [114].

Scheller *et al.* (2004) combined a synthetic spider silk protein (SO1) and the elastin like polypeptide (ELP) (V-P-G-X-G)$_{100}$ (X stands for G, V or A) in a fusion protein, which was expressed in transgenic tobacco and potato plants. Through the targeting to and retention in the ER, best producer plants accumulated the spider silk-elastin fusion protein up to 4% of TSP. The SO1-100xELP protein could be purified by a simple method and extraction of one kg tobacco leaf material leads to a yield of 80mg pure recombinant spider silk-elastin protein [115]. The proportion of 4% TSP reached is twice as high as the level of the spider silk protein alone produced in plants by Scheller *et al.* (2001).

This enhancement could base on the fact, that endoplasmic reticulum-targeted elastin-like polypeptide fusions induce the formation of a novel type of protein body [116]. This exclusion of heterologous proteins from normal physiological turnover is possibly responsible for positive effect of elastin like polypeptides on recombinant protein accumulation.

The biocompatibility of the transgenic protein SO1-100xELP was demonstrated by successful growth of fixation-dependent mammalian cells on spider SO1-100xELP coated culture plates [115]. The fusion of spider silk and elastin leads to a new biomaterial. Foils and layers of SO1-100xELP prepared by casting and spin-coating were more elastic and harder than the thermoplastic polymers polyethylenterephthalate (PET) or polyetherimide (PEI) [117], which are currently used for industrial applications like PET bottles for soft drinks, food packaging and thermal isolation. Thus, spider silk-elastin polymer has been suggested as a promising starting point for the development of novel layers and membranes [112, 117].

Also Patel *et al.* (2007) used a 11.3 kDa elastin-like polypeptide $(V-P-G-X-G)_{27}$ as fusion partner for the natural *N. clavipes* silk protein MaSp2 and expressed it in tobacco. The concentration of the recombinant protein MaSp2 increased from 0.0125% TSP (MaSp2 alone) by a factor 60 to 0.75% TSP in case of MaSp2-ELP fusion [118].

Efficient expression of large proteins, which contain a high proportion of only a few amino acids, in eukaryotes is probably limited by the availability of these amino acids or the corresponding tRNAs. Therefore, high level production of such proteins in plants may be dependent on additional genetic engineering, in particular modifications in the amino acid biosynthetic pathways and/or tRNA pools [88].

The fibrous protein synthesis in plants could also be improved by targeting to subcellular compartments and tissues that are optimal for protein synthesis and storage [2].

Assuming that expression and processing limitations can be resolved, heterologous expression in plants would enable production on much larger scale than it is possible in bioreactors [2] and could be a more realistic approach for resource-poor areas with a low level of technological infrastructure than the current alternatives of bacteria, yeast, insect and mammalian cells [112].

6.2.2   Polyhydroxyalkanoates

Polyhydroxyalkanoates (PHAs) are biological polyesters that are synthesized in over 100 different genera of bacteria [88]. These PHAs consist of 3-hydroxy fatty acids with chain length of 4–16 carbons and serve as carbon and energy storage compounds that accumulate in the form of granules [2].

Depending on composition and resulting properties PHAs have many and wide-ranging potential applications [2]. These molecules have material properties that are similar to those of some common plastics, such as polypropylene [87]. Therefore, PHAs have been proposed as alternatives to conventional petroleum-based plastics [3].

The homopolymer poly(3-hydroxybutyrate) (PHB), the simplest, most common and well studied PHA, is a relatively stiff and brittle plastic with limited commercial applications [2, 88].

In contrast, PHA co-polymers, composed primarily of 3-hydroxybutyrate with a fraction of longer chain monomers, such as 3-hydroxyvalerate, 3-hydroxyhexanoate or 3-hydroxyoctanoate, are more flexible and tougher plastics [88].

Medium-chain-length PHAs (mclPHAs), PHAs consisting of higher-molecular-weight monomers (C6-C16), are typically rubber-like materials with an amorphous soft/sticky consistency [2, 119].

Since PHA products can be decomposed using a PHA-depolymerase detected in several bacteria and fungi, they are favorable raw materials to produce disposable items that often end up in the environment like flowerpots, fishing lines, nets, bags and foils, which should decompose if lost and other items like disposable tableware that can be composted after using (reviewed by [120]).

Earlier applications of PHAs were mainly in the area of packaging e.g. shampoo bottles, cosmetic containers, cover for paper and cardboards, milk cartons and films, moisture barriers in nappies and sanitary towels etc. [121]. More recently attention has focused on the medical applications of PHAs e.g. in tissue engineering. They are widely used as bone plates, osteosynthetic materials and surgical sutures [120].

For poly(3-hydroxybutyrate) (PHB) synthesis three enzymes are necessary [122] (Figure 6.8):

(i) β-ketoacetyl-CoA thiolase (PhbA) that catalyses the reversible condensation of two acetyl-CoA monomers to form acetoacetyl-CoA,

(ii) acetoacetyl-CoA reductase (PhbB) which reduces acetoacetyl-CoA to (R)-3-hydroxybutyryl-CoA, and

(iii) PHA synthase (PhbC) which finally polymerizes (R)-3-hydroxybutyryl-CoA monomers into PHB.

Despite their interesting characteristics, commercial PHB production is limited by the high costs of bacterial fermentation, which are reported to be five times more expensive than chemical synthesis of polyethylene [87]. Production of PHA in transgenic plants has been proposed as an economically viable alternative to bioreactor-based or chemical synthesis [3]. It was calculated that polymer concentration in plants will need to reach at least 15% of dry weight for economically useful production [87].

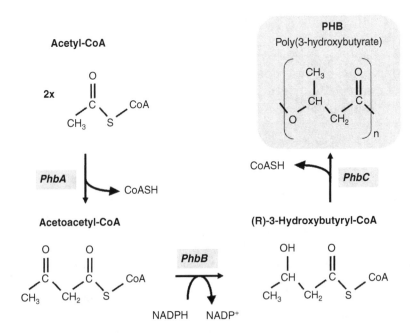

Figure 6.8 Biosynthetic pathway for PHB. PHB is synthesized in three enzymatic steps catalyzed by ß-ketoacetyl-CoA thiolase (PhbA); NADPH-dependent acetoacetyl-CoA reductase (PhbB) and PHA-synthase (PhbC) (modified after [122]).

The general feasibility of the production of PHB in plants was shown 1992 in pilot experiments with the model plant *Arabidopsis thaliana*. Since the β-ketothiolase is already present in cytoplasm of plants, the cytoplasmatic expression of the acteoactetyl-CoA reductase and PHA synthase of *Ralstonia eutropha* led to PHB accumulation of 0.1% of leaf dry weight (dw). Despite this low PHB content the plants showed strong growth retardation and reduced seed production. Nevertheless, the plant-produced PHB formed granules comparable to those formed by bacteria [123] and also the analysis of the resulting polymer indicated that its chemical structure is identical to that of bacterial produced PHB [124].

The deleterious effects of PHB production in the cytoplasm of plants might be caused by the diversion of acetyl-CoA and acetoacetyl-CoA away from the endogenous flavonoid and isoprenoid pathways, which are responsible for the synthesis of a range of plant hormones and sterols [2].

Similar low accumulation was observed when PHB synthesis was established in the cytoplasm of tobacco (0.01% dw; [125]), oilseed rape (0.1 % dw, [126]) and in cotton fiber cells (0.34% dw, [127]).

Production of higher amounts of PHB in plants was achieved by moving the biosynthetic enzymes to the plastids. Therefore, the three enzymes (PhbA, PhbB and PhbC) were fused to a transit peptide and combined in one plant by sexual crossing. In this way the polymer accumulation in *A. thaliana* leaves could be increased up to 14% of leaf dw without significant effects on plant growth, despite evidence of leaf chlorosis [128]. By combination of all three genes in one vector plasmid, Bohmert *et al.* 2000 generated transgenic *A. thaliana* plants with PHB accumulation as high as 40% of leaf dw. Nevertheless, these plants exhibited stunted growth and a loss in fertility [129].

These results indicate that although chloroplasts could accommodate a higher production of PHAs with minimal impact on plant growth compared to the cytoplasm, polymer accumulation above a critical point had a negative impact on the chloroplast function [130].

In general, plastids have proven to be the most promising cellular compartments for PHB synthesis in a wide range of plant species. Alfalfa (0,18% dw [131] and tobacco (0,32% [132]) typically yielded lower amounts, while intermediate results were obtained with sugar cane leaves (1,9% dw [133]) sugar beet hairy roots (5,5% dw [134]) corn leaves (5.7% dw [126]) and switchgrass (3,7% leaf dw [135]) (Table 6.4). PHB producing switchgrass, a dominant species of the central North American tall-grass prairie that is seen as

Table 6.4 Polyhydroxyalkanoates produced in transgenic crop plants (modified after [2]).

Produced PHA		Plant species	Tissue	Subcellular localization	PHA yield (% dry weight)	Reference
PHB	Thale cress	*Arabidopsis thaliana*	Shoot	Cytoplasm	0.1	[123]
			Shoot	Plastid	14–40	[128, 129]
	Switchgrass	*Panicum virgatum*	Leaves	Plastid	3.72	[135]
	Alfalfa	*Medicago sativa*	Shoot	Plastid	0.18	[131]
	Corn	*Zea mays*	Shoot	Plastid	5.7	[126]
	Cotton	*Gossypium hirsutum*	Vascular bundles	Cytoplasm	0.34	[127]
	Oilseed rape	*Brassica napus*	Shoot	Cytoplasm	0.1	[126]
			Seed	Plastid	9	[136]
	Tobacco	*Nicotiana tabacum*	Shoot	Cytoplasm	0.01	[125]
			Shoot	Plastid	0.32–1.7	[132, 138]
	Sugar beet	*Beta vulgaris*	Hairy roots	Plastid	5.5	[134]
	Sugar cane	*Saccharum spp.*	Leaves	Plastids	1.88	[133]
	Flax	*Linum usitatissimum*	Stem	Plastids	0.005	[139, 140]
P(HB-co-HV)	Oilseed rape	*Brassica napus*	Seed	Plastid	2.3	[141]
mclPHA	Thale cress	*A. thaliana*	Whole plant	Peroxisome	0.4	[142]
	Potato	*Solanum tuberosum*	Shoot	Plastid	0.03	[143]
	Tobacco	*N. tabacum*	Leaves	Plastid	0.48	[130]

PHA, polyhydroxyalkanoates; PHB, poly(3-hydroxybutyrate); P(HB-co-HV), poly(3-hydroxybutyrate-co-3-hydroxyvalerate); mclPHA, medium-chain-length PHA valuable biomass crop [3], developed normally and accumulated biomass at similar levels compared to wild-type [135].

In addition to leaf chloroplasts, seed leucoplasts in rapeseed were used for PHB production. The highest producing line accumulated 7.7% of fresh seed weight (fsw) of mature seeds [136]. Since such seeds still contain approximately 15% (w/w) water [137] 7.7% fsw corresponds to 9% of dry weight.

Since nuclear encoded proteins are expressed to a lesser extent then those encoded by plastidic genes, it was supposed, that the direct expression of the PHB pathway in the plastid genome might increase the PHB yield without increasing the deleterious effects. Nevertheless, in tobacco this strategy only leads to relatively low amounts up to 1.7% dw, accompanied by growth deficiency and male sterility [138].

In order to improve the physical properties of PHB, extensive efforts have been made to synthesize co-polymers with better properties like poly(3-hydroxybutyrate-co-3-hydroxyvalerate) (P(HB-co-HV)) and medium-chain-length PHA (mclPHA) [2].

P(HB-co-HV) produced by inclusion of 3-hydroxyvalerate in PHB is less stiff and tougher than PHB and also easier to process, which renders it a good target for commercial application. This copolymer marketed under the tradename Biopol™, was first synthesized by bacterial fermentation [88]. The biochemical pathway for PHB and P(HB-co-HV) are essentially identical, differing only in the initial metabolites. P(HB-co-HV) synthesis requires a condensation of acetyl-CoA with propionyl-CoA instead of condensation of two acetyl-CoA molecules as in the case of PHB [141]. Although acetyl-CoA is present in the cytosol and plastids, propionyl-CoA is restricted to the mitochondria [3]. Slater *et al.* (1999) tried to overcome this problem by coexpression of a threonine deaminase from *E. coli* along with the three PHB biosynthetic proteins targeted to plastids. The threonine deaminase leads to an accumulation of 2-ketobutyrate, which can be converted, admittedly at low efficiency, to propionyl-CoA by the plant pyruvate hydroxygenase complex. This strategy led to P(HB-co-HV) accumulation up to 2.3% of seed weight in seeds of oil seed rape and 1.6% plant tissue dw in *A. thaliana* [141]. Levels of P(HB-co-HV) copolymer production are lower than for PHB, indicating a potential metabolic penalty of propionyl-CoA formation and/or threonine deaminase expression in the plastid [88].

Recent approaches analyze the possibility of plastidic PHB accumulation in oil palm. Oil palm seems to be especially suitable for PHB production because this plant has a high concentration of the PHB precursor acetyl-CoA. Various PHB and P(HP-co-HV) multiple gene vectors driven by oil palm specific promoters are available [144, 145]. Ismail *et al.* produced 2010 transgenic oil palm plants by *Agrobacterium*-mediated transformation. Nevertheless, PHB contents are not published up to now [146].

MclPHBs represent a broad family of PHAs containing 3-hydroxyacids ranging from six to 16 carbons. Mcl-PHAs are generally attractive as a source for elastomers, rubber and glue [147].

By integration of a peroxisome targeted PHA synthase a mclPHA was produced up to 0.4% dw in peroxisomes of *A. thaliana* seedlings [142]. The PHA synthase used 3-hydroxyacyl-CoA intermediates generated in the peroxisome by the fatty acid β-oxidation pathway for PHA polymerization. The mclPHA polymers consisted of 40–50 mol % of C12 and longer monomers and contained a significant amount of unsaturated monomers, which provides the polymer soft- and stickiness [2].

Similar mclPHA levels (0.48% dw) were achieved in transplastomic tobacco plants harboring the PHA synthase and the 3-hydroxyacyl-acyl carrier protein-CoA transferase. This transferase converted 3-hydroxyacyl-carrier protein (ACP) of the fatty acid biosynthetic pathway to 3-hydroxyacyl-CoAs, which subsequently were polymerized by the PHA synthase [130]. The monomers of accumulated mclPHAs were mainly 3-hydroxyoctanoate and 3-hydroxydecanoate. The transplastomic tobacco plants showed beside leaf chlorosis normal morphology just as the wild type tobacco [130].

In potato leaves the same set of genes targeted to the plastids led to only low mclPHA accumulation (below 0.03% dw) accompanied by abnormal phenotypes [143].

Although most efforts on the metabolic engineering of plant for PHA synthesis have been focused towards using the polymer as a source for plastics for consumer products, a novel use could be the modification of plant fiber properties.

For example, low level PHB production (0.3% dw) in transgenic cotton fibers cause measurable changes in their thermal properties. The rate of heat uptake and cooling was slower in transgenic fibers, resulting in higher heat capacity, what suggested enhanced insulation characteristics [127]. And even low PHB accumulation in vascular bundle of flax resulted in the substantial improvement of elastic properties of the harvested flax fibers [139, 140]. Wrobel-Kwiatkowska and colleagues expected that thus-engineered bio-fibers will be a good starting point for the generation of novel composites for biomedical and industrial applications [148].

Even though the great efforts to optimize the PHA production in plants led to considerable success, the PHA composition is still not ideal and the synthesis, regulation of precursors (like acetyl-CoA, propionyl-CoA or 3-hydroxyacyl-ACP) and the efficacy to channel them towards PHA without deleterious effects on plant growth needs more investigation [1].

The PHA accumulation level in bacteria reaches up to 85 % for PHB and 50% for mclPHA, which is probably unachievable for plants. Also producing a wide range of PHA types is unrealistic in the context of transgenic crops. The PHA synthesis in crop plants must be limited to few types of PHAs that would be used for large-scale, low-cost bulk applications, like substitutes for plastics used in common consumer products. In contrast, bacterial fermentation allows an enormous flexible production of a wide spectrum of PHAs with different physical properties. Therefore bacterial PHA would be used for lower-volume, higher-value applications [2].

Possibly, the transient expression [149] of PHA synthase genes in the last phase of the plant life cycle might be an approach to increase PHA production without deleterious effects on the plant. Nevertheless, this would mean that transgenic agrobacteria have to be applied in the field on fully developed plants and transformation has to occur without laborious wounding or vacuuminfiltration. Also biosafety aspects of this bacterial release remain to be ascertained. In addition, public acceptance might limit the use of transient expression in the field.

6.2.3 Cyanophycin

Cyanophycin (multi-L-arginyl-poly-L-aspartic acid) is a branched cyanobacterial polymer composed of equimolar amounts of aspartic acid and arginine arranged as a poly-α-aspartic acid backbone to which arginine residues are linked via their α-amino group to the β-carboxyl group of each aspartate residue [150, 151]. Lysine can be incorporated as well, possibly replacing arginine [152, 153].

The variable polymer in length (25–125 kDa) [150] is deposited in the cytoplasm in membrane-less granules [154] in many Cyanobacteria and some other non-photosynthetic bacteria [155, 156].

It functions as a temporary nitrogen, energy and possibly also carbon reserve [157, 158]. Cyanophycin contains five nitrogen atoms in every building block, representing an ideal intracellular nitrogen reserve [159].

Only one enzyme, the cyanophycin synthetase encoded by *cph*A, is necessary to catalyze the ATP-dependent elongation of a cyanophycin primer by the consecutive addition of L-aspartic acid and L-arginine [152, 156, 160] using nonribosomal peptide synthesis [161].

Cyanophycin is insoluble under physiological conditions (neutral pH), soluble in diluted acid or base and is insensitive to commercially available proteases [151, 159]. Only degradation by specific cyanophycin hydrolases leads mainly to the release of arginine-aspartate dipeptides [158].

Although cyanophycin per se is not a polymer with interesting material properties, it is of interest as a source of its single components.

Mild hydrolysis of cyanophycin results in homo- and copolymers of polyaspartate and L-arginine [162].

Polyaspartate is a soluble, non-toxic and biodegradable polycarboxylate [163], which could replace the non-biodegradable petrochemical polyacrylates in a number of industrial, agricultural, and medical applications [162, 164–166]. Since polyaspartate-producing organism have not been identified so far, polyaspartate can only be produced chemically [167].

The amino acid L-arginine has been suggested to be a regulator of several physiological and immunological processes, for instance being a growth inductor [168–170], an immune system stimulator [171–178] or an inhibitor of tumor cell growth [179–181].

Current investigations concentrate on the production and application of the dipeptides aspartate-arginine and aspartate-lysine resulting from enzymatic degradation of cyanophycin. Since dipeptides are more efficiently utilized than free amino acids, because of a higher absorption rate, cyanophycin derived dipeptides could be an effective approach for the oral administration of the constituting amino acids as therapeutic and/or nutritional agent [158, 182, 183].

Furthermore, the pure amino acids aspartate and arginine could serve as starting point for the synthesis of a range of chemicals (Figure 6.9).

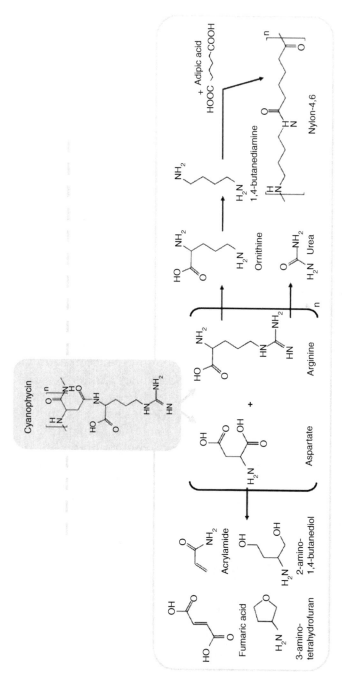

Figure 6.9 Potential products derived from the polymer cyanophycin (modified after [185]).

Arginine can be transformed to 1,4-butanediamine, which can be used for the production of nylon-4,6 and aspartate into numerous chemicals like 2-amino-1,4-butanediol, 3-aminotetrahydrofuran (used in the polymer industry as analogs of high-volume chemicals), fumaric acid (used for polyester resins) and acrylamide (used as thickener, manufacture of dyes or papermaking) [184].

For large scale production, cyanobacterial *cph*A genes were heterolougously expressed in *Escherichia coli, Pseudomonas putida, Corynebacterium glutamicum, Ralstonia eutropha* as well as yeast. Recombinant cyanophycin exhibits a molecular mass of 25–30kDa and contains small amounts of lysine replacing arginine [156, 158, 186].

To establish the production of cyanophycin in plants, the cyanophycin synthetase gene from *Thermosynechococcus elongatus* BP-1 (*cph*A$_{Te}$) was integrated into tobacco and potato plants by *Agrobacterium*-mediated transformation. The *cph*A$_{Te}$ gene was constitutively expressed under the control of the *Cauliflower Mosaic Virus* (CaMV) 35S promoter. Maximum polymer contents of 1.14% per dry weight (dw) in tobacco and 0.24% of dw in potato leaves were achieved when the enzyme was expressed in the cytosol [187].

The isolated polymer was polydisperse with maximum size of about 35kDa. Digestion trials with the cyanophycin-specific, hydrolytic enzyme cyanophycinase showed rapid degradation of the plant derived polymer. Also amino acid analysis validated the successful production of cyanophycin in plants, which also contained lysine in small quantities. However, cyanophycin producing plants of both species exhibited phenotypical changes like thickened cell walls, variegated leaves and growth retardation. Tubers of the potato clone with highest cyanophycin accumulation could not be propagated because of missing eye development, although the cyanophycin content was so low that it could only be demonstrated by electron microscopy.

Like already observed for PHB, targeting the cyanophycin synthase to plastids of *Nicotiana tabacum* improved the cyanophycin production drastically. For the translocation different transit peptides were tested but only the signal sequence of the integral protein of photosystem II (PsbY) worked in combination with the *cph*A$_{Te}$ coding region.

Yields of up to 6.8% dw in leaves of the T$_2$-progenies were obtained, without significant disturbance of plant growth and development. However, the line producing the highest amounts of cyanophycin produced fewer seeds [188].

To decrease the phenotypic abnormalities and to increase polymer production in case of potato, cyanophycin accumulation was restricted to the potato tubers by using the $cphA_{Te}$ gene in combination with the tuber-specific class 1 promoter (B33).

Tuber-specific cytosolic as well as tuber-specific plastidic expression resulted in significant polymer accumulation exclusively in the tubers. In plants with cytosolic expression, both cyanophycin synthetase and cyanophycin were detected in the cytoplasm leading to an increase up to 2.3% cyanophycin of dw. Unfortunately, these plants developed small and deformed tubers. While targeting cyanophycin synthetase to the amyloplasts led to polymer accumulation in tubers of up to 7.5% dw with only minimal effects on growth and morphology of the plants, potato tubers tend to develop stress symptoms, especially those with high cyanophycin content. According to green house experiments, the starch content in the transgenic lines was reduced to 76.4 to 93.6% of the starch measured in the control lines [189]. These adverse effects require further attention and might possibly be overcome by manipulation of the L-aspartate and/or L-arginine biosynthetic pathway.

Nevertheless, the commercial use of cyanophycin producing potatoes requires several properties of the transgenic lines: firstly similar polymer amounts have to be produced stably in the field, secondly, the tubers capacity to propagate after a storage period of around eight month, and thirdly, the development of cheap and efficient isolation methods.

If this is guaranteed, production of cyanophycin as co-product seems to be a possible way to succeed in economic terms.

6.3   Conclusion

Based on gene technology, great progress has been made in the production of biopolymers in plants. This is due to the careful selection of production compartments like chloroplast or endoplasmic reticulum (ER), production organs like seeds or tubers as well as the selection and combination of genes involved in the synthesis. Nevertheless, application is not in sight except for the amylopectin potato *Amflora*. This is because of a number of limitations that have to be addressed.

One important challenge is, the production efficiency which still to be improved, possibly by an improved supply of precursors, tissue specific expression and the selection of optimized genes. In addition, it is of major importance to analyze the causes for yield limitations, reduced seed production and other deleterious effects observed in several high production lines.

The selection of the optimal production system is also of great significance. Besides the ability to cope with the high level expression of foreign compounds, the optimal plant should allow the usage of areas with low agricultural value to reduce the pressure on food production. Unfortunately, arable land area is restricted and will in future probably be further reduced due to urbanization, pollution and climate changes [190, 191]. On the other hand, world population increases constantly [192] and the global food security will be more than ever a challenge [193]. Consequently, there is an urgent need to develop improved, strategies for sustainable agriculture, which provides sufficient amounts and quality of plant biomass.

Hence, species able to produce high biomass in less fecund areas like *Miscanthus* and switchgrass are potential targets for polymer synthesis. Another possibility to reduce the amount of arable land used for non-food purposes is the creation of a double-use plant. Van Beilen and Poirier conclude that competitive PHA production in plants must assume that PHA is only one of several valuable products obtained from the crop [2]. This idea was also taken into account by the development of starch potatoes that in addition produces the polypeptide cyanophycin. Since the reduction in starch content was low despite of high amounts of the transgene encoded protein [189], it may be assumed that this approach is feasible and plants still posses unexplored production reserves. Nonetheless, it remains to be ascertained whether this holds true for all possible biopolymers, and which plant species can be used as a double-use plant. Industrially processed crops already used for biomass, carbohydrate or fatty acid production like sugar beet, sugar cane, certain starch potato varieties and fiber crops, as well as tobacco and biomass crops like *Miscanthus* and switchgrass could be promising candidates. This approach might in addition increase the outcome for the farmer, and diminish CO_2 emission.

It should also be possible to further explore the plant potential by the usage of normally unexploited parts of the plant to produce the biopolymer such as the stover of corn or leaves of sugar beet and sugar cane [2].

One of the main challenges that has to be met is the development of cost effective and environmental friendly isolation protocols for the biopolymers that are most suitably adapted to the isolation procedure of possible primary compounds.

Beside the changes in the production system that should reduce production costs, some features of the market have to change to allow the application of transgenic biopolymer crops. The economic success of these plants should replace fuel based products will certainly depend on the development of fuel prices. In addition, the political will to change the regulatory framework and to invest in the development of optimized plants is inevitable. Last but not least, public acceptance of transgenic crops is an essential prerequisite for the application. In contrast to North America where the use of transgenic plants is more accepted, it is presently undesired in other regions of the world, especially in Europe. The choice of non-food crops, if possible those lacking close relatives in the cultivation area, as production system might help to overcome these reservations.

References

1. Börnke F, Broer I: Tailoring plant metabolism for the production of novel polymers and platform chemicals. *Current Opinion in Plant Biology* 2010, 13:354–362.
2. van Beilen JB, Poirier Y: Production of renewable polymers from crop plants. *Plant Journal* 2008, 54:684–701.
3. Mooney BP: The second green revolution? Production of plant-based biodegradable plastics. *Biochemical Journal* 2009, 418:219–232.
4. Copeland L, Blazek J, Salman H, Tang MCM: Form and functionality of starch. *Food Hydrocolloids* 2009, 23:1527–1534.
5. Keeling PL, Myers AM: Biochemistry and Genetics of Starch Synthesis. *Annual Review of Food Science and Technology, Vol 1* 2010, 1:271–303.
6. Davis JP, Supatcharee N, Khandelwal RL, Chibbar RN: Synthesis of novel starches in planta: Opportunities and challenges. *Starch-Starke* 2003, 55:107–120.
7. Sestili F, Janni M, Doherty A, Botticella E, D'Ovidio R, Masci S, Jones HD, Lafiandra D: Increasing the amylose content of durum wheat through silencing of the SBEIIa genes. *Bmc Plant Biology* 2010, 10:144.
8. Smith AM: Prospects for increasing starch and sucrose yields for bioethanol production. *Plant Journal* 2008, 54:546–558.
9. Jane JL, Kasemsuwan T, Leas S, Zobel H, Robyt JF: Anthology of Starch Granule Morphology by Scanning Electron-Microscopy. *Starch-Starke* 1994, 46:121–129.

10. Cai LM, Shi YC: Structure and digestibility of crystalline short-chain amylose from debranched waxy wheat, waxy maize, and waxy potato starches. *Carbohydrate Polymers* 2010, 79:1117–1123.

11. Tester RF, Karkalas J, Qi X: Starch - composition, fine structure and architecture. *Journal of Cereal Science* 2004, 39:151–165.

12. Slattery CJ, Kavakli IH, Okita TW: Engineering starch for increased quantity and quality. *Trends in Plant Science* 2000, 5:291–298.

13. Halford NG, Curtis TY, Muttucumaru N, Postles J, Mottram DS: Sugars in crop plants. *Annals of Applied Biology* 2011, 158:1–25.

14. BASF. AMFLORA User Guide. Edited by BASF plant Science Company GmBH. 1–15. 2010. Ref Type: Pamphlet

15. Burrell MM: Starch: the need for improved quality or quantity - an overview. *Journal of Experimental Botany* 2003, 54:451–456.

16. Ellis RP, Cochrane MP, Dale MFB, Duffus CM, Lynn A, Morrison IM, Prentice RDM, Swanston JS, Tiller SA: Starch production and industrial use. *Journal of the Science of Food and Agriculture* 1998, 77:289–311.

17. Weatherwax P: A rare carbohydrate in waxy maize. *Genetics* 1922, 7:568–572.

18. Collins GN: A new type of Indian Corn from China. *Bureau of Plant Industry (Bulletin)* 1909, 161:1–30.

19. Nelson OE, Rines HW: Enzymatic Deficiency in Waxy Mutant of Maize. *Biochemical and Biophysical Research Communications* 1962, 9:297-300.

20. Hovenkamp-Hermelink JHM, Jacobsen E, Ponstein AS, Visser RGF, Vosscheperkeuter GH, Bijmolt EW, Devries JN, Witholt B, Feenstra WJ: Isolation of An Amylose-Free Starch Mutant of the Potato (Solanum-Tuberosum-L). *Theoretical and Applied Genetics* 1987, 75:217–221.

21. Nakamura T, Yamamori M, Hirano H, Hidaka S, Nagamine T: Production of Waxy (Amylose-Free) Wheats. *Molecular & General Genetics* 1995, 248:253–259.

22. Li XQ: Molecular characterization and biotechnological improvement of the processing quality of potatoes. *Canadian Journal of Plant Science* 2008, 88:639–648.

23. Blennow A, Wischmann B, Houborg K, Ahmt T, Jorgensen K, Engelsen SB, Bandsholm O, Poulsen P: Structure function relationships of transgenic starches with engineered phosphate substitution and starch branching. *International Journal of Biological Macromolecules* 2005, 36:159–168.

24. Raemakers K, Schreuder M, Suurs L, Furrer-Verhorst H, Vincken JP, de Vetten N, Jacobsen E, Visser RGF: Improved cassava starch by antisense inhibition of granule-bound starch synthase I. *Molecular Breeding* 2005, 16:163–172.

25. Kull B, Salamini F, Rhode W: genetic engineering of potato starch composition. Inhibition of amylose biosynthesis in tubers from transgenic potato lines by the expression of antisense sequences of the gene for granule-bound-starch synthase. *Journal of Genetics and Breeding* 1995, 49:69–76.

26. Wandelt C: Implementation of general surveillance for amflora potato cultivation - Data management. *Journal fur Verbraucherschutz und Lebensmittelsicherheit-Journal of Consumer Protection and Food Safety* 2007, 2:70–71.

27. EFSA report. Opinion of the Scientific panel on genetically modified organisms on an application (Reference EFSA-GMO-UK-2005-14) for the placing on the market of genetically modified potato EH92-527-1 with altered starch composition, for production of starch and food/feed uses, under regulation

(EC) No 1829/2003 from BASF Plant Science. The EFSA Journal 324, 1–20. 2005. Ref Type: Report.

28. Bundessortenamt. Beschreibende Sortenliste Kartoffeln. 2010. Ref Type: Pamphlet.

29. Böhme T. BASF Plant Science Company GmbH. 2011. Ref Type: Personal Communication.

30. Boerjan W, Ralph J, Baucher M: Lignin biosynthesis. *Annual Review of Plant Biology* 2003, 54:519–546.

31. Collinson SR, Thielemans W: The catalytic oxidation of biomass to new materials focusing on starch, cellulose and lignin. *Coordination Chemistry Reviews* 2010, 254:1854–1870.

32. Siro I, Plackett D: Microfibrillated cellulose and new nanocomposite materials: a review. *Cellulose* 2010, 17:459–494.

33. Reddy N, Yang Y: Biofibers from agricultural byproducts for industrial applications. *Trends in Biotechnology* 2005, 23:22–27.

34. Spinelli R, Hartsough BR, Pottle SJ: On-site veneer production in short-rotation hybrid poplar plantations. *Forest Products Journal* 2008, 58:66–71.

35. Abramson M, Shoseyov O, Shani Z: Plant cell wall reconstruction toward improved lignocellulosic production and processability. *Plant Science* 2010, 178:61–72.

36. Novaes E, Kirst M, Chiang V, Winter-Sederoff H, Sederoff R: Lignin and Biomass: A Negative Correlation for Wood Formation and Lignin Content in Trees. *Plant Physiology* 2010, 154:555–561.

37. Hisano H, Nandakumar R, Wang ZY: Genetic modification of lignin biosynthesis for improved biofuel production. *In Vitro Cellular & Developmental Biology-Plant* 2009, 45:306–313.

38. Li X, Weng JK, Chapple C: Improvement of biomass through lignin modification. *Plant Journal* 2008, 54:569–581.

39. O'Connell A, Holt K, Piquemal J, Grima-Pettenati J, Boudet A, Pollet B, Lapierre C, Petit-Conil M, Schuch W, Halpin C: Improved paper pulp from plants with suppressed cinnamoyl-CoA reductase or cinnamyl alcohol dehydrogenase. *Transgenic Research* 2002, 11:495–503.

40. Halpin C, Holt K, Chojecki J, Oliver D, Chabbert B, Monties B, Edwards K, Barakate A, Foxon GA: Brown-midrib maize (bm1) - a mutation affecting the cinnamyl alcohol dehydrogenase gene. *Plant Journal* 1998, 14:545–553.

41. Vignols F, Rigau J, Torres MA, Capellades M, Puigdomenech P: The Brown Midrib 3 (Bm3) Mutation in Maize Occurs in the Gene Encoding Caffeic Acid O-Methyltransferase. *Plant Cell* 1995, 7:407–416.

42. Baucher M, Halpin C, Petit-Conil M, Boerjan W: Lignin: Genetic engineering and impact on pulping. *Critical Reviews in Biochemistry and Molecular Biology* 2003, 38:305–350.

43. Vanholme R, Demedts B, Morreel K, Ralph J, Boerjan W: Lignin Biosynthesis and Structure. *Plant Physiology* 2010, 153:895–905.

44. Chen F, Reddy MSS, Temple S, Jackson L, Shadle G, Dixon RA: Multi-site genetic modulation of monolignol biosynthesis suggests new routes for formation of syringyl lignin and wall-bound ferulic acid in alfalfa (Medicago sativa L.). *Plant Journal* 2006, 48:113–124.

45. Wrobel-Kwiatkowska M, Starzycki M, Zebrowski J, Oszmianski J, Szopa J: Lignin deficiency in transgenic flax resulted in plants with improved mechanical properties. *Journal of Biotechnology* 2007, 128:919–934.

46. Hu WJ, Harding SA, Lung J, Popko JL, Ralph J, Stokke DD, Tsai CJ, Chiang VL: Repression of lignin biosynthesis promotes cellulose accumulation and growth in transgenic trees. *Nature Biotechnology* 1999, 17:808–812.

47. Leple JC, Dauwe R, Morreel K, Storme V, Lapierre C, Pollet B, Naumann A, Kang KY, Kim H, Ruel K, Lefebvre A, Joseleau JP, Grima-Pettenati J, De Rycke R, ndersson-Gunneras S, Erban A, Fehrle I, Petit-Conil M, Kopka J, Polle A, Messens E, Sundberg B, Mansfield SD, Ralph J, Pilate G, Boerjan W: Downregulation of cinnamoyl-coenzyme a reductase in poplar: Multiple-level phenotyping reveals effects on cell wall polymer metabolism and structure. *Plant Cell* 2007, 19:3669–3691.

48. Coleman HD, Park JY, Nair R, Chapple C, Mansfield SD: RNAi-mediated suppression of p-coumaroyl-CoA 3 '-hydroxylase in hybrid poplar impacts lignin deposition and soluble secondary metabolism. *Proceedings of the National Academy of Sciences of the United States of America* 2008, 105:4501–4506.

49. Shadle G, Chen F, Reddy MSS, Jackson L, Nakashima J, Dixon RA: Downregulation of hydroxycinnamoyl CoA: Shikimate hydroxycinnamoyl transferase in transgenic alfalfa affects lignification, development and forage quality. *Phytochemistry* 2007, 68:1521–1529.

50. Huntley SK, Ellis D, Gilbert M, Chapple C, Mansfield SD: Significant increases in pulping efficiency in C4H-F5H-transformed poplars: Improved chemical savings and reduced environmental toxins. *Journal of Agricultural and Food Chemistry* 2003, 51:6178–6183.

51. Simmons BA, Logue D, Ralph J: Advances in modifying lignin for enhanced biofuel production. *Current Opinion in Plant Biology* 2010, 13:313–320.

52. Coleman HD, Samuels AL, Guy RD, Mansfield SD: Perturbed Lignification Impacts Tree Growth in Hybrid Poplar-A Function of Sink Strength, Vascular Integrity, and Photosynthetic Assimilation. *Plant Physiology* 2008, 148:1229–1237.

53. Wagner A, Donaldson L, Kim H, Phillips L, Flint H, Steward D, Torr K, Koch G, Schmitt U, Ralph J: Suppression of 4-Coumarate-CoA Ligase in the Coniferous Gymnosperm Pinus radiata. *Plant Physiology* 2009, 149:370–383.

54. Ralph J, Akiyama T, Kim H, Lu FC, Schatz PF, Marita JM, Ralph SA, Reddy MSS, Chen F, Dixon RA: Effects of coumarate 3-hydroxylase down-regulation on lignin structure. *Journal of Biological Chemistry* 2006, 281:8843–8853.

55. Reddy MSS, Chen F, Shadle G, Jackson L, Aljoe H, Dixon RA: Targeted downregulation of cytochrome P450 enzymes for forage quality improvement in alfalfa (Medicago sativa L.). *Proceedings of the National Academy of Sciences of the United States of America* 2005, 102:16573–16578.

56. Pilate G, Guiney E, Holt K, Petit-Conil M, Lapierre C, Leple JC, Pollet B, Mila I, Webster EA, Marstorp HG, Hopkins DW, Jouanin L, Boerjan W, Schuch W, Cornu D, Halpin C: Field and pulping performances of transgenic trees with altered lignification. *Nature Biotechnology* 2002, 20:607–612.

57. Voelker SL, Lachenbruch B, Meinzer FC, Jourdes M, Ki CY, Patten AM, Davin LB, Lewis NG, Tuskan GA, Gunter L, Decker SR, Selig MJ, Sykes R, Himmel ME, Kitin P, Shevchenko O, Strauss SH: Antisense Down-Regulation of 4CL

Expression Alters Lignification, Tree Growth, and Saccharification Potential of Field-Grown Poplar. *Plant Physiology* 2010, 154:874–886.

58. Pedersen JF, Vogel KP, Funnell DL: Impact of reduced lignin on plant fitness. *Crop Science* 2005, 45:812–819.

59. Baucher M, Chabbert B, Pilate G, VanDoorsselaere J, Tollier MT, PetitConil M, Cornu D, Monties B, VanMontagu M, Inze D, Jouanin L, Boerjan W: Red xylem and higher lignin extractability by down-regulating a cinnamyl alcohol dehydrogenase in poplar. *Plant Physiology* 1996, 112:1479–1490.

60. Wadenbäck J, von Arnold S, Egertsdotter U, Walter MH, Grima-Pettenati J, Goffner D, Gellerstedt G, Gullion T, Clapham D: Lignin biosynthesis in transgenic Norway spruce plants harboring an antisense construct for cinnamoyl CoA reductase (CCR). *Transgenic Research* 2008, 17:379–392.

61. Liang HY, Frost CJ, Wei XP, Brown NR, Carlson JE, Tien M: Improved sugar release from lignocellulosic material by introducing a tyrosine-rich cell wall peptide gene in poplar. *Clean-Soil Air Water* 2008, 36:662–668.

62. Mooibroek H, Cornish K: Alternative sources of natural rubber. *Applied Microbiology and Biotechnology* 2000, 53:355–365.

63. Cornish K: Similarities and differences in rubber biochemistry among plant species. *Phytochemistry* 2001, 57:1123–1134.

64. Ko JH, Chow KS, Han KH: Transcriptome analysis reveals novel features of the molecular events occurring in the laticifers of Hevea brasiliensis (para rubber tree). *Plant Molecular Biology* 2003, 53:479–492.

65. van Beilen JB, Poirier Y: Guayule and Russian dandelion as alternative sources of natural rubber. *Critical Reviews in Biotechnology* 2007, 27:217–231.

66. van Beilen JB, Poirier Y: Establishment of new crops for the production of natural rubber. *Trends in Biotechnology* 2007, 25:522–529.

67. Bushman BS, Scholte AA, Cornish K, Scott DJ, Brichta JL, Vederas JC, Ochoa O, Michelmore RW, Shintani DK, Knapp SJ: Identification and comparison of natural rubber from two Lactuca species. *Phytochemistry* 2006, 67:2590–2596.

68. Schmidt T, Hillebrand A, Wurbs D, Wahler D, Lenders M, Gronover CS, Prufer D: Molecular Cloning and Characterization of Rubber Biosynthetic Genes from Taraxacum koksaghyz. *Plant Molecular Biology Reporter* 2010, 28:277–284.

69. Lieberei R: South American leaf blight of the rubber tree (Hevea spp.): New steps in plant domestication using physiological features and molecular markers. *Annals of Botany* 2007, 100:1125–1142.

70. Metcalfe CR: Distribution of Latex in Plant Kingdom. *Economic Botany* 1966, 21:115-127.

71. Kang HS, Kim YS, Chung GC: Characterization of natural rubber biosynthesis in Ficus benghalensis. *Plant Physiology and Biochemistry* 2000, 38:979–987.

72. Coffelt TA, Nakayama FS: Determining optimum harvest time for guayule latex and biomass. *Industrial Crops and Products* 2010, 31:131–133.

73. Siler DJ, Cornish K, Hamilton RG: Absence of cross-reactivity of IgE antibodies from subjects allergic to Hevea brasiliensis latex with a new source of natural rubber latex from guayule (Parthenium argentatum). *Journal of Allergy and Clinical Immunology* 1996, 98:895–902.

74. Hamilton RG, Cornish K: Immunogenicity studies of guayule and guayule latex in occupationally exposed workers. *Industrial Crops and Products* 2010, 31:197–201.

75. Foster MA, Coffelt TA: Guayule agronomics: establishment, irrigated production, and weed control. *Industrial Crops and Products* 2005, 22:27–40.

76. Coffelt TA, Ray DT, Nakayama FS, Dierig DA: Genotypic and environmental effects on guayule (Parthenium argentatum) latex and growth. *Industrial Crops and Products* 2005, 22:95–99.

77. Coffelt TA, Ray DT: Cutting height effects on guayule latex, rubber, and resin yields. *Industrial Crops and Products* 2010, 32:264–268.

78. Ray DT, Coffelt TA, Dierig DA: Breeding guayule for commercial production. *Industrial Crops and Products* 2005, 22:15–25.

79. Dong N, Montanez B, Creelman RA, Cornish K: Low light and low ammonium are key factors for guayule leaf tissue shoot organogenesis and transformation. *Plant Cell Reports* 2006, 25:26–34.

80. Pan ZQ, Ho JK, Feng Q, Huang DS, Backhaus RA: Agrobacterium-mediated transformation and regeneration of guayule. *Plant Cell Tissue and Organ Culture* 1996, 46:143–150.

81. Veatch ME, Ray DT, Mau CJD, Cornish K: Growth, rubber, and resin evaluation of two-year-old transgenic guayule. *Industrial Crops and Products* 2005, 22:65–74.

82. Parmenter G. *Taraxacum officinale* - Common dandelion, Lion's tooth. 2002. New Zealand Institute for Crop & Food Research Ltd, Mana Kai Rangahau. Ref Type: Pamphlet.

83. Wahler D, Gronover CS, Richter C, Foucu F, Twyman RM, Moerschbacher BM, Fischer R, Muth J, Prufer D: Polyphenoloxidase Silencing Affects Latex Coagulation in Taraxacum Species. *Plant Physiology* 2009, 151:334–346.

84. Asawatreratanakul K, Zhang YW, Wititsuwannakul D, Wititsuwannakul R, Takahashi S, Rattanapittayaporn A, Koyama T: Molecular cloning, expression and characterization of cDNA encoding cis-prenyltransferases from Hevea brasiliensis - A key factor participating in natural rubber biosynthesis. *European Journal of Biochemistry* 2003, 270:4671–4680.

85. Schmidt T, Lenders M, Hillebrand A, van Deenen N, Munt O, Reichelt R, Eisenreich W, Fischer R, Prufer D, Gronover CS: Characterization of rubber particles and rubber chain elongation in Taraxacum koksaghyz. *Bmc Biochemistry* 2010, 11.

86. Arokiaraj P, Yeang HY, Cheong KF, Hamzah S, Jones H, Coomber S, Charlwood BV: CaMV 35S promoter directs beta-glucuronidase expression in the laticiferous system of transgenic Hevea brasiliensis (rubber tree). *Plant Cell Reports* 1998, 17:621–625.

87. Scheller J, Conrad U: Plant-based material, protein and biodegradable plastic. *Current Opinion in Plant Biology* 2005, 8:188–196.

88. Moire L, Rezzonico E, Poirier Y: Synthesis of novel biomaterials in plants. *Journal of Plant Physiology* 2003, 160:831–839.

89. Holland C, Terry AE, Porter D, Vollrath F: Natural and unnatural silks. *Polymer* 2007, 48:3388–3392.

90. Nagapudi K, Brinkman WT, Leisen J, Thomas BS, Wright ER, Haller C, Wu XY, Apkarian RP, Conticello VP, Chaikof EL: Protein-based thermoplastic elastomers. *Macromolecules* 2005, 38:345–354.

91. Altman GH, Diaz F, Jakuba C, Calabro T, Horan RL, Chen J, Lu H, Richmond J, Kaplan DL: Silk-based biomaterials. *Biomaterials* 2003, 24:401–416.

92. Huang J, Foo CWP, Kaplan DL: Biosynthesis and applications of silk-like and collagen-like proteins. *Polymer Reviews* 2007, 47:29–62.
93. Kon'kov AS, Pustovalova OL, Agapov II: Biocompatible Materials from Regenerated Silk for Tissue Engineering and Medicinal Therapy. *Applied Biochemistry and Microbiology* 2010, 46:739–744.
94. Rising A, Nimmervoll H, Grip S, Fernandez-Arias A, Storckenfeldt E, Knight DP, Vollrath F, Engstrom W: Spider silk proteins--mechanical property and gene sequence. *Zoolog.Sci.* 2005, 22:273–281.
95. Heim M, Keerl D, Scheibel T: Spider Silk: From Soluble Protein to Extraordinary Fiber. *Angewandte Chemie-International Edition* 2009, 48:3584–3596.
96. Hinman MB, Jones JA, Lewis RV: Synthetic spider silk: a modular fiber. *Trends in Biotechnology* 2000, 18:374–379.
97. Xia XX, Qian ZG, Ki CS, Park YH, Kaplan DL, Lee SY: Native-sized recombinant spider silk protein produced in metabolically engineered Escherichia coli results in a strong fiber. *Proceedings of the National Academy of Sciences of the United States of America* 2010, 107:14059–14063.
98. Vendrely C, Scheibel T: Biotechnological production of spider-silk proteins enables new applications. *Macromolecular Bioscience* 2007, 7:401–409.
99. Hayashi CY, Lewis RV: Evidence from flagelliform silk cDNA for the structural basis of elasticity and modular nature of spider silks. *Journal of Molecular Biology* 1998, 275:773–784.
100. Scheller J, Guhrs KH, Grosse F, Conrad U: Production of spider silk proteins in tobacco and potato. *Nature Biotechnology* 2001, 19:573–577.
101. Menassa R, Hong Z, Karatzas CN, Lazaris A, Richman A, Brandle J: Spider dragline silk proteins in transgenic tobacco leaves: accumulation and field production. *Plant Biotechnology Journal* 2004, 2:431–438.
102. Yang JJ, Barr LA, Fahnestock SR, Liu ZB: High yield recombinant silk-like protein production in transgenic plants through protein targeting. *Transgenic Research* 2005, 14:313–324.
103. Barr LA, Fahnestock SR, Yang JJ: Production and purification of recombinant DP1B silk-like protein in plants. *Molecular Breeding* 2004, 13:345–356.
104. Lazaris A, Arcidiacono S, Huang Y, Zhou JF, Duguay F, Chretien N, Welsh EA, Soares JW, Karatzas CN: Spider silk fibers spun from soluble recombinant silk produced in mammalian cells. *Science* 2002, 295:472–476.
105. Teule F, Cooper AR, Furin WA, Bittencourt D, Rech EL, Brooks A, Lewis RV: A protocol for the production of recombinant spider silk-like proteins for artificial fiber spinning. *Nature Protocols* 2009, 4:341–355.
106. Stein H, Wilensky M, Tsafrir Y, Rosenthal M, Amir R, Avraham T, Ofir K, Dgany O, Yayon A, Shoseyov O: Production of Bioactive, Post-Translationally Modified, Heterotrimeric, Human Recombinant Type-I Collagen in Transgenic Tobacco. *Biomacromolecules* 2009, 10:2640–2645.
107. Ruggiero F, Exposito JY, Bournat P, Gruber V, Perret S, Comte J, Olagnier B, Garrone R, Theisen M: Triple helix assembly and processing of human collagen produced in transgenic tobacco plants. *Febs Letters* 2000, 469:132–136.
108. Canty EG, Kadler KE: Procollagen trafficking, processing and fibrillogenesis. *Journal of Cell Science* 2005, 118:1341–1353.
109. Scheibel T: Protein fibers as performance proteins: new technologies and applications. *Current Opinion in Biotechnology* 2005, 16:427–433.

110. Kyle S, Aggeli A, Ingham E, McPherson MJ: Production of self-assembling biomaterials for tissue engineering. *Trends in Biotechnology* 2009, 27:423–433.
111. Merle C, Perret S, Lacour T, Jonval V, Hudaverdian S, Garrone R, Ruggiero F, Theisen M: Hydroxylated human homotrimeric collagen I in Agrobacterium tumefaciens-mediated transient expression and in transgenic tobacco plant. *Febs Letters* 2002, 515:114–118.
112. Floss DM, Schallau K, Rose-John S, Conrad U, Scheller J: Elastin-like polypeptides revolutionize recombinant protein expression and their biomedical application. *Trends in Biotechnology* 2010, 28:37–45.
113. Zhang XD, Letham DS, Zhang R, Higgins TJV: Expression of the isopentenyl transferase gene is regulated by auxin in transgenic tobacco tissues. *Transgenic Research* 1996, 5:57–65.
114. Guda C, Lee SB, Daniell H: Stable expression of a biodegradable protein-based polymer in tobacco chloroplasts. *Plant Cell Reports* 2000, 19:257–262.
115. Scheller J, Henggeler D, Viviani A, Conrad U: Purification of spider silk-elastin from transgenic plants and application for human chondrocyte proliferation. *Transgenic Research* 2004, 13:51–57.
116. Conley AJ, Joensuu JJ, Menassa R, Brandle JE: Induction of protein body formation in plant leaves by elastin- like polypeptide fusions. *Bmc Biology* 2009, 7.
117. Junghans F, Morawietz M, Conrad U, Scheibel T, Heilmann A, Spohn U: Preparation and mechanical properties of layers made of recombinant spider silk proteins and silk from silk worm. *Applied Physics A-Materials Science & Processing* 2006, 82:253–260.
118. Patel J, Zhu H, Menassa R, Gyenis L, Richman A, Brandle J: Elastin-like polypeptide fusions enhance the accumulation of recombinant proteins in tobacco leaves. *Transgenic Research* 2007, 16:239–249.
119. Witholt B, Kessler B: Perspectives of medium chain length poly(hydroxyalkanoates), a versatile set of bacterial bioplastics. *Current Opinion in Biotechnology* 1999, 10:279–285.
120. Philip S, Keshavarz T, Roy I: Polyhydroxyalkanoates: biodegradable polymers with a range of applications. *Journal of Chemical Technology and Biotechnology* 2007, 82:233–247.
121. Keshavarz T, Roy I: Polyhydroxyalkanoates: bioplastics with a green agenda. *Current Opinion in Microbiology* 2010, 13:321–326.
122. Madison LL, Huisman GW: Metabolic engineering of poly(3-hydroxyalkanoates): From DNA to plastic. *Microbiology and Molecular Biology Reviews* 1999, 63:21–+.
123. Poirier Y, Dennis DE, Klomparens K, Somerville C: Polyhydroxybutyrate, a biodegradable thermoplastic, produced in transgenic plants. *Science* 1992, 256:520–523.
124. Poirier Y, Somerville C, Schechtman LA, Satkowski MM, Noda I: Synthesis of High-Molecular-Weight Poly([R]-(-)-3-Hydroxybutyrate) in Transgenic Arabidopsis-Thaliana Plant-Cells. *International Journal of Biological Macromolecules* 1995, 17:7–12.
125. Nakashita H, Arai Y, Yoshioka K, Fukui T, Doi Y, Usami R, Horikoshi K, Yamaguchi I: Production of biodegradable polyester by a transgenic tobacco. *Bioscience Biotechnology and Biochemistry* 1999, 63:870–874.

126. Poirier Y, Gruys KJ: Production of PHAs in transgenic plants. *In: Biopolyesters (Doi, Y. and Steinbüchel, A., eds.) Weinheim: Wiley-VCH*: 2001, 401–435.
127. John ME, Keller G: Metabolic pathway engineering in cotton: Biosynthesis of polyhydroxybutyrate in fiber cells. *Proceedings of the National Academy of Sciences of the United States of America* 1996, 93:12768–12773.
128. Nawrath C, Poirier Y, Somerville C: Targeting of the polyhydroxybutyrate biosynthetic-pathway to the plastids of *Arabidopsis thaliana* results in high-levels of polymer accumulation. *Proceedings of the National Academy of Sciences of the United States of America* 1994, 91:12760–12764.
129. Bohmert K, Balbo I, Kopka J, Mittendorf V, Nawrath C, Poirier Y, Tischendorf G, Trethewey RN, Willmitzer L: Transgenic Arabidopsis plants can accumulate polyhydroxybutyrate to up to 4% of their fresh weight. *Planta* 2000, 211:841–845.
130. Wang YH, Wu ZY, Zhang XH, Chen GQ, Wu Q, Huang CL, Yang Q: Synthesis of medium-chain-length-polyhydroxyalkanoates in tobacco via chloroplast genetic engineering. *Chinese Science Bulletin* 2005, 50:1113–1120.
131. Saruul P, Srienc F, Somers DA, Samac DA: Production of a biodegradable plastic polymer, poly-beta-hydroxybutyrate, in transgenic alfalfa. *Crop Science* 2002, 42:919–927.
132. Bohmert K, Balbo I, Steinbüchel A, Tischendorf G, Willmitzer L: Constitutive expression of the beta-ketothiolase gene in transgenic plants. A major obstacle for obtaining polyhydroxybutyrate-producing plants. *Plant Physiology* 2002, 128:1282–1290.
133. Petrasovits LA, Purnell MP, Nielsen LK, Brumbley SM: Production of polyhydroxybutyrate in sugarcane. *Plant Biotechnology Journal* 2007, 5:162–172.
134. Menzel G, Harloff HJ, Jung C: Expression of bacterial poly (3-hydroxybutyrate) synthesis genes in hairy roots of sugar beet (*Beta vulgaris* L.). *Applied Microbiology and Biotechnology* 2003, 60:571–576.
135. Somleva MN, Snell KD, Beaulieu JJ, Peoples OP, Garrison BR, Patterson NA: Production of polyhydroxybutyrate in switchgrass, a value-added co-product in an important lignocellulosic biomass crop. *Plant Biotechnology Journal* 2008, 6:663–678.
136. Houmiel KL, Slater S, Broyles D, Casagrande L, Colburn S, Gonzalez K, Mitsky TA, Reiser SE, Shah D, Taylor NB, Tran M, Valentin HE, Gruys KJ: Poly(beta-hydroxybutyrate) production in oilseed leukoplasts of *Brassica napus*. *Planta* 1999, 209:547–550.
137. Valentin HE, Broyles DL, Casagrande LA, Colburn SM, Creely WL, DeLaquil PA, Felton HM, Gonzalez KA, Houmiel KL, Lutke K, Mahadeo DA, Mitsky TA, Padgette SR, Reiser SE, Slater S, Stark DM, Stock RT, Stone DA, Taylor NB, Thorne GM, Tran M, Gruys KJ: PHA production, from bacteria to plants. *International Journal of Biological Macromolecules* 1999, 25:303–306.
138. Lössl A, Eibl C, Harloff HJ, Jung C, Koop HU: Polyester synthesis in trans-plastomic tobacco (*Nicotiana tabacum* L.): significant contents of polyhydroxybutyrate are associated with growth reduction. *Plant Cell Reports* 2003, 21:891–899.
139. Wrobel-Kwiatkowska M, Zebrowski J, Starzycki M, Oszmianski J, Szopa J: Engineering of PHB synthesis causes improved elastic properties of flax fibers. *Biotechnology Progress* 2007, 23:269–277.

140. Wrobel M, Zebrowski J, Szopa J: Polyhydroxybutyrate synthesis in transgenic flax. *Journal of Biotechnology* 2004, 107:41–54.
141. Slater S, Mitsky TA, Houmiel KL, Hao M, Reiser SE, Taylor NB, Tran M, Valentin HE, Rodriguez DJ, Stone DA, Padgette SR, Kishore G, Gruys KJ: Metabolic engineering of Arabidopsis and Brassica for poly(3-hydroxybutyrate-co-3-hydroxyvalerate) copolymer production. *Nature Biotechnology* 1999, 17:1011–1016.
142. Mittendorf V, Robertson EJ, Leech RM, Kruger N, Steinbuchel A, Poirier Y: Synthesis of medium-chain-length polyhydroxyalkanoates in Arabidopsis thaliana using intermediates of peroxisomal fatty acid beta-oxidation. *Proceedings of the National Academy of Sciences of the United States of America* 1998, 95:13397–13402.
143. Romano A, van der Plas LHW, Witholt B, Eggink G, Mooibroek H: Expression of poly-3-(R)-hydroxyalkanoate (PHA) polymerase and acyl-CoA-transacylase in plastids of transgenic potato leads to the synthesis of a hydrophobic polymer, presumably medium-chain-length PHAs. *Planta* 2005, 220:455–464.
144. Masani MYA, Parveez GKA, Izawati AMD, Lan CP, Akmar ASN: Construction of PHB and PHBV multiple-gene vectors driven by an oil palm leaf-specific promoter. *Plasmid* 2009, 62:191–200.
145. Yunus AMM, Ho CL, Parveez GKA: Construction of Phb and Phbv Transformation Vectors for Bioplastics Production in Oil Palm. *Journal of Oil Palm Research* 2008, 37–55.
146. Ismail I, Iskandar NF, Chee GM, Abdullah R: Genetic transformation and molecular analysis of polyhydroxybutyrate biosynthetic gene expression in oil palm (Elaeis guineensis Jacq. var Tenera) tissues. *Plant Omics* 2010, 3:18–27.
147. Poirier Y: Production of new polymeric compounds in plants. *Current Opinion in Biotechnology* 1999, 10:181–185.
148. Wrobel-Kwiatkowska M, Skorkowska-Telichowska K, Dyminska L, Maczka M, Hanuza J, Szopa J: Biochemical, Mechanical, and Spectroscopic Analyses of Genetically Engineered Flax Fibers Producing Bioplastic (Poly-beta-Hydroxybutyrate). *Biotechnology Progress* 2009, 25:1489–1498.
149. Gleba Y, Klimyuk V, Marillonnet S: Viral vectors for the expression of proteins in plants. *Current Opinion in Biotechnology* 2007, 18:134–141.
150. Simon RD: Biosynthesis of multi-L-arginyl-poly(L-aspartic acid) in filamentous cyanobacterium *Anabaena cylindrica*. *Biochimica et Biophysica Acta* 1976, 422:407–418.
151. Simon RD, Weathers P: Determination of structure of novel polypeptide containing aspartic acid and arginine which is found in cyanobacteria. *Biochimica et Biophysica Acta* 1976, 420:165–176.
152. Ziegler K, Diener A, Herpin C, Richter R, Deutzmann R, Lockau W: Molecular characterization of cyanophycin synthetase, the enzyme catalyzing the biosynthesis of the cyanobacterial reserve material multi-L-arginyl-poly-L-aspartate (cyanophycin). *European Journal of Biochemistry* 1998, 254:154–159.
153. Aboulmagd E, Oppermann-Sanio FB, Steinbuchel A: Molecular characterization of the cyanophycin synthetase from Synechocystis sp strain PCC6308. *Archives of Microbiology* 2000, 174:297–306.

154. Allen MM, Weathers PJ: Structure and Composition of Cyanophycin Granules in the Cyanobacterium Aphanocapsa 6308. *Journal of Bacteriology* 1980, 141:959–962.

155. Krehenbrink M, Oppermann-Sanio FB, Steinbuchel A: Evaluation of non-cyanobacterial genome sequences for occurrence of genes encoding proteins homologous to cyanophycin synthetase and cloning of an active cyanophycin synthetase from Acinetobacter sp strain DSM 587. *Archives of Microbiology* 2002, 177:371–380.

156. Ziegler K, Deutzmann R, Lockau W: Cyanophycin synthetase-like enzymes of non-cyanobacterial eubacteria: Characterization of the polymer produced by a recombinant synthetase of *Desulfitobacterium hafniense*. *Zeitschrift fur Naturforschung C-A Journal of Biosciences* 2002, 57:522–529.

157. Krehenbrink M, Steinbuchel A: Partial purification and characterization of a non-cyanobacterial cyanophycin synthetase from Acinetobacter calcoaceticus strain ADP1 with regard to substrate specificity, substrate affinity and binding to cyanophycin. *Microbiology-Sgm* 2004, 150:2599–2608.

158. Sallam A, Steinbüchel A: Cyanophycin-degrading bacteria in digestive tracts of mammals, birds and fish and consequences for possible applications of cyanophycin and its dipeptides in nutrition and therapy. *Journal of Applied Microbiology* 2009, 107:474–484.

159. Simon RD: Cyanophycin Granules from Blue-Green Alga Anabaena-Cylindrica - Reserve Material Consisting of Copolymers of Aspartic Acid and Arginine. *Proceedings of the National Academy of Sciences of the United States of America* 1971, 68:265–267.

160. Berg H, Ziegler K, Piotukh K, Baier K, Lockau W, Volkmer-Engert R: Biosynthesis of the cyanobacterial reserve polymer multi-L-arginyl-poly-L-aspartic acid (cyanophycin) - Mechanism of the cyanophycin synthetase reaction studied with synthetic primers. *European Journal of Biochemistry* 2000, 267:5561–5570.

161. Simon RD: Effect of Chloramphenicol on Production of Cyanophycin Granule Polypeptide in Blue-Green-Alga Anabaena-Cylindrica. *Archiv fur Mikrobiologie* 1973, 92:115–122.

162. Joentgen W, Groth TSA, Hai T, Oppermann FB: Polyasparaginic acid homopolymers and copolymers, biotechnical production and use thereof. *US Patent* 2001, 6180752.

163. Tabata K, Abe H, Doi Y: Microbial degradation of poly(aspartic acid) by two isolated strains of Pedobacter sp and Sphingomonas sp. *Biomacromolecules* 2000, 1:157–161.

164. Oppermann-Sanio FB, Steinbüchel A: Occurrence, functions and biosynthesis of polyamides in microorganisms and biotechnological production. *Naturwissenschaften* 2002, 89:11–22.

165. Schwamborn M: Chemical synthesis of polyaspartates: a biodegradable alternative to currently used polycarboxylate homo- and copolymers. *Polymer Degradation and Stability* 1998, 59:39–45.

166. Zotz RJ, Schenk S, Kuhn A, Schlunken S, Krone V, Bruns W, Genth S, Schuler G: Safety and efficacy of LK565 - a new polymer ultrasound contrast agent. *Zeitschrift fur Kardiologie* 2001, 90:419–426.

167. Schwamborn M: Polyasparaginsäuren. *Nachr. Chem. Techn. Lab* 1996, 44:1167–1179.
168. Lenis NP, van Diepen HTM, Bikker P, Jongbloed AG, van der Meulen J: Effect of the ratio between essential and nonessential amino acids in the diet on utilization of nitrogen and amino acids by growing pigs. *Journal of Animal Science* 1999, 77:1777–1787.
169. Roth FX, Fickler J, Kirchgessner M: Effect of Dietary Arginine and Glutamic-Acid Supply on the N-Balance of Piglets .5. Communication on the Importance of Nonessential Amino-Acids for Protein Retention. *Journal of Animal Physiology and Animal Nutrition-Zeitschrift fur Tierphysiologie Tierernahrung und Futtermittelkunde* 1995, 73:202–212.
170. Wu GY, Bazer FW, Davis TA, Jaeger LA, Johnson GA, Kim SW, Knabe DA, Meininger CJ, Spencer TE, Yin YL: Important roles for the arginine family of amino acids in swine nutrition and production. *Livestock Science* 2007, 112:8–22.
171. Cen Y, Luo X, Liu X. Effect of arginine on deep partial thickness burn in rats. Hua Xi Yi Ke Da Xue Xue Bao. 30[2], 198–201. 1999. Ref Type: Generic
172. de Jonge WJ, Kwikkers KL, te Velde AA, van Deventer SJH, Nolte MA, Mebius RE, Ruijter JM, Lamers MC, Lamers WH: Arginine deficiency affects early B cell maturation and lymphoid organ development in transgenic mice. *Journal of Clinical Investigation* 2002, 110:1539–1548.
173. Li P, Yin YL, Li D, Kim SW, Wu GY: Amino acids and immune function. *British Journal of Nutrition* 2007, 98:237–252.
174. Nieves C, Jr., Langkamp-Henken B: Arginine and immunity: a unique perspective. *Biomed. Pharmacother.* 2002, 56:471–482.
175. Popovic PJ, Zeh HJ, Ochoa JB: Arginine and immunity. *Journal of Nutrition* 2007, 137:1681S–1686S.
176. Taheri F, Ochoa JB, Faghiri Z, Culotta K, Park HJ, Lan MS, Zea AH, Ochoa AC: L-arginine regulates the expression of the T-cell receptor zeta chain (CD3 zeta) in Jurkat cells. *Clinical Cancer Research* 2001, 7:958S–965S.
177. Tapiero H, Mathe G, Couvreur P, Tew KD: I. Arginine. *Biomed. Pharmacother.* 2002, 56:439–445.
178. Yeramian A, Martin L, Arpa L, Bertran J, Soler C, McLeod C, Modolell M, Palacin M, Lloberas J, Celada A: Macrophages require distinct arginine catabolism and transport systems for proliferation and for activation. *European Journal of Immunology* 2006, 36:1516–1526.
179. Amber IJ, Hibbs JB, Taintor RR, Vavrin Z: The L-Arginine Dependent Effector Mechanism Is Induced in Murine Adenocarcinoma Cells by Culture Supernatant from Cyto-Toxic Activated Macrophages. *Journal of Leukocyte Biology* 1988, 43:187–192.
180. Caso G, McNurlan MA, McMillan ND, Eremin O, Garlick PJ: Tumour cell growth in culture: dependence on arginine. *Clinical Science* 2004, 107:371–379.
181. Flynn NE, Meininger CJ, Haynes TE, Wu G: Dossier: Free amino acids in human health and pathologies - The metabolic basis of arginine nutrition and pharmacotherapy. *Biomedicine & Pharmacotherapy* 2002, 56:427–438.
182. Sallam A, Kast A, Przybilla S, Meiswinkel T, Steinbüchel A: Biotechnological Process for Production of beta-Dipeptides from Cyanophycin on a Technical Scale and Its Optimization. *Applied and Environmental Microbiology* 2009, 75:29–38.

183. Sallam A, Steinbüchel A: Dipeptides in nutrition and therapy: cyanophycin-derived dipeptides as natural alternatives and their biotechnological production. *Applied Microbiology and Biotechnology* 2010, 87:815–828.

184. Mooibroek H, Oosterhuis N, Giuseppin M, Toonen M, Franssen H, Scott E, Sanders J, Steinbuchel A: Assessment of technological options and economical feasibility for cyanophycin biopolymer and high-value amino acid production. *Applied Microbiology and Biotechnology* 2007, 77:257–267.

185. Hühns M, Broer I: Biopolymers. In *Biotechnology in Agriculture and Foresty Vol 64: Genetic Modification of Plants - Agriculture, Horticulture and Foresty*. Edited by Kempken F, Jung C. Springer; 2010:237–252.

186. Steinle A, Oppermann-Sanio FB, Reichelt R, Steinbuchel A: Synthesis and accumulation of cyanophycin in transgenic strains of Saccharomyces cerevisiae. *Applied and Environmental Microbiology* 2008, 74:3410–3418.

187. Neumann K, Stephan DP, Ziegler K, Hühns M, Broer I, Lockau W, Pistorius EK: Production of cyanophycin, a suitable source for the biodegradable polymer polyaspartate, in transgenic plants. *Plant Biotechnology Journal* 2005, 3:249–258.

188. Hühns M, Neumann K, Hausmann T, Ziegler K, Klemke F, Kahmann U, Staiger D, Lockau W, Pistorius EK, Broer I: Plastid targeting strategies for cyanophycin synthetase to achieve high-level polymer accumulation in *Nicotiana tabacum*. *Plant Biotechnol. J.* 2008, 6:321–336.

189. Hühns M, Neumann K, Hausmann T, Klemke F, Lockau W, Kahmann U, Kopertekh L, Staiger D, Pistorius EK, Reuther J, Waldvogel E, Wohlleben W, Effmert M, Junghans H, Neubauer K, Kragl U, Schmidt K, Schmidtke J, Broer I: Tuber-specific *cph*A expression to enhance cyanophycin production in potatoes. *Plant Biotechnology Journal* 2009, 7:883–898.

190. Döös BR: Population growth and loss of arable land. *Global Environmental Change-Human and Policy Dimensions* 2002, 12:303–311.

191. Fischer G, Shah M, van Velthuizen H. Climate Changes and Agricultural Vulnerability. 2002. International Institute for Applied Systems Analysis. Ref Type: Report.

192. Lutz W, Samir KC: Dimensions of global population projections: what do we know about future population trends and structures? *Philosophical Transactions of the Royal Society B-Biological Sciences* 2010, 365:2779–2791.

193. Baulcombe D: Reaping Benefits of Crop Research. *Science* 2010, 327:761.

7

Polyesters, Polycarbonates and Polyamides Based on Renewable Resources

Bart A. J. Noordover

Eindhoven University of Technology, Faculty of Chemical Engineering and Chemistry, Laboratory of Polymer Chemistry, Eindhoven, The Netherlands

Abstract

In this chapter, we have given an overview of a range of biomass-based polymers prepared through polycondensation chemistry. This is by no means an exhaustive summary of available starting materials and polymers. Still, it is clear that this rapidly expanding field of research opens up new synthetic pathways to performance polymers with exciting novel properties. Some of these bio-based polycondensates can rival conventional polymers from petro-chemistry in terms of mechanical and chemical performance and a limited number have already found their way to the market place. Further development of bio-refining processes is expected to improve the availability and economics of the discussed starting materials and polymers. Several of the presented polymers can be produced in existing process equipment with only minor adjustments. Typically, improved temperature control and an inert atmosphere are required to prevent degradation and discoloration of the bio-based products. In addition, more sustainable and milder synthetic routes as well as more effective catalysis are necessary to make optimal use of biomass, constituting stimuli for new research into the field of step-growth polymerization.

Keywords: Saccharides, polycarbonates, aliphatic polyamides, polyesters and 1,4:3,6-dianhydrohexitols, bio-based polycondensates, biomass

Abbreviations and Symbols: 1,3-PD: 1,3-propanediol, 1,4-BD: 1,4-butanediol, ^{13}C NMR: carbon-13 Nuclear Magnetic Resonance spectroscopy, ^{1}H NMR: hydrogen-1 Nuclear Magnetic Resonance

Vikas Mittal (ed.) Renewable Polymers, (305–354) © Scrivener Publishing LLC

spectroscopy, 2,3-BD: 2,3-butanediol, *AV:* acid value, determined by titration [mg KOH/g]; BPA: isphenol-A; CA: citric acid, DAH: 1,4:3,6-dianhydrohexitol, DAII: diamino isoside, DBTDL: dibutyltin dilaurate, DMA: Dynamic Mechanical Analysis, DPC: diphenyl carbonate, DSC: Differential Scanning Calorimetry, EG: ethylene glycol, FDA: 2,5-furandicarboxylic acid, GLY: glycerol, GPC: Gel Permeation Chromatography, HCl: hydrochloric acid, HFIP: 1,1,1,3,3,3-hexafluoro-2-propanol; HMDI: hexamethylene diisocyanate, HMF: 5-hydroxy-methylfurfural, II: 1,4:3,6-dianhydro-L-iditol (or: isodide), IM: 1,4:3,6-dianhydro-D-mannitol (or: isomannide), IPDI: isophorone diisocyanate, IS: 1,4:3,6-dianhydro-D-glucitol (or: isosorbide), MALDI-ToF-MS: Matrix-Assisted Laser Desorption/Ionization Time-of-Flight Mass Spectrometry, MDI: methylene diphenyl isocyanate, MeOH: methanol, M_n: number average molecular weight [g/mol], M_w: weight average molecular weight [g/mol], *N:* number of repeating units in a polymer chain, NCO: isocyanate group, NMP: N-methyl-2-pyrrolidone, NMR: Nuclear Magnetic Resonance spectroscopy, NPG: neopentyl glycol, *OHV:* hydroxyl value, determined by titration [mg KOH/g], PA: polyamide, PBT: poly(butylene terephthalate), PDI: polydispersity index, PET: poly(ethylene terephthalate): PMMA: poly(methyl methacrylate), PS: polystyrene, SA: succinic acid, SEC: Size Exclusion Chromatography, SSP: Solid State Polymerization, *t:* time [s], *T:* temperature [°C], T_c: crystallization temperature, T_{flow}: flow temperature, T_g: glass transition temperature, T_m: melting temperature, TBD: 1,5,7-triazabicyclo [4.4.0]dec-5-ene, TGA: ThermoGravimetric Analysis, TGIC: triglycidyl isocyanurate, TMA: trimellitic anhydride, TMP: trimethylolpropane, TPA: terephthalic acid, TPh: triphosgene, UV: ultraviolet

7.1 Introduction

Some of the most important industrial polymer products are step-growth polymers, produced by a range of polycondensation chemistries and processes. Polyesters, for example, are used in countless applications ranging from engineering plastics such as poly(ethylene terephthalate) (PET) and poly(butylene terephthalate) (PBT) to relatively low molar mass, functional polyester polyols applied in polyurethane elastomers and coatings. Polycarbonates, on the other hand, are used where toughness and excellent optical transparency are required. Another major class of step-growth polymers are polyamides. Since the invention of polyamide 6.6 by Wallace H. Carothers in 1935, these materials have gained enormously in industrial importance as a result of their outstanding mechanical performance, thermal stability and insulating properties. Such performance polymers, prepared through polycondensation, are typically based on monomers from petrochemical origins, such as 1,6-diaminohexane, 1,4-butanediol, neopentyl glycol, bisphenol-A, terephthalic acid and a range of other aliphatic and aromatic polyhydric alcohols, dicarboxylic acids and diamines. The polymers need to comply with stringent requirements in terms of polymer composition, end-group functionality, processability as well as thermal, hydrolytic and UV stability. Also, the production cost is obviously one of the key parameters determining the success of the material in the marketplace. The desire to develop new polymers from alternative feedstock has led to a rapidly increasing number of scientific publications concerning all kinds of polymers from renewable resources. Also in industry, an increasing interest for alternative feedstocks and materials exists. One approach towards new materials is to try to mimic nature by learning from its impressively elaborate design of functional polymers. In nature, polyesters and polypeptides are some of the most common types of biopolymers, in addition to polyether species such as the polysaccharides. While polyesters such as polyhydroxyalkanoates are typically used to store carbon and energy, polypeptides have a wide range of functions in living organisms, including hormonal, signaling, antioxidant and structural proteins (wool, silk). Another way of working is to extract or derivatize monomeric building blocks from biomass, which can be used to produce polymers according to our own specifications in developed production processes. In this chapter, we will focus on such man-made polyesters, polycarbonates and polyamides prepared using polycondensation chemistry.

7.2 Biomass-Based Monomers

As was already discussed in previous chapters of this book, a number of biomass-based raw material platforms exist, providing various types of monomers for polymer synthesis. Here, we give a brief overview of the platforms which are of interest in connection with polycondensation chemistry in general and polyester, polycarbonate and polyamide synthesis in particular. There is a preference to start from non-edible feedstock, avoiding conflicts with the human and animal food chains. Therefore, cellulose, lignin, cork and non-edible vegetable oils such as castor oil are increasingly studied [1, 2]. From a polymer chemistry and material science point of view, on the other hand, all monomers leading to new polymer structures and properties are worth investigating. It should be noted here that, although synthetic routes towards the described monomers have been demonstrated, some of these routes are still in their infancy and require further development before large scale production is feasible.

7.2.1 Monomers from Saccharides

By fermentation of glucose, a broad range of important monomers suitable for polycondensation reactions can be produced. Several of these monomers feature on the US Department of Energy (DOE) list of 15 molecules which may be produced from carbohydrates [3]. Glucose itself is mainly produced by enzymatic hydrolysis of starch, obtained from corn or potatoes. However, sources such as sucrose or cellulose can also yield glucose. The main products obtained from glucose fermentation are lactic acid, succinic acid, itaconic acid, 2,3-butanediol, 3-hydroxypropionic acid, citric acid [4] and glutamic acid, all of which can in principle be directly used in esterification and amidation reactions. Lactic acid can be transformed into other interesting building blocks, including lactide and 1,2-propanediol. Succinic acid (SA) is the starting point to prepare other useful compounds such as 1,4-butanediol, dimethyl succinate and succinic anhydride [5]. By hydrogenation of 3-hydroxypropionic acid, 1,3-propanediol can be synthesized, whereas oxidation yields malonic acid. Itaconic acid derivatives include 2-methyl-1,4-butanediamine and 2-methyl-1,4-butanediol, whereas glutamic acid serves as a basis to produce glucaric acid, 1,5-pentanediol and some other polyfunctional monomers of potential interest such as glutaminol and 4-amino-5-hydroxypentanoic acid [6, 7]. Glucose is also the source of another very interesting class of monomers: the 1,4:3,6-dianhydrohexitols (DAH) are produced through hydrogenation of D-glucose to form sorbitol (suitable for use as a monomer itself), followed by dehydration (Figure 7.1) [8].

Isosorbide (1,4:3,6-dianhydro-D-glucitol, IS), one of the DAH isomers, is currently being produced on an industrial scale by e.g. Roquette Frères in France and has been the subject of many studies into biomass-based polymer synthesis, as will be discussed later on in this chapter.

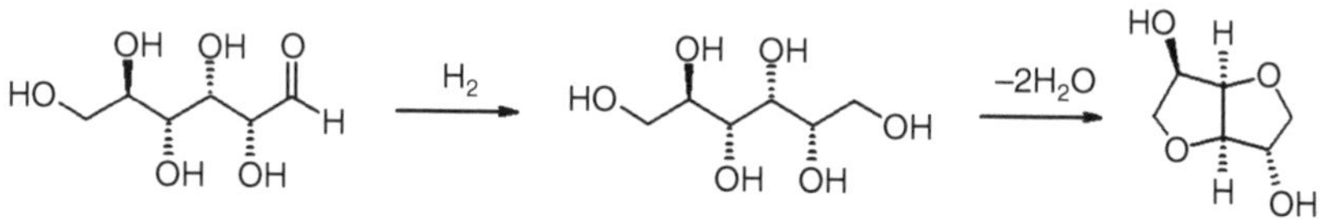

Figure 7.1 Preparation of isosorbide from D-glucose.

Another important monomer platform is obtained through the dehydration of monosaccharides including glucose, fructose and xylose. Depending on the number of carbon atoms present in these sugars, compounds such as furfural and 5-hydroxymethylfurfural (HMF) can be produced. The former is utilized to prepare difuran monomers, including diamine, diisocyanate and dicarboxylic acid species [9, 10]. The latter is the intermediate required to make 2,5-furandicarboxylic acid (FDA) or 2,5-bis(hydroxymethyl)-tetrahydrofuran. Especially FDA receives much attention, as it may serve as a replacement for commonly used compounds such as terephthalic or isophthalic acid. Polymers derived from HMF are discussed elsewhere in this book. An overview of the most important saccharide-derived monomers is given in Figure 7.2.

7.2.2   Monomers from Vegetable Oils and Other Sources

Vegetable oils have a long history in polymer chemistry, especially in the coating field, starting long before the development of the petrochemical industry. Drying oils such as linseed oil, tung oil and soybean oil have been used in oil paints and varnishes already for centuries [11]. In recent years, a revival of the triglyceride chemistry has taken place and a plethora of starting materials has become available. Several of these compounds can be used to prepare polyesters and polyamides. Hydrolysis (or alcoholysis) of the triglyceride oils (as well as animal fats) yields three fatty acid (or ester) molecules and glycerol. For example, transesterification of triglycerides with methanol yields bio-diesel and glycerol. Glycerol itself is a useful monomer, especially when preparing branched polyesters with increased functionality for, e.g., coating applications (*vide infra*). In addition, glycerol serves as a starting point to prepare other compounds, among which glycerol carbonate, glycidol, diglycerols, 1,3-propanediol, 1,2-propanediol and potentially ethylene glycol. The fatty acids resulting from hydrolysis of triglycerides differ in structure depending on the type of oil used.

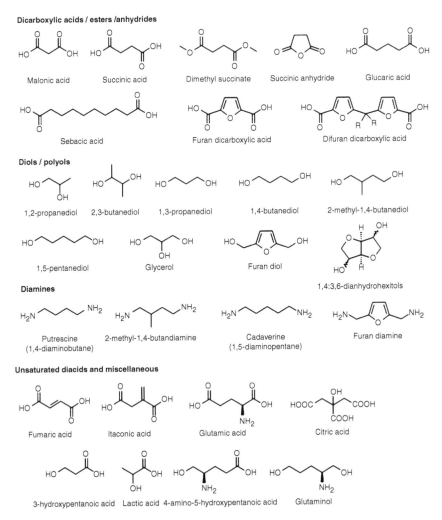

Figure 7.2 A selection of monomers derived from biomass.

Soybean oil, for example, mainly consists of the unsaturated linoleic, oleic and α-linolenic acids as well as the saturated stearic and palmitic acids. The non-edible castor oil, on the other hand, consists mainly of ricinoleic acid (approx. 90%) and smaller amounts of oleic and linoleic acids. A convenient route to prepare dicarboxylic acid monomers is the dimerization of fatty acids. The resulting flexible diacids are suitable for polyamide and polyester synthesis. In addition, they may be derivatized to form diol, diamine or diisocyanate monomers [6]. The double bonds present in the long fatty acid chains can be epoxidized, after which a range of chemistries can be applied to yield, e.g., hydroxyl or cyclic carbonate functionalities. Alternatively, hydroformylation of the unsaturations can introduce carbonyl groups onto these chains, serving as a starting point for the formation of carboxylic acids. Also, metathesis can be used to obtain a range of bifunctional monomers suitable for polycondensation reactions [12].

Several fatty acids, long chain dicarboxylic acids and hydroxy-fatty acid can be derived from suberin, an abundant resource found in plant bark. Suberin consists of a cross-linked network of aromatic and aliphatic building blocks, which can be hydrolyzed to prepare the mentioned compounds [13]. Lignin, another highly cross-linked material mainly built up out of phenolic moieties, is being considered as a source for aromatic monomers such as p-hydroxybenzoic acid [2, 14, 15].

7.3   Polyesters Based on Renewable Resources

7.3.1   Polyesters from Saccharide-Derivatives

Linear and branched, fully aliphatic polyesters can be synthesized by combining a range of monomers form renewable resources. Examples include poly(1,3-propylene succinate) [16, 17] and a broad spectrum of functional polyesters from alditol monomers (such as sorbitol, xylitol and arabinitol derivatives, of which part of the secondary OH-groups are optionally transformed into the corresponding methyl esters prior to polycondensation using linear dicarboxylic acids) [18, 19]. Such polyesters, typically having rather low glass transition temperatures, are especially interesting for biomedical applications and as biodegradable materials. Also, they are often applied as soft segments in poly(ester urethane)s, as shown in the example in Figure 7.3 [20, 21].

Apart from the fully aliphatic polymers, aromatic polyesters partially based on renewable resources are known, such as poly(propylene terephthalate), which is commercialized by DuPont under the trade name Sorona.

Self-condensation of 5-hydroxylevulinic acid, obtained from cellulose via the intermediate levulinic acid, yields a polyester product which is susceptible to hydrolytic degradation. An important disadvantage of this monomer is that it is difficult to obtain polymer of high molecular weight. Interestingly, this polymer (Figure 7.4) has a surprisingly high T_g of up to 120°C, even at low molecular weights of approximately 2,600 g/mol. Inter- and intramolecular hydrogen

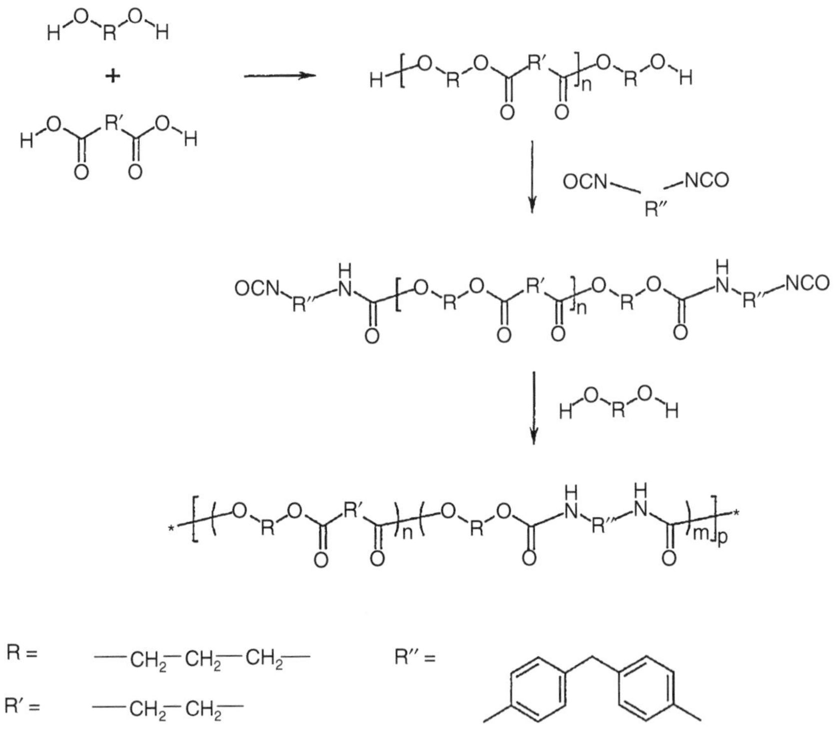

Figure 7.3 Synthesis of dihydroxy-terminated oligo(1,3-propylene succinate) through thermal polycondensation, followed by a chain-extension reaction using MDI to form a poly(ester urethane). Figure taken from reference [21]. Copyright John Wiley and Sons. Reproduced with permission.

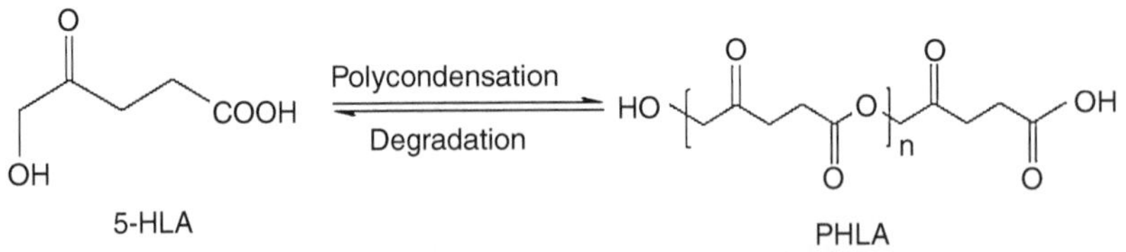

Figure 7.4 Synthesis and degradation of poly(5-hydroxylevulinic acid). Figure taken from reference [22]. With kind permission from Springer Science & Business Media.

bonding involving the carbonyls and enol hydroxyl groups formed through keto-enol tautomerism are thought to cause this high T_g [22].

Unsaturated polyesters based on, for example, itaconic acid in combination with succinic acid, maleic acid and several polyols including sorbitol, 1,4-butanediol and trimethylolpropane may be suitable as photo-curable biomaterials or hydrogels [23, 24].

7.3.2 Aliphatic DAH-Based Polyesters

Much of the information discussed in this paragraph is taken from references [25] and [26]. Adapted with permission. Copyright 2006/2007 American Chemical Society.

Many commercial polyesters, such as poly(ethylene terephthalate) (PET), poly(butylene terephthalate) (PBT) and several types of polyester resins used in thermosetting coating systems [27], are successful in the marketplace as a result of their relatively high glass transition temperatures (T_g). To achieve such high T_g values, rigid molecules are incorporated into the polymer chain. Typically, aromatic structures including terephthalic acid, isophthalic acid, trimellitic anhydride and bisphenol-A are used. Apart from the aromatic structures present in lignin, which up to now have proven to be difficult to harvest and purify, not many rigid molecules are available from renewable resources. This is one of the reasons why the 1,4:3,6-dianhydrohexitols (DAH, Figure 7.5) have received much attention in the polymer community. These bifunctional, aliphatic molecules consist of two nearly planar, *cis*-fused, five-membered rings. These rings, which are at an angle of 120°, do not undergo conformational changes as those occurring in cyclohexane moieties. The hydroxyl groups of the DAHs, located at the C_2 and C_5 positions, are either *exo*- or *endo*-oriented, indicating their configurations with respect to the V-shape formed by the two anhydro-rings. Isosorbide (IS) thus has one *endo*-OH and one

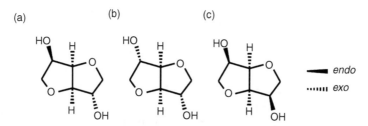

Figure 7.5 The 1,4:3,6-dianhydrohexitols: a) isosorbide, b) isoidide and c) isomannide.

exo-OH, while isoidide (II) has two *exo*-oriented hydroxyls and iso-mannide (IM) has two hydroxyl groups in the *endo* configuration [28–30]. The orientation of these OH-groups has consequences for their relative reactivities depending on the reaction conditions and on the chemistry they are submitted to (*vide infra*) [8, 30–35].

Linear aliphatic oligo- and polyesters based on DAHs and dicarboxylic acids can be prepared through either the non-activated diacids or their dialkyl esters in melt polycondensations or by using activated monomers such as acid chlorides in the melt or in solution reactions [36, 37]. Unsaturated polyesters prepared from isosorbide, copolymerized with 1,2-propanediol, maleic anhydride and phthalic anhydride were described for the first time in a patent dating from 1963 and were applied as lacquers after radical curing with styrene [38]. Apart from some other patent applications from the 1960s [39, 40], little was written about DAH-based polyesters until the 1990s, when more extensive studies were published in the open literature [41–49].

Thermal Properties of DAH-Based Polyesters

The simplest series of polyesters are copolymers of the DAHs with linear dicarboxylic acids, ranging from succinic acid up to longer chain diacids such as tetradecanedioic acid. In most cases, isosorbide was used as the DAH isomer in such series, for availability reasons. In Figures 7.6 and 7.7, the T_g and T_m data for several series of DAH-based polyesters reported by three different research groups are summarized.

Data in Figures 7.6 and 7.7 from: □ Okada *et al.*; ○ Braun *et al.*; △ Noordover *et al.* Data taken from references [46, 49–51]. Copyright John Wiley and Sons/Wiley-VCH Verlag GmbH & Co. KGaA. Reproduced with permission. Figure adapted from reference [37]. Copyright (2010), with permission from Elsevier.

The T_g of these linear polyesters decreases with increasing dicarboxylic acid length, as may be expected as a result of the corresponding increased chain flexibility. Isoidide-based polymers appear to have slightly higher T_g values at the same number of CH$_2$ units in the aliphatic diacid residues, which may result from a denser packing of the more extended II-based polymer chains compared to their IS- and IM-based counterparts. When looking at the T_m data, the different DAHs clearly display deviant behavior. The IS-based polyesters are amorphous when relatively short chain dicarboxylic

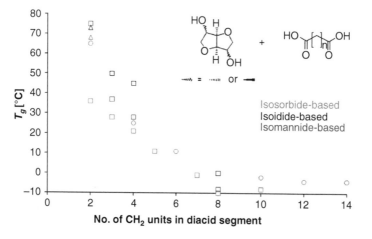

Figure 7.6 DAH-based polyester T_g as a function of the dicarboxylic acid length.

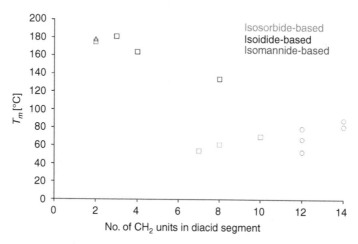

Figure 7.7 DAH-based polyester T_m as a function of the dicarboxylic acid length.

acids are used, but melting temperatures are reported for polyesters based on dicarboxylic acids longer than suberic acid (i.e. more than 6 methylene units between the ester linkages). In addition, the melting temperature increases with the dicarboxylic acid length and is expected to level off for very long chain diacids. In the case of II-based polyesters, the T_m is highest at short diacid length and decreases afterwards, presumable towards a similar value as for the IS-based polymers, as the influence on T_m of the DAH residues is diminished and the melting temperature is mainly determined by the diacid component. The difference in crystallization behavior between IS and II-based polyesters is caused by their different chain conformations: II-based polyesters are thought to have a more extended chain conformation than there IS-based counterparts. Isomannide also facilitates semi-crystalline behavior in polyesters, but this is only observed at short dicarboxylic acid lengths. In addition, it was observed by several authors that isomannide is not sufficiently stable to be polymerized at high temperatures, leading to ring opening of its ether rings and, thus, to branched or cross-linked polymer chains [32, 52–54]. In Figure 7.8, the second heating traces of Differential Scanning Calorimetry (DSC) measurements for poly(isosorbide succinate), poly(isoidide succinate) and poly(isomannide succinate) are displayed.

Depending on the envisioned application of DAH-based polyesters, the T_g as well as the degree of crystallinity, the melting temperature (T_m) and the crystallization temperature (T_c) of these polymers need to be controlled. This is achieved by incorporating additional comonomers (Figure 7.9).

If these comonomers impart more flexibility to the polyester backbone, the T_g of the formed terpolyester is decreased. At the same time, such comonomers disturb the regularity of the polyester chain, thus lowering its T_m. When discussing properties such as the glass transition temperature, the molecular weight of the compared polymers should be similar. After all, the T_g of a polymer depends on its M_n value, especially in the low molecular weight regime. The Fox-Flory equation describes this phenomenon, stating that the polymer T_g increases with molecular weight, leveling off at a certain plateau value [55]. For example, a copolymer based on succinic acid and isosorbide with a molecular weight of $M_n = 3,100$ g/mol (relative to PS standards) has a T_g of 68°C. For virtually the same copolymer, having an M_n of 2,400 g/mol, the T_g is only 57°C (copolyester 1b) and decreases to 53°C at an M_n of 2,200 g/mol. Figure 7.10 shows the effect on T_g and M_n of the systematic replacement of neopentyl glycol (NPG) by isosorbide in a succinic acid-based polyester synthesized at constant reaction times and temperature (but not necessarily at constant conversion).

The plot shows a decrease of M_n with increasing isosorbide content. This is caused by the lower reactivity of the secondary hydroxyl groups of this monomer, compared to the primary OH-groups of NPG. Nevertheless, a significant increase in T_g with

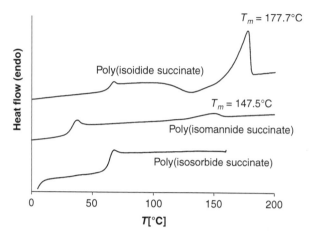

$T_m = 177.7°C$

Heat flow (endo)

Poly(isoidide succinate)

$T_m = 147.5°C$

Poly(isomannide succinate)

Poly(isosorbide succinate)

0   50   100   150   200

$T[°C]$

Figure 7.8 DSC thermograms (2nd heating curves) of polyesters based on succinic acid and the three DAH isomers.

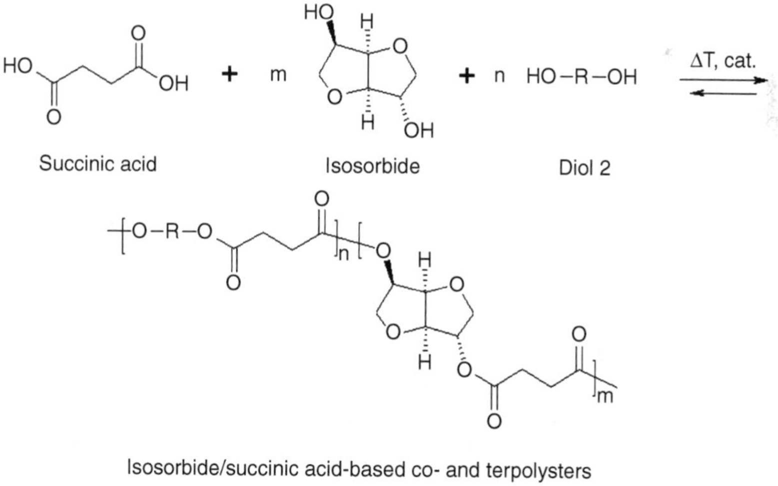

HO

O

Succinic acid   +   m   Isosorbide   +   n   HO–R–OH ΔT, cat.

Diol 2

O–R–O ... O ... O

Isosorbide/succinic acid-based co- and terpolysters

Figure 7.9 Example of a polyesterification reaction between succinic acid, isosorbide and a second diol species. Reprinted with permission from reference [25]. Copyright 2006 American Chemical Society.

increasing isosorbide content can be observed. Taking into account the T_g decreasing effect of lowering the molecular weight, one can expect that the increase in T_g should be even more pronounced when comparing materials of equal molecular weights. Figure 7.10 does therefore not provide maximum values for the glass transition

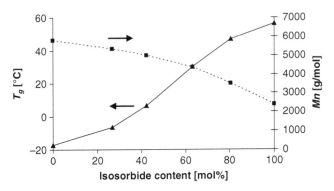

Figure 7.10 T_g (▲) and M_n (■) as a function of isosorbide content for a polyester based on succinic acid, neopentyl glycol and isosorbide (isosorbide mol% relative to total diol content). *The lines in this figure are meant only as guides to the eye.* Reprinted with permission from reference [25]. Copyright 2006 American Chemical Society.

temperatures of these polyesters. The influence of the incorporation of several different comonomers such as NPG and the bio-based 2,3-butanediol (2,3-BD) and 1,3-propanediol (1,3-PD) in SA/IS-based polyesters is shown in Figure 7.11. The number average molecular weights of the polyesters described in Figure 7.11 are in the same range to make a fair comparison.

The terpolyesters in Figure 7.11 have lower T_g values than poly(isosorbide succinate), as a results of the increase in chain flexibility. 1,3-PD and NPG affect the T_g more than 2,3-BD, which is caused by the more flexible nature of 1,3-PD and NPG as compared to 2,3-BD. In semi-crystalline polyesters, the incorporation of comonomers leads to lower melting and crystallization temperatures, as displayed for isoidide-based polyesters in Table 7.1. Also, the degree of crystallinity is reduced.

The polyesters prepared from the DAH isomers are thermally stable up to 250°C, as determined by Thermogravimetric Analysis (TGA). They thus offer sufficient thermal stability to be used in a range of applications, including thermosetting coating or engineering plastics.

Functionality of DAH-Based Polyesters

For applications in which biomass-based polyesters are used as reactive polyfunctional compounds, such as thermosetting polymer systems [56, 57], the number and type of end-groups

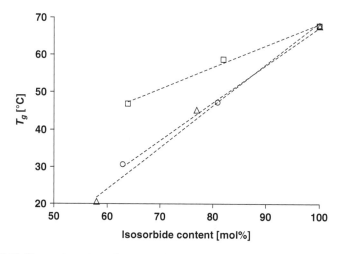

Figure 7.11 T_g as a function of isosorbide content for copolyesters: SA/IS/diol 2, with diol 2 = 1,3-propanediol (Δ), neopentyl glycol (O) or 2,3-butanediol (☐). Reprinted with permission from reference [25]. Copyright 2006 American Chemical Society.

Table 7.1 Terpolyesters based on succinic acid and isoidide.

Composition (NMR)	T_g [°C]	T_m [2] [°C]	M_n [g/mol]	PDI
SA/II [1:1.08]	73	175	3,900	2.2
SA/II/GLY [1:0.98:0.06]	63	166	6,200	3.4
SA/II/2,3-BD [1:0.89:0.14]	58	151	6,700	2.0
SA/II/1,3-PD [1:0.82:0.23]	44	137	5,900	2.0

[1] GLY = glycerol.
[2] The melting temperatures were determined from the second heating curves of DSC thermograms.

(i.e. hydroxyl or carboxylic acid groups) are crucial parameters. In general, the type of end-group resulting from a polycondensation reaction can be determined by controlling the reaction stoichiometry. An excess of diol/polyols species relative to the amount of carboxylic acids/esters will yield a predominantly OH-functional polymer, suitable for further reaction with diisocyanate chain extenders or polyisocyanate curing agents. Inversely, an excess of carboxylic acid groups will afford an acid-functional polyester, which can be reacted with epoxy compounds, for example. In DAH-based polyesters, the control over functionality is not as straightforward. The secondary, sterically hindered OH-groups of the DAH moieties are only moderately reactive and limit the overall conversion during the polyesterification reaction. In melt reactions, using non-activated diacid species, especially the *endo*-oriented DAH OH-groups are hard to convert. Obtaining well-defined, linear polyesters with telechelic functionality is therefore troublesome. A detailed analysis of the functionality of DAH-based polyesters, using NMR spectroscopy, end-group titration and special mass spectroscopy techniques such as matrix-assisted laser desorption/ionization time-of-flight mass spectroscopy (MALDI-ToF-MS) reveals the presence of several polymeric species. Still, almost exclusively OH-functional polyester can be made using IS or II. For example, a linear poly(isosorbide succinate), with a M_n of 2,400 g/mol (PDI = 1.8) prepared using a 0.14 molar excess of IS relative to SA, has an OH-value (measure for the number of hydroxyl groups present in 1 gram of sample) of 1.16 mmol/g, whereas its acid value is only 0.03 mmol/g. Although some cyclic oligo-ester chains and some carboxylic and-groups are detected using MALDI-ToF-MS, this polyester can be regarded as a useful diol for, e.g., thermosetting coating applications. The situation even improves when the more reactive isoidide is used instead of isosorbide. Applying a 0.14 excess of diol yields a poly(isoidide succinate) with a M_n of 3,400 g/mol (PDI = 1.9) and an OH-value of 0.81 mmol/g, while an acid value was not detected. This is supported by the MALDI-ToF-MS spectrum shown in Figure 7.12, in which only OH-telechelic chains and some cyclic species are visible. The average number of OH-functionalities per chains can be further enhanced by incorporating small amounts of polyhydric alcohols, such the bio-based monomer glycerol.

The situation is quite different when acid-functionalized polymers are targeted. Even after using an 0.2 molar excess of succinic acid, a combination of all possible end-groups can be identified in the MALDI-ToF-MS spectrum. Although the majority of the high intensity peaks can be assigned to carboxylic acid-functionalized chains, peaks of moderate to high intensity corresponding to hydroxyl end-groups are present as well (Figure 7.13).

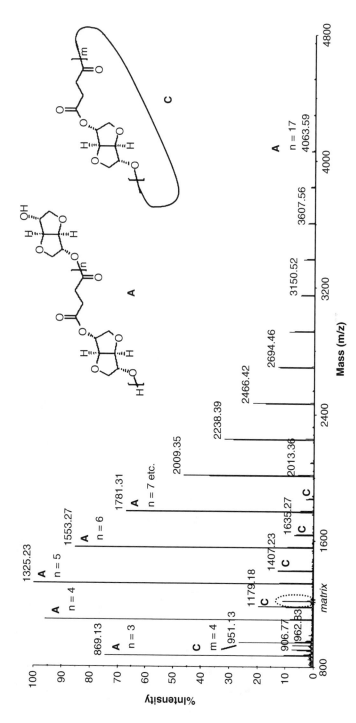

Figure 7.12 MALDI-ToF-MS spectrum of OH-functional poly(isoidide succinate).

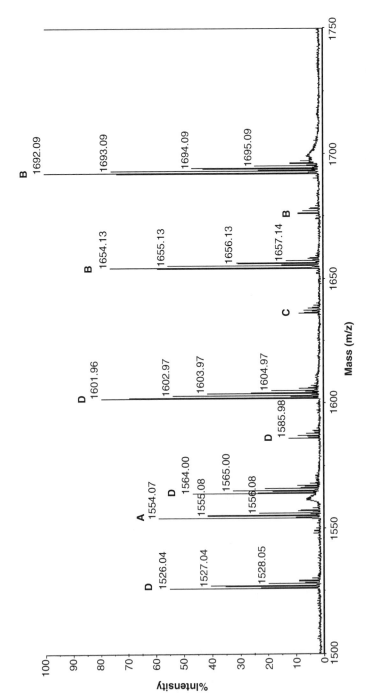

Figure 7.13 Section of the MALDI-ToF-MS spectrum of poly(isosorbide succinate), obtained after polymerization with an excess of succinic acid: (**A**) linear chains having two hydroxyl end-groups; (**B**) linear chains having one carboxylic acid and one hydroxyl end-group; (**C**) cyclic chains; (**D**) linear chains having two carboxylic acid end-groups. *N.B. Acid-functionalized species produce several peaks due to proton replacement by sodium and potassium, originating from the ionization agent.* Adapted with permission from reference [25]. Copyright 2006 American Chemical Society.

Titration confirms that the amount of hydroxyl groups present in this polymer is of a comparable order of magnitude as the amount of acid groups. This observation can only be explained by a reduced reactivity of the secondary hydroxyl groups in the chain growth steps, in particular at chain ends when additional steric constraints are present (i.e. *endo*-orientation of the OH-group). Therefore, the synthesis of fully acid-functional linear isosorbide polymers cannot be achieved under these reaction conditions, when using non-activated diacid species. This problem can be resolved by the addition of carboxylic acid anhydrides, such as succinic anhydride, which is a more reactive species and will more readily convert the secondary OH-groups at the chain ends.

Another attractive method exists to prepare carboxylic acid functional polymers based on the DAHs. Starting from a (mainly) OH-functional polymer, the reaction with the cheap bio-based monomer citric acid yields a COOH-terminated product with an enhanced average functionality. Although citric acid as such contains non-activated carboxylic acid groups, the reaction with moderately reactive secondary OH-groups does proceed as a result of a special feature of the citric acid molecule. Upon heating to around its melting temperature (T_m = 153°C) citric acid anhydride species are formed which are more reactive towards the DAH hydroxyl groups. In the ideal case, only citric acid anhydride is formed through the dehydration of CA. The formation of this asymmetric cyclic anhydride (step **1**) and its subsequent esterification with DAH-terminated polyester chains (reactions **2** and **3**) are shown in Figure 7.14. The citrate end-group resulting from reaction **3** can in principle again form a cyclic anhydride (step **4**) and react once more with a hydroxyl-terminated polyester chain (reaction **5**).

Due to the limited thermal stability of citric acid, this modification should be performed under well-controlled conditions to avoid excessive decomposition through dehydration and/or decarboxylation at temperatures in excess of 180°C.

Unsaturated DAH-Based Polyesters

Apart from saturated DAH/succinic acid-based polyesters, unsaturated structures can be introduced by replacing part of the succinic acid by, for example, maleic anhydride. This approach results in terpolyesters having similar thermal properties as those described

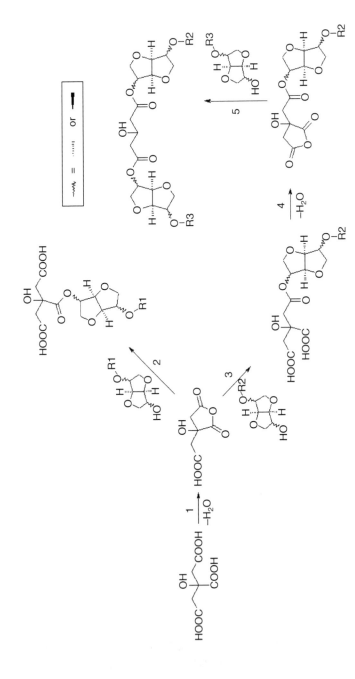

Figure 7.14 Reaction scheme of the citric acid-modification of OH-functional, DAH-based polyesters. Adapted with permission from reference [26]. Copyright 2007 American Chemical Society.

previously, i.e. T_g values between 50 and 65°C and no melting transitions as a result of the amorphous character of such polymers. These poly(isosorbide maleate-*co*-succinate)s, of which the maleic acid units isomerize during the polymer synthesis to form fumaric acid structures (Figure 7.15), offer the attractive feature of being curable through radical copolymerization with monomers such as 2-hydroxyethyl methacrylate or acrylic acid, a.o. Moreover, waterborne polyesters can be prepared from copolyesters of isosorbide, maleic anhydride and poly(ethylene glycol), which can also be cross-linked via radical mechanisms [58–60].

Application of DAH-Based Polyesters in Coating Systems

The aliphatic polyesters based on DAHs are especially suitable for coating applications. In such systems, moderate molecular weights are sufficient, provided that the T_g and functionality are in the right range. Subsequent curing using polyisocyanates, polyepoxides or

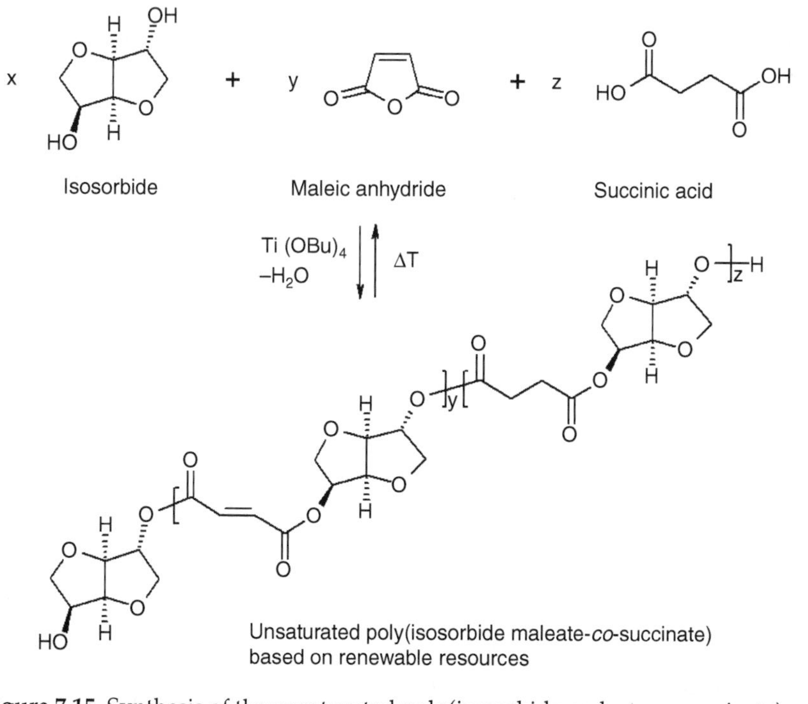

Figure 7.15 Synthesis of the unsaturated poly(isosorbide maleate-*co*-succinate)s. Figure taken from reference [58]. Copyright John Wiley and Sons. Reproduced with permission.

radical chemistry yields mechanically and chemically stable poly-mer networks. The hydroxyl-functional as well as the carboxylic acid-functional polyesters (obtained through the citric acid modifi-cation) were successfully tested as binder resins for powder coating applications, leading to mechanically and chemically stable coat-ings after curing with isocyanate- or epoxy-functional cross-linkers, a.o. (Figure 7.16, Table 7.2) [25, 26, 61].

Aliphatic polyesters can also positively influence the outdoor durability of coatings, as they are less susceptible to degradation by exposure to UV radiation and show reduced yellowing compared to the aromatic polyesters currently used in, for example, powder coatings. Such polyesters can be produced in currently available processing equipment with only minor adjustments in processing conditions. In particular, reaction temperatures should not exceed 250°C and reactions should be performed in an inert atmosphere (i.e. nitrogen purge).

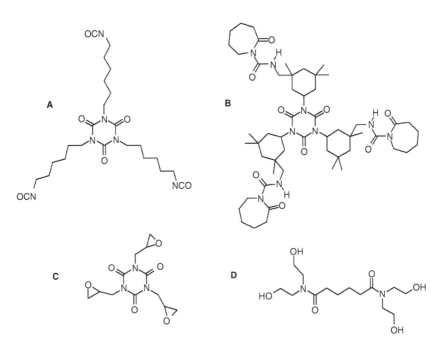

Figure 7.16 Curing agents used to form poly(ester urethane) and polyester networks from DAH-based polyesters: (**A**) trimer of hexamethylene diisocyanate, (**B**) ε-caprolactam-blocked trimer of isophorone diisocyanate, (**C**) triglycidyl isocyanurate and (**D**) N,N,N′,N′-tetrakis(2-hydroxyethyl)adipamide.

Table 7.2 Coating properties of DAH-based poly(ester urethane) and polyester networks.

Polyester Topology	Curing Agent [1]	Solvent Resistance [2]	Impact Resistance [3]
OH-functional			
linear	A	good	good
	B	moderate	poor
branched	A	good	good
	B	good	moderate
COOH-functional			
linear/branched	C	good	good
	D	good	moderate

[1] Curing agents are depicted in Figure 7.16.
[2] Determined by assessing the damage upon rubbing the coating with an acetone-drenched cloth.
[3] Determined by reverse impact testing at 100 kg×cm.

7.3.3   Aromatic DAH-Based Polyesters

Aromatic polyesters are used in a wide variety of applications. Some of the most important commercial products are poly(ethylene terephthalate) (PET), used in food and beverage packaging as well as in various engineering applications. Poly(butylene terephthalate) (PBT) is another example of an aromatic engineering polyester, used in plastic furniture, electronics applications etc. In addition, most powder coating resins are functional aromatic polyesters of moderate molar mass, which are cured to form poly(ester urethane) or polyester networks. In the previous paragraph, the use of aliphatic DAH-based polyesters as replacements for such aromatic coating resins was discussed. Here, we focus on the incorporation of DAH moieties into aromatic engineering polyesters to yield partially bio-based polymers with enhanced properties [36].

Engineering Polyesters Containing DAHs

Thiem and Lüders made copolyesters of terephthaloyl dichloride and the three DAH isomers, yielding polymers with T_g values above

150°C at moderate molecular weights. Poly(isoidide terephthalate) was found to be a semi-crystalline material [41]. Later, Storbeck and Ballauff prepared copolyesters of terephthaloyl dichloride, ethylene glycol and isosorbide, showing good thermal stability and increased T_g values [42, 52]. Such copolyesters were patented by DuPont de Nemours [62–64]. Melt polymerizations of isosorbide or isoidide in combination with ethylene glycol and terephthalic acid, performed at 270°C, demonstrate that up to 25 mol% of isosorbide is lost from the reactor through evaporation. Isoidide losses are significantly lower (up to 10 mol%), confirming its higher reactivity in melt reactions compared to IS [37]. Figure 7.17 shows the conversion of the carboxylic acid groups of terephthalic acid (TPA) as a function of time and of the relative amount of isosorbide present in the ethylene glycol/isosorbide diol mixture. These observations again demonstrate the difficulties encountered in obtaining high molecular weight products when applying high DAH loadings.

Kricheldorf c.s. studied the influence on T_g and T_m of the incorporation of the DAH in PBT. The authors demonstrated that melt polycondensation of different ratios of isosorbide and 1,4-butanediol with dimethyl terephthalate, performed at temperatures up to 250°C using Ti(OBu)$_4$ as a transesterification catalyst, yields low molecular weight products. This means that the conventional process to

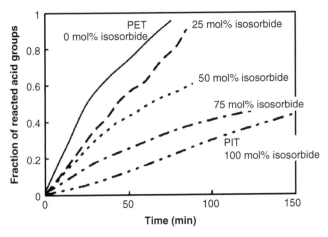

Figure 7.17 Conversion of the carboxylic acid groups of terephthalic acid (TPA) during the copolymerization of ethylene glycol, isosorbide and TPA in the melt, at an initial diol/diacid ratio of 1.2. Figure taken from reference [37]. Copyright (2010), with permission from Elsevier.

make PBT cannot be used when incorporating significant amounts of the DAHs. However, the solution copolymerization of isosorbide and 1,4-butanediol was successful when terephthaloyl dichloride was used. With increasing IS content, the T_g of PBT increased from around 45°C to 92°C at a 1:1 ratio of IS:1,4-BD. Complete replacement of 1,4-BD by IS even yielded a T_g close to 200°C, confirming the results published by Thiem and Lüders. On the other hand, the crystallinity and the T_m of the modified PBT decrease with increasing DAH loading. When the isosorbide content exceeds 36%, the polymer becomes amorphous [65]. As an alternative method to improve the T_g of PBT by building in DAH moieties, while retaining as much as possible the high melting temperature and degree of crystallinity of the original polymer, solid state polymerization (SSP) technology can be applied. The DAHs as such are insufficiently reactive and too volatile to be incorporated through SSP, so a DAH-based macromonomer with primary hydroxyl end-groups was synthesized, as shown in Figure 7.18.

This macromonomer was mixed into PBT in hexafluoroisopropanol solution, after which the solvent was evaporated. The obtained powder mixture was heated to temperatures approx. 30°C below the melting temperature of PBT (T_m = 215–225°C), allowing for transesterification reactions to occur in the mobile amorphous phase, resulting in the incorporation of the macromonomer into the PBT backbone. The SSP modified PBT product has a more blocky structure than its counterpart prepared in the melt. Up to 25 mol% of isosorbide (relative to 1,4-BD) was built into the polymer, affording a T_g of 78°C while the melting temperature of the crystalline phase was decreased by only 20–25°C [66]. In this respect, the SSP

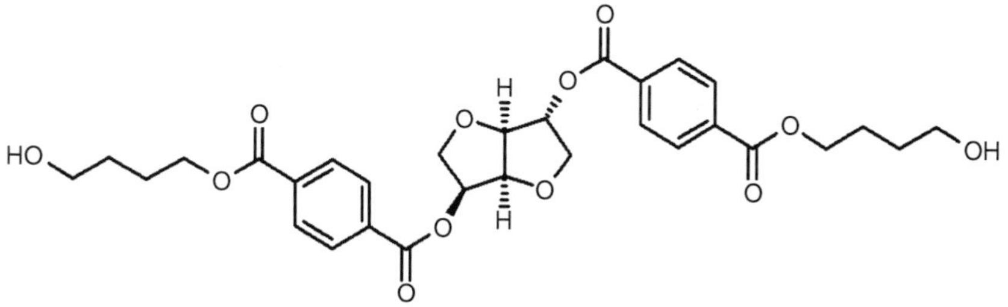

Figure 7.18 Isosorbide-based macromonomer: isosorbide 2,5-bis [4-(hydroxybutyl) terephthalate]. Reproduced with permission from reference [66]. Copyright 2008 American Chemical Society.

results are comparable to the data presented for solution polymerizations using terephthaloyl dichloride.

Liquid Crystalline Polyesters Containing DAHs

In addition to modified engineering plastics, several aromatic polyesters containing DAHS have been studied to assess the effects of
the chirality of these sugar diols. Several types of linear and photo-
setting or thermosetting cholesteric polyesters, poly(ester imide)s
and poly(ester carbonate)s were described, as summarized in excellent reviews by Kricheldorf and Fenouillot [36, 37].

7.3.4   Polyesters Based on Furan Monomers

The use of furan-2,5-dicarboxylic acid (FDA) to prepare bio-based
polyesters was described by Moore *et al.* in 1978 [67]. Copolymerization of FDA and 1,6-hexanediol, 2,5-bis(hydroxymethyl)furan
or 2,5-bis(hydroxymethyl)tetrahydrofuran yielded polyesters of
moderate molecular weight and limited thermal stability. In the
past decade, a renewed interest in FDA has led to very interesting findings, notably by Gandini and his co-workers. For example,
poly(ethylene 2,5-furandicarboxylate) (Figure 7.19) was found to
be a potential alternative for PET, showing a T_g of 75–80°C and a
T_m of 210–215°C (which is approximately 45°C lower than the T_m of
PET) [68].

The fully bio-based poly(propylene 2,5-furandicarboxylate) has
a T_g of 65°C and a T_m of 174°C. Copolyesters of FDA and the three
1,4:3,6-dianhydrohexitols, reported for the first time by Storbeck
and Ballauff in the early 1990s, have high T_g values of up to 196°C

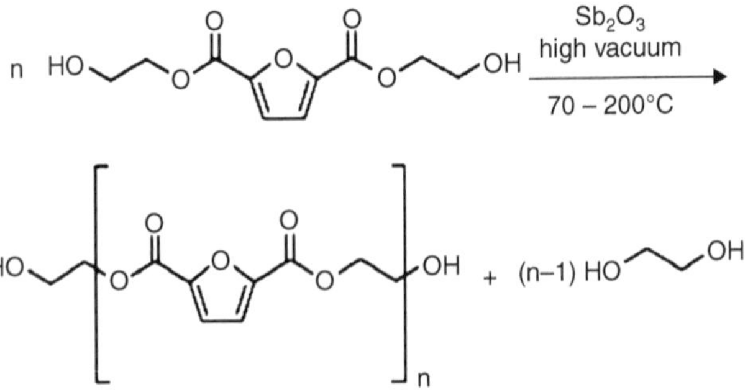

Figure 7.19 Synthesis of poly(ethylene 2,5-furandicarboxylate). Figure taken from
reference [68]. Copyright John Wiley and Sons. Reproduced with permission.

as a result of their highly rigid polymer backbones [69]. The FDA-based polyesters mentioned here are all thermally stable up to approximately 300–350°C [9, 10, 70].

Apart from polymers based on FDA several polyesters from difuran monomers, including difuran dicarboxylic acid and difuran diols, were described [19, 70, 71]. Polymers based on 5-hydroxymethylfurfural are discussed in detail in Chapter 10.

7.3.5 Vegetable Oil-Based Polyesters

As mentioned previously, vegetable oils are used as drying oils through auto-oxidation and polymerization of the unsaturations present in their fatty acid residues. However, the oils can also be seen as a source of individual monomers, which can be purified and modified if necessary. Castor oil, for example, consists mainly of the fatty acid ricinoleic acid (approx. 88–90%), which can be obtained by hydrolysis of the triglyceride [12]. Alternatively, the corresponding methyl ester can be made through methanolysis. Upon hydrogenation of the double bond, methyl-12-hydroxystearate is obtained. This material is already available as a commercial product and can be used to prepare polyesters, for example through enzymatic catalysis (Figure 7.20) [72].

An alternative approach was published by Bakare c.s., who prepared monoglycerides from rubber seed oil, which were subsequently polymerized with phthalic anhydride to yield an aromatic polyester with long fatty acid side-groups. These partially bio-based polyesters are suitable for use as binders in coatings and composites [73]. Fully aliphatic polyesters based on glycerol and divinyl sebacate having unsaturated fatty acid side-chains, which were subsequently cured through oxidation of the double bonds, were also investigated [74]. Can c.s. studied partially vegetable

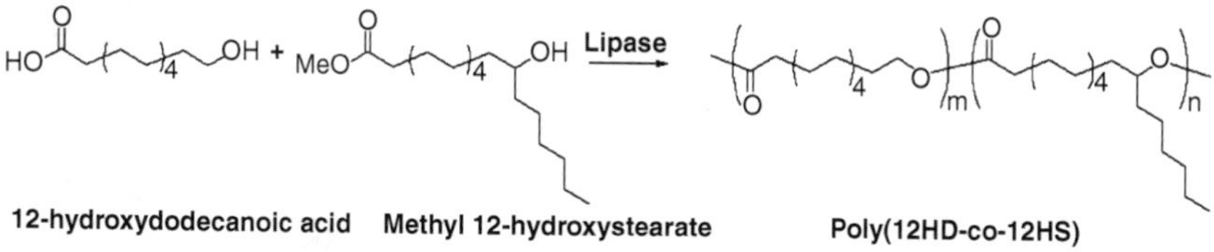

12-hydroxydodecanoic acid   Methyl 12-hydroxystearate   Poly(12HD-co-12HS)

Figure 7.20 Enzymatic copolymerization of 12-hydroxydodecanoic acid and methyl 12-hydroxystearate. Figure taken from reference [72]. Copyright John Wiley and Sons. Reproduced with permission.

oil-based, unsaturated oligo-esters, combining soybean and castor oil-derivates with conventional monomers such as pentaerythritol and bisphenol-A. These oligomers were subsequently copolymerized with styrene. The properties of the resulting materials can be tuned over a wide range in terms of, e.g., flexural modulus (0.8–2.5 GPa) and T_g (70–150°C) [75]. In the examples described so far, aliphatic pendant groups are always present as a result of the fact that the fatty acid derivatives do not have functional groups at their chain ends. Recently, a few solutions to this potential problem were presented, for example by Quinzler and Mecking, who successfully converted internal unsaturations into linear carboxylic acid esters. In this way, 1,19-nonadecanedioate was prepared from methyl oleate. Furthermore, the corresponding long chain diol was synthesized by reduction of the diester. These two monomers were polymerized using titanium alkoxides as catalysts, yielding semi-crystalline polyesters having molecular weights comparable to commercial polyesters [76]. Alternatively, castor oil is a source of methyl 10-undecenoate, which can be used as a starting point to prepare a range of monomers suitable for polycondensation reactions, including diols and dicarboxylic acid methyl esters, which were successfully polymerized within the group of Meier [77].

7.4   Polycarbonates Based on Renewable Resources

Conventional polycarbonates, such poly(bisphenol-A carbonate), are known for their high toughness, excellent transparency and good solvent resistance. In addition, the carbonate links are more hydrolytically stable than esters. The synthesis of polycarbonates requires a carbonyl-donor of sufficient reactivity, such as the industrially applied phosgene or diphenyl carbonate (DPC). Phosgene is a highly toxic compound, whereas using DPC yields phenol which can only be removed from the reaction mixture at high temperature and low pressure. Therefore, much effort is currently put in developing less toxic and potentially bio-based alternatives, such as dialkyl carbonates. Conceptually, these compounds can be prepared from carbon dioxide (or ureum) and low molecular weight aliphatic alcohols. Another very interesting option is to use carbon dioxide directly in copolymerization reactions with oxiranes

[78–81] or to apply cyclic carbonate monomers as carbonyl sources [82–84]. A fully bio-based example of the former route was published by the group of Coates, demonstrating the alternating copolymerization of limonene oxide with CO_2 [85]. In this discussion, however, the main topic will be polycarbonates prepared through polycondensation reactions of biomass-derived polyols with several carbonyl-donors.

7.4.1 Polycarbonates Based on 1,4:3,6-dianhydrohexitols

Much of the information discussed in this paragraph is taken from reference [86]. Copyright John Wiley and Sons. Reproduced with permission.

Polycarbonates and copolycarbonates based on the three DAH isomers isosorbide, isoidide and isomannide have been described in several publications. The first mention of poly(isosorbide carbonate), prepared through the transesterification reaction between isosorbide and diphenyl carbonate, was made in 1964 in a patent application assigned to Courtaulds [39]. In 1985, Bayer AG patented (co)polycarbonates based on the DAHs, prepared using DPC, phosgene or the bis-chloroformate of the DAHs [87]. Some properties of these polycarbonates are also described by Braun and Bergmann, who prepared poly(isosorbide carbonate) by phosgenation using gaseous phosgene or the liquid diphosgene. The resulting aliphatic polycarbonate has a glass transition temperature of 155°C at a molecular weight of M_n = 15,000 g/mol (PDI = 2.0). Interestingly, the T_g and mechanical properties of the DAH-based polymer are comparable to those of poly(bisphenol-A carbonate), the most commonly used commercial polycarbonate [51]. Bisphenol-A is under increased pressure, as it is suspected to have adverse health effects [88]. It is therefore no surprise that the polycarbonates based on all three non-toxic DAH isomers feature in several patents and patent applications, filed and published during the past decades [89–91]. Several synthetic methods have been investigated, yielding M_n values of up to 50,000 g/mol and T_g values of up to 175°C [37]. Apart from the DAH-based homopolycarbonates, a wide range of copolycarbonates can be made to tune the thermal, mechanical and chemical properties. For example, liquid-crystalline polycarbonates are obtained when copolymerizing isosorbide with photoreactive moieties such as 4,4'-dihydrochalcone [36]. Here, we focus on fully aliphatic copolycarbonates based exclusively on biomass-derivatives. Yokoe *et al.* prepared copolymers of DAHs and linear aliphatic diols, targeting biodegradable polycarbonates. These polymers were synthesized by reacting the primary dihydric alcohols with the bis(phenyl carbonate) [92] or the bis(nitrophenoxy carbonate) derivatives of the DAHs, catalyzed by zinc acetate. As expected, the T_g decreases strongly with the increasing length of the linear diol comonomer. In addition, the materials show faster biodegradation with decreasing T_g (and thus with increasing linear diol chain length) [93].

As discussed previously, aliphatic polymer resins with a relatively high T_g and reactive chain-ends are very interesting intermediates to prepare thermosetting polymeric networks for coating applications. Combining excellent optical appearance and transparency with very good toughness and hydrolytic stability, polycarbonates are of interest for such applications. However, the conventional aromatic polycarbonates do not qualify for (exterior) coating applications due to their poor UV stability and their high cost. Aliphatic, bio-based polycarbonates with reactive end-groups are therefore highly interesting candidates for such thermosetting materials, provided that the economics improve with the further development of the biomass-derived monomers. Different polymerization routes can be considered to synthesize linear and branched copolycarbonates with the required characteristics with respect to molecular weight, T_g and functionality. Depending on the desired properties of the final cured network, the molecular weight between cross-links (i.e. the molecular weight between the functional groups of the polymer resin and the curing agent, respectively) as well as the end-group structure can be influenced by carefully controlling the reaction stoichiometry and the average functionality of the applied monomers. In polycarbonate systems, the most logical approach is to target OH-functional polymers, which can subsequently be reacted with (blocked) polyisocyanates. Alternative possibilities include the end-group modification of such OH-functional resins with anhydrides to facilitate carboxylic acid/epoxide curing or to introduce acrylate functionalities for UV- or electron beam-initiated cross-linking.

Phosgene-Route

When using highly reactive compounds such as phosgene to prepare polycarbonates, controlling the molecular weight and the structure of the polymer end-groups is challenging. The required ratio of phosgene to polyhydric alcohols needs to be tuned for each different polymer composition. When using phosgene derivatives such as the liquid diphosgene or the solid triphosgene, matters are complicated even further, as the effective amount of phosgene formed *in situ*, is strongly dependent on the composition of the reaction mixture, the solvent and the applied base. Polymerization of the DAHs using di- and/or triphosgene in dichloromethane solution (Figure 7.21), using pyridine as a catalyst and HCl scavenger, results in polycarbonate chains having hydroxyl end-groups as well as in cyclic polycarbonate species, as observed by MALDI-ToF mass spectrometry and ^{1}H NMR spectroscopy. In addition, chloroformate end-groups can be formed, which may react further during work-up or processing of the product [86, 94, 95].

An advantage of preparing polycarbonates through phosgenation is that controlling the polymer composition is straightforward, meaning that the feed ratio of, e.g., two different diols is reflected in the final polymer composition. Some examples of DAH-based homo- and copolycarbonates, prepared using triphosgene, are listed in Table 7.3. NMR studies also provide information concerning the relative reactivities of the DAH hydroxyl groups (i.e. *endo-* or *exo*-oriented) under different reaction conditions. Contrary to what was observed previously for isosorbide-based polyesters prepared by melt polycondensation, it appears that the *endo*-oriented OH-groups of isosorbide are more reactive (conversion = 97%) than the *exo*-oriented hydroxyls (conversion = 88%). This confirms observations described in literature concerning the reactivity of these different types of OH-groups in the presence of, for example, pyridine hydrochloride [31–33].

The MALDI-ToF-MS spectra of the three isomeric poly(DAH carbonate)s (Figure 7.22) show signals at the same molecular masses, but the relative intensities of these peaks differ for the different materials. Especially, the ratios of linear versus cyclic chains appear

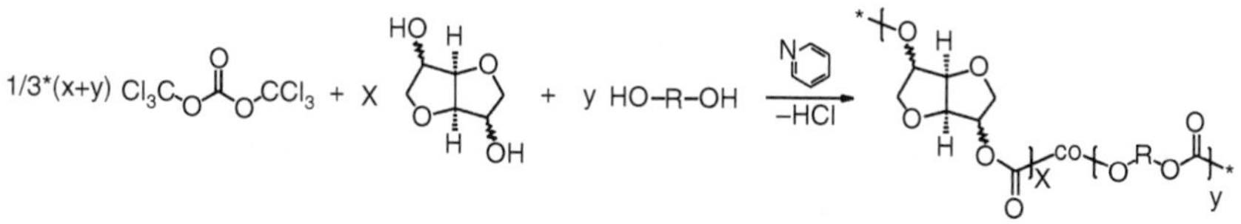

Figure 7.21 Synthesis of DAH-based polycarbonates, using triphosgene as a carbonyl-donor. Figure taken from reference [86]. Copyright John Wiley and Sons. Reproduced with permission.

Table 7.3 Compositions and properties of DAH-based (co)polycarbonates prepared using triphosgene.

Composition[1, 2]	T_g [3] [°C]	M_n [4] [g/mol]	M_w [g/mol]	PDI
IS/TPh [1:0.33]	134.8	3,500	6,100	1.8
II/TPh [1:0.33]	130.9	3,200	6,200	1.9
IM/TPh [1:0.33]	120.9	3,400	10,700	3.1
IS/1,3-PD [0.53:0.47]	57.2	1,800	2,500	1.4
IS/1,3-PD [0.50:0.50]	83.0	2,600	4,200	1.6
II/1,3-PD [0.47:0.53]	48.1	1,800	2,300	1.3
IM/1,3-PD [0.54:0.46]	46.7	1,900	3,200	1.7
IS/GLY [0.9:0.1]	109.8	3,900	12,900	3.3

[1] Reaction conditions: addition of pyridine solution to diol/TPh solution at 0°C, followed by stirring at room temperature for 20 hrs.

[2] TPh = triphosgene, IS = isosorbide, II = isoidide, IM = isomannide, 1,3-PD = 1,3-propanediol, GLY = glycerol. Composition determined by ^{1}H NMR spectroscopy.

[3] Determined from the second heating curve of a DSC thermogram.

[4] Determined by SEC in HFIP, using PMMA calibration.

to be different for the three isomeric polymers based on isosorbide (IS), isoidide (II) or isomannide (IM), respectively.

Peaks corresponding to linear chains (**A** and **C**) are abundant in the spectra of the II and IS-based polymers, while signals attributed to cyclic polycarbonate chains (**B**) dominate in the IM-based polycarbonate. The II-based polycarbonate shows only very low intensity signals corresponding to cyclic species. Although a quantitative interpretation of the MALDI-ToF-MS spectra in terms of relative concentrations of linear and cyclic species is not reliable, the trend

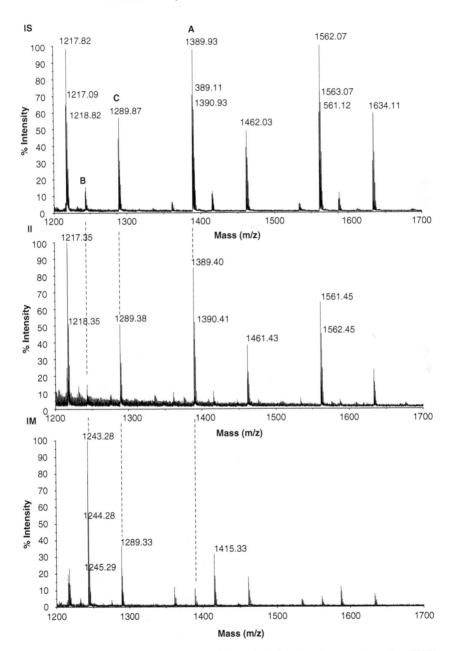

Figure 7.22 Sections of MALDI-ToF-MS spectra of polycarbonates based on IS, II and IM, having (**A**) linear chains with two hydroxyl end-groups, (**B**) cyclic chains and (**C**) linear chains containing one ether linkage and having methyl carbonate or ether end-groups. Figure taken from reference [86]. Copyright John Wiley and Sons. Reproduced with permission.

observed here is as may be expected when considering the orientations of the OH-groups of the different 1,4:3,6-dianhydrohexitols. Possibly, the conformation of poly(isomannide carbonate) is such that cycle formation is favored. The observation that polymerization of isoidide, having two *exo*-oriented hydroxyl groups, apparently yields even less cyclic polymer chains than are observed in poly(isosorbide carbonate), may indicate that the *exo*-orientation of the hydroxyl groups of the 1,4:3,6-dianhydrohexitols favors the formation of more extended, linear polycarbonate chains. Similar results were published for polycarbonates prepared from IS and IM using diphosgene. MALDI-ToF-MS spectra of homopolycarbonates of isosorbide hardly show signals corresponding to cyclic species, while the MALDI-ToF-MS spectra of copolycarbonates containing both isosorbide and isomannide feature high intensity signals of cyclic chains [96].

DPC-Route

The reaction between an alcohol and DPC is a type of transesterification reaction (diphenyl carbonate being the phenolic diester of carbonic acid) which is also referred to as transcarbonation. As an example, Figure 7.23 depicts the reaction between diphenyl carbonate (DPC) and a 1,4:3,6-dianhydrohexitol [97]. It is well known that in such reactions, primary alcohols are more reactive than their secondary counterparts. Therefore, longer reaction times and/or higher temperatures are required to achieve high conversions when using secondary alcohols such as the DAHs. In addition, the removal of phenol from the increasingly viscous reaction mixture is facilitated by high temperature and low pressure.

As was already observed in polyesterification reactions, isoidide is the most reactive DAH isomer in melt reaction using DPC

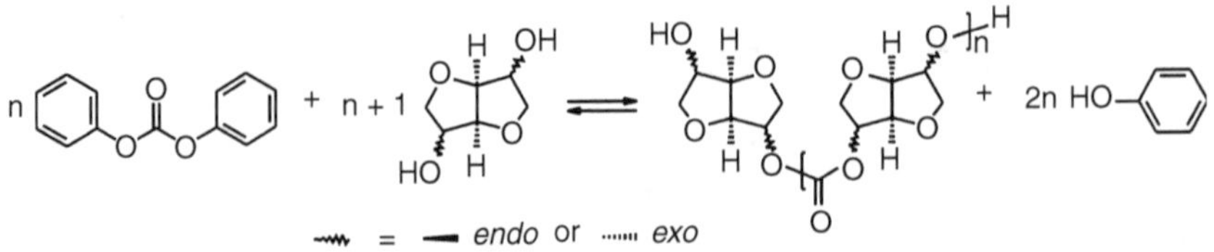

Figure 7.23 Synthesis of a poly(DAH carbonate) from DPC and a DAH, through a transcarbonation mechanism. Figure taken from reference [86]. Copyright John Wiley and Sons. Reproduced with permission.

as the carbonyl source, affording higher molecular weights than for example isosorbide (Table 7.4). The end-group structure of the polycarbonates prepared using DPC can be controlled relatively easily by adjusting the stoichiometry of the reaction mixture. As expected, cyclic polycarbonate chains are formed during these polycondensation reactions, in addition to the linear species. From MALDI-ToF-MS spectra, evidence was also found for decarboxylation, resulting in ether linkages in the polycarbonate main chain. The presence of ether linkages in coating resins is disadvantageous, since they are expected to be labile towards UV radiation. Isomannide again proved to be insufficiently stable to withstand the high reaction temperatures applied. Volatile bio-based comonomers such as 1,3-propanediol may be partially lost during the production of polycarbonates at high temperatures, making it more difficult to control the reaction stoichiometry, the final polymer composition and end-group structure.

DAH bis(alkyl/aryl carbonate)-Route

An alternative synthetic strategy was developed to avoid the high reaction temperatures (up to 250°C), needed in the DPC route to convert the secondary DAH OH-groups of moderate reactivity into carbonate links. Highly reactive alkyl or aryl chloroformates can be used to convert the 1,4:3,6-dianhydrohexitols into their respective bis(alkyl/aryl carbonate)s [92, 93]. Examples of such monomers are shown in Figure 7.24, depicting isoidide bis(phenyl carbonate) and isoidide bis(ethyl carbonate) which can easily be synthesized at high purity.

These monomers can subsequently be reacted with species containing primary OH-groups, such as flexible biomass-based alkylene diols, glycerol or trimethylolpropane, in the presence of a transesterification catalyst. As a result of the higher reactivity of such primary alcohols, the reaction temperature can be reduced (compared to DPC-based polymerizations). At the same time, the control over the build-in of the 1,4:3,6-dianhydrohexitols and its comonomers into the polycarbonates is improved. This approach is similar to the previously discussed method applied by Sablong et al. [66] to achieve incorporation of isosorbide into poly(butylene terephthalate). Table 7.4 lists the properties of a range of linear and branched copolycarbonates prepared using the mentioned DAH derivatives. These copolycarbonates are typically colorless

Table 7.4 Compositions and properties of DAH-based (co)polycarbonates prepared using DPC or DAH-based bis(alkyl/aryl carbonates).

Composition[1]	T_g [2] [°C]	M_n [3] [g/mol]	M_w [g/mol]	PDI	OHV [4] [mg KOH/g]
Polycarbonates prepared using DPC[5]					
IS [-]	138.0	3,200	7,200	2.2	55.0
II [-]	148.2	4,600	10,300	2.2	34.0
IS/1,3-PD [0.57:0.43]	48.1	3,200	5,300	1.7	57.2
Polycarbonates prepared using DAH bis(alkyl/aryl carbonates)[6]					
IS/1,3-PD [0.57:0.43]	89.7	4,200	8,100	1.9	47.3
IS/1,3-PD [7] [0.67:0.33]	64.4	2,200	3,700	1.7	*n.d.*
II/1,3-PD [0.44:0.56]	35.6	3,100	6,400	2.1	76.4
II/1,3-PD [7] [0.83:0.17]	67.1	1,900	3,100	1.6	*n.d.*
IS/1,3-PD/TMP [1:0.28:0.06]	78.8	3,600	11,600	3.2	50.4
II/1,3-PD/TMP [1:0.24:0.05]	83.4	2,900	6,600	2.3	56.8

[1] IS = isosorbide, II = isoidide, IM = isomannide, 1,3-PD = 1,3-propanediol, TMP = trimethylolpropane. Composition determined by ¹H NMR spectroscopy.

[2] Determined from the second heating curve of the DSC thermogram.

[3] Determined by SEC in HFIP, using PMMA calibration.

[4] Hydroxyl-value, determined by potentiometric titration. Such measurements were only performed on materials available in sufficient quantities and at full solubility of the resins in NMP.

[5] Prepared in the melt using diphenyl carbonate, T_{max} = 250°C.

[6] Prepared in the melt using the corresponding DAH bis(phenyl carbonate)s (unless stated otherwise), T_{max} = 210°C.

[7] Prepared using the corresponding DAH bis(ethyl carbonate)s, T_{max} = 210°C.

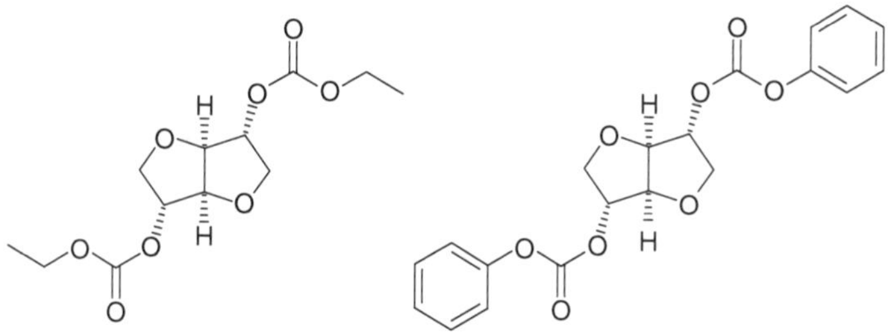

Figure 7.24 (**A**) isoidide bis(ethyl carbonate) and (**B**) isoidide bis(phenyl carbonate).

or pale yellow, transparent polymers. Their molecular weights and OH-functionalities are in an appropriate range for application in thermosetting polymer systems. The T_g values depend on the molecular weight as well as on the composition of the copolycarbonates. According to the copolymer compositions determined by ^{1}H NMR, the monomers are generally built-in into the polymer chains in the same ratio as the monomer feed. Significant losses of volatile diol comonomers were not observed in the reactions with DAH bis(alkyl/aryl carbonate)s, indicating that the reaction temperatures were low enough to control the reaction stoichiometry and the polymer composition. This is confirmed by MALDI-ToF-MS spectra, in which the main signals can be attributed to linear chains having two hydroxyl end-groups. In order to improve the polycarbonate functionality, branched polycarbonates can also be prepared, using glycerol or the more effective yet nonrenewable trimethylolpropane.

Thermal Stability and Rheological Properties of DAH-Based (co)Polycarbonates

Homopolycarbonates based on the DAH isomers are thermally stable up to approx. 300°C. The copolycarbonates, on the other hand, show a reduced thermal stability. It is possible that the copolycarbonates are more prone to degradation by alcoholysis, since primary hydroxyl end-groups are present in these polymers. For these polycarbonates, no significant weight loss is observed up to 250°C (Figure 7.25).

In applications such as powder coating systems, flow plays a crucial role during film formation and coalescence of the resin

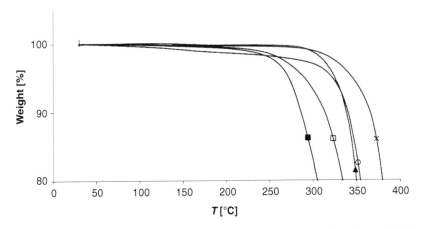

Figure 7.25 Thermogravimetric analysis (TGA) of poly(isosorbide carbonate) (O), poly(isoidide carbonate) (▲), poly(isomannide carbonate) (×), poly(isosorbide-*co*-butylene carbonate) (□) and poly(isosorbide-*co*-propylene-*co*-TMP carbonate) (■). Figure taken from reference [86]. Copyright John Wiley and Sons. Reproduced with permission.

particles. If the melt viscosity of the polymeric resin is too high, this process does not occur properly and gelation of the formulation will occur before a smooth film is obtained. The melt viscosity of a polymer depends on the T_g, the molecular weight, the degree of branching and on intermolecular interactions such as hydrogen bonding. Obviously, the addition of additives such as pigments and flow agents will also strongly affect the behavior of the formulated powder paint during film formation. In Figure 7.26, the complex viscosities of several of the discussed bio-based polycarbonates are plotted as a function of temperature. In the same plot, the complex viscosity of a typical commercially available polyester resin, used in powder coatings, is shown.

Poly(isosorbide-*co*-propylene carbonate) has a similar viscosity profile as the commercial polyester, showing a rapid drop in viscosity between approximately 70°C and 170°C to viscosity values between 10^1 and 10^2 Pa·s. The viscosity of poly(isoidide-*co*-butylene carbonate) starts to decrease at slightly lower temperatures, *viz.* around 60°C, which is probably due to its relatively low T_g (36.3°C). Poly(isoidide-*co*-propylene carbonate) has a significantly lower melt viscosity over the whole temperature range, which is thought to be caused by its low T_g (35.6°C) in combination with its moderate molecular weight. The branched poly(isosorbide-*co*-propylene-*co*-TMP

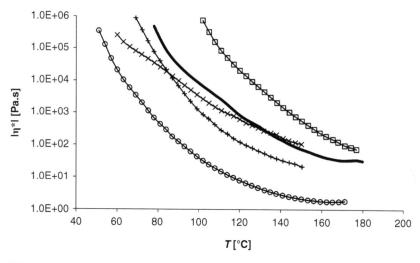

Figure 7.26 Complex viscosity as a function of temperature for poly (isosorbide-*co*-propylene carbonate) (+), poly(isoidide-*co*-propylene carbonate) (O), poly(isoidide-*co*-butylene carbonate) (×), poly(isosorbide-*co*-propylene-*co*-TMP carbonate) (□) and a commercially available polyester resin (──), used in thermosetting powder coatings. Figure taken from reference [86]. Copyright John Wiley and Sons. Reproduced with permission.

carbonate) has the highest melt viscosity and its viscosity only starts to decrease at approximately 100°C.

Curing of the bio-based linear and branched aliphatic (co) polycarbonates using conventional polyisocyanate cross-linkers (Figure 7.16) yields colorless to pale yellow poly(carbonate urethane) networks. These transparent, glossy materials show excellent resistance against solvents and withstand falling dart reverse impact testing [98]. In Figure 7.27, an example of such a coating, applied onto an aluminum substrate and subjected to reverse impact and solvent resistance testing, is shown.

7.4.2   Other Biomass-Based Polycarbonates

Apart from the DAH-derived polymers, not many bio-based polycarbonates prepared from diol monomers have been described. Copolycarbonates of arabinitol- or xylitol-based diols (Figure 7.28) with bisphenol-A are mostly amorphous polymers, with T_g-values ranging between 30 and 149°C. These polymers are susceptible to degradation in water in the presence of enzymes [99].

Figure 7.27 Poly(carbonate urethane) coating based on isoidide.

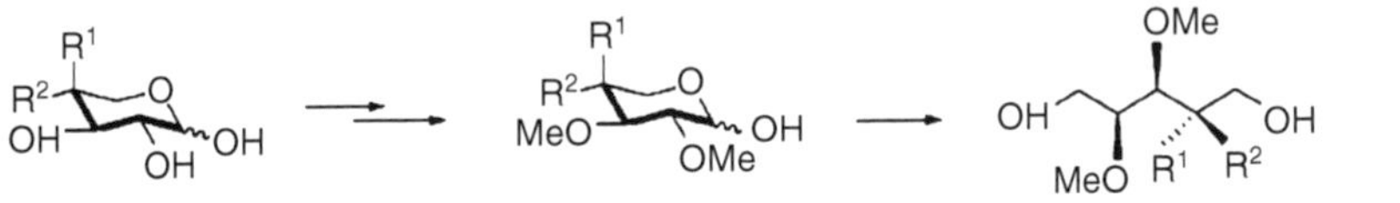

Sugar	R¹	R²
L-Arabinose	OH	H
D-Xylose	H	OH

Sugar	R¹	R²
L-Arabinitol (Ar)	OMe	H
Xylitol (Xy)	H	OMe

Figure 7.28 Synthesis of the L-Arabinitol- and Xylitol-derived diol monomers. Reprinted with permission from reference [99]. Copyright 2005 American Chemical Society.

Poly(ester carbonate)s can be prepared from low molecular weight, dihydroxy-terminated poly(1,3-propylene succinate)s followed by chain extension facilitated by phosgenation. These interesting, semi-crystalline materials have T_g values in the range of –35 to –28°C and melting temperatures of around 46°C [100]. The use of phosgene can be avoided if diethyl carbonate is used, as demonstrated by Chandure *et al*. They made poly(1,3-propylene adipate carbonate)s of moderate number-average molecular weights of up to 4,400 g/mol with similar properties as mentioned for the poly(1,3-propylene succinate carbonate)s. These poly(ester carbonate)s are also biodegradable [101].

7.5 Polyamides Based on Renewable Resources

7.5.1 Linear, Aliphatic Polyamides

While many biomass-based polyesters and polycarbonates, discussed in the previous paragraphs, are still in the early stages of their development in terms of industrial production and commercialization, several (partially) bio-based polyamides (PA) are already being sold commercially. The aliphatic diamines and dicarboxylic acids available from renewable resources can be used to produce a large number of different polyamides, spanning a vast range of properties.

In 1931, Carothers polymerized 11-aminoundecanoic acid (Figure 7.29), an amino-acid which is obtained from ricinoleic acid, the main component of castor oil [102]. One of the main advantages of using castor oil as a feedstock is the fact that this oil is not edible, preventing conflicts with the food chain. The resulting polyamide 11, currently a commercial product sold by Arkema (France), has a melting point of 188°C, a tensile strength of 38 MPa at a tensile modulus of 1.4 GPa and it has a high elongation at break of 250%. It displays excellent toughness at low temperature and its relatively apolar molecular structure (i.e. low amide bond density) leads to a low moisture uptake compared to the well-known polyamides 6 and 6.6. In addition, PA 11 exhibits excellent chemical stability towards, e.g., hydrocarbons, a low density, good thermal and weathering stability and is easily processable. Therefore, this polymer is applied in tubing and hoses for automotive brake and fuel systems, pipes for the petroleum and gas industries, food packaging and a variety of engineering, medical and sports applications.

Another important castor oil-derivative is sebacic acid, which can be used in both polyester and polyamide syntheses. Initially, this dicarboxylic acid was combined with the readily available, non-renewable monomer 1,6-diaminohexane to form PA 6.10. PA 6.10 is very suitable for use in toothbrush bristles, thanks to its high resistance against deformation under load in humid conditions. It is also used in precision injection molding parts such as electrical connectors and industrial parts, as well as in tubing and profiles and is sold by BASF and Rhodia, a.o. [103]. Combining sebacic acid with 1,4-diaminobutane (or: putrescine, a naturally occurring aliphatic diamine) yields a potentially fully biomass-based polyamide 4.10. PA 4.10 can be synthesized through a prepolymerization step of the salt of 1,4-diaminobutane and sebacic acid in water, followed

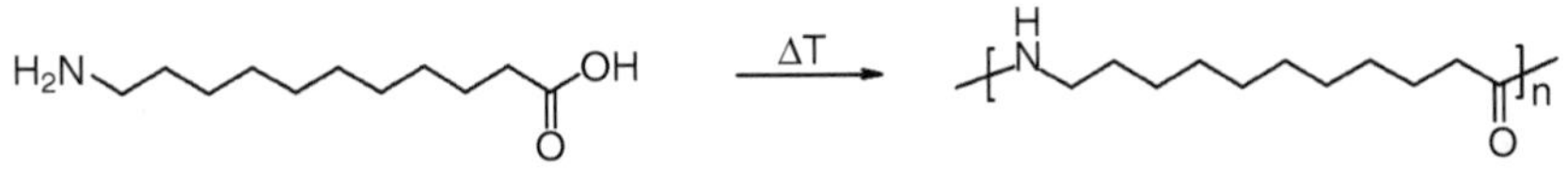

Figure 7.29 Polymerization of 11-aminoundecanoic acid to form polyamide 11.

by the removal of water through flashing. The prepolymer is then subjected to solid state post condensation to form the high molecular weight polymer (Figure 7.30) [104].

PA 4.10 has a high melting point of 249°C, a low moisture uptake compared to PA 6 and PA 6.6 as well as good tensile characteristics (tensile strength = 82 MPa, tensile modulus = 3 GPa, elongation at break = 10%). In addition, this polyamide is hydrolysis resistant and crystallizes rapidly from the melt, leading to short injection molding cycle times. PA 4.10 was only recently commercialized by DSM (The Netherlands). PA 4.10 is suitable for demanding applications in the automotive and electronics markets. Another natural diamine is cadaverine (1,5-diaminopentane), which could therefore be used to make another polyamide, fully derived from renewable resources. PA 5.10 has a lower melting point than PA 4.10 due to its lower amide bond density (i.e. a longer distance between amide linkages) and as a result of the well-known odd-even effect. Yet another sebacic acid-based, linear polyamide is PA 10.10, produced by Evonik. This hydrophobic polymer takes up low amounts of moisture and is hydrolysis resistant. It is therefore used in applications in which substrates must be protected from water, such as cable jacketing and as a thermoplastic powder coating material.

Azaleic acid (or: nonanedioic acid) is produced from oleic acid, a mono-unsaturated fatty acid present in several types of vegetable oils, by ozonolysis. Polycondensation of azaleic acid with 1,6-diaminohexane yields PA 6.9, a polymer with a melting temperature of approx. 210°C. This polyamide, a former commercial product, has a moderate degree of crystallinity and is applied in tubing, bristles and various injection molding parts [103, 105].

7.5.2   Fatty Acid-Based Polyamides

Polyamides with much lower amide bond densities can be prepared from dimerized fatty acids. Such flexible, mainly aliphatic

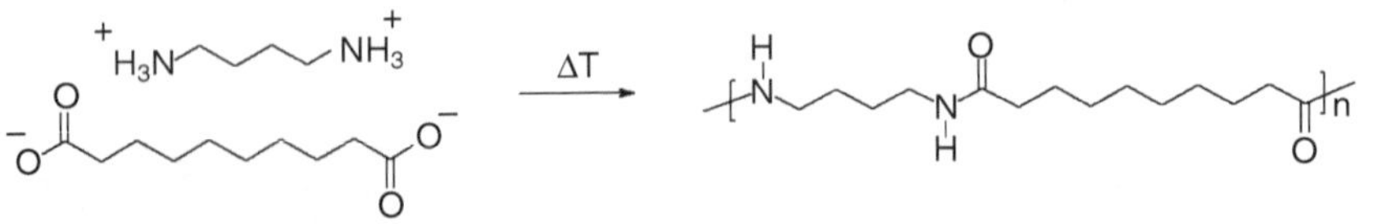

Figure 7.30 Synthesis of polyamide 4.10 from the corresponding salt.

dicarboxylic acids are derived from vegetable oils such as soybean or rapeseed oil. Diamine monomers used in such systems include 1,2-ethylenediamine [106], 1,6-diaminohexane, dimer fatty diamines and piperazine. Also, other dicarboxylic acids may be used [107–109]. Such polyamides typically have glass transition temperatures below 0°C and melting temperatures up to 110°C for short diamines such as ethylene diamine [110, 111]. This class of bio-based polyamides is applicable as pressure sensitive, hot melt adhesives or as matrices for biocomposites [112].

Acyclic diene metathesis can be applied to couple two units of methyl-10-undecenoate from castor oil, forming an unsaturated dimethyl ester, which can subsequently be polymerized with diamines (catalyzed by a strong base such as 1,5,7-triazabicyclo [4.4.0]dec-5-ene, TBD). Alternatively, two units of methyl-10-undecenoate can first be reacted with a diamine, to form a α,ω-diene, which can then be polymerized using metathesis (Figure 7.31). These routes were compared by Mutlu et al., showing that the former route yields a very interesting range of novel polyamides, with number average molecular masses up to 15,000 g/mol. Depending on the diamine used, melting temperatures of up to 226°C can be achieved (for the unsaturated PA 2.20) [113].

7.5.3 Other Biomass-Based Polyamides

DAH-based polyamides were described by Thiem and Bachmann, who transformed the three isomeric diols into their corresponding diamine derivatives. These diamine hydrochloride compounds were subsequently polymerized with aliphatic as well as aromatic acid chlorides. Polymerizations of these diamines with terephthaloyl or isophthaloyl chloride yields polyamides with number average molecular weights up to 25,000 g/mol and high melting temperatures of up to 310°C. During melting, decomposition occurs as well. Fully aliphatic polyamides, obtained through reactions of the DAH diamine hydrochlorides with C10–C16 dicarboxylic acid chlorides, have T_g values in the range of 50 to 70°C and melting temperatures between 130 and 200°C. In addition, polyamides based on 1,4-anhydroalditols were reported [114]. Recently, fully bio-based, semi-crystalline copolyamides were described prepared from sebacic acid, putrescine and diaminoisoidide (DAII). The crystals present in these new polymers contain both polyamide 4.10 and DAII.10-based repeat units. The polarities and melting

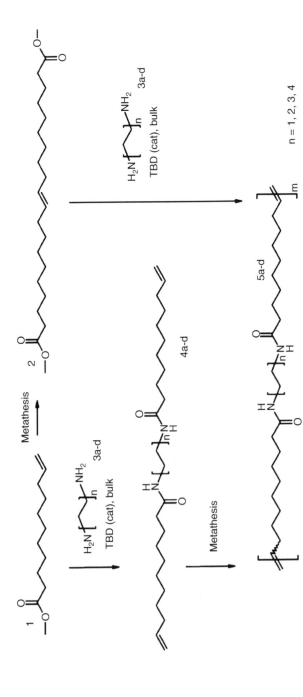

Figure 7.31 Schematic representation of two different routes to prepare castor oil-derived polyamides x.20 through metathesis and TBD-catalyzed bulk polymerization. Figure taken from reference [113]. Copyright Wiley-VCH Verlag GmbH & Co. KGaA. Reproduced with permission.

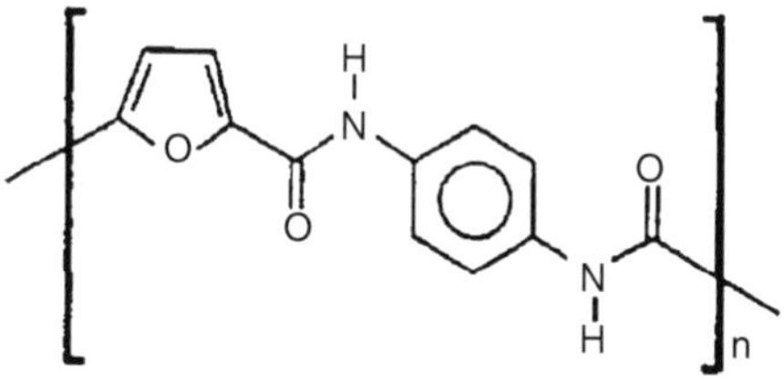

Figure 7.32 Polyamide from 2,5-furandicarboxylic acid and 1,4-phenylene diamine. Figure taken from reference [70]. Copyright (1997), with permission from Elsevier.

temperatures of such polymers can be tuned depending on the DAII content [115].

A range of polyamides can also be prepared based on furan derivatives. One of the most interesting examples is the aromatic polymer of 2,5-furandicarboxylic acid (2,5-FDA) and 1,4-phenylene diamine (Figure 7.32), having a T_g of 325°C and an onset of decomposition at 385°C [70].

Reactions between the dimethyl ester of 2,5-FDA and linear aliphatic diamines such as hexamethylene diamine yield amorphous polymers with glass transition temperatures around 70–110°C, depending on the diamine chain length, at moderate molar mass [116].

Another range of interesting polyamides are based on L-lysine in combination with carboxylic acid-functional compounds such as citric acid, malic acid or tartaric acid. Depending on the type of comonomers, these polymers may be applicable as water-soluble polyelectrolytic drug carriers [117]. Polymerizing L-lysine with tartaric acid yields non-crystalline yet optically active polymers with T_g values up to 105°C, which are thermally stable up to 250–300°C [118].

7.6   Conclusions

In this chapter, we have given an overview of a range of biomass-based polymers prepared through polycondensation chemistry. This is by no means an exhaustive summary of available starting materials and polymers. Still, it is clear that this rapidly expanding field of research opens up new synthetic pathways to performance polymers with exciting novel properties. Some of these bio-based polycondensates can rival conventional polymers from petro-chemistry in terms of mechanical and chemical performance and a limited number have already found their way to the market place. Further development of bio-refining processes is expected to improve the availability and economics of the discussed starting materials and polymers. Several of the presented polymers can be produced in existing process equipment with only minor adjustments. Typically, improved temperature control and an inert atmosphere are required to prevent degradation and discoloration of the bio-based products. In addition, more sustainable and milder synthetic routes as well as more effective catalysis are necessary to make optimal use of biomass, constituting stimuli for new research into the field of step-growth polymerization.

References

1. A. Gandini and N. M. Belgacem, *Polym. Int.*, Vol. 47, p. 267, 1998.
2. A. Gandini, *Green Chem.*, 2011.
3. J. J. Bozell and G. R. Petersen, *Green Chem.*, Vol. 12, p. 539, 2010.
4. H. Ikram-ul, S. Ali, M. A. Qadeer and J. Iqbal, *Bioresour. Technol.*, Vol. 93, p. 125, 2004.
5. I. Bechthold, K. Bretz, S. Kabasci, R. Kopitzky and A. Springer, *Chem. Eng. Technol.*, Vol. 31, p. 647, 2008.
6. A. Corma, S. Iborra and A. Velty, *Chem. Rev.*, Vol. 107, p. 2411, 2007.
7. J. Y. Qin, Z. J. Xiao, C. Q. Ma, N. Z. Xie, P. H. Liu and P. Xu, *Chin. J. Chem. Eng.*, Vol. 14, p. 132, 2006.
8. P. Stoss and R. Hemmer, *Adv. Carbohydr. Chem. Biochem.*, Vol. 49, p. 93, 1991.
9. A. Gandini, D. Coelho, M. Gomes, B. Reis and A. Silvestre, *J. Mater. Chem.*, Vol. 19, p. 8656, 2009.
10. A. Gandini, *Polym. Chem.*, Vol. 1, p. 245, 2010.
11. J. T. P. Derksen, F. P. Cuperus and P. Kolster, *Prog. Org. Coat.*, Vol. 27, p. 45, 1996.
12. H. Mutlu and M. A. R. Meier, *Eur. J. Lipid Sci. Tech.*, Vol. 112, p. 10, 2010.
13. A. Gandini, *Macromolecules*, Vol. 41, p. 9491, 2008.
14. M. Kleinert and T. Barth, *Chem. Eng. Technol.*, Vol. 31, p. 736, 2008.
15. J. Zakzeski, P. C. A. Bruijnincx, A. L. Jongerius and B. M. Weckhuysen, *Chem. Rev.*, Vol. 110, p. 3552, 2010.
16. M. S. Lindblad, Y. Liu, A. C. Albertsson, E. Ranucci and S. Karlsson, "Polymers from renewable resources" in A. C. Albertsson, ed. *Degradable Aliphatic Polyesters*, Springer-Verlag, Berlin, p. 139, 2002.
17. Y. Xu, J. Xu, D. Liu, B. Guo and X. Xie, *J. Appl. Polym. Sci.*, Vol. 109, p. 1881, 2008.
18. J. A. Galbis and M. G. García-Martín, "Synthetic Polymers from Readily Available Monosaccharides" in A. P. Rauter, P. Vogel and Y. Queneau, eds., *Carbohydrates in Sustainable Development II*, Springer Verlag, Berlin, p. 147, 2010.

19. M. Okada, *Prog. Polym. Sci.*, Vol. 27, p. 87, 2002.

20. M. F. Sonnenschein, S. J. Guillaudeu, B. G. Landes and B. L. Wendt, *Polymer*, Vol. 51, p. 3685, 2010.

21. Y. Liu, Soderquist Lindblad, M., Ranucci, E., Albertsson, A.-C., *J. Polym. Sci., Part A: Polym. Chem.*, Vol. 39, p. 630, 2001.

22. L. Wu, Y. Zhang, H. Fan, Z. Bu and B.-G. Li, *J. Polym. Environ.*, Vol. 16, p. 68, 2008.

23. N. Teramoto, M. Ozeki, I. Fujiwara and M. Shibata, *J. Appl. Polym. Sci.*, Vol. 95, p. 1473, 2005.

24. D. G. Barrett, T. J. Merkel, J. C. Luft and M. N. Yousaf, *Macromolecules*, Vol. 43, p. 9660, 2010.

25. B. A. J. Noordover, V. G. Van Staalduinen, R. Duchateau, C. E. Koning, R. A. T. M. Van Benthem, M. Mak, A. Heise, A. E. Frissen and J. Van Haveren, *Biomacromolecules*, Vol. 7, p. 3406, 2006.

26. B. A. J. Noordover, R. Duchateau, R. A. T. M. Van Benthem, W. Ming and C. E. Koning, *Biomacromolecules*, Vol. 8, p. 3860, 2007.

27. T. A. Misev, *Powder coatings-chemistry and technology*, New York, John Wiley & Sons, 1991.

28. L. W. Wright and J. D. Brandner, *J. Org. Chem.*, Vol. 29, p. 2979, 1964.

29. A. C. Cope and T. Y. Shen, *J. Am. Chem. Soc.*, Vol. 78, p. 3177, 1956.

30. C. Cecutti, Z. Mouloungui and A. Gaset, *Bioresour. Technol.*, Vol. 66, p. 63, 1998.

31. W. Szeja, *J. Chem. Soc., Chem. Commun.*, Vol. 5, p. 215, 1981.

32. G. Fleche and M. Huchette, *Starch*, Vol. 38, p. 26, 1986.

33. K. W. Buck, J. M. Duxbury, A. B. Foster, A. R. Perry and J. M. Webber, *Carbohydr. Res.*, Vol. 2, p. 122, 1966.

34. D. Abenhaim, A. Loupy, L. Munnier, R. Tamion, F. Marsais and G. Queguiner, *Carbohydr. Res.*, Vol. 261, p. 255, 1994.

35. G. Le Lem, P. Boullanger, G. Descotes and E. Wimmer, *Bull. Soc. Chim. Fr.*, Vol. 3, p. 567, 1988.

36. H. R. Kricheldorf, *J.M.S.-Rev. Macromol. Chem. Phys.*, Vol. C37, p. 599, 1997.

37. F. Fenouillot, A. Rousseau, G. Colomines, R. Saint-Loup and J.-P. Pascault, *Prog. Polym. Sci.*, Vol. 35, p. 578, 2010.

38. V. F. Jenkins, A. Mott and R. J. Wicker, Unsaturated polyesters and polyester compositions, patent no. 927,786, assigned to Howards of Ilford Limited, 1963.

39. J. R. Collins, Polyesters, patent no. GB1079686, assigned to Courtaulds Ltd, 1964.

40. Atlas, Fibrous products bound with polyester resins, patent no. US1012563, assigned to Atlas Chemical Industries, 1962.

41. J. Thiem and H. Lueders, *Polym. Bull.*, Vol. 11, p. 365, 1984.

42. R. Storbeck and M. Ballauff, *J. Appl. Polym. Sci.*, Vol. 59, p. 1199, 1996.

43. H. R. Kricheldorf, S. Chatti, G. Schwarz and R. P. Kruger, *J. Polym. Sci., Part A: Polym. Chem.*, Vol. 41, p. 3414, 2003.

44. H. R. Kricheldorf, M. Berghahn, N. Probst, M. Gurau and G. Schwarz, *React. Func. Polym.*, Vol. 30, p. 173, 1996.

45. H. R. Kricheldorf and Z. Gomourachvili, *Macromol. Chem. Phys.*, Vol. 198, p. 3149, 1997.

46. M. Okada, Y. Okada, A. Tao and K. Aoi, *J. Appl. Polym. Sci.*, Vol. 62, p. 2257, 1996.

47. M. Okada, K. Tachikawa and K. Aoi, *J. Polym. Sci., Part A: Polym. Chem.*, Vol. 35, p. 2729, 1997.
48. M. Okada, K. Tachikawa and K. Aoi, *J. Appl. Polym. Sci.*, Vol. 74, p. 3342, 1999.
49. M. Okada, K. Tsunoda, K. Tachikawa and K. Aoi, *J. Appl. Polym. Sci.*, Vol. 77, p. 338, 2000.
50. M. Okada, Y. Okada and K. Aoi, *J. Polym. Sci., Part A: Polym. Chem.*, Vol. 33, p. 2813, 1995.
51. D. Braun and M. Bergmann, *J. Prakt. Chem.*, Vol. 334, p. 298, 1992.
52. R. Storbeck, M. Rehahn and M. Ballauff, *Makromol. Chem.*, Vol. 194, p. 53, 1993.
53. S. Chatti, M. Bortolussi, A. Loupy, J. C. Blais, D. Bogdal and P. Roger, *J. Appl. Polym. Sci.*, Vol. 90, p. 1255, 2003.
54. B. A. J. Noordover, *PhD thesis: Biobased step-growth polymers-chemistry, functionality and applicability* Eindhoven University of Technology, 2008.
55. T. G. J. Fox and P. J. Flory, *J. Appl. Phys.*, Vol. 21, p. 581, 1950.
56. J. Van Haveren, E. A. Oostveen, F. Micciche, B. A. J. Noordover, C. E. Koning, R. A. T. M. Van Benthem, A. E. Frissen and J. G. J. Weijnen, *J. Coat. Tech. Res.*, Vol. 4, p. 177, 2007.
57. J. M. Raquez, M. Deléglise, M. F. Lacrampe and P. Krawczak, *Prog. Polym. Sci.*, Vol. 35, p. 487, 2010.
58. L. Jasinska and C. E. Koning, *J. Polym. Sci., Part A: Polym. Chem.*, Vol. 48, p. 2885, 2010.
59. L. Jasinska and C. E. Koning, *J. Polym. Sci., Part A: Polym. Chem.*, Vol. 48, p. 5907, 2010.
60. M. Reinecke, Ritter, H., *Makromol. Chem.*, Vol. 194, p. 2385, 1993.
61. B. A. J. Noordover, A. Heise, P. Malanowski, D. Senatore, M. Mak, L. Molhoek, R. Duchateau, C. E. Koning and R. A. T. M. Van Benthem, *Prog. Org. Coat.*, Vol. 65, p. 187, 2009.
62. D. J. Adelman, L. F. Charbonneau and S. Ung, Process for making poly(ethylene-co-isosorbide) terephthalate polymer, patent no. US2003232959, assigned to Du Pont de Nemours & Company, 2003.
63. D. J. Adelman, R. N. Greene and D. E. Putzig, Poly(1,3-propylene-co-1,4:3,6-dianhydro-D-sorbitol terephthalate) and manufacturing process, patent no. US7049390, assigned to E.I. Du Pont de Nemours and Company, 2006.
64. L. F. Charbonneau, Process for maing low color poly(ethylene-co-isosorbide) terephthalate polymer, patent no. US2006173154, assigned to E.I. Du Pont de Nemours and Company, 2006.
65. H. R. Kricheldorf, G. Behnken and M. Sell, *J. Macromol. Sci. A*, Vol. 44, p. 679, 2007.
66. R. Sablong, R. Duchateau, C. E. Koning, G. De Wit, D. Van Es, R. Koelewijn and J. Van Haveren, *Biomacromolecules*, Vol. 9, p. 3090, 2008.
67. J. A. Moore and J. E. Kelly, *Macromolecules*, Vol. 11, p. 568, 1978.
68. A. Gandini, A. J. D. Silvestre, C. P. Neto, A. F. Sousa and M. Gomes, *J. Polym. Sci., Part A: Polym. Chem.*, Vol. 47, p. 295, 2009.
69. R. Storbeck and M. Ballauff, *Polymer*, Vol. 34, p. 5003, 1993.
70. A. Gandini and M. N. Belgacem, *Prog. Polym. Sci.*, Vol. 22, p. 1203, 1997.
71. A. Khrouf, Boufi, S., El Gharbi, R., Gandini, A., *Polym. Int.*, Vol. 48, p. 649, 1999.
72. H. Ebata, K. Toshima and S. Matsumura, *Macromol. Biosci.*, Vol. 7, p. 798, 2007.

73. I. O. Bakare, C. Pavithran, F. E. Okieimen and C. K. S. Pillai, *J. Appl. Polym. Sci.*, Vol. 100, p. 3748, 2006.
74. T. Tsujimoto, H. Uyama and S. Kobayashi, *Biomacromolecules*, Vol. 2, p. 29, 2001.
75. E. Can, R. P. Wool and S. Kusefoglu, *J. Appl. Polym. Sci.*, Vol. 102, p. 1497, 2006.
76. D. Quinzler and S. Mecking, *Angewandte Chemie International Edition*, Vol. 49, p. 4306, 2010.
77. O. Turunc and M. A. R. Meier, *Macromol. Rapid Commun.*, Vol. 31, p. 1822, 2010.
78. S. Inoue, H. Koinuma and T. Tsuruta, *Polymer Letters*, Vol. 7, p. 287, 1969.
79. D. J. Darensbourg and M. W. Holtcamp, *Macromolecules*, Vol. 28, p. 7577, 1995.
80. W. J. Van Meerendonk, R. Duchateau, C. E. Koning and G.-J. M. Gruter, *Macromol. Rapid Commun.*, Vol. 25, p. 382, 2004.
81. D. J. Darensbourg, *Chem. Rev.*, Vol. 107, p. 2388, 2007.
82. G. Rokicki and T. Kowalczyk, *Polymer*, Vol. 41, p. 9013, 2000.
83. W. Kuran, M. Sobczak, T. Listos, C. Debek and Z. Florjanczyk, *Polymer*, Vol. 41, p. 8531, 2000.
84. P. Y. W. Dankers, Z. Zhang, E. Wisse, D. W. Grijpma, R. P. Sijbesma, J. Feijen and E. W. Meijer, *Macromolecules*, Vol. 39, p. 8763, 2006.
85. C. M. Byrne, S. D. Allen, E. B. Lobkovsky and G. W. Coates, *J. Am. Chem. Soc.*, Vol. 126, p. 11404, 2004.
86. B. A. J. Noordover, D. Haveman, R. Duchateau, R. A. T. M. van Benthem and C. E. Koning, *J. Appl. Polym. Sci.*, Vol. 121, p. 1450, 2011.
87. H. Medem, M. Schreckenberg, R. Dhein, W. Nouvertne and H. Rudolph, Thermoplastische Polycarbonate, Verfahren zu ihrer Herstellung, Bis-chlorkohlensaureester von hexahydro-furo(3,2-b) furan-3,6-diolen, sowie Polycarbonatmassen patent no. EP0025937, assigned to Bayer AG, 1980.
88. A. Schecter, N. Malik, D. Haffner, S. Smith, T. R. Harris, O. Paepke and L. Birnbaum, *Environ. Sci. Tech.*, Vol. 44, p. 9425, 2010.
89. A. Ono, K. Toyohara, H. Minematsu and Y. Kageyama, Polycarbonate and process for producing the same, patent no. EP1640400, assigned to Teijin Limited, 2006.
90. P. Schuhmacher, H. R. Kricheldorf and S.-J. Sun, Chiral nematic polycarbonates, patent no. US6156866, assigned to BASF A.G., 2000.
91. D. Dhara, A. A. G. Shaiki, G. Chatterjee and C. Seetharaman, Aliphatic diol polycarbonates and their preparation, patent no. WO2005066239, assigned to General Electric Company, 2005.
92. H. R. Kricheldorf, S. J. Sun, A. Gerken and T. C. Chang, *Macromolecules*, Vol. 29, p. 8077, 1996.
93. M. Yokoe, K. Aoi and M. Okada, *J. Polym. Sci., Part A: Polym. Chem.*, Vol. 41, p. 2312, 2003.
94. H. R. Kricheldorf, S. Bohme and G. Schwarz, *Macromol. Chem. Phys.*, Vol. 206, p. 432, 2005.
95. H. R. Kricheldorf, S. Bohme and G. Schwarz, *Macromolecules*, Vol. 39, p. 3210, 2006.
96. S. Chatti, G. Schwarz and H. R. Kricheldorf, *Macromolecules*, Vol. 39, p. 9064, 2006.
97. J.-P. Hsu and J.-J. Wong, *Ind. Eng. Chem. Res.*, Vol. 45, p. 2672, 2006.
98. B. A. J. Noordover, R. Duchateau, C. E. Koning and R. A. T. M. Van Benthem, Polycarbonate and process for producing the same, patent no. WO2009033934, assigned to Stichting Dutch Polymer Institute, 2009.

99. M. G. Garcia-Martin, R. R. Perez, E. B. Hernandez, J. L. Espartero, S. Munoz-Guerra and J. A. Galbis, *Macromolecules*, Vol. 38, p. 8664, 2005.

100. E. Ranucci, Liu, Y., Soderquist Lindblad, M., Albertsson, A.C., *Macromol. Rapid Commun.*, Vol. 21, p. 680, 2000.

101. A. S. Chandure, S. S. Umare and R. A. Pandey, *Eur. Polym. J.*, Vol. 44, p. 2068, 2008.

102. W. H. Carothers, *Chem. Rev.*, Vol. 8, p. 353, 1931.

103. I. B. Page, *Polyamides as engineering thermoplastic materials*, Rapra Technology Ltd, 2000.

104. C. Koning, L. Teuwen, R. de Jong, G. Janssen and B. Coussens, *High Perform. Polym.*, Vol. 11, p. 387, 1999.

105. M. I. Kohan, "Commercial Nylon plastics and their applications" in M. I. Kohan, ed. *Nylon Plastics Handbook* Hanser, Munich, Vienna, New York, p. 556, 1995.

106. J. Heidarian, N. M. Ghasem and W. Daud, *J. Appl. Polym. Sci.*, Vol. 92, p. 2504, 2004.

107. H. S. Vedanayagam and V. Kale, *Polymer*, Vol. 33, p. 3495, 1992.

108. C. R. Frihart, *Int. J. Adhes. Adhes.*, Vol. 24, p. 415, 2004.

109. X. D. Fan, Y. L. Deng, J. Waterhouse and P. Pfromm, *J. Appl. Polym. Sci.*, Vol. 68, p. 305, 1998.

110. H. J. Manuel and R. J. Gaymans, *Polymer*, Vol. 34, p. 4325, 1993.

111. E. Hablot, B. Donnio, M. Bouquey and L. Avérous, *Polymer*, Vol. 51, p. 5895, 2010.

112. E. Hablot, R. Matadi, S. Ahzi and L. Avérous, *Comp. Sci. Tech.*, Vol. 70, p. 504, 2010.

113. H. Mutlu and M. A. R. Meier, *Macromol. Chem. Phys.*, Vol. 210, p. 1019, 2009.

114. J. Thiem and F. Bachmann, *Makromol. Chem.*, Vol. 192, p. 2163, 1991.

115. L. Jasinska, M. Villani, J. Wu, D. Van Es, E. Klop, S. Rastogi and C. E. Koning, *Macromolecules*, 2011.

116. O. Grosshardt, U. Fehrenbacher, K. Kowollik, B. Tubke, N. Dingenouts and M. Wilhelm, *Chem. Ing. Tech.*, Vol. 81, p. 1829, 2009.

117. A.-C. Couffin-Hoarau, M. Boustta and M. Vert, *J. Polym. Sci., Part A: Polym. Chem.*, Vol. 39, p. 3475, 2001.

118. M. A. Majó, A. Alla, J. J. Bou, C. Herranz and S. Muñoz-Guerra, *Eur. Polym. J.*, Vol. 40, p. 2699, 2004.

8

Succinic Acid: Synthesis of Biobased Polymers from Renewable Resources

Stephan Kabasci and Inna Bretz

Fraunhofer Institute for Environmental, Safety, and Energy Technology UMSICHT, Oberhausen, Germany

Abstract

Succinic acid can be derived from renewable resources in a biotechnological production process. Because of its chemical structure as linear C4 dicarboxylic acid it can be used as biobased monomer in different polycondensation reactions. An overview of the resulting biopolymers (e.g. polyesters, polyamides and poly(ester amide)s) based on succinic acid and its derivatives is given in this contribution. With biotechnological succinic acid production on the rise these polymers can contribute considerably to the substitution of fossil based plastics.

Keywords: Succinic acid, biopolymers, polyamides, polyesters, poly(ester amide)s, renewable resources

8.1 Introduction

8.1.1 General

The resources of fossil raw materials for the chemical industry are limited. In addition, the use of fossil resources leads to severe environmental problems and increasing CO_2 concentrations in the atmosphere. Therefore, new CO_2 neutral processes based on renewable resources become increasingly important in the chemical industry. A variety of biobased platform chemicals was recognized as a significant research topic by politics and industry in the last decade [1, 2].

The technologies for the production of biobased chemicals [3, 4, 5, 6, 7] or biobased polymers [8, 9] and their possible economical

---

Vikas Mittal (ed.) Renewable Polymers, (355–380) © Scrivener Publishing LLC

potentials were reviewed in several reports in the last decade. In all these publications succinic acid was identified as one of the most important platform chemicals. It can rather easily be produced biotechnologically and a huge variety of chemicals can be derived from it via different chemical conversion pathways.

Succinic acid, a linear four carbon dicarboxylic acid, can be used as a starting material for the production of e.g. succinate salts and esters, 1,4-butanediol, tetrahydrofuran, γ-butyrolactone, 2-pyrrolidone or N-methyl pyrrolidone, which are used in a number of industries including polymers and fibres, food, surfactants and detergents, flavours and fragrances. The current worldwide annual production of succinic acid is around 20,000 to 30,000 tonnes [10].

Succinic acid based on carbohydrates by bacterial fermentation is mostly used for the food industry. Several reviews have presented the fermentation of succinic acid [11, 12, 13, 14]. In the following sections a brief review of the biotechnological production of succinic acid and some chemical conversion pathways will be given with the focus on monomers for homo- and copolymers of succinic acid.

8.1.2   Biotechnological Production of Succinic Acid

Up to now succinic acid is mainly produced by chemical processes from n-butane or butadiene via maleic anhydride using the C4-fraction of naphtha. Its price range is estimated to be 6-9 $/kg. Derivatives of succinic acid are predicted to have a potential of hundreds of thousand tons annually [6, 7, 15].

Succinic acid is an intermediate of the tricarboxylic acid (TCA) cycle and one of the fermentation end-products of anaerobic metabolism. In the course of bacterial succinic acid production carbon dioxide (CO_2) is assimilated which makes this process very interesting as a CO_2 capturing technology.

A wide variety of natural succinic acid overproducing bacteria, including *Bacteroides ruminicola, Bacteroides amylophilus, Enterococcus faecalis, Anaerobiospirillum succiniciproducens, Actinobacillus succinogenes and, Mannheimia succiniciproducens* were already identified [16, 17, 18, 19]. Using metabolic engineering methods *Echerichia coli* and *Corynebacterium glutamicum* were modified to produce succinic acid. Further, several fungi, including different *Aspergillus* strains, *Byssochlamys nivea, Lentinus degener, Paecilomyces varioti and,*

Penicillium viniferum were screened and studied for succinic acid production [20, 21].

Anaerobiospirillum succiniciproducens [17, 22, 23] and *Actinobacillus succinogenes* [24, 25, 26] are well analyzed due to their ability to produce large amounts of succinic acid.

Different renewable resources were tested as substrates for succinic acid production by *A. succinogenes*, like whey [27], cane molasses [28], straw hydrolysates [25], corn fiber [29], crop stalk wastes [26], wheat [30] and hydrolysates of spent yeast cells [31]. *A. succinogenes* metabolizes glucose to succinic acid, acetic acid and formic acid and has a high acid tolerance [32, 33, 34]. Main problems of *A. succinogenes* fermentation are the high concentrations of byproducts like acetic acid, propionic acid and pyruvic acid which lead to high costs in product separation.

A. succiniciproducens can utilize glucose, glycerol, sucrose, maltose, lactose, fructose, untreated whey, wood hydrolysate and corn steep liquor as carbon sources [35, 36, 37].

More recently, *Mannheimia succiniciproducens* MBEL55E was isolated as a natural succinic acid overproducer from bovine rumen by Lee *et al.* [38]. Furthermore, the efficient and economical production of succinic acid was possible by fermentation of *M. succiniciproducens* using a whey-based medium containing corn steep liquor instead of yeast extract [23]. Metabolic engineering led to increased succinic acid production and yield with simultaneous reduction of byproduct formation [39]. To achieve a higher productivity and evaluate the industrial applicability, a recent study used *Mannheimia succiniciproducens* LPK7 (knock-out: *ldh*A, *pfl*B, *pta-ack*A), which was designed to enhance the productivity of succinic acid and reduce by-product secretion [40].

Additionally, there have been great efforts in developing recombinant *Escherichia coli* strains, which are capable of enhanced succinic acid production under aerobic and anaerobic conditions [41]. Recently, a metabolically engineered *E. coli* strain capable of aerobically producing succinic acid through the glyoxylate pathway and the oxidative branch of the TCA cycle was developed [42, 43].

Raab *et al.* [44] reported the suitability of high acid- and osmotolerant yeast *Saccharomyces cerevisiae* as a succinic acid production host. They implemented a metabolic engineering strategy for the oxidative production of succinic acid in yeast by deletion of the genes SDH1, SDH2, IDH1 and IDP1. Succinic acid is not accumulated intracellularly. This makes the yeast *S. cerevisiae* a suitable and promising candidate for the biotechnological production of succinic acid on an industrial scale.

The purification cost for fermentation based products usually determines the total production costs. In average 80% of the overall production costs are caused by down stream processing [15]. Efficient separation of succinic acid from by-products like acetic, formic, and pyruvic acids is necessary for an economically viable process. Various recovery techniques including crystallization [45, 46], extraction [47, 48], adsorption [49] and electrodialysis [50, 51] have been reported.

8.1.3   Chemical Conversion

Succinic acid is often being described as a "Green" platform chemical because it can be transformed into several commodity or specialty chemicals (see Figure 8.1) [7, 52]. A general chemistry of succinic acid is summarized by Fumagalli *et al.* [53].

Direct hydrogenation of succinic acid esters, succinic anhydride, or succinic acid leads to the product family consisting of 1,4-butanediol (BDO), tetrahydrofuran (THF), and gamma-butyrolactone (GBL). These compounds are large-scale commodities used in a variety of industrial applications and as monomer precursors for

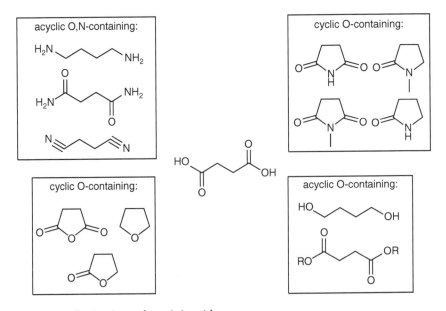

Figure 8.1 Derivatives of succinic acid.

polybutylene terephthalate (PBT) and poly(tetramethylene ether glycol) (PTMEG) [54].

1,4-Butanediol (BDO) is a raw material for a number of chemical syntheses. BDO can be produced from succinic acid anhydride using Raney-type catalysts in the form of hollow spheres. A route from succinic acid to 1,4-butanediole (BDO) via GBL is also mentioned [55].

Tetrahydrofuran (THF) is a widely used high performance solvent and a precursor for a range of chemical syntheses. Tetrahydrofuran synthesis from 1,4-butanediol is carried out via dehydration in high-temperature liquid water (HTW) without added catalyst at 200–350°C [56]. THF can be polymerized by strong acids to receive a linear polymer called poly(tetramethylene ether) glycol (PTMEG), also known as poly(tetramethylene oxide) (PTMO). The primary use of this polymer is the production of elastomeric polyurethane fibers like Spandex®.

The purified GBL can be catalytically hydrogenated over bi- or trimetallic catalysts (Ni and Pd coupled with Rh) supported by carbon [57, 58, 59]. Processes using a hydrogenation catalyst comprising one or more active hydrogen catalyst components on rutile supports for the production of GBL, THF and BDO are described by Bhattacharyya and Manila [60], Roesch et al. [61], Wood et al. [62].

Pyrrolidones are excellent solvents (liquid at low temperatures) for numerous applications. They are also used as starting materials for a wide range of chemical conversions in the pharmaceutical and food industries. 2-Pyrrolidone is an important solvent, extractant and starting material for the production of pharmaceuticals (e.g., dopaminergics and antibiotics) and N-vinylpyrrolidone (NVP). N-Methylpyrrolidone (NMP) is a powerful aprotic solvent that is applied in polymer syntheses, in the manufacture of different resins and fibers, as well as in numerous organic syntheses [14]. The production of 2-pyrrolidone (2-PDO) and N-methylpyrrolidone by hydrogenation of ammonium succinate using active metal catalysts has been described by Werpy et al. [52, 63, 64].

8.2 Polymerization

In addition to succinic acid the production of polyesters, polyamides (PA) and poly(ester amide)s requires components like diamines or diols. In the case of using products with four carbon atoms, they could be obtained by chemical conversion of succinic acid (compare Chapter 8.1.3).

8.2.1   Polyesters

Biodegradable and biobased polymers are becoming more and more important. They will join the market-established materials and partially replace them either to reduce the required use of fossil resources or to promote environmentally friendly waste disposal. Predominantly, aliphatic polyesters are used in packaging, hygiene and medical applications, where they could be replaced by new polymers based on succinic acid. In the literature biodegradable linear aliphatic polyesters, poly(alkylene succinate)s, are described, namely poly(ethylene succinate) PES, poly(propylene succinate) PPS and poly(butylene succinate) PBS and their copolyesters. The general molecular formula of these polymers is given in Figure 8.2.

Carothers and Berchet first studied the synthesis of aliphatic polyesters by polycondensation reactions of diols with different dicarboxylic acids or their esters in 1930s [65]. The low melting points of most of these polyesters in combinations with the difficulty to obtain high molecular weight materials has prevented their usage for long-time.

Polyesters of succinic acid can be manufactured by polycondensation of succinic acid diesters or succinic acid with diols. High-molecular weight polyesters can be obtained e.g. by chain extensions from low-molecular weight polymers with hexamethylene diisocyanate as chain-extender.

The polycondensations are carried our using different metal compounds as catalysts. Various authors successfully applied tin(II) chloride [$SnCl_2$], titanium butoxyde [$Ti(OBu)_4$], titanium tetraisopropoxide [$Ti(OiPr)_4$], or scandium compounds like $Sc(CF_3SO)_3$ or $Sc(NTf_2)_3$ to receive low or high molecular weight polyesters. Compounds of antimony, tin and titanium are the most active catalysts for these polycondensation reactions [66, 67, 68, 69, 70, 71].

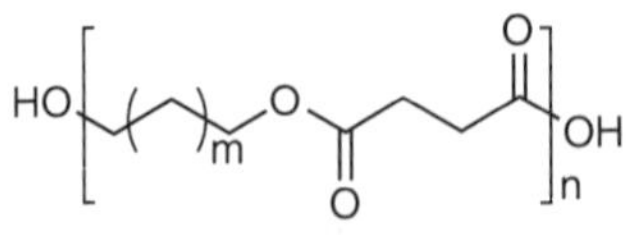

Figure 8.2 Structure of poly(alkylene succinate)s (PAS).

Poly(ethylene succinate) (PES) and poly(butylene succinate) (PBS) are two commercial polyesters with relatively high melting temperatures of 103–106°C and 112–114°C respectively. They display controllable biodegradation rate and good processability. Their mechanical properties like elongation at break and tensile strength are comparable with those of PP and LDPE and their crystallization behavior is similar to that of polyethylene with well formed lamellar morphologies [72]. PBS exhibits a lower biodegradation rate than PES due to its higher crystallization rate and crystallinity. Poly(propylene succinate) PPS biodegrades faster than PES or PBS.

Low molecular weight PBS and its copolymers, poly(butylene succinate-co-adipate) PBSA and poly(butylene-co-hexylene succinate) PBHS, were synthesized by direct polyesterification of corresponding diols and dicarboxylic acids with dimethyl benzene as solvent and water-removing agent [67]. $SnCl_2$ was used as catalyst. A faster reaction rate was achieved and the molecular weight of some polyesters surpassed 30,000 g/mol, if a 4-Å molecular sieve water trap was used. The authors described a decreasing trend of M_w/M_n ratio with increasing reaction time. The DSC analyses showed decreasing glass-transition temperature T_g and melting point T_m if a third monomer unit was incorporated into PBS.

Various metal alkoxides as catalysts can be used to receive high molecular weight PBS whereas titanium tetraisopropoxide [$Ti(OiPr)_4$] gave the best results with respect to highest molecular weight and yield [68]. The molecular weight of the obtained polyesters (M_n = 15,900–26,000 g/mol) can be increased (M_n = 34,700–56,000 g/mol) by a chain extension reaction using hexamethylene diisocyanate (HDI) as a chain extender.

Linear high molecular poly(butylene succinate) (PBS) can be produced in a coupling reaction of hydroxyl and carboxylic acid group terminated PBS. Branched PBS was produced by addition of glycerol to carboxylic group terminated PBS [69]. The author mentions average molecular weights of 100,000–220,000 g/mol for the linear and branched polymers.

Similar to PBS a high molecular weight poly(ethylene succinate) (PES) and its copolyesters containing 7, 10, or 48 mol% butylene succinate (BS) can be synthesized by direct polycondensation of succinic acid, ethylene glycol and 1,4-butanediol with titanium tetraisopropoxide as catalyst [70]. The molecular weight of the obtained polymers was determined by intrinsic viscosity. Viscosities of 1.08–1.27 dL/g reveal that high molecular weight polyesters were formed with random distributions of ethylene succinate and butylene succinate units determined from ^{1}H and ^{13}C-NMR.

Additionally, the production of different aliphatic polyesters - poly(ethylene succinate) (PES), poly(propylene succinate) (PPS), poly(butylene succinate) (PBS) and poly(hexamethylene succinate) (PHS) – was studied using the two-stage melt polycondensation method (esterification and polycondensation) in a glass batch reactor [71, 73, 74, 75]. Thermogravimetry (TG) and differential TG (DTG) showed a decomposition temperature of 399°C for PBS, 404°C for PPS and 413°C for PES. These results reveal PES as the most stable polyester and further indicate that the chemical structure influences the thermal decomposition process.

On the contrary, regarding biodegradability the molecular structure of the polymers has a much smaller effect than the physical structure of the material. Biodegradation rates of the polymers decreased following the order PPS > PES ≥ PBS. This was attributed to the lower crystallinity of PPS compared to the other polyesters [75]. Franco and Puiggali [71] have studied the crystallization kinetics of poly(hexamethylene succinate) and determined a heterogeneous nucleation and a spherulitic growth mechanism in the performed isothermal experiments.

A rapid synthesis of poly(butylene succinate) (PBS) using microwaves was developed in the presence of 1,3-dichloro-1,1,3,3-tetrabutyldistannoxane as catalyst by Velmathi et al. [76]. PBS with a weight-average molecular weight (M_w) of 23,500 g/mol was obtained in 20 min under optimal conditions.

In addition to classical chemical catalysis, enzymatic processes for the production of polyesters have been investigated. The polymerization of divinyl adipate and 1,4-butanediol was performed with physically immobilized Candida antarctica Lipase B (Novozym 435) yielding a weight average molecular weight (M_w) of 23,200 g/mol [77]. However, activated monomers are expensive and therefore limit the potential impact of the polymerization method. Yet a review of literature from recent years shows an important progress in lipase-catalyzed condensation polymerization of conventional (unactivated) diacids and diols. Lipase-catalyzed polycondensation reactions have been explored at temperatures of 90°C and below [78, 79, 80]. Candida antarctica Lipase B-catalyzed synthesis of poly(butylene succinate) was investigated by Azim et a. [81]. Dimethyl succinate and 1,4-butanediol were used as monomers. After 24 h PBS with M_n of 2,000, 4,000, 8,000, and 7,000 g/mol was obtained at 60, 70, 80, and 90°C using diphenyl ether as organic solvent. No significant increase of M_n was achieved by extending the   reaction time to 72 h. However, increasing the reaction time produced PBS with extraordinarily low M_w/M_n due to diffusion and reaction between low-molecular weight oligomers and chains occurring with higher frequency than interchain transesterification. Time-course studies and visual observation showed a limited chain growth of PBS at 80°C due to the formation of precipitates after 5 to 10 hours. A monophasic reaction mixture could be maintained by increasing the reaction temperature from 80 to 95°C after 21 h. The result was a PBS with increased molecular weight of M_w 38,000 g/mol.

Furthermore, different methods for synthesis and characterisation of several other succinic acid copolyesters are described [82, 83, 84].

Ahn et al. [82] produced and examined poly(butylene succinate) (PBS) and poly(butylene adipate) (PBA) homopolyesters and poly(butylene succinate-co-butylene adipate) (PBSA) copolyesters with a two-step process. The polyester had a higher molecular weight with narrow molecular weight distributions M_w/M_n. The melting point (T_m) of copolyesters decreased gradually with increasing butylene adipate content. The glass-transition temperature (T_g) of copolyesters decreased linearly as the fraction of adipoyl units increased.

Functional aliphatic copolyesters of succinic acid (SA) and citric acid (CA) were synthesized via direct copolycondensation in the presence of 1,4-butanediol, with titanium(IV) butoxide as a catalyst [83]. The free functional hydroxyl and carboxyl groups of the resulting polyester provide a better compatibility and processability, particularly when blending with natural polymers like starch and protein. The melting temperature was very sensitive to the molar ratio of the SA–CA comonomer units. The chain extension of this poly(butylene succinate citrate) was carried out with hexamethylene diisocyanate. The materials with increased molecular weight are proposed to be used as impact modifiers and compatibilizers in polyester/starch/protein blends.

Lee et al. [84] investigated the synthesis and chain extension of poly(L-lactic acid-co-succinic acid-co-1,4-butenediol) PLASB. Poly(L-lactic acid) (PLLA) homopolymer obtained from a direct condensation polymerization of LA had molecular weights (M_w) lower than 41,000 g/mol and was very brittle. Addition of SA and 1,4-BDO to LA produced PLASB with M_w above 140,000 g/mol exhibiting tensile properties comparable to a commercially available high-molecular-weight PLLA. Intermolecular linking reactions at the unsaturated 1,4-BED units in PLASB with benzoyl peroxide further increased the molecular weight and made PLASB more ductile and flexible to show elongation at break as high as 450%.

The properties of succinic acid polyesters can be changed by copolymerization with aromatic monomers or through side branching by polymerization of succinic acid with 1,4-butanediol in the presence of 7-octene-1,2-diol [85, 86, 87, 88]. The increasing content of the phenyl side chain reduces the melting temperature and crystallinity, but increases the glass transition temperature of the aliphatic polyesters. However, the increase of the phenyl content increased elongation at break and tear resistance of poly(butylene succinate) considerably. Jin *et al.* [88] describes a decreased glass-transition temperature (T_g) with increased branching density. The differential scanning calorimetry (DSC) measurements showed that T_g increased due to the in situ cross linking of the unsaturated groups. N-hexenyl side branches decreased melting temperature (T_m) more significantly than ethyl side branches, but the effect was on par with that by n-octyl branches. Epoxidation of the double bonds decreased T_m and melting enthalpy (ΔH_m), but increased T_g of the aliphatic polyester.

Some homo- and copolyesters of succinic acid are already produced industrially and being currently introduced in the polymer market. The major manufacturers of PBS are summarized in Table 8.1.

Mitsubishi Chemical Corporation (Tokyo, Japan) has developed a new biodegradable poly(butylene succinate) with the trade name GS Pla®. Their goal is to produce a "Green Sustainable Plastic" using succinic acid produced by bacterial fermentation and 1,4-butanediol [89]. GS Pla® is flexible, has high heat-seal strength and is permeable to oxygen and moisture. According to Mitsubishi this material has a higher tensile elongation and notched izod impact in comparison to PLA. It was introduced in Japan, where it is currently being used in agricultural applications. It can be processed to fibers, suggesting non-wovens and synthetic papers as applications.

Bionolle® (Showa Denko) is being produced by polycondensation of glycols (ethylene glycol, 1,4-butanediol) together with aliphatic dicarboxylic acids (succinic acid, adipic acid) [90]. Bionolle® is a typical thermoplastic polyester with good manufacturing properties, flexibility and tenacity, which are comparable to those of

Table 8.1 Manufacturers of poly(butylene succinate) [66].

Manufacturer	Product	Monomers
Hexing Chemical, China	PBS and its copolymers	Succinic acid, butanediol, branched alkanedicarboxylic acid
Xinfu Pharmaceutical, China	PBS, PBSA	Succinic acid, adipic acid, butanediol
Jinfa Tech, China	PBSA	Succinic acid, adipic acid, butanediol
Showa, Japan	Bionolle	Succinic acid, adipic acid, ethylene glycol, butanediol
Mitsubishi Chemical, Japan	GS Pla	Succinic acid, butanediol
Nippon Shokubai, Japan	Lunare	Succinic acid, adipic acid, ethylene glycol

LDPE. Bionolle® is in the introductory market phase with a production capacity of 10,000 t/a [91].

In 2006 Hexing Chemical (Anhui, China) established a 3,000 t/year manufacturing line for direct melt polycondensation of PBS, which was developed by Tsinghua University in Beijing, China. A facility capable of manufacturing PBS in a quantity of 10,000 tons per year is presently under construction by Hexing Chemical. In October 2007, Xinfu Pharmactical (Hangzhou, China) built a PBS production line with one-step polymerization technology, which was supported by Chinese Academy of Sciences [66].

8.2.2 Polyamides

Polyamides are technical polymers on a high performance level. Their properties are characterized by high durability with simultaneous large hardness and rigidity.

Carothers [92, 93] was the first researcher performing systematic investigations on the polycondensation of dicarboxylic acids with diamines. A wide variety of methods for the production of polyamides by polycondensation of aliphatic diamines and dicarboxylic acids or the polyaddition of lactams is described in literature.

PA 44 is a polyamide based on succinic acid and its derivative 1,4-butanediamine. It was manufactured only in laboratory scale [94, 95]. A technical synthesis of PA 44 has not been described in literature yet. Ambiguous melting temperatures are mentioned in the academic literature for PA 44. In the fundamental publication about PA 44 by Dreyfuss [95] a melting temperature of >260°C is reported. Other papers state that PA 44 - as well as PA 42 - will decompose before reaching the melting point [96]. However, Koning and Buning [97] reported a melting temperature of PA 42 of 390°C.

The polycondensation of succinic acid and 1,4-butanediamine was investigated thoroughly in a reserch project at Fraunhofer UMSICHT. The polymerization procedure was optimized with regard to the resulting molecular weight and the thermal properties of the polymers. First, a catalyzed polycondensation was investigated using organic compounds of titanium and inorganic compounds of tin as catalysts. The type of catalyst did not influence the molecular and thermal properties of the polyamide.

Furthermore, PA 44 was synthesized in a two-step process from the salt of 1,4-butanediamine and succinic acid, similar to the PA 66 production from "Nylon-salt". The first step was the synthesis of prepolymers and the second step was the solid state polycondensation (SSP) of prepolymers. The molecular weight M_n of the synthesized PA 44 was less than 5,000 g/mol. Infrared spectroscopic analysis revealed side reactions like imide and pyrrolidine formation. The prepared PA 44 samples were further characterized by DSC and TGA. An average melting point of 300°C was estimated by DSC. The decomposition of PA 44 was found to start at temperatures directly above the melting point as shown in Figure 8.3.

For this reason, PA 44 cannot be used in conventional thermoplastic processing.

The literature describes interesting and unexpected properties for other polyamide types produced from 1,4-butanediamine or succinic acid (PA 4n or PA n4) [93, 98, 99, 100,101, 102, 103, 104, 105].

For example PA 24 possesses the same properties like silk proteins and commercial polyamides regarding IR absorption, solubility and thermal behavior [105]. A short-chain aliphatic PA 42 manufactured by polycondensation of 1,4-butanediamine with diethyl oxalate could be interesting for use as fibers [99].

PA 46 was described the first time in 1938 by Carothers [93] and is now being produced and distributed by DSM under the trade

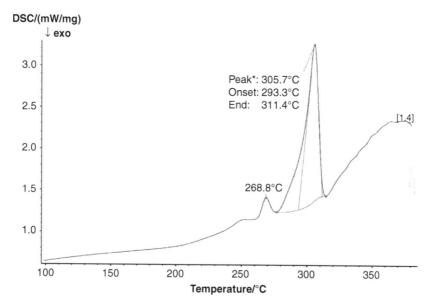

Figure 8.3 DSC thermogram of PA 44.

name Stanyl®. Carothers reported a melting point of 278°C, which later was confirmed by Coffman *et al.* [101].

Katsarava *et al.* [103, 104] suggested a new polycondensation method for the synthesis of high-molecular weight polysuccinamides on mild reaction conditions. They demonstrated that their formation is linked to a low rate of chain-terminating formation of a cyclic imide and its nearly complete suppression by chain-growth aminolysis reaction.

Morgan and Kwolek [106] have described aromatic polyamides from phenylenediamines and aliphatic diacids. High molecular weight polyamides were made from o-, m- and p-phenylenediamines and 4- to 12-carbon aliphatic diacid chlorides by interfacial and solution methods. Melting temperatures increased from ortho to meta to para structures and decreasing chain length of the aliphatic acid. Solubility decreased in the above order. The melting temperatures of polyamides from phenylenediamines and succinic acid were detected at 195°C, 325°C and at >400°C, if o-, m- and p-phenylenediamine respectively were used.

Nylon 4I is an amorphous polyamide from 1,4-butanediamine and isophtalic acid which is difficult to synthesize. The melt polymerization of nylon 4I was studied starting with nylon-salt and nylon prepolymers (η_{inh} = 0.25). With nylon-salt only low molecular polymers were obtained. The polymer is glassy and could easily be melt pressed [102].

The synthesis of nylon 4T from 1,4-butanediamine and a terephthalic acid derivative was studied by Gaymans [107] in a two step-process: prepolymerization, followed by postcondensation in solid state (4 h, 290°C). The prepolymers were prepared by the nylon salt method, ester polymerization method, interfacial method, and a low temperature solution method. Nylon 4T has a high amide content and a regular spacing of the amide groups. The polymer is highly crystalline and has a high melting point.

The properties of the homo polyamides are determined by the density and the sequence of the polar amide groups along the polymer chain. The water absorption depends on the structure and the processing of the polyamides.

Gaymans et al. [100] have examined the water adsorption of cast films from PA 42 and showed, that the material was very hygroscopic (7.5% at 65% RH). PA 42 with a crystallinity of 70% has exhibited lower water absorption (3.1% by 50% RH) [99]. From these results it can be assumed that PA 44 will show a similar behavior concerning water absorption.

Another derivative of succinic acid, 2-pyrrolidone, can be used for the preparation of polyamides. Nylon 4 is the polymer of butanolactam, also known as butyrolactam, 2-pyrrolidinone or α- or 2-pyrrolidone (further named 2-pyrrolidone).

Heterogeneous anionic polymerization of 2-pyrrolidone yields a fiber-forming polyamide (Nylon 4, Polyamide 4). The main challenge in its processing is the low thermal stability of the polymer because melting and decomposition points are very close. Polyamide 4 is highly crystalline and contains a large number of closely spaced hydrogen bonding sites. Due to the high concentration of amide groups per unit length, this polyamide can adsorb up to 30% of its weight of water [108].

Most of the publications were published by the working group of Jan Roda [109, 110, 111, 112, 113, 114, 115, 116, 117, 118, 119, 120, 121, 122, 123, 124, 125, 126, 127, 128, 129]. The authors and co-authors investigated the activated and non-activated anionic polymerization of 2-pyrrolidone initiated with a wide variety of catalysts, e.g. metallic potassium, potassium tert-butoxide, potassium 2-pyrrolidonate or pentamethylguanidin. N-acyllactams, N-iminolactams or carbon dioxide can be used as activators.

Some polymerization conditions for the activated anionic polymerization of 2-pyrrolidone differ from those of 6-caprolactam. The polymerization proceeds at lower temperatures (25–70°C). The lower polymerization temperatures certainly influence the character and scope of the side reactions. The polymerization is heterogeneous except in the very early stages. This may lead to a burial of the propagating chain ends in the precipitated phase and thus to an apparent termination reaction. Fixation in the crystalline structure also may prevent condensations of the CO-N-CO- groups, owing to the reduced contact between the two groups. The incorporation of segments of polypyrrolidone in the crystal structure suppresses side reactions with these parts of the polymer molecule [109].

The non-activated polymerization of 2-pyrrolidone initiated with potassium tert-butoxide (conc. range 0.24–0.97 mole-%) was studied at 40°C by Roda et al. [110]. In accordance with theory, conversion increases linearly with polymerization time and with initiator concentration. The linear dependence on conversion of the content of basic groups in the polymer and, above all, of the number of polymer molecules and its independence on the initiator concentration up to 30% conversion indicate that a stationary state between formation and destruction of growth centers is very rapidly reached.

Patil et al. [118] studied the effect of transient cooling of the polymerization mixture on the anionic polymerization of 2-pyrrolidone initiated with potassium 2-pyrrolidinonate (2-oxo-1-pyrrolidinylpotassium) prepared in situ and activated with N-acetyl-2-pyrrolidone. An increase in conversion and molecular weight could be observed which is controlled by the time interval between the onset of polymerization and the starting of transient cooling of the mixture, which is different to isothermal arrangements of the reaction.

The following limit (ceiling) temperatures of polymerization were extrapolated from kinetic data for the anionic polymerization of 2-pyrrolidone initiated with its potassium salt and the initiation systems of optimum compositions: 66°C for the non-activated polymerization, 68°C and 73°C for the polymerization accelerated with 1-(1-pyrrolin-2-yl)-2-pyrrolidone and carbon dioxide, respectively, 73°C and 76°C for the polymerization activated with N-benzoyl-2-pyrrolidone and N-acetyl-2-pyrrolidone, respectively [126].

Poly(2-pyrrolidone)(PPD,Polyamide4)is most frequently obtained by means of alkali salts, of which 2-oxo-1-pyrrolidinylpotassium

(KPD) is the most efficient, optimally at 40–55°C. In this non-activated polymerization the formation of propagation sites proceeds via a slow disproportionation reaction between the lactam and its anion [115].

In the temperature interval between the ceiling temperature (ca. 70°C) and 110°C polymerization mixtures of 2-pyrrolidone, i.e. the mixtures of monomer, polymer and their potassium salts prepared by non-activated polymerization under optimal conditions (45°C), exhibit an increase in polymer yield, whereas the yield and intrinsic viscosity of the polymer decreases at higher temperatures up to 160°C [127].

Kawasaki *et al.* [130] described the synthesis of branched polyamide 4 types and their properties. Polybasic acid chlorides were effective initiators for the synthesis of the branched polyamide 4. The melting points of polyamide 4 in the high molecular weight region were near 265°C and showed no significant difference depending on their chain structure. On the other hand, it was found that the branched polyamide 4 showed remarkable increase of tensile strength, compared to similar molecular weight of the linear polyamide 4 (e.g. four-branched type M_w 28,000 g/mol, tensile strength 72 MPa). The biodegradation of branched polyamide 4 was evaluated using a standard activated sludge (e.g. four-branched type $M_w = 8.25*10^4$, biodegradation 41%).

PA 4 membranes showed some interesting dialysis properties which were comparable to those of commercial cellophane and cellulose acetate membranes.

8.2.3   Poly(ester amide)s

Although polyesters are nowadays the most important family of biodegradable polymers, their thermal and mechanical properties are not optimal for industrial applications. Attempts have been made to regulate the physical properties and biodegradation rate of polyesters by copolymerization with various monomeric units. Especially, the properties of the polymers can be improved significantly by the introduction of amide groups into the main chain, since these comonomers give rise to strong intermolecular hydrogen-bond interactions [131, 132, 133].

The physical properties, such as melting temperature and tensile strength of poly(ester amide)s show a strong dependence on both the content and the distribution of the amide units in the polyester chains.

The first works on poly(ester amide)s were carried out in 1979 and were based on the amide-ester interchange reaction that takes place if a polyamide/polyester mixture is heated at temperatures near 270°C [134]. Their major application in industry nowadays is as hot-melt adhesives. These are materials which are optimized by incorporating special long-chain carboxylic acids, e.g. dimeric fatty acids or mixtures of cycloaliphatic diamines to produce low melting points and delayed crystallization [135].

Insertion of ester groups in a polyamide chain is a well known strategy to make the polymer more vulnerable to hydrolysis. A number of aliphatic poly(ester amide)s varying either in the chemical structure and/or microstructure has been described by various authors [136, 137 138, 139].

Succinic acid and its derivatives can also be used for the synthesis of poly(ester amide)s. Novel biodegradable copolymers with a periodic sequential structure of ester and amide groups were synthesized by two-step polycondensation reactions from dimethyl succinate, 1,4-butanediol and 1,4-butanediamine [132]. These periodic copolyester-amides have a high thermal stability, comparable with the PBS homopolyester. This suggests that the copoly(ester amide)s create different chain packing structures due to the formation of the intermolecular hydrogen bonds coming from the periodic amide units introduced into the polyester chains. The melting temperature of the periodic copolymers increased with an increase of the amide content.

On the other hand, the introduction of ester segments can lead to the reduction of crystallinity of the polyamide depending on size and concentration of the ester segments. The combination of good physical properties, biocompatibility and controlled degradability of poly(ester amide)s makes it possible to use these materials in medicine as absorbable wound seams or temporary implants.

Different biodegradable poly(ester amide)s were commercialized in the 1990's (e.g. BAK®, Bayer AG) with a wide range of applications. For example, BAK1095 was manufactured from ε-caprolactam, adipic acid and 1,4-butanediol. BAK2195 was a mixture from a polyamide on the basis of 1,6-hexamethylenediamine and adipic acid with a polyester from adipic acid and 1,4-butanediol [140]. Economic reasons stopped the production of BAK in 2001.

8.3   Conclusions

A wide range of polymers like polyamides, polyesters and poly(ester amide)s can be produced based on succinic acid and its derivatives. In order to be competitive to fossil-based materials, these materials like all biopolymers need to have a good processibility on conventional machines and a broad and excellent range of characteristics.

Presently, aliphatic polyesters are mainly used in packaging, hygiene and medical applications. Most of these materials can be replaced by new polymers based on succinic acid. Depending on the utilization profile, homo- or co-polyesters of succinic acid with ethylene, propylene or butylene diol can be prepared to meet the needed product specifications.

Polyamides are technical plastics with high performance level. Their properties are characterized by high durability with simultaneous large hardness and rigidity. Basically, the preparation of polyamides based on succinic acid is possible, with increasing ease using long chain aliphatic diamines. On the other hand, polymers based on the succinic acid derivative 1,4-butanediamine cannot be prepared easily, since side reactions inhibiting chain propagation cannot be suppressed in most instances.

Succinic acid and its derivatives are not only basic molecules for polyesters and polyamides, they can also be used for the synthesis of poly(ester amide)s. The introduction of ester segments can lead to the reduction of crystallinity of the polyamide as a function of size and concentration of the ester segments. The combination of good physical properties, biocompatibility and controlled degradability of poly(ester amide)s makes it possible to use these materials in different applications, e.g. in medicine as absorbable seams or temporary implants.

References

1. T. M. Carole, J. Pellegrino, and M. D. Paster, *Applied Biochemistry and Biotechnology*, Vol. 113–116, p. 871, 2004.
2. R. Busch, T. Hirth, A. Liese, S. Nordhoff, J. Puls, D. Sell, C. Syldatk, and R. Ulber, *Biotechnology Journal*, Vol. 1, p. 770, 2006.
3. É. Archambault, in *et al. Towards a Canadian R&D Strategy for Bioproducts and Bioprocesses Prepared for National Research Council of Canada*

(NRC) by Science-Metrix http://www.science-metrix.com/pdf/SM_2003_014_ NRC_Canadian_R&D_Strategy.pdf, 2004.

4. B. Hüsing, G. Angerer, S. Gaisser, F. Marscheider-Weidemann, *Biotechnologische Herstellung von Wertstoffen unter besonderer Berücksichtigung von Energieträgern und Biopolymeren*. Forschungsbericht 20066301, Umweltbundesamt, Berlin, 2003.

5. M. D. Paster, J. Pellegrino and T. M. Carole in *Industrial Bioproducts: Today and Tomorrow. Washington, DC: prepared by Energetics, Inc. for the Office of the Biomass Program, Office of Energy Efficiency and Renewable Energy, US Department of Energy*, Washington, DC, 2003.

6. M. Patel, M. Crank, V. Dornburg and B. Hermann, in *Medium and Long-term Opportunities and Risks of the Biotechnological Production of Bulk Chemicals from Renewable Resources - The Potential of White Biotechnology, "The BREW Project", Final report, Prepared under the European Commission's GROWTH Programme (DG Research)*, Utrecht, http://www.chem.uu.nl/brew/BREW_Final_ Report_September_2006.pdf, 2006.

7. T. Werpy and G. Petersen, *Top Value Added Chemicals From Biomass Volume 1: Results of Screening for Potential Candidates from Sugar and Synthesis Gas. Golden, Co,: National Renewable Energy Laboratory (NREL)*, available electronically on: http://www.osti.gov/bridge, 2004.

8. M. Crank, M. Patel, F. Marscheider-Weidemann, J. Schleich, B. Hüsing and G. Angerer, in *Techno-economic Feasibility of Large-scale Production of Bio-based Polymers in Europr (Pro-BIP), Final Report, Prepared for the European Commissions´s Institute for Prospective Technological Studies (IPTS)*, Sevilla, Spain, 2004.

9. L. Shen, J. Haufe, and M. K. Patel, *Product overview and market projection of emerging bio-based plastics (PRO-BIO), Commissioned by European Polysaccharide Network of Excellence (EPNOE) and European Bioplastics. Group Science, Technology and Society (STS), Copernicus Institute for Sustainable Development and Innovation*, Utrecht University, Utrecht, the Netherlands, Report No: NWS-E-2009–32, 2009.

10. H. Kidwell, *Bio-succinic acid to go commercial*. http://www.in-pharmatechnol-ogist.com/Materials-Formulation/Bio-succinic-acid-to-go-commercial, 2008.

11. H. Song, and S. Y. Lee, *Enzyme and Microbial Technology*, Vol. 39, p. 352, 2006.

12. J. B. McKinlay, C. Vieille, and J. G. Zeikus, *Applied Microbiology and Biotechnology*, Vol. 76, p. 727, (2007).

13. I. Bechthold, K. Bretz, S. Kabasci, R. Kopitzky, and A. Springer, *Chemical Engineering & Technology*, Vol. 31, p. 647, 2008.

14. A. Cukalovic, and C. V. Stevens, *Biofuels, Bioproducts & Biorefining*, Vol. 2, p. 505, 2008.

15. J. G. Zeikus, M. K. Jain, and P. Elankovan, *Applied Microbiology and Biotechnology*, Vol. 51, p. 545, 1999.

16. C. Andersson, D. Hodge, K. A. Berglund, and U. Rova, *Biotechnology progress*, Vol. 23, p. 381, 2007.

17. C. P. Davis, D. Cleven, J. Brown, and E. Balish, *International Journal of Systematic Bacteriology*, Vol. 26, p. 498, 1976.

18. T. Y. Kim, H. U. Kim, H. Song, and S. Y. Lee, *Journal of Biotechnology*, Vol. 144, p. 184, 2009.

19. S. Okino, M. Inui, and H. Yukawa, *Applied Microbiology and Biotechnology*, Vol. 68, p. 475, 2005.

20. E. T. Ling, J. T. Dibble, M. R. Houston, L. B. Lockwood, and L. P. Elliott, *Applied and Environmental Microbiology*, Vol. 35, p. 1213, 1978.

21. C. Rossi, J. Hauber, and T. P. Singer, *Nature*, Vol. 204, p. 167, 1964.

22. P. C. Lee, W. G. Lee, S. Kwon, S. Y. Lee, and H. N. Chang, *Enzyme and Microbial Technology*, Vol. 24, p. 549, 1999.

23. P. C. Lee, S. Y. Lee, S. H. Hong, and H. N. Chang, *Bioprocess and Biosystems Engineering*, Vol. 26, p. 63, 2003.

24. M. V. Guettler, D. Rumler, and M. K. Jain, *International Journal of Systematic Bacteriology*, Vol. 49, p. 207, 1999.

25. P. Zheng, J. J. Dong, Z. H. Sun, Y. Ni, and L. Fang, *Bioresource Technology*, Vol. 100, p. 2425, 2009.

26. Q. Li, M. Yang, D. Wang, W. Li, Y. Wu, Y. Zhang, J. Xing, and Z. Su, *Bioresource Technology*, Vol. 101, p. 3292, 2010.

27. C. Wan, Y. Li, A. Shahbazi, and S. Xiu, *Applied Biochemistry and Biotechnology: Part A: Enzyme Engineering and Biotechnology*, Vol. 145, p. 111, 2008.

28. Y.-P. Liu, P. Zheng, Z.-H. Sun, Y. Ni, J.-J. Dong, and L.-L. Zhu, *Bioresource Technology*, Vol. 99, p. 1736, 2008.

29. K. Chen, M. Jiang, P. Wei, J. Yao, and H. Wu, *Applied Biochemistry and Biotechnology*, Vol. 160, p. 477, 2010.

30. M. P. Dorado, S. K. C. Lin, A. Koutinas, C. Du, R. Wang, and C. Webb, *Journal of Biotechnology*, Vol. 143, p. 51, 2009.

31. K. Q. Chen, J. Li, J. F. Ma, M. Jiang, P. Wei, Z. M. Liu, and H. J. Ying, *Bioresource Technology*, Vol. 102, p. 1704, 2011.

32. M. J. Van der Werf, M. V. Guettler, M. K. Jain, and Z. J. G, *Archives of Microbiology*, Vol. 167, p. 332, 1997.

33. M. V. Guettler, M. K. Jain and D. Rumler, Method for making succinic acid, bacterial variants for use in the process, and methods for obtaining variants, US Patent 5573931, assigned to Michigan Biotechnology Institute, November 12, 1996.

34. M. V. Guettler, M. K. Jain, and B. K. Soni, Process for making succinic acid, microorganisms for use in the process and methods of obtaining the microorganisms, US Patent 5504004, assigned to Michigan Biotechnology Institute, April 2, 1996.

35. P. C. Lee, W. G. Lee, S. Y. Lee, and H. N. Chang, *Biotechnology and Bioengineering*, Vol. 72, p. 41, 2001.

36. P. C. Lee, S. Y. Lee, S. H. Hong, H. N. Chang, and S. C. Park, *Biotechnology Letters*, Vol. 25, p. 111, 2003.

37. P. C. Lee, W. G. Lee, S. Y. Lee, H. N. Chang, and Y. K. Chang, *Biotechnology and Bioprocess Engineering*, Vol. 5, p. 379, 2000.

38. P. C. Lee, S. Y. Lee, S. H. Hong, and H. N. Chang, *Applied Microbiology and Biotechnology*, Vol. 58, p. 663, 2002.

39. S. J. Lee, H. Song, and S. Y. Lee, *Applied and Environmental Microbiology*, Vol. 72, p. 1939, 2006.

40. I. J. Oh, H. W. Lee, C. H. Park, S. Y. Lee, and J. W. Lee, *Journal of Microbiology and Biotechnology*, Vol. 18, p. 908, 2008.

41. G. N. Vemuri, M. A. Eiteman, and E. Altman, *Journal of Industrial Microbiology and Biotechnology*, Vol 28, p. 325, 2002.

42. H. Lin, G. N. Bennett, and K. Y. San, *Metabolic Engineering*, Vol. 7, p. 116, 2005.

43. H. Lin, G. N. Bennett, and K.-Y. San, *Biotechnology and Bioengineering*, Vol. 90, p. 775, 2005.

44. A. M. Raab, G. Gebhardt, N. Bolotina, D. Weuster-Botz, and C. Lang, *Metabolic Engineering*, Vol. 12, p. 518, 2010.

45. R. Datta, D. A. Glassner, M. K. Jain, R. Vick, and R. John, Fermentation and purification process for succinic acid, US Patent 5168055, December 1, 1992.

46. K. A. Berglund, S. Yedur, and D. D. Dunuwila, Succinic acid production and purification. US Patent 5958744, assigned to Applied Carbochemicals, September 28, 1999.

47. Y. S. Huh, Y.-S. Jun, Y. K. Hong, H. Song, S. Y. Lee, and W. H. Hong, *Process Biochemistry*, Vol. 41, p. 1461, 2006.

48. Y.-S. Jun, E. Z. Lee, Y. S. Huha, Y. K. Hong, W. H. Honga, and S. Y. Lee, *Biochemical Engineering Journal*, Vol. 36, p. 8, 2007.

49. B. H. Davison, N. P. Nghiem, and G. L. Richardson, *Applied Biochemistry and Biotechnology*, Vol. 114, p. 653, 2004.

50. D. A. Glassner, and R. Datta, Process for the production and purification of succinic acid. US Patent 5143834, September 1, 1992.

51. I. Meynial-Salles, S. Dorotyn, and P. Soucaille, *Biotechnology and Bioengineering*, Vol. 99, p. 129, 2008.

52. T. A. Werpy, J. G, Frye and J. Holladay, "Succinic acid – a model building block for chemical production from renewable resources", in B. Kamm, P. R. Gruber and M. Kamm, eds, *Biorefineries - Industrial Processes and Products. Status Quo and Future Directions*, Wiley-VCH, Weinheim, Germany, Bd. 2, chapter 13, pp. 367–379, 2006.

53. C. Fumagalli, "Succinic acid and succinic anhydride", in *Kirk-Othmer Encyclopedia of Chemical Technology*, 5th Ed, Vol. 23, Wiley, pp 416–435, 2006.

54. S. Varadarajan, and D. J. Miller, *Biotechnology Progress*, Vol. 15, p. 845, 1999.

55. D. Ostgard, M. Berweiler and S. Roeder, Method for producing alcohols by hydrogenating carbonyl compounds, WO Patent 2002051779, assigned to Degussa, July 4, 2002.

56. S. E. Hunter, C. E. Ehrenberger, and P. E. Savage, *The Journal of Organic Chemistry*, Vol. 71, p. 6229, 2006.

57. N. Nghiem, B. H. Davison, M. I. Donnelly, S.-P. Tsai, and J. G. Frye, "An integrated process for the production of chemicals from biologically derived succinic acid", in J. J. Bozell, eds, *Chemicals and Materials from Renewable Resources*. American Chemical Society, Washington DC, Vol. 784, chapter 13, pp. 160–173, 2001.

58. D. Campos, Ruthenium-molybdenum catalyst for hydrogenation in aqueous solution, US Patent 2004122242, June 24, 2004.

59. D. Campos and G. M. Sisler, Platinum-rhenium-tin catalyst for hydrogenation in aqueous solution, US Patent 6670490, assigned to DuPont, December 23, 2003.

60. A. Bhattacharyya, and M. D. Manila, Catalysts for maleic acid hydrogenation to 1,4-butanediol, US Patent 2006004212, January 5, 2006.

61. M. Roesch, R. Pinkos, M. Hesse, S. Schlitte, H. Junicke, O. Schubert, A. Weck and G. Windecker, Method for the production of defined mixtures of THF, BDO and GBL by gas phase hydrogenation, WO Patent 2005058853, assigned to BASF AG, June 30, 2005.

62. M. A. Wood, S. P. Crabtree and D. V. Tyers, Homegeneous process for the hydrogenation of dicarboxylic acids and/or anhydrides thereof, WO Patent 2005051875, assigned to Davy Process Techn Ltd [GB], June 9, 2005.

63. T. A. Werpy, J. G. Frye, Y. Wang and A. H. Zacher, Methods of making pyrrolidones, US Patent 6632951, assigned to BATTELLE MEMORIAL INSTITUTE [US], October 14, 2003.

64. T. A. Werpy, J. G. Frye, Y. Wang and A. H. Zacher, Methods of making pyrrolidones, WO Patent WO2002102772, assigned to BATTELLE MEMORIAL INSTITUTE [US], June 17, 2002.

65. W. H. Carothers, *Chemical Review*, Vol. 8, p. 353, 1931.

66. J. Xu and B.-H.Guo, "Microbial Succinic acid, its polymer poly(butylene succinate), and applications", in G.-O. Chen, ed., *Plastisc from Bacteria: Natural Functions and Applications*, Springer-Verlag Berlin Heidelberg, pp. 347–388, 2010.

67. C. Zhu, Z. Zhang, Q. Liu, Z. Wang, and J. Jin, *Journal of Applied Polymer Science*, Vol. 90, p. 982, 2003.

68. H. Shirahama, Y. Kawaguchi, M. S. Aludin, and H. Yasuda, *Journal of Applied Polymer Science*, Vol. 80, p. 340, 2001.

69. Y. K. Han, S. R. Kim, and J. Kim, *Macromolecular Research*, Vol. 10, p. 108, 2002.

70. C.-H. Chen, H.-Y. Lu, M. Chen, J.-S. Peng, C.-J. Tsai, and C.-S. Yang, *Journal of Applied Polymer Science*, Vol. 111, p. 1433, 2009.

71. L. Franco, and J. Puiggalı, *European Polymer Journal*, Vol. 39, p. 1575, 2003.

72. R. Ishioka, E. Kitakuni and Y. Ichikawa, "Aliphatic polyesters: "Bionole"", in Y. Doi and A. Steinbüchel, eds, *Biopolymers polyesters III - applications and commercial products*, Vol. 4, Wiley, New York, pp. 275–297, 2002.

73. K. Chrissafis, K. M. Paraskevopoulos, and D. N. Bikiaris, *Thermochimica Acta*, Vol. 435, p. 142, 2005.

74. K. Chrissafis, K. M. Paraskevopoulos, and D. N. Bikiaris, *Polymer Degradation and Stability*, Vol. 91, p. 60, 2006.

75. D. N. Bikiaris, G. Z. Papageorgiou, and D. S. Achilias, *Polymer Degradation and Stability*, Vol. 91, p. 31, 2006.

76. S. Velmathi, R. Nagahata, J. Sugiyama, and K. Takeuchi, *Macromolecular Rapid Communications*, Vol. 26, p. 1163, 2005.

77. A. K. Chaudhary, J. Lopez, E. J. Beckman, and A. J. Russell, *Biotechnology Progress*, Vol. 13, p. 318, 1997.

78. A. Mahapatro, B. Kalra, A. Kumar, and R. A. Gross, *Biomacromolecules*, Vol. 4, p. 544, 2003.

79. A. Mahapatro, A. Kumar, B. Kalra, and R. A. Gross, *Macromolecules*, Vol. 37, p. 35, 2004.

80. Y. Mei, L. Miller, W. Gao, and R. A. Gross, *Biomacromolecules*, Vol. 4, p. 70, 2003.

81. H. Azim, A. Dekhterman, Z. Jiang, and R. A. Gross, *Biomacromolecules*, Vol. 7, p. 3093,2006.

82. B. D. Ahn, S. H. Kim, Y. H. Kim, and J. S. Yang, *Journal of Applied Polymer Science*, Vol. 82, p. 2808, 2001.

83. R. Mani, M. Bhattacharya, C. Leriche, L. Nie, and S. Bassi, *Journal of Polymer Science Part A: Polymer Chemistry*, Vol. 40, p. 3232, 2002.

84. C. M. Lee, H. S. Kim, and J. S. Yoon, *Journal of Applied Polymer Science*, Vol. 95, p. 1116, 2005.

85. S.-W. Kim, J.-C. Lim, D.-J. Kim, and K.-H. Seo, *Journal of Applied Polymer Science*, Vol. 92, p. 3266, 2004.

86. H.-J. Jin, B.-Y. Lee, M.-N. Kim, and J.-S. Yoon, *European Polymer Journal*, Vol. 36, p. 2693, 2000.

87. H.-J. Jin, B.-Y. Lee, M.-N. Kim, and J.-S. Yoon, *Journal of Polymer Science Part B: Polymer Physics*, Vol. 38, p. 1504, 2000.

88. H.-J. Jin, D.-S. Kim, M.-N. Kim, I.-M. Lee, H.-S. Lee, and J.-S. Yoon, *Journal of Applied Polymer Science*, Vol. 81, p. 2219, 2001.

89. M. Baumann, ed., PolymerPlace Notes. A plastics technology newsletter, http://www.polymerplace.com/newsletters/August%2007_newsletter.pdf, 2007.

90. T. Fujimaki, *Polymer Degradation and Stability*, Vol. 59, p. 209, 1998.

91. P. Drachman, *Developments in biodegradable plastics for packaging - industry insight*. Intertech Pira, 2007.

92. W. H. Carothers, and J. W. Hill, *Journal of the American Chemical Society*, Vol. 54, p. 1566, 1932.

93. W. H. Carothers, Synthetic fiber, US Patent 2130948, assigned to DuPont, September 20, 1938.

94. R. G. Beaman, P. W. Moran, C. R. Koller, E. L. Wittbecker, and E. E. Magat, *Journal of Polymer Science*, Vol. XL, p. 329, 1959.

95. P. Dreyfuss, *Journal of Polymer Science: Polymer Physics Edition*, Vol. 11, p. 201, 1973.

96. N. A. Jones, E. D. Atkins, and M. J. Hill, *Macromolecules*, Vol. 33, p. 2642, 2000.

97. C. E. Koning, and G. W. Buning, *Recent Research Developments in Macromolecules Research*, Vol. 4, p. 1, 1999.

98. G. Pipper and E. M. Koch, Continuous preparation of copolyamides, US Patent 5030709, assigned to BASF AG, July 9, 1991.

99. R. J. Gaymans, V. S. Venkatraman, and J. Schuijer, *Journal of Polymer Science: Polymer Chemistry Edition*, Vol. 22, p. 1373, 1984.

100. R. J. Gaymans, T. E. C. van Utteren, J. W. A. van den Berg, and J. Schuyer, *Journal of Polymer Science: Polymer Chemistry Edition*, Vol. 15, p. 537, 1977.

101. D. D. Coffman, G. J. Berchet, W. R. Peterson, and E. W. Spanagel, *Journal of Polymer Science*, Vol. 2, p. 306, 1947.

102. R. J. Gaymans, and A. G. J. Van Der Ham, *Polymer (Guildford)*, Vol. 25, p. 1755, 1984.

103. R. D. Katsarava, D. P. Kharadze, and L. M. Aualkhuili, *Makromolecular Chemistry*, Vol. 181, p. 2053, 1986.

104. R. D. Katsarava, D. P. Kharadze, T. M. Bendiashvili, Y. G. Urman, I. Y. Slonim, S. G. Alekseeva, P. Gefelin, and V. Janout, *Acta Polymerica*, Vol. 39, p. 523, 1988.

105. Q. Wang, Z. Shao, and T. Yu, *Polymer Bulletin*, Vol. 36, p. 659, 1996.

106. P. W. Morgan, and S. L. Kwolek, *Macromolecules*, Vol. 8, p. 104, 1975.

107. R. J. Gaymans, *Journal of Polymer Science: Polymer Chemistry Edition*, Vol. 23, p. 1599, 1985.

108. M. A. Bellinger, C.-W. A. Ng, and W. J. MacKnight, *Acta Polymerica*, Vol. 46, p. 361, 1995.

109. J. Roda, M. Kuskova, and J. Králiček, *Die Makromolekulare Chemie*, Vol. 178, p. 3203, 1977.

110. J. Roda, I. Kminek, and J. Králiček, *Die Makromolekulare Chemie*, Vol. 179, p.345, 1978.

111. J. Roda, I. Kminek, and J. Králiček, *Die Makromolekulare Chemie*, Vol. 179, p. 353, 1978.

112. J. Roda, Z. Hula, M. Kuskova, and J. Králiček, *Die Angewandte Makromolekulare Chemie*, Vol. 70, p. 159, 1978.

113. J. Roda, O. Kucera, and J. Králiček, *Die Makromolekulare Chemie*, Vol. 180, p. 89, 1979.

114. J. Roda, J. Brožek, and J. Králiček, *Die Makromolekulare Chemie Rapid Communications*, Vol. 1, p. 165, 1980.

115. J. Roda, Z. Votrubcová, J. Králiček, J. Stehlícek, and S. Pokorný, *Die Makromolekulare Chemie*, Vol. 182, p. 2117, 1981.

116. J. Roda, P. Sysel, and J. Králiček, *Polymer Bulletin*, Vol. 5, p. 609, 1981.

117. J. Roda, J. Králiček, V. Subr, and J. Stehlicek, *Die Makromolekulare Chemie*, Vol. 184, p. 1987, 1983.

118. V. S. Patil, J. Roda, and J. Králiček, *Die Makromolekulare Chemie*, Vol. 182, p. 3119, 1981.

119. S. S. Mahajan, J. Roda, and J. Králiček, *Die Angewandte Makromolekulare Chemie*, Vol. 75, p. 63, 1979.

120. J. Brožek, P. Řehák, J. Roda, and J. Králiček, *Polymer Bulletin*, Vol. 11, p. 353, 1984.

121. J. Brožek, J. Roda, and J. Klalicek, *Die Makromolekulare Chemie*, Vol. 189, p. 1, 1988.

122. J. Brožek, P. Řehák, M. Marek Jr., J. Roda, and J. Králiček, *Die Makromolekulare Chemie*, Vol. 189, p. 9, 1988.

123. J. Brožek, M. Marek Jr., J. Roda, and J. Králiček, *Die Makromolekulare Chemie*, Vol. 189, p. 17, 1988.

124. F. Chuchma, A. Bostiska, J. Roda, and J. Králiček, *Die Makromolekulare Chemie*, Vol. 180, p. 1849, 1979.

125. F. Chuchma, J. Roda, and J. Králiček, *Die Makromolekulare Chemie*, Vol. 184, p. 1781, 1983.

126. T. Fries, J. Bělohlávková, J. Roda, and J. Králiček, *Polymer Bulletin*, Vol. 12, p. 87, 1984.

127. F. Fries, B. Ciekerová, J. Roda, and J. Králiček, *Die Makromolekulare Chemie*, Vol. 186, p. 1995, 1985.

128. T. Fries, O. Nováková, J. Roda, and J. Králiček, *Die Makromolekulare Chemie*, Vol. 186, p. 2005, 1985.

129. M. Kusková, J. Roda, and J. Králiček, *Die Makromolekulare Chemie*, Vol. 179, p. 337, 1978.

130. N. Kawasaki, A. Nakayama, N. Yamano, S. Takeda, Y. Kawata, N. Yamamoto, and S.-i. Aiba, *Polymer*, Vol. 46, p. 9987, 2005.

131. M. Okada, *Progress in Polymer Science*, Vol. 27, p. 87, 2002.

132. H. Abe, and Y. Dui, *Macromolecular Rapid Communications*, Vol. 25, p. 1303, 2004.

133. V. De Simone, G. Maglio, R. Palumbo, and V. Scardi, *Journal of Applied Polymer Science*, Vol. 46, p. 1813, 1992.

134. Y. Tokiwa, T. Suzuki, and T. Ando, *Journal of Applied Polymer Science*, Vol. 24, p. 1701, 1979.

135. M. Bolze and M. Drawert, Polyester-amide based melting adhesives and their use in bonding organic and inorganic substrates, EP Patent 0027944, assigned to Schering AG, May 6, 1981.

136. L. Castaldo, F. de Candia, G. Maglio, R. Palumbo, and G. Strazza, *Journal of Applied Polymer Science*, Vol. 27, p. 1809, 1982.

137. I. Goodman, and R. J. Sheahan, *European Polymer Journal*, Vol. 26, p. 1081, 1990.

138. R. J. Gaymans, and J. L. de Haan, *Polymer*, Vol. 34, p. 4360, 1993.

139. H. R. Stapert, A.-M. Bouwens, P. J. Dijkstra, and J. Feijen, *Macromolecular Chemistry and Physics*, Vol. 200, p. 1921, 1999.

140. E. Grigat, R. Koch, and R. Timmermann, *Polymer Degradation and Stability*, Vol. 59, 223, 1998.

9

5-Hydroxymethylfurfural Based Polymers

Ananda S. Amarasekara

*Department of Chemistry, Prairie View A&M University,
Prairie View, TX, USA*

Abstract

5-Hydroxymethylfurfural (HMF) is a renewable resource based potential platform chemical useful in the synthesis of polymers. This versatile six carbon furan system with primary alcohol and aldehyde groups can be prepared from the triple dehydration of hexoses, which are produced from depolymerization of the major cellulosic fraction of abundant lignocellulosic biomass. A verity of monomers suitable for polymer synthesis can be prepared from HMF by adjusting the oxidation states of the functional groups and simple derivatization methods. Some of the well studied derivatives of HMF includes; 2,5-bis(formyl)furan, 2,5-furandicarboxylic acid, 2,5-furandicarboxylicacid dichloride, 2,5-bis(hydroxymethyl)furan, 2,5-bis(aminomethyl)furan, furanic diesters and furanic diisocyanates. Recent progress in the synthesis, characterization, and physical properties studies of the poly-Schiff-base, polyester, polyamide, polyurethane, polybenzoimidazole, and polyoxadiazole type furanic polymers from these HMF derived monomers will be discussed in this chapter.

Keywords: 5-Hydroxymethylfurfural, furanic polymers,
2,5-furandicarboxylic acid, furanic polyesters, furanic polyamides

9.1 Introduction

Human civilization has utilized natural polymeric materials such as wood, plant fibers, animal hides, cotton, and silk for utensils, clothes, and various other purposes for several millenniums. Since the discovery of crude oil in the 19th century and the development

Vikas Mittal (ed.) Renewable Polymers, (381–428) © Scrivener Publishing LLC

of chemical technologies, petroleum based synthetic polymers have played a major role in all aspects of our lives. Starting from simple day to day routines like using a plastic tooth brush and a comb in the morning, modern human civilization is totally dependent on the materials derived from these petroleum based raw materials. Now with declining petroleum resources, combined with increased demand for petroleum from emerging economies such as China and India, political and environmental concerns about petroleum based non renewable raw materials, it is imperative to develop renewable resources based novel polymeric materials to replace petroleum based polymers. In essence, the major challenge for the human civilization in the 21st century is the development of energy sources, and materials that can substitute for the petroleum resources. Seeing that bulk of natural polymeric materials have limited processability and applications range, the search for sustainable renewable resources based platform chemicals and technologies for the post-petroleum era materials is a high priority interest in contemporary science and engineering. The majority of non-renewable resources based synthetic polymers used today are derived from a handful of monomers such as ethylene, propylene, styrene, vinyl chloride, acrylate esters, caprolactam, terephthalic acid, and ethylene glycol. As the main fraction of plant based organic carbon occurs as highly oxygenated polysaccharide forms like cellulose and hemicellulose, it is fundamentally impossible to derive all the current major polymer industry monomers in the present form directly from renewable resources. Nevertheless, some present monomers like acrylates could be derived from the three carbon platform of plant based glycerol. In spite of the fact that some of the current monomers or closely related materials will be derived from plant sources, carbohydrate based new platform chemicals are expected to take over or play a major role in the future chemical and polymer industry.

9.2   5-Hydroxymethylfurfural

9.2.1   Preparation of 5-Hydroxymethylfurfural from Hexoses

Biomass derived potential platform chemicals can be classified into major classes like fatty acids, small carboxylic acid related compounds, lignin aromatics, terpenoids, alcohols/polyols and furans [1]. Furan-2-carboxaldehyde (furfural) (2) and 5-hydroxy-methylfurfural (HMF) (1) are the two most important biomass derived furan compounds in the furan class and the disubstituted furan derivative 5-hydroxymethylfurfural (1) is quite attractive due to the fact that it has six carbons and two substituent groups attached to the ring structure. HMF (1) is known since late 19th century. In 1895 Dull [2] and Kiermayer [3] working independently reported the first synthesis of this important bifunctional five membered heterocyclic system, which they called "oxymethylfurfurol". Even though, HMF was known for over a century, the potential of this renewable resources based platform chemical has been realized only in very recent times. Renewed interest on HMF is evident in the exponential growth in the number of publications on HMF in last 5 years, and there are about 500 research articles published on HMF in 2010. Several review articles in recent years have summarized the progress of HMF chemistry, and in 2010 Yan [4], and Tong [5] published two separate reviews on the general dehydration of carbohydrates for HMF production, and Zakrzewska [6] reviewed the use of ionic liquids in HMF preparation. Comprehensive reviews covering HMF have also appeared in literature [7], [8] and the review by Lewkowoski [8] gives a complete account of synthesis, chemistry and applications of HMF up to 2001. Additionally, Shiyong [9] reviewed the dehydration of fructose in various solvents including bi-phase solvent systems, sub/supercritical fluid solvent systems, and ionic liquids in 2009. The significance of this six carbon molecule lies in the fact that it can be derived from the major fraction of the renewable lignocellulosic biomass, which is the most abundant organic substance on earth. Typical percentage composition of lignocellulosic biomass in the plant material composed of cellulose, hemicellulose, and lignin are; 35–50% cellulose, 20–35% hemicellulose, and 5–30% lignin by dry weight [10].

The dehydration reaction of five carbon sugars or pentoses, which is mainly xylose in the lignocellulosic biomass leads to the formation of furfural (2-furancarboxaldehyde) (2), whereas the dehydration of six carbon sugars (hexoses) like glucose and fructose leads to the formation of HMF (1) as shown in Figure 9.1. Furthermore, rehydration of HMF in acidic medium can lead to the formation of open chain compound levulinic acid (3). Furfural has been known as one of the leading biomass derived platform chemicals for more than fifty years, and being used in the preparation of polymeric resins and used as a solvent as well. However, HMF has started

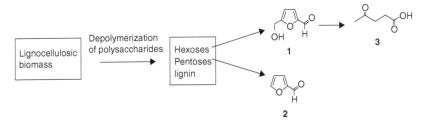

Figure 9.1 Lignocellulosic biomass derived furan derivatives.

gaining attention only in recent times, particularly in the last couple of years that it has come to the verge of becoming an industrial chemical. Nevertheless, there are a number of complications associated with HMF, like difficulties in purification and storage for long periods due to susceptibility to polymerization, rehydration to levulinic acid (3), and proneness to degradation even under relatively mild conditions. Therefore, it is more practical to convert HMF into its more stable and useful derivatives for purification, long term storage, and transportation.

The dehydration of carbohydrates is generally carried out in an acid catalyzed reaction using Lewis or mineral acids, but at high temperatures, hexoses like fructose can be dehydrated to HMF without any added acid as well. HMF is formed in the triple dehydration of hexoses and this acid catalyzed dehydration of hexoses is a very complex process, and usually leads to a number of by products as well. Van Dam [11] showed that aqueous and nonaqueous dehydration processes leads to about 37 products. These by products are formed due to a series of side reactions such as isomerization, fragmentation, and condensation in the acidic medium, and these processes strongly influence the HMF yield. There is a large number of reports on the studies of the mechanisms of dehydration of hexoses to HMF. Earliest mechanistic studies by Van Dam [11] Antal [12] and Kuster [13] based on the kinetics of the formation of products, (fructose and glucose) and suggested that dehydration could occur through two possible pathways (Figure 9.2.). The most widely studied monosaccharide used for the HMF production is fructose, as it is the easiest precursor to use and gives high yields of HMF under variety of conditions. In general, the dehydration of ketohexoses is easier than the dehydration of aldohexoses. This is due to the fact that aldohexose enolyzes to a very low degree, compared to ketohexoses like fructose, and the dehydration process initiates in the fructofuranose form as shown in the mechanism.

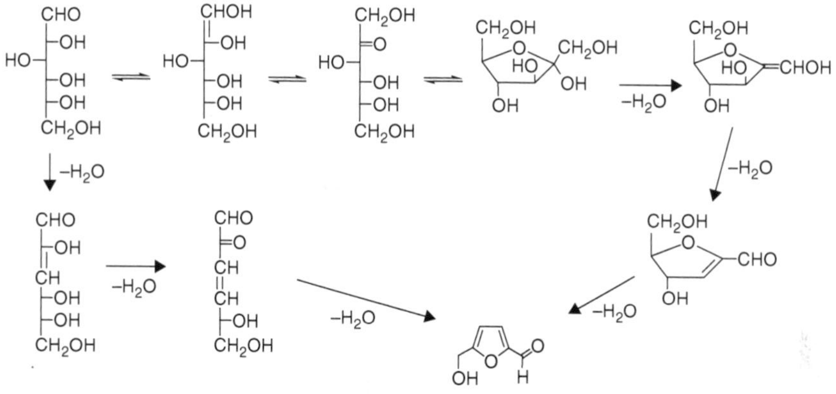

Figure 9.2 Two possible pathways for the dehydration of hexoses to HMF.

According to the proposed mechanism, aldohexoses like glucose can also be used as a precursor for HMF after isomerization to keto-hexose form, and in general the acid catalyzed dehydration of glucose gives a much lower yield of HMF.

Dehydration can be carried out in aqueous, non aqueous, organic solvents and in ionic liquid mediums using a verity of catalysts. In some cases, the solvent medium itself behaves as the catalyst in the dehydration reaction. The common classes of catalysts used in HMF production are shown in the Table 9.1.

9.2.1.1 Dehydration of Fructose using Acid Catalysts

The dehydration of fructose in the aqueous acidic medium is a very convenient process from an ecological point of view, but the aqueous medium reaction can lead to numerous side reactions and gives

Table 9.1 Groups of catalysts used in the preparation of HMF from fructose.

Mineral Acids	Organic acids	Salts	Lewis acids	Zeolites	Ionic Liquids
H_2SO_4, HCl, H_3PO_4	p-TsOH, Oxalic acid, Levulinic acid, Maleic acid	Cr, Al, Ti, Zr, Th salts, $Vo(SO_4)_2$, TiO_2, Zr, Cr, Ti-porphyrine, Pyrid/HCl, Pyrid/PO_4^{-3} $(NH_4)_2SO_4/SO_3$	BF_3, $AlCl_3$, $ZnCl_2$	ZSM-5, LZY	[BMIM]Cl [EMIM]Cl

poor yield of about 30%. The major problem in the aqueous system is the degradation of HMF to levulinic and formic acid. Van Dam [11] and Kuster [14] independently studied the dehydration of fructose in water under homogeneous catalysis and found that the HMF yield can be improved to about 75% by using methyl isobutyl ketone as a simultaneous extraction solvent. This way the HMF formed in the reaction is removed from the system before conversion into levulinic and formic acids. Later, Dumesic further improved the biphase extraction process for HMF production [15] [16] in which, fructose was dehydrated, in a two-phase reactor system, using hydrochloric acid or an acidic ion-exchange resin as the catalyst. In their process, dimethyl sulfoxide and/or poly(1-vinyl-2-pyrrolidinone) was added to suppress undesired side reactions, and the HMF product was continuously extracted into an organic phase (methyl isobutyl ketone) modified with 2-butanol to enhance partitioning from the reactive aqueous solution. Non aqueous solvent systems such as dimethyl sulfoxide (DMSO) [17], acetonitrile [18] and poly(glycol ether) [19] have also been used in the HMF synthesis.

9.2.1.2   Dehydration of Fructose without a Catalyst

Fructose can be dehydrated to HMF in dimethyl sulfoxide at high temperatures in the range of 150–160°C without adding any extra catalyst. Recently, Amarasekara et al. [20] has studied the mechanism of the high temperature (150°C) dehydration of fructose in DMSO-d_6, without any added acid catalyst, using ^{1}H and ^{13}C NMR spectroscopy and proposed a mechanism suggesting that DMSO catalyzes the dehydration. Furthermore, A key intermediate in the reaction was identified as (4R,5R)-4-hydroxy-5-hydroxymethyl-4,5-dihydro-furan-2-carbaldehyde by using in situ ^{1}H and ^{13}C NMR spectroscopy. The main advantage of using DMSO as a solvent is that it prevents the formation of levulinic and humic acids; however, the disadvantage of using DMSO is that this solvent is difficult to separate from the HMF product.

9.2.1.3   Dehydration of Fructose in Ionic Liquids

Ionic liquids have gained attention as a solvent medium as well as a catalyst for the dehydration of carbohydrates in the past few years. These low melting organic salts are excellent solvents for carbohydrates, especially for carbohydrate polymers, which are difficult to dissolve in common solvents. In some of the work ionic liquid are used to play the dual role of catalyst as well as the solvent medium. The first report of the use of molten organic salts for the dehydration of carbohydrates dates back to the year 1983. [21] In this initial work molten tetraethylammonium or pyridinium salts with various anions like chloride or bromide were used as solvents and yields as high as 70–75% were obtained for the dehydration of fructose. The next development in the use of ionic liquid appears after 20 years, in which Lansalot-Matras *et al.* [22] reported the acid-catalyzed dehydration of fructose in 1-butyl-3-methylimidazolium tetrafluoroborate ([BMIM]BF$_4$) and 1-butyl-3-methylimidazolium hexafluorophosphate ([BMIM]PF$_6$) with DMSO as a co solvent. In this publication they claimed that 50% yield of HMF could be obtained in 3 hr. by dehydration of fructose at 80°C using Amberlyst-15 as catalyst, in ionic liquid-DMSO mixtures as the solvent. Then in 2006, the same research group reported [23] that 92% yield HMF could be achieved within 15–45 min. using 1-*H*-3-methyl imidazolium chloride, acting both as a solvent and catalyst. Furthermore, they reported that under similar operating conditions, sucrose is nearly quantitatively transformed into HMF and unreacted glucose. Ionic liquid - ethyl acetate bi phase system was introduced by Hu [24] for the easy separation of the HMF. By using DFT calculations Lai [25] demonstrated that ionic liquid solvent "switches" the dehydration from thermodynamically unfavorable into a thermodynamically favorable reaction.

Efficient dehydration of fructose in ionic liquid with the use of added acid catalyst under mild conditions has been reported by several groups, and various Brönsted and Lewis acids have been used for this purpose including, HCl [26] ion exchange resins [27] [28], 12-molybdophosphoric acid [29]. In another approach, solid acid catalyst prepared by immobilizing sulfonic group bearing ionic liquid 1-*H*-3-(3-sulfonic acid) propylimidazolium chloride on micro-ball silica-gels were used for the dehydration process [30]. The results indicated that these sulfonic acid functionalized imidazole modified silicas are efficient catalysts for fructose dehydration to HMF. Fructose solution of 15%~25% (wt) could be dehydrated with these heterogeneous catalysts to give over 80% yield of HMF. The catalysts could be reused conveniently, but the yield of HMF dropped gradually with the times of reuse, especially after fourth reuse. Then there are examples of the use of ionic liquid as a catalyst in aprotic solvent systems. This includes the use of N-methylmorpholinium methyl sulfonate in N,N-dimethylformamide-lithium bromide system [31] and results showed that N-methyl-2-pyrrolidonium methyl sulfonate and N-methyl-2-pyrrolidonium hydrogen sulfate in dimethyl sulfoxide [32] efficiently catalyzes the dehydration of fructose under mild conditions.

9.2.1.4   Inulin in Ionic Liquids

Polysaccharides can also be used as feedstock for HMF, and among these, inulin is the most favorable carbohydrate for the HMF production. Inulins are polymers composed mainly of fructose units, and typically have a terminal glucose, therefore hydrolysis gives mostly fructose. In 2009 Hu [33] showed that inulin is soluble in choline chloride ionic liquid and can be converted to HMF using oxalic acid or citric acid as the catalyst. A one pot, two-step process is also known [34] for the preparation of 5-hydroxymethylfurfural from inulin in ionic liquids. In this method, first, 1-butyl-3-methyl imidazolium hydrogen sulfate ([BMIM]HSO$_4$) was used as both solvent and catalyst for the rapid hydrolysis of inulin into fructose with 84% fructose yield in 5 min. reaction time. In the second step, 1-butyl-3-methyl imidazolium chloride ([BMIM]Cl) and a strong acidic cation exchange resin were added to the mixture to selectively convert fructose into HMF, giving a HMF yield of 82% in 65 min. at 80°C, which is the highest HMF yield reported by thus far for an inulin feedstock. In another recent report [35] glycerol has been used as a co solvent with ionic liquids, and in this work, they demonstrate that the amount of the ionic liquid [BMIM]Cl required for the acid-catalyzed hydrolysis of inulin into HMF, over Amberlyst-70 solid acid catalyst, can be reduced by substituting it up to 90-wt% with glycerol or glycerol carbonate.

9.2.1.5   Dehydration of Glucose in Ionic Liquids

As discussed earlier, glucose normally gives poor yields of HMF in the direct dehydration as it requires a less favorable isomerization to a fructofuranose form. In 2007 Zhao et al. [36] reported the first application of metal halides as effective catalysts in 1-alkyl-3-methylimidazolium chlorides for the conversion of carbohydrates into HMF. In this publication in Science they described the use of various metal halides and different ionic liquids, 1-ethyl-3-methylimidazolium ([EMIM]Cl), butylmethyl imidazolium chloride ([BMIM]Cl), and 1-octyl-3-methylimidazolium chloride ([OMIM]Cl)). This landmark study involves the conversion of fructose and more difficult glucose in to HMF under catalytic action of various metal ions. Catalytic effect on the dehydration of fructose was seen by the addition of 6 mole percent loading (based on sugar) of $CrCl_2$, $CrCl_3$, $FeCl_2$, $FeCl_3$, $CuCl$, $CuCl_2$, VCl_3, $MoCl_3$, $PdCl_2$, , $PtCl_2$, $PtCl_4$, $RuCl_3$, and $RhCl_3$, HMF yields ranging from 63 to 83% were achieved in 3 hr at 80°C. The experiment was repeated with glucose as feed but raised the temperature to 100°C because of its lower reactivity. Twelve of the metal halides tested showed less than 40% conversion of glucose, but only one catalyst, $CrCl_2$, gave high 70% yield of HMF. In order to explain the unprecedentedly high conversion of glucose to HMF they propose that the $CrCl_3^-$ anion plays a role facilitating mutarotation of glucose in [EMIM]Cl, as shown in the mechanism in Figure 9.3. Zhang suggested that mutarotation of the α-glucopyranosyl anomer to the β-glucopyranosyl anomer is a critical step of the reaction. The role of the chromium chloride species ($CrCl_3^-$) was identified as the proton transfer by forming hydrogen bonds with the hydroxyl groups, and was confirmed using ^1H NMR spectroscopy methods. Then the β-glucopyranose anomer undergoes a ring opening with [EMIM]$CrCl_3$, and the chromium enolate intermediate created, thus enabled the conversion of glucose into fructose. Once fructose is formed, dehydration of fructofuranose is rapid in the presence of the catalyst in the ionic liquid solvent.

Later NHC = N-heterocyclic carbine coordinated Cr species [37] are also shown to be effective catalyst for the dehydration of fructose, as well as glucose when used as the starting material. The use of $SnCl_4$ as a catalyst for the isomerization of glucose to fructose and subsequent dehydration was reported by Han and co-workers [38], who investigated a system composed of tin(IV) chloride and [EMIM]BF_4. This system gave a 60% yield of HMF. They screened various metal chlorides and ionic liquids, and only chromium(III), aluminum(III), and tin(IV) chlorides exhibited any activity, and tin(IV) was found to be the most efficient of these catalysts. The tin(IV)chloride system was also suitable for the conversion of fructose, sucrose, inulin, cellobiose and starch.

9.2.1.6   Cellulose in Ionic Liquids

The use of cellulose or abundant raw cellulosic biomass for the preparation of HMF is a very attractive proposition, as this will provide a more highly economical route than using a purified expensive monosaccharide such as fructose and glucose. However, the major challenge with cellulose and raw lignocellulosic biomass is poor

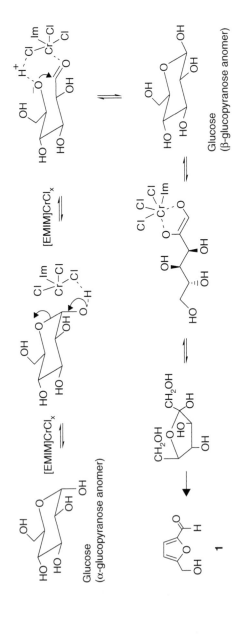

Figure 9.3 Mechanism of the CrCl₃ catalyzed isomerization of α-glucopyranosyl anomer of glucose to the β-glucopyranosyl anomer, and then conversion to HMF (1).

solubility and the difficulty of the hydrolysis into monosaccharides. Initial attempts with metal catalyzed degradation of cellulose under pyrolysis conditions have led to poor yields of HMF, for example, cellulose containing 0.5 mol of $ZnCl_2$/mol of glucose unit of cellulose was found to degrade at 200°C when heated for more than 60 s in air gave only 9%, yield of 5-hydroxymethylfurfural [39]. With the development of ionic liquids as a cellulose solvent and metal catalyzed isomerizations of glucose, a number of researchers have recognized that HMF can be obtained from abundant cellulosic feed stocks. In 2009 two research groups reported the metal catalyzed direct conversion of cellulose into HMF. Zhang and co workers reported [40] that the pair of metal chlorides $CuCl_2$ and $CrCl_2$ dissolved in 1-ethyl-3-methylimidazolium chloride ([EMIM]Cl) at temperatures of 80–120°C collectively catalyze the single-step process of converting cellulose to HMF with an unrefined 96% purity among recoverable products at 55% HMF yield. Furthermore, they found that, after extractive separation of HMF from the solvent, the catalytic performance of recovered ionic liquid and the catalysts was not affected in repeated uses. Then, Li and coworkers reported [41] [42] the use of 10% catalytic amount $CrCl_3$ in [BMIM]Cl for the hydrolysis and conversion of α–cellulose to hydroxymethylfurfural (HMF) in 62% under microwave irradiation, using 400W power for 2 min. Brönsted acidic ionic liquids with sulfonic acid group covalently attached to the imidazole structure can also be used efficiently in the first step of hydrolysis of cellulose to glucose [43]. In another recent report $FeCl_2$ [44] was used as the catalyst in 1-(4-sulfonic acid) butyl-3-methylimidazolium hydrogen sulfate acidic ionic liquid. In these experiments microcrystalline cellulose (MCC) was hydrolyzed to an appreciable extent (70%) by using 1-(4-sulfonic acid) butyl-3-methylimidazolium hydrogen sulfate as an effective catalyst producing 5-hydroxymethyl furfural and furfural in relatively high yields (15% and 7%, respectively), in addition to glucose. Then with the introduction of $FeCl_2$ as the catalyst a higher conversion of MCC (84%) and higher yields of HMF and furfural (34% and 19%, respectively) were achieved under the same experimental conditions.

Continuous flow reactors are an attractive alternative to the batch reactor process in an industrial scale production of HMF. Even though, this process gives moderate yields, McNeff [45] recently described a promising fixed bed catalytic bed flow reactor process for the production of HMF from both simple and complex sugars such as fructose, glucose, starch and cellulose, which involved a liquid phase process giving HMF up to 29% yield. The catalysts studied were spherical, porous zirconia and titania, which were both found to be stable over a variety of operating conditions and reusable through a simple pyrolytic regeneration process and the solvent system consists of a mixture of water and methyl isobutyl ketone.

The use of raw biomass forms such as corn stalk, poplar and switch grass is the next step in the development of a truly industrial process for large scale economical production of HMF as a platform chemical. HMF chemistry has now matured to this stage and several researchers have reported their finding in the use of raw lignocellulosic biomass for HMF production in 2010. Zhang [46] and co workers reported their experiments on raw lignocellulosic biomass in ionic liquid in the presence of $CrCl_3$ under microwave irradiation. Corn stalk, rice straw, and pine wood were treated under typical reaction conditions, and produced HMF and furfural in yields of 45–52% and 23–31%, respectively, within 3 min. In another approach, the use of a mixture of 1-ethyl-3-methylimidazolium chloride ([EMIM]Cl), ionic liquid, N,N-dimethylacetamide (DMA) and lithium chloride (LiCl) is also known as the solvent medium [47], where 10 mol % $CrCl_2$ and 10 mol % HCl was used as the catalyst. In a typical experiment untreated raw biomass was heated at 140°C for 2–3 hr in this complex solvent-salt mixture with the catalyst, and reported the production of HMF from corn stover and pine saw dust in 48 and 19% yields respectively. Other biomass components, such as lignin and protein, did not interfere substantially in the process, as yields of HMF based on the cellulose content of the biomass were comparable to those from purified cellulose. Recent shift in the interest of HMF production starting materials towards raw lignocellulosic biomass forms such as cornstover, switchgrass, saw dust and moving away from edible sugars fructose and glucose is a promising trend. Still there are a number of vast challenges of using these raw materials, such as use of expensive ionic liquids in a large scale industrial production of HMF. The major challenge with ionic liquids is the efficient recycling, and nearly 100% reuse of ionic liquid is essential in any industrial setup due to the cost of the current generation of ionic liquids. The use of chromium salts as a catalyst is another concern in a scale up process due to the toxicity of chromium salts. Some research has indicated that other metals such as iron or lanthanides can also be used in the glucose to fructose isomerization process.

9.3   5-Hydroxymethylfurfural Derivatives

9.3.1   Monomers Derived from 5-Hydroxymethylfurfural

The presence of two functional groups on opposing sides of the heterocyclic ring is an attractive feature in the 5-hydroxymethyl-furfural molecule, and makes it a proper platform molecule for the development of monomers. The oxidation states of these primary alcohol and aldehydes groups can be adjusted and derivatized to a verity of functionalities that undergoes rapid condensation as well as addition reactions that are suitable for polymerization processes. The use of HMF itself as a direct monomer is rare and being used in the preparation of Novolak type phenolic resins [48], and as a substituent for formaldehyde. Employing various types of oxidations, reductions and derivatizations, HMF can be converted into a number of versatile monomers suitable for poly-mer synthesis as shown in Figure 9.4. Reduction of the aldehyde function gives a diol: 2,5-bis(hydroxymethyl)furan (BHF) (4), and an alternative route to this diol is the hydroxymethylation of fur-fural alcohol. This furan diol can be used in the preparation of the

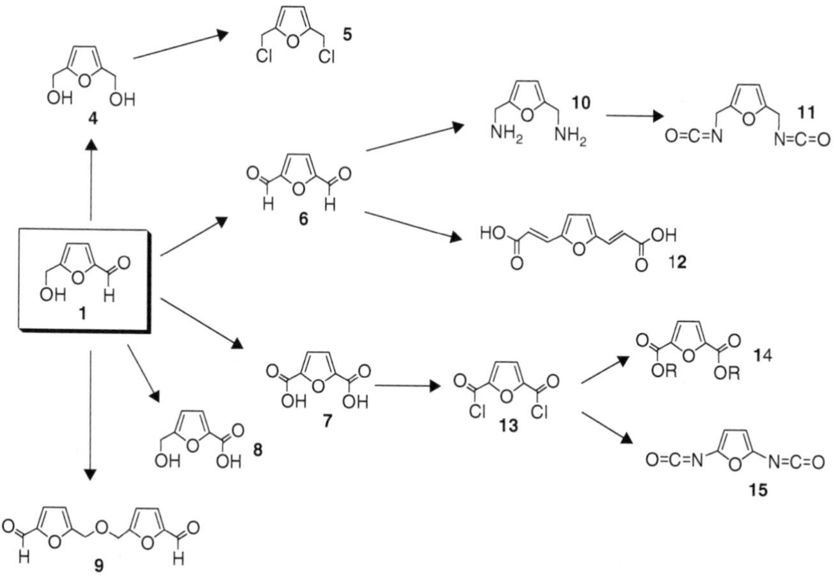

Figure 9.4 5-Hydroxymethylfurfural derivatives.

dichloride: 2,5-bis(chloromethyl)furan (BCF) (**5**). Through oxidation, HMF can be transformed into the dialdehyde: 2,5-bis(formyl)
furan (BFF) (**6**), dicarboxylic acid: 2,5-furandicarboxylic acid
(FDCA) (**7**) or 5-(hydroxymethyl)-2-furancarboxylic acid (HFCA)
(**8**) depending on the oxidation conditions used in the reaction.
These oxidized derivatives can be further transformed into;
2,5-bis(aminomethyl)furan (BAF) (**10**), 2,5-furandiacrylicacid
(FDAA) (**12**) and 2,5-furandicarboxylicacid dichloride (FDCC) (**13**).
Furthermore, this diacid dichloride (**13**) can be used in the synthesis of furanic diesters (**14**) or furanic diisocyanate (**15**). Whereas
the furanic bis methylene diisocyanate (**11**) can be prepared from
the diamine: 2,5-bis(aminomethyl)furan (BAF) (**10**). Dimerization
of HMF with the loss of a water molecule leads to the twelve carbon dialdehyde; 5,5′(oxy-bis(methylene))bis-2-furfural (OBMF)
(**9**). The use of HMF derivatives in the preparation of polymers has
been discussed in several excellent review articles on furan system
containing polymers, written by the pioneer in the field of renewable resources based furanic polymers, Alessandro Gandini [49]
[50] [51], the progress till 1997 is particularly discussed in detail in
a review published in 1997.

9.3.2   Synthesis of 5-Hydroxymethylfurfural Derivatives

9.3.2.1   *Synthesis of 2,5-bis(hydroxymethyl)furan*

2,5-bis(hydroxymethyl)furan (BHF) (**4**) is produced by the high
pressure catalytic reduction of HMF, using copper chromite catalyst
[52]. An alternative approach to this furanic diol is the hydroxymethylation of furfural alcohol with formaldehyde in the presence
of an acid catalyst [53].

9.3.2.2   *Synthesis of 2,5-bis(formyl)furan*

Oxidation of the primary alcohol group in HMF gives the symmetrical dialdehyde 2,5-bis(formyl)furan (BFF) (**6**), which is a stable solid
of melting point 108–9°C. A number of common oxidation methods
have been used [8] for the oxidation of HMF to BFF. The use of
oxygen or air as a primary oxidant is the most economical type of
oxidation available and vanadyl phosphate (VOP) based catalysts
have been identified as best catalytic systems for the oxidation of
HMF to BFF. The best performances in the oxidation have been

obtained when VOP in *N,N*-dimethylformamide (DMF) was used as a heterogeneous catalyst at 100°C [54]. The more recent development is a one-pot, two-step method involving dehydration of fructose, followed by in situ air-oxidation using a vanadium pentoxide catalyst in DMSO [55]. Room temperature oxidation of HMF in a buffered sodium hypochlorite-methylene chloride biphase system using Mn(III)-salen catalysts is also known to give good yields [56] as shown in the Figure 9.5.

9.3.2.3   Synthesis of 2,5-furandicarboxylic Acid

2,5-furandicarboxylic acid (FDCA) (**7**) is the symmetrical diacid resulting from the oxidation of both primary alcohol and the aldehydes group of HMF to carboxylic acid groups. Unlike HMF the diacid is a very stable compound, a crystalline solid at room temperature with a melting point >300°C. This is probably the most important derivative of HMF and this symmetrical aromatic diacid has been looked as a potential replacement and the renewable resources based equivalent of terephthalic acid, [57] which is a monomer in polyethylene terephthalate (PET) plastics. Furthermore, FDCA is listed in a 2004 US Department of Energy National Renewable Energy Laboratory (NREL) report [58] as one of the twelve building blocks that can be subsequently converted to a number of high-value bio-based chemicals or materials.

Many of the common stoichiometric oxidation reagents have been used for the conversion of HMF to FDCA [8]. Then, there are a number of reports on catalytic oxidation methods for the conversion of HMF to FDCA, and Van Bekkum [59] and Vinke [60], [61] introduced the use of oxygen and noble metal catalysts. Lew patented [62] a very efficient oxidation method using platinum adsorbed on activated charcoal as a catalyst and reported the isolation of FDCA

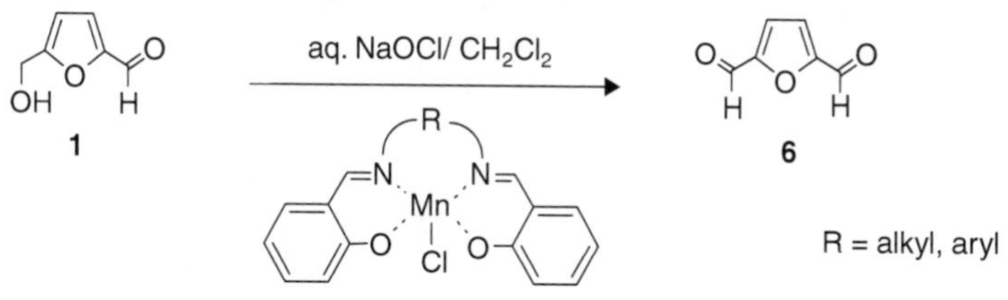

Figure 9.5 Mn(III)-salen catalyzed oxidation of HMF to 2,5-bis(formyl)furan (BFF) (**6**) using NaOCl as the primary oxidant.

in 95% yield (Figure 9.6.). However, when the Pt/C/CuO-Ag$_2$O mixture was used as the catalyst, FDCA was obtained in 99% yield. Lew suggested that HMF was first oxidized to 5-hydroxymethyl-furoic acid with CuO-Ag$_2$O pair and the latter is subsequently oxidized to FDCA (7) with charcoal-on-platinum catalyst.

Another catalytic oxidation method used a platinum/lead catalyst deposited on carbon in a strongly alkaline aqueous medium and in the presence of oxygen, the reaction was quantitative in less than 2 hr and the purity of the diacid was >99%. [63]. Additionally, encapsulation of the catalyst in silica beads is also known [64].

The use of metal/bromide catalysts [Co/Mn/Br, Co/Mn/Zr/Br, Co/Mn/Br/(Co+Mn)] for the oxidation of HMF to BFF and then to FDCA was described by Partenheimer [65] and the use of gold nanoparticles supported on nanoparticulate ceria (Au-CeO$_2$) and titania (Au-TiO$_2$) was reported by Casanova and co workers [66]. In another recent report the use of titania-supported gold-nanoparticle catalyst in water at ambient temperature was reported by Gorbanov [67]. In 2010 the results of a study on application of a flow reactor [68] for the oxidation of HMF to FDCA appeared, with the potential of scaling up to an industrial process. In addition to the common oxidation of HMF, direct use of sugars like glucose with insitu generation of HMF and then oxidation is also proposed for the large scale production of FDCA [69].

9.3.2.4   Synthesis of 5,5′(oxy-bis(methylene))bis-2-furfural

5,5′(oxy-bis(methylene))bis-2-furfural (OBMF) (9) is the dimer of 5-Hydroxymethylfurfural produced as a result of a dehydration. This twelve carbon symmetrical dialdehyde is a stable crystalline solid with a melting point of 111–2°C. Two different routes are known for the synthesis of OBMF: acid catalyzed dehydration of HMF, and Williamson reaction between HMF and 5-chloromethyl-2-furfural (16) in the presence of a base [70], [71]. These two approaches are shown in Figure 9.7.

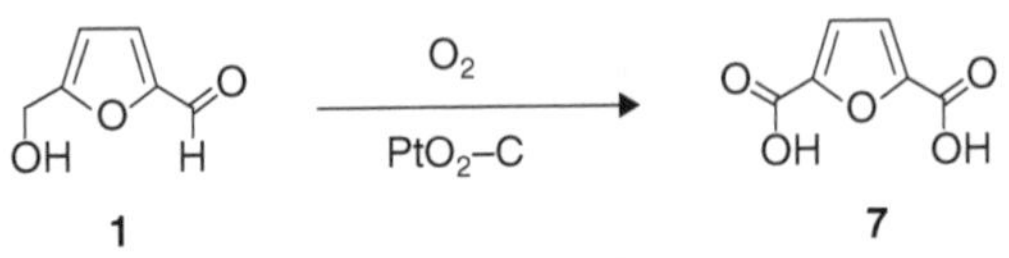

Figure 9.6 Oxidation of HMF to 2,5-furandicarboxylic acid (FDCA) (7) using oxygen or air and a platinum oxide catalyst.

Recently, Corma *et al.* showed [71] that etherification of HMF to OBMF can be performed in high conversion and selectivity using molecular sieves with Brönsted and Lewis acid sites. Among them, mesoporous materials (Al-MCM-41 and Sn-MCM-41) gave better performance than zeolites. They claimed that this is due to the existence of diffusional constraints in the microporous materials. Particularly, Al-MCM-41 bearing Brönsted acid sites gives better results than a homogeneous acid catalyst, such as *p*-toluenesulphonic acid. These solid acid catalysts have the added advantage that they can be easily recovered and reused.

9.3.2.5 Synthesis of 2,5-bis(aminomethyl)furan

Symmetrical furanic diamine: 2,5-bis(aminomethyl)furan (BAF) (**10**) is a useful polymer building block for the preparation of Schiff base polymers as well as polyamides with furan rings. This useful monomer has been prepared *via* furan bis aldoxime **17**. Catalytic reduction of the aldoxime **17** gave 2,5-bis(aminomethyl)furan (BAF) (**10**) as shown in Figure 9.8.

The furanic diamine (**10**), is a useful precursor for the furanic diisocyanate, where **10** can be converted to furanic diisocyanate (**11**) (Figure 9.4.), by reaction with triposgene [72].

9.3.2.6 Synthesis of 2,5-bis(chloromethyl)furan

The furanic dichloride, 2,5-bis(chloromethyl)furan (BCF) (**5**) is prepared from the furanic diol (BHF) (**4**) by standard chlorination methods.

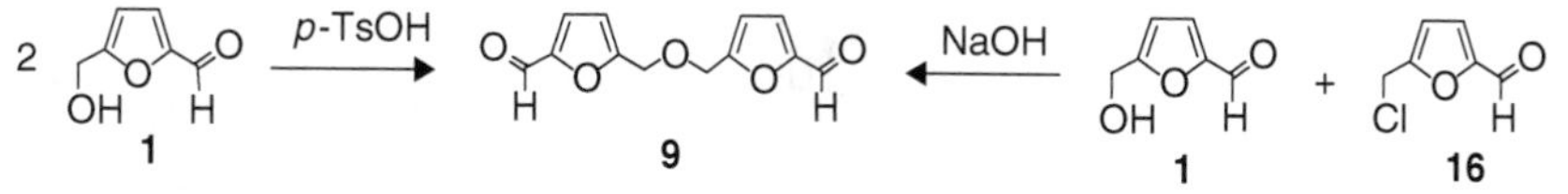

Figure 9.7 Two paths for the preparation of 5,5′(oxy-bis(methylene)) bis-2-furfural.

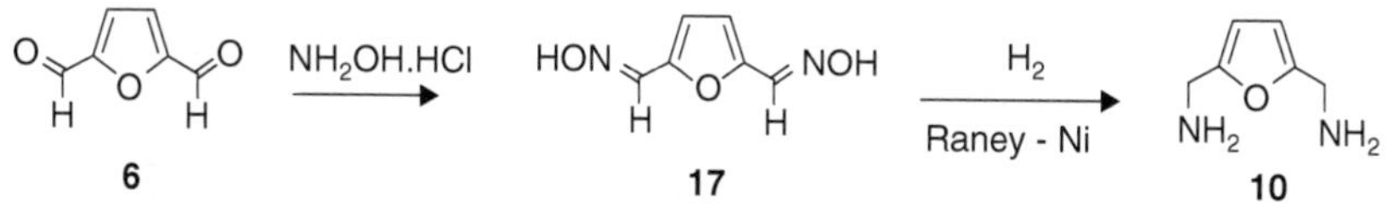

Figure 9.8 Synthesis of 2,5-bis(aminomethyl)furan (**10**).

9.3.2.7   Synthesis of 2,5-furandicarboxylicacid Dichloride and Related Compounds

2,5-Furandicarboxylicacid dichloride (FDCC) (13) is the acid dichloride of FDCA (7). This symmetrical acid chloride is a stable compound, and can be used in the synthesis of polyesters, polyamides, and polyurethanes. The acid dichloride can be prepared by reaction of FDCA with PCl_5 or thionyl chloride ($SOCl_2$) [73]. Additionally, FDCC serves as the intermediate in the preparation of diesters (14) by reaction with alcohols, and diisocyanate (15) by Curtius reaction. This furanic diisocyanate (15) is highly reactive compared to other furanic diisocyanate (11), and is susceptible to react rapidly with external agents and must therefore be stored in an inert atmosphere in a dark and cold place.

9.4   Polymers from 5-Hydroxymethylfurfural Derivatives

9.4.1   Furanic Poly Schiff Bases

Symmetrical dialdehyde, 2,5-bis(formyl)furan (BFF) (6) has been used as a co-monomer component in poly Schiff base polymers, where condensation of BFF with different diamines (18); 1,4-phenylenediamine, 1,6-diaminohexane and 2,5-bis(aminomethyl)furan gave low molecular weight polymers (19) of M_n in the range of 1500–2500, and these polymers are soluble in common organic solvents [74] (Figure 9.9). In particularly, polymer resulting from BFF and 1,4-phenylenediamine showed good thermal stability and semiconduction properties after doping with iodine.

In a follow up investigation, the same research group prepared [75] furanic polyazine (20) using hydrazine. The resulting materials possessed the expected structure, showed molecular weights

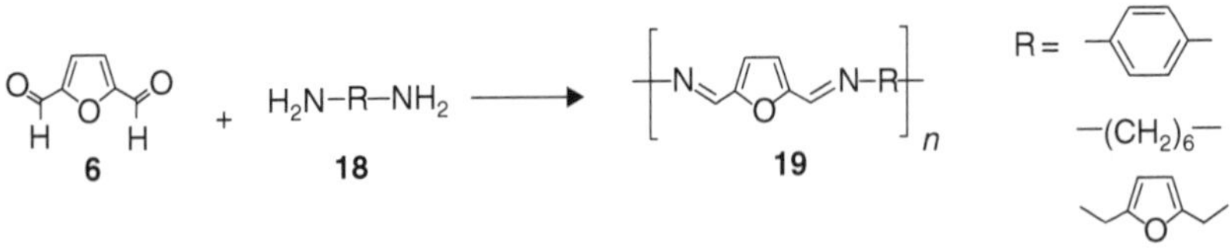

Figure 9.9 Synthesis of furanic poly Schiff bases (19) from 2,5-bis(formyl)furan (BFF) (6).

reaching 2000, and these materials underwent thermal decomposition starting at 270°C without any sign of melting, and showed conductivity in the range of 10^{-5} to 10^{-4} S cm^{-1} after doping with iodine, indicating semiconducting properties, as in the case of the aromatic/furanic counterparts (Figure 9.10). The low molecular weight fractions of the polyazine (20) were soluble in CH$_2$Cl$_2$ and showed an absorption peak at 440 nm, confirming the conjugated nature of the polymer.

Furthermore, attempts have been made to synthesize more amenable structures, in terms of the processability of solution or melt processability, by introducing soft segments in the form of oligo(ethylene oxide) moieties within the main chain. In these studies first, two trimeric precursors were prepared by condensation of excess of BFF with hydrazine and the second using BFF with excess of oligo(ethylene oxide)-diamine (Jeffamine, M$_n$ = 600) [75]. Then the two macromonomer precursors were condensed to yield a plastic like soluble polymer with a molecular weight of about 20000. This product was therefore made up of short conjugated sequences like 20 alternated with short poly(ethylene oxide) blocks like 21 (Figure 9.11).

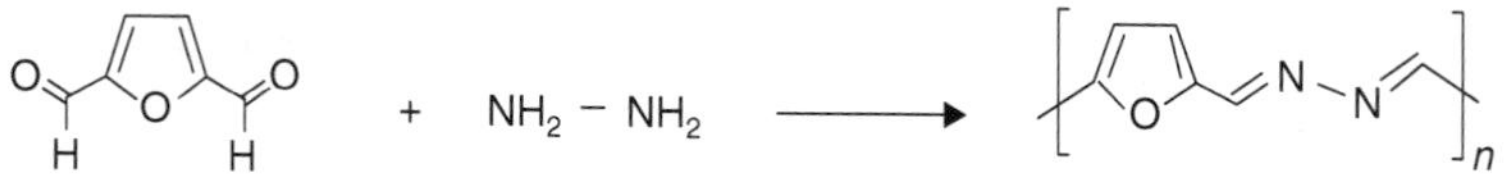

Figure 9.10 Synthesis of furanic polyazine (20).

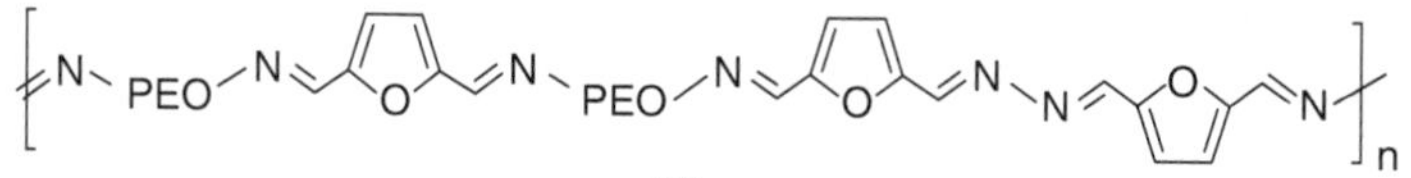

Figure 9.11 Poly(ethylene oxide) blocks in the polymer prepared using BFF, oligo(ethylene oxide)-diamine (Jeffamine, M$_n$ = 600), and hydrazine In particular, the combination.

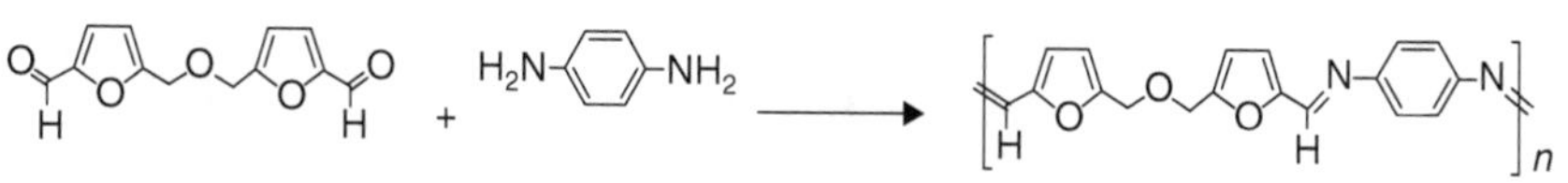

Figure 9.12 Synthesis of Schiff base polymer 22 from 5,5′(oxy-bis(methylene)) bis-2-furfural (OBMF) and p-phenylenediamine.

Similar to the reactions of 2,5-bis(formyl)furan (BFF) (6), difuranic dialdehyde; 5,5′(oxy-bis(methylene))bis-2-furfural (OBMF) (9) can also be used for furanic Schiff base polymer synthesis [74].

Thus condensation of 9 with p-phenylenediamine gave the corresponding Schiff base polymer 22, (Figure 9.12) and this polymer showed a broad UV absorption around 375 nm, closer to the UV spectrum of the model compound of this polymer. This material failed to exhibit any transition in the DSC analysis between room temperature and 300°C, but decomposition started around 270°C, and the latter feature was confirmed by TGA as well. Even though this is an unconjuagated polymer, this material showed modest semi-conductivity of about 10^{-6} S cm^{-1} at room temperature. This semi-conductivity behavior may be due to an inter chain hopping mechanism between long Fu-CH=N-Ph-N=Ch-Fu conjugated units in the polymer chains [74].

9.4.2   Furanic Poly Esters

The first record of a preparation of a polyester from 2,5-furandicarboxylic acid (FDCA) (7) is found in a 1951 US Patent [76]. In this patent they claimed that a polymer with a melting point in the range of 205–10° could be obtained by heating equal parts of 2,5-furandicarboxylic acid and ethylene glycol for 3 hr at 180° and 4 hr at 210°C, followed by further heating under reduced pressure. Next, in 1978 Moore and Kelly published a detailed study of polyester preparation using FDCA [77]. They used a *trans*-esterification technique for the polymerization reaction. A 2:1 molar ratio mixture of 1,6-hexanediol, dimethyl-2,5-furan dicarboxylate (23), with a mixture of calcium acetate, and antimony oxide added as a catalyst was used. First the reaction mixture was heated in a fluidized sand bath (190–200°C), under nitrogen, for 6 hr as shown in Figure 9.13. Then the nitrogen flow was discontinued and the system was evacuated to less than 1 Torr at 200°C. The temperature was maintained at 200°C for 1 hr and was then raised to 275°C over the course of 2 hr.

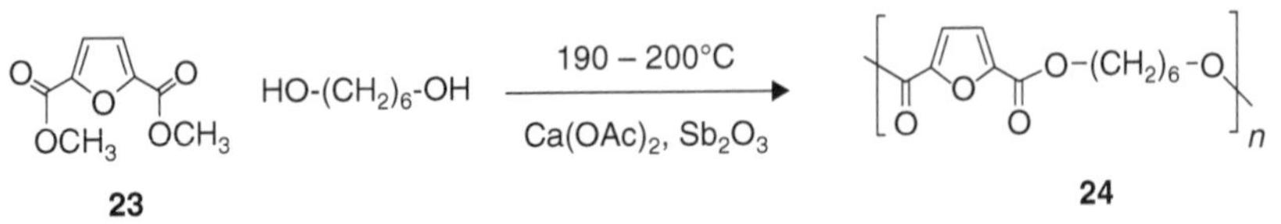

Figure 9.13 Synthesis of furanic polyester **24** by *trans*-esterification.

Then the system was heated at 275°C for an additional 2 hr., and upon cooling, the polyester was obtained as a hard, greenish solid which was found to be soluble in halogenated hydrocarbons. The crude polymer was dissolved in chloroform, and purified by re-precipitation in light petroleum ether to give the poly(2,5-furandi-ylcarbonyloxy-hexamethyleneoxycarbonyl) (24) as a white, fibrous solid, in 50% yield. Molecular weight data measured using GPC showed to have an apparent M_n = 7438, M_w = 18860, and polydispersity of 2.54. Brittle fibers could be drawn from the polymer melt and transparent, flexible films could be cast from a 20% (by weight) solution of the polymer in chloroform.

Later in 2009, Fehrenbacher and co workers [78] following a procedure similar to Moore and Kellys trans-esterification method, prepared a series of polyesters. In this work Fehrenbacher's group compared the molecular weights and thermal properties of four polyesters derived from four different diols: 1,3-propane diol, 1,6-hexane diol, 1,12-dodecane diol, and 1,18-octadecane diol, in trans-esterification with dimethyl-2,5-furan dicarboxylate (23). A mixture containing 0.13 mol% of calcium acetate and 0.1 mol% of antimony (III) oxide was used as the catalyst. Molecular weight and melting point data of the polyesters prepared are shown in Table 9.2.

As shown in Table 9.2, this method produced polyesters with low to moderate molecular weights, and melting points of the series illustrated the effect of increasing the carbon chain of the diol [78].

Difuranic diesters derived from furan 2-carboxylic acid has also been used for the preparation of polyesters bearing furan moieties using similar trans-esterification techniques [79] [80] [81].

Table 9.2 Molecular weight and melting point data of polyesters prepared from various diols and dimethyl-2,5-furan dicarboxylate (23), via trans-esterification method. A mixture containing 0.13 mol% of calcium acetate and 0.1 mol% of antimony (III) oxide was used as the catalyst, at 140–240°C for 8 hr.

Diol comonomer	M_n(g/mol)	M_w(g/mol)	T_m(°C) from DSC
1,3-propane diol	13901	30640	176.5
1,6-hexane diol	13394	22347	147.2
1,12-dodecane diol	25354	51613	109.2
1,18-octadecane diol	22118	46780	98.1

Recently, Gandini [57] and co workers compared the synthesis, spectroscopic, and thermal stability properties of polyester poly(ethylene-2,5-furandicarboxylate) (PEF) (26) derived from 2,5-furandicarboxylic acid (FDCA) and ethylene glycol with polyethylene terephthalate (PET). According to this brief but interesting study, when compared to the earlier approaches used for the preparation of PEF (26) (solution polycondensation between the 2,5-furandicarboxylicacid dichloride (FDCC) (13) and ethylene glycol (EG), trans-esterification of the FDCA dimethyl ester with an excess of EG, followed by the poly-trans-esterification of the ensuing product), their approach of poly-trans-esterification of the diester diol 25 proved the most rewarding. This diester diol 25 used as the monomer was prepared in 98% yield by reacting FDCA with a excess of EG (100 fold) for 6 h at 75°C in the presence of catalytic amount of HCl under a vacuum, and then removing the excess ethylene glycol after neutralization. Then 25 was subjected to Sb_2O_3 (5×10^{-3} to 2×10^{-2} M), catalyzed trans-esterification polymerization in a high-vacuum system under magnetic stirring, while the temperature was raised progressively from 70 to 220°C in the course of several hours, with the concurrent trapping of the released ethylene glycol at liquid nitrogen temperature as shown in Figure 9.14.

The crude product PEF could be obtained as a white solid and found to dissolve only in trifluoroacetic acid (TFA) and in hot tetrachloroethane (TCE), among the numerous potential solvents tested. The polymer was purified by precipitation from the TFA solutions using an excess of an EtOH/Et$_3$N mixture. After filtering, and washing with the same solvent mixture, the polymers were dried and characterized using IR and NMR spectroscopy. Figure 9.15 shows the typical FTIR (a), and ^{1}H NMR (b) spectrums of PEF produced. The peaks at 1716 and 1264 cm^{-1} in the IR spectrum were attributed to the ester carbonyl and C-O moieties. The 2,5-disubstituted furan heterocycle was indicated by peaks at 3123, 1578, 1015, 960, 834, and 761 cm^{-1}. The very weak OH absorption around 3400 cm^{-1} suggested that the PEF had reached a reasonably high molecular weight [57].

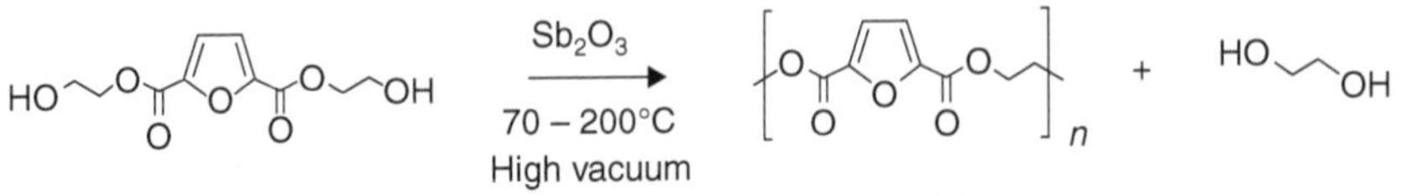

Figure 9.14 Synthesis of poly(ethylene-2,5-furandicarboxylate) (PEF) (26) using trans-esterification polymerization of diester diol (25).

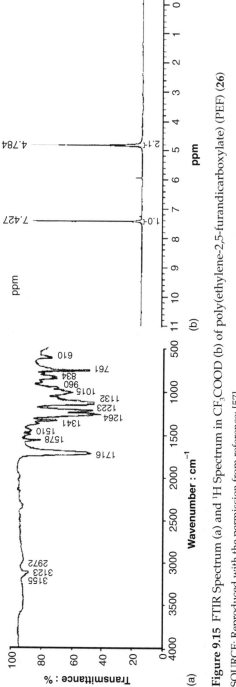

Figure 9.15 FTIR Spectrum (a) and ^{1}H Spectrum in CF$_3$COOD (b) of poly(ethylene-2,5-furandicarboxylate) (PEF) (**26**)

SOURCE: Reproduced with the permission from reference: [57].

The ¹H NMR spectra of the PEF in CF₃COOD (Figure 9.15 b) showed a striking resemblance to that of PET in the same solvent. This spectrum showed the resonance of the H3 and H4 furan protons at 7.43 ppm, and that of the ester CH₂ at 4.78 ppm with the expected 1:2 integration ratio. The ¹³C NMR spectra in the same solvent exhibited the peaks associated with the furan ring (C2/C5 at 147.1 ppm and C3/C4 at 121.1 ppm), with the methylene groups at 64.7 ppm and with the carbonyl moieties at 161.0 ppm. Elemental analyses of these polymers were in agreement with a high molecular weight polymer, with the calculated DP in the order of about 200. TGA of these PEFs showed that they are thermally stable up to ~300°C and degraded thereafter with a major decomposition step, which left an approximately 20% residue at 400°C, and a slower complete volatilization ending at about 580°C. This behavior is similar to that displayed by PET [82] albeit with a somewhat lower thermal stability.

The DSC thermograms of the precipitated polymer poly(ethylene-2,5-furandicarboxylate) (PEF) (26) (Figure 9.16) indicated a high degree of crystallinity, with a melting temperature of 210–215°C depending on the sample tested, which is about 45°C lower than that of a typical sample PET. After quenching the melted PEF (26) in liquid nitrogen, the tracings of the ensuing amorphous

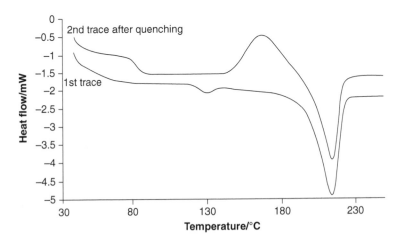

Figure 9.16 DSC tracing of the precipitated polymer poly(ethylene-2,5-furandicarboxylate) (PEF) (26) - first trace. After quenching the melt of 26 in liquid nitrogen – second trace.

SOURCE: Reproduced with permission from reference: [57].

morphologies (Figure 9.16) displayed a glass transition at 75–80°C (similar to that of PET) and a crystallization exotherm with a maximum at ~165°C, followed by the same melting pattern as that of the precipitated precursor.

In theory, one should be able to use the differentially functionalized 5-(hydroxymethyl)-2-furancarboxylic acid (HFCA) (8) (Figure 9.4) as the monomer to make an all furanic polyester. The main obstacle for this approach is the difficulty in the preparations of acid-alcohol 8, which requires the selective conversion of the aldehyde group in HMF to a carboxylic acid group without affecting the primary alcohol. This selective oxidation is a considerable challenge and there are very few reports on the preparation of this interesting furan derivative. For example chloroperoxidase (CPO) enzyme catalyzed the oxidation of 5-hydroxymethylfurfural with hydrogen peroxide as the oxidant [83] to give 26–40% yield of 5-(hydroxymethyl)-2-furancarboxylic acid (8) as the main byproduct, whereas the main product was FDCA. There were a few attempts to use 5-(hydroxymethyl)-2-furancarboxylic acid (8) for the polyester synthesis, which in some cases resulted oligoesters. In these earlier experiments [84], oligoesters of 8 were prepared by polycondensation of 5-(hydroxymethyl)-2-furancarboxylic acid using 2-chloro-1-methylpyridinium iodide as a polycondensation agent. Solution polycondensation in pyridine at 60°C gave the macrocyclic crown-oligoesters in 91% total yield. The product was found to be a mixture of oligomers with cyclic trimer, tetramer, pentamer and higher oligomers. Polycondensation in *n*-hexane in place of pyridine with tri-*n*-butylamine as a scavenger for hydrogen halide produced more of the linear oligoesters and less of the cyclic oligoesters.

Later, Moore [85] succeeded in making a linear polyester with all furanic units by *trans*-esterification polymerization of the methyl ester of 8 as shown in Figure 9.17. In this case,

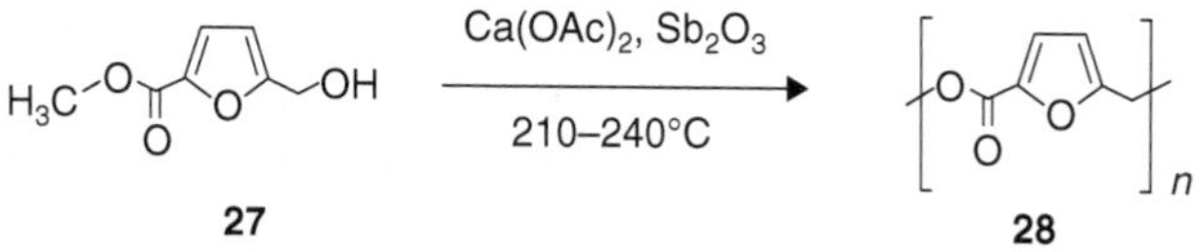

Figure 9.17 Synthesis of all furanic polyester: poly(2,5- Furan diylcarbonyloxymethylene) (28) by *trans*-esterification of methyl-5-hydroxymethyl-2-furoate (27).

methyl-5-hydroxymethyl-2-furoate (27) was used to produce the
linear polyester: poly(2,5-furan diylcarbonyloxymethylene) (28),
where a mixture of calcium acetate and antimony oxide was used as
the catalyst, heating at the 210–240°C temperature range, first under
nitrogen and then under a vacuum, produced the crude polymer.
The product was purified by dissolving in chloroform, and precipitation into light petroleum ether, which was then dried to give the
pure polymer (28) in low 38% yield. The polymer produced from
the hydroxyester was a brown powder, showed moderate molecular weight, and had no film-forming or fibrous properties.

Even though the most widely used polyesterification catalysts are Sb or Sn oxide based systems, lanthanide salts are gaining attention in many catalytic applications in recent years. For
instance, another interesting catalyst used in the polyesterification is scandium triflate; thus, 2,5-furandicarboxylic acid could
be directly polymerized with ethylene glycol at 160°C, producing
poly(ethylene-2,5-furandicarboxylate) with M_n 4.9×10^4 g/mol [86].
In another recently patented method, [87] FDCA polyester preparation is described using titanium based catalysts, and in this procedure, diester-diol of 2,5-furandicarboxylate was prepared by reacting
FDCA with excess of diol in the presence of hydrochloric acid or
zinc acetate catalyst, and heating to 75–110°. Then, the diester-diol of
2,5-furandicarboxylate was subjected to a polyesterification reaction
at a temperature of 140–160° using *tetra*-butyltitanate as the catalyst.
A variety of diols, including ethylene glycol, 1,3-propylene glycol,
1,4-butanediol or 1,6-hexanediol could be used, producing furanic
polyesters in high yield. The use of aromatic diols and long chain
aliphatic diols for the polyester preparation with FDCA is an active
research area in the last couple of years, especially in the patented
literature. There are claims for the preparation of FDCA based polyesters with relatively high molecular weights as high as 110,000–
190,000 [88]. Polyesterification with polyols for the preparation of
resins are also known, and according to the claims in one patent
[89], these novel polyester resins displayed excellent heat resistance
properties, and could be used for producing molded products.

One can envision that any diol can be used in the polyesterification reaction with FDCA or its ester/acid dichloride for the preparation of furanic polyesters, and the use of a renewable resources
based diol as the co-monomer would make the process even more
attractive, as shown in the next example. In 1993, Storbeck and
Ballauff reported [90] an elegant preparation of polyesters based

on all renewable resources based monomers, namely FDCA acid dichloride and 1,4:3,6-anhydrohexitols. Three polyesters **29–31** were prepared by the reaction of 1,4:3,6-anhydro-D-sorbitol (DAS) (**32**), 1,4:3,6-anhydro-D-mannitol (DAM) (**33**), and 1,4:3,6-anhydro-L-iditol (DAI) (**34**) as shown in Figure 9.18. In the synthesis of these polymers, equimolar mixtures of 2,5-furandicarboxylicacid dichloride (FDCC) (**13**) and 1,4:3,6-anhydrohexitols in 1,1,2,2-tetrachloroethane were treated with an excess of pyridine and stirred at room temperature for four days. The polymers were isolated by pouring the reaction mixtures into large excess of methanol and the products were purified by re-dissolving in 1,1,2,2-tetrachloroethane and re-precipitation in methanol. Polyesters were obtained in 75–85% yield after drying in a vacuum. Molecular weight, viscosity, and thermogravimetry data for the three polyesters **29–31** prepared are shown in Table 9.3. These polyesters showed glass transition temperatures (Tg) in the range of 173–196°C, and these values are

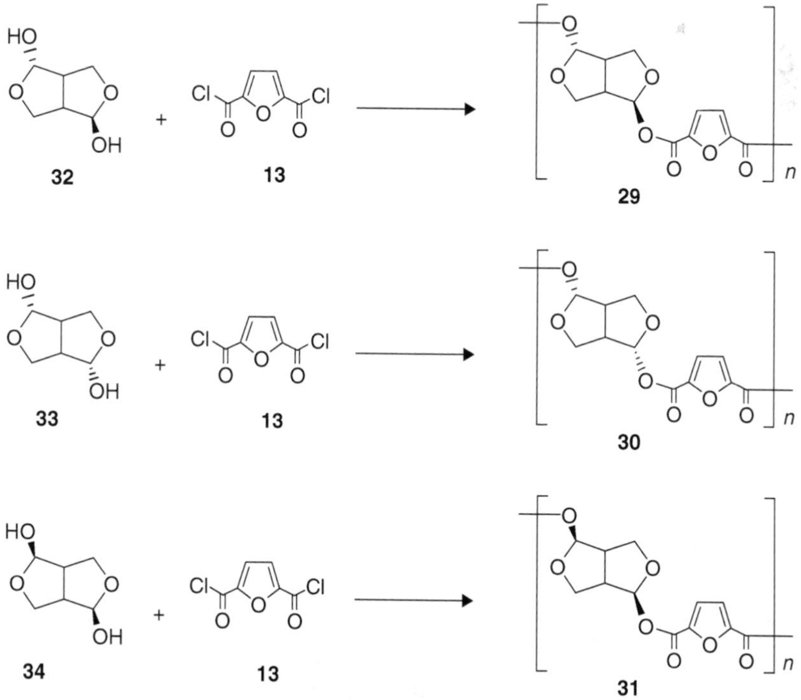

Figure 9.18 Synthesis of polyesters **29–31** using all renewable based monomers; 2,5-Furandicarboxylic acid dichloride (FDCC) (**13**) and 1,4:3,6-anhydrohexitols (**32–34**) in 1,1,2,2-tetrachloroethane.

Table 9.3 Molecular weight, viscosity and glass transition temperature data for the three polyesters **29–31** prepared from 2,5-Furandicarboxylicacid dichloride (FDCC) (**13**) and 1,4:3,6-anhydrohexitols in 1,1,2,2-tetrachloroethane. M_n was determined by membrane osmometry in 1,1,2,2-tetrachloroethane, at 30°C.

Polyester	M_n (g/mol)	DP	[η] (dL/g)	T_g (°C)
29	9000	34	0.11	173
30	14900	56	0.22	187
31	21500	81	0.22	196

relatively high compared to the furanic polyesters prepared from diols with long carbon chains. A similar range of Tg values are known in the case of polyesters prepared with these hexitols and terephthalic acid as well [91], and this may be due to the intrinsic stiffness of the bicyclic hexitol units in these polyesters.

¹H NMR Spectrum of typical polyester, **29** is shown in Figure 9.19. The two protons in the furan ring appear as two peaks in the spectrum, pointing to slightly different environments due to the secondary hydroxyl groups of chiral 1,4:3,6-anhydro-D-sorbitol (DAS) (**32**) unit and the possibility of having different connections to the adjacent hexitol moieties.

In another example for using all renewable resources based monomers for the synthesis of polyesters, 2,5-bis(hydroxymethyl)furan (BHF) (**4**) (Figure 9.4) produced from the reduction of HMF has been used as the diol monomer. This all furanic polyester; poly(2,5-furandimethylene 2,5-furandicarboxylate) (**35**) [92] was prepared in 94% yield at low temperature condensation of 5-bis(hydroxymethyl) furan (BHF) (**4**) with 2,5-furandicarboxylicacid dichloride (FDCC) (**13**) in acetonitrile in the presence of pyridine or triethylamine as the catalyst as shown in Figure 9.20. The polymer produced with the triethylamine as the catalyst showed a higher yield, reduced viscosity, and melting point when compared with polymer formed in the presence of pyridine catalyst. The poly(2,5-furandimethylene 2,5-furandicarboxylate) thus prepared had a density in the range of 1.476–1.495 g/cm³ and was compatible with 15% dibutyl phthalate or tricresyle phosphate plasticizer and the antioxidant diphenylolpropane. Additionally, this all furanic polyester showed a thermomechnical curve with a highly elastic zone [92].

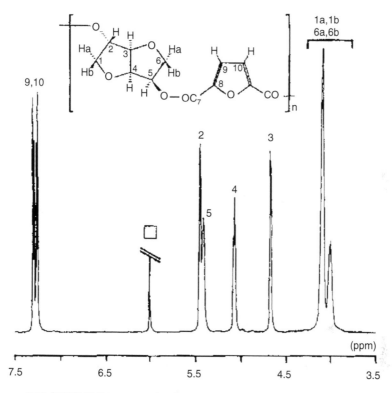

Figure 9.19 ^{1}H NMR Spectrum of polyester **29** prepared from all renewable resources based monomers; 1,4:3,6-anhydro-D-sorbitol (DAS) (**32**) and 2,5-furandicarboxylicacid dichloride (FDCC) (**13**), recorded in $C_2D_2Cl_4$
SOURCE: Reproduced with the permission from reference: [90].

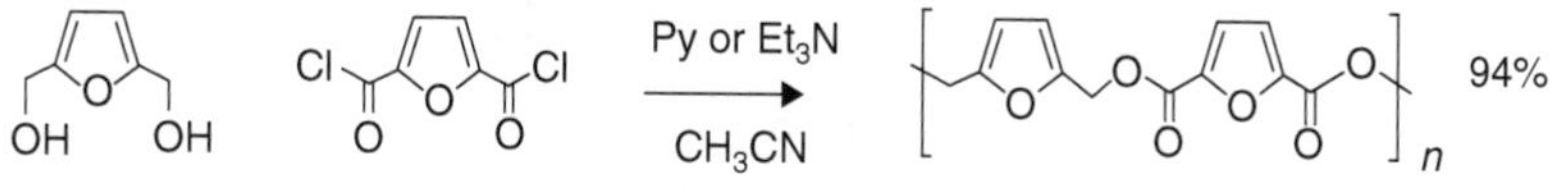

4   **13**   **35**

Figure 9.20 Synthesis of all furanic polyester; poly(2,5-furandimethylene 2,5-furandicarboxylate) (**35**) from 2,5-bis(hydroxymethyl)furan (BHF) (**4**) and 2,5-furandicarboxylic acid dichloride (FDCC) (**13**).

The same research group who prepared the polyester poly(2,5-furandimethylene 2,5-furandicarboxylate) (**35**) reported the use of 5-bis(hydroxymethyl)furan (**4**) for the synthesis of polyesters with adipoyl and sebacoyl, terephthaloyl chlorides [93], [94]. In these reactions polymerizations were carried out by interfacial, emulsion, and solution polycondensation at low temperatures in the presence of pyridine. Furthermore, they reported that these polyesters were susceptible to thermal oxidative degradation.

In a more recent work, natural dicarboxylic acids; succinic acid, fumaric acid, and maleic acid were used to prepare all renewable resources based polyesters with furanic diol; 5-bis(hydroxymethyl) furan (4) [95]. In these experiments polymerizations were carried out by two methods; one using acid chloride and the other one using a condensing reagent. Polycondensation of 5-bis(hydroxymethyl) furan (4) and succinic acid gave a polymer with an average DP of 30, and NMR and IR spectra of these polymers indicated that they have linear and regular polyester structures. Furthermore, copolymerizations were also performed using mixtures of two kinds of dicarboxylic acids with furanic diol 4 giving various furanic copolyesters [95].

9.4.3   Furanic Polyamides

The earliest reports on the use of HMF derived monomers for the preparation of furanic polyamides goes as far back as 1960's. These early works by Malyshevskaya and co-workers involved the studies [96] [97], [98] on the polycondensation of 2,5-furandicarboxylic acid (FDCA) (7) with hexamethylenediamine. A number of polycondensation techniques has been attempted, and among those, melting the salt of 7 and hexamethylenediamine for 30 min. at 260°C and polycondensation 6 hr at 220°C in sealed ampuls in vacuum gave polymers with DP of about 34. In another method involving interfacial polycondensation, a mixture of 0.15 mole/L each of 7 and hexamethylenediamine in $C_2H_2Cl_2$ was treated with 0.6 mole/L solution of aqueous NaOH to give a white polyamide with DP of 57. The polyamide thus prepared was soluble in cresol, N,N-dimethylformamide, and sulfuric acid, and swelled in ethanol and acetone. These polyamides could be formed into films and fibers, but failed to display satisfactory stability.

In the next stage of development, the use of acid dichloride; 2,5-furandicarboxylicacid dichloride (FDCC) (13). [99] as well as the furanic diesters (14) [100] are reported for the synthesis of furanic polyamides, but these investigations did not pursue the details related to the physical properties of the products.

Subsequently, in a fairly comprehensive study by Gandinis group [101], [102], a number of polyamides bearing 2,5-furandicarbonyl and aromatic diamino structures were prepared and characterized by spectroscopic methods. In these studies, two distinct types of furanic polyamides were synthesized, where the first group (36–39) was synthesized from the acid dichloride; 2,5-furandicarboxylicacid dichloride (FDCC) (13) and various aromatic diamines, and another polyamide (40) was prepared by condensation of acid dichloride 13 with 2,5-bis(aminomethyl)furan (10) to produce the all-furanic system (Figure 9.21). Furthermore, they showed that these polyamides possess regular structures, high molecular weights, and good thermal stability (Table 9.4). It is interesting to note that the 2,5-bis aminomethylfuran required as the co monomer was also prepared from HMF as described in section 9.3.2.5 [102].

After experimenting with several synthetic routes and the optimization of the procedures, the furanic-aromatic polyamides had higher molecular weights and better thermal stability than their all-furanic counterparts. A number of polymerization conditions were tested in these studies and the best operating conditions were found to be the use of a solvent mixture of N-methylpyrrolidone (NMP) and hexamethylphosphortriamide (HMPT) (1/1, v/v), total monomer concentrations of 2–3% (w/v), temperature of about 100°C, and reaction times of 10–20 hr. Furthermore, it was found that better yields can be obtained with the use of 1–2% excess of

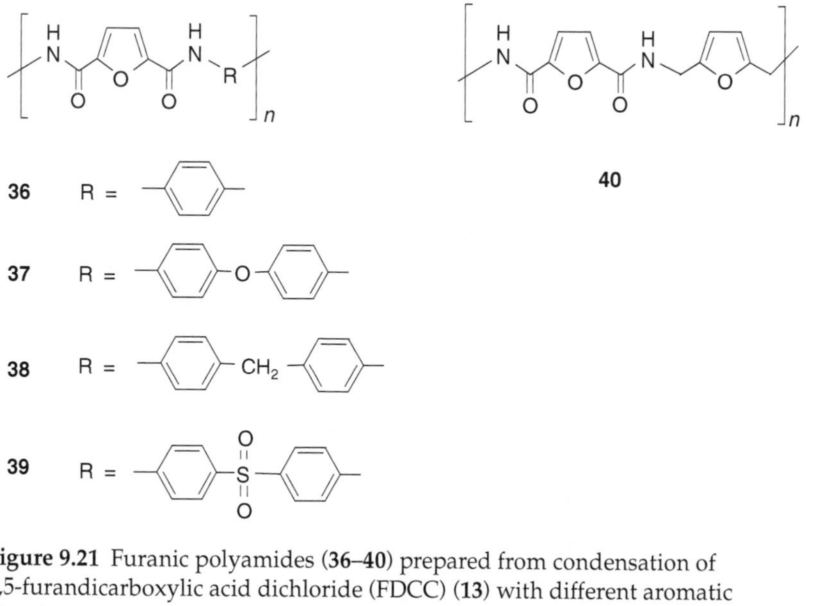

Figure 9.21 Furanic polyamides (36–40) prepared from condensation of 2,5-furandicarboxylic acid dichloride (FDCC) (13) with different aromatic diamines.

acid dichloride with respect to the stoichiometric amount relative to the diamine. They reported that during these experiments, the reaction mixture became very viscous, but the polymer did not precipitate. Another technique used was the so called "direct poly-condensation", which implies that the diamine (aromatic) was used in conjunction with the furanic diacids themselves instead of the dichlorides. This technique was developed for fully aromatic poly-amides. Structures of all the polyamides were confirmed by IR, ^{1}H and ^{13}C NMR spectroscopy.

Another approach to make the amide link is the reaction of an ester with an amine in the presence of a Lewis acid type catalyst, and many *trans*-esterification Lewis acid catalysts are good amidation catalysts as well, and this route is shown in the next example. In 2009 Fehrenbacher's group published [78] their findings on the use of dimethyl-2,5-furan dicarboxylate (23) for the polycondensation reactions. They studied the use of 23 for the preparation of polyesters as well as polyamides by reacting with various diols and diamines. In the preparation of polyamides, aliphatic C-6 to C-12 diamines were used in the presence of *n*-butyltinchloride dihydroxide as the catalyst as shown in the Figure 9.22.

For instance, in the polyamide synthesis, a 1:1 mixture of furan-2,5-dicarboxylic acid dimethylester and diamine with 0.13 mol% of

Table 9.4 Viscosity and molecular weight data obtained for representative furanic polyamides **36–38**. Viscosity measured in 98% H_2SO_4 at 30°C with $c = 0.5$ g/dL.

Polyamide	Viscosity (dL/g)	M$_w$
36	0.55	29500
37	0.95	37000
38	0.28	12300

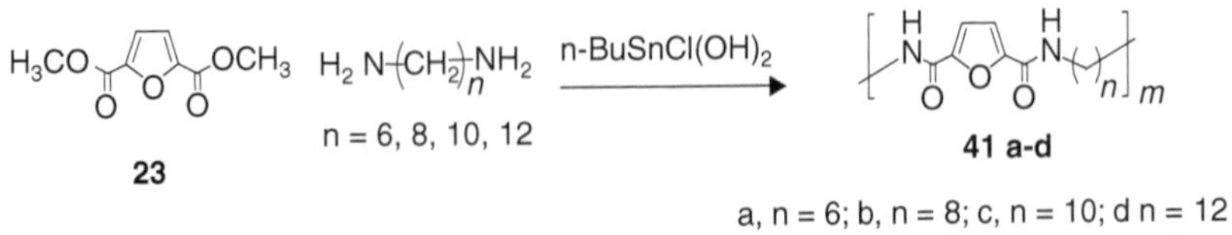

n = 6, 8, 10, 12

23

41 a-d

a, n = 6; b, n = 8; c, n = 10; d n = 12

Figure 9.22 Synthesis of furanic polyamides (41 a-d) from aliphatic diamines, using *n*-butyltinchloride dihydroxide as the catalyst.

n-butyltinchloride dihydroxide catalyst was first heated in an inert atmosphere at 140°C for 2hr, then the temperature was raised to 180–230°C range, and a vacuum (<100 Pa) was applied to remove the methanol formed and polymerization. They have studied the effect of temperature programming as well as isothermal heating under vacuum on the molecular weight of the polyamide formed, and the summary of the results for a representative example of the polycondensation of **23** with 1,6-diaminohexane is shown in Table 9.5.

Furthermore, the catalyst loading was optimized and it was found that even 0.06 mol% n-butyltinchloride dihydroxide is sufficient for the polymerization. Even with the optimization of polymerization conditions, the polymers showed relatively low number average molecular weights and high polydispersity values for all the diamines studied. M_n and M_w values and the glass transition temperatures of the polymer, obtained from C-6 to C-12 aliphatic diamines are shown in Table 9.6. The Tg values measured using DSC showed the expected decreasing trend as the diamine carbon chain length increases.

The use of FDCA and aromatic polyamine for the preparation of polyamine amide containing pendant amine groups and use of this polymer as a curative for polyureas, hybrid epoxy-urethanes, hybrid urea-urethanes, chain extenders for polyurethane and polyurea elastomers, and also for reaction injection molding (RIM) products are claimed in two recent patents [103], [104].

Table 9.5 The effect of the temperature program on the number and weight average molecular weights of polymer **41a**, during the n-butyltinchloride dihydroxide catalyzed polycondensation of dimethyl-2,5-furan dicarboxylate (**23**) and 1,6-diaminohexane.

Catalyst (Mol%)	Polycondensation P ≤ 0.3 MPa	Polycondensation P < 0.3 MPa	M_n (g/mol)	M_w (g/mol)
0.13	2 h, 140°C	2 h, 180°C 2 h, 200°C 1 h, 230°C	3313	8018
0.13	2 h, 140°C	2 h, 190°C	2803	5267
0.13	2 h, 140°C	2 h, 230°C	5115	13436

Table 9.6 The effect of carbon chain length of the aliphatic diamine on the number and weight average molecular weights of polymer **41a-d**, during the *n*-butyltinchloride dihydroxide catalyzed polycondensation of dimethyl-2,5-furan dicarboxylate (**23**) and C6-C12 aliphatic diamines.

Comonomer	M_n (g/mol)	M_w (g/mol)	T_g °C
1,6-diaminohexane	5160	13845	110.4
1,8-diaminooctane	4299	11130	82.8
1,10-diaminodecane	5266	12748	70.9
1,12-diaminododecane	6970	18837	68.2

9.4.4   Furanic Polyurethanes

Synthesis of polyurethanes by polycondensation of HMF derived 2,5-bis(hydroxymethyl)furan (BHF) (**4**) and 5,5'(oxy-bis(methylene))bis-2-furfural (OBMF) (**9**) with various diisocyanates are described in a series of papers published from Gandinis group in the early 1990's [72], [105], [106], [107], [108]. These work involved the synthesis and characterization of model urethanes and kinetic studies of the polyurethane formations. Typically, the polymerizations were carried out in dimethylacetamide (DMA) under a nitrogen atmosphere with a monomer concentration 0.5 mol/L. After mixing the reactant, the temperature was raised to 70°C under stirring and the reaction was allowed to proceed for 24 hr. The polyurethane polymer formed was isolated by pouring the clear viscous solution into an excess of ether. These polymers in the form of powders or elastomers were filtered and vacuum-dried to constant weight. The characterization of the product involved FTIR and NMR spectroscopy, elemental analysis, molecular mass determination by viscosimetry, GPC, end-group enhancement, light scattering, DCS, and TGA. Typical structures of liner polyurethanes prepared from 2,5-bis(hydroxymethyl)furan (**4**) and 5,5 (oxy-bis(methylene))bis-2-furfural (OBMF) (**9**) are shown in the Figure 9.23.

The hexamethylene diisocyanate (**42**) used in the first reaction is a commercially available monomer. Additionally, it is interesting to note that 2,5-furyl diisocyanate (**15**) was not used because of its

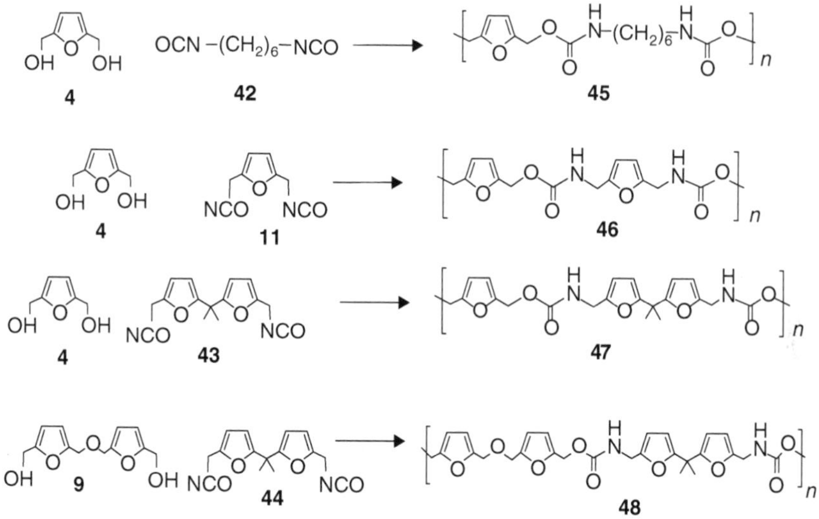

Figure 9.23 Synthesis of furanic polyurethanes **45–48**, using 2,5-bis(hydroxymethyl)furan (BHF) (**4**) and 5,5′(oxy-bis(methylene))bis-2-furfural (OBMF) (**9**) with various diisocyanates.

excessive sensitivity to resinification and the urethanes derived from it were not very stable. Instead, they utilized its homologue diisocyanate bearing methylene groups between the ring and each NCO function: 2,5-furfuryl diisocyanate (**11**), which was synthesized from 2,5-bis(aminomethyl)furan, by reacting with bis(trichloromethyl) carbonate (triphosgene) in chlorobenzene to insert the CO moiety. 5,5′-Isopropylidenebis(2-furfuryl isocyanate) (**43**) was the other furanic diisocyanates used in this work. This was synthesized from the corresponding amine using a similar procedure [107]. ^{1}H and ^{13}C NMR spectrums of these polyurethanes indicated that they all have expected linear structures; ^{1}H NMR spectrum of the polyurethane **48** is shown in Figure 9.24 as an example.

Preparation and studies on the properties of thermoplastic elastomers with furan moieties are also known [108]. These segmented furanic polyurethanes were based on the three typical components: macrodiol, diisocyanate and diol chain extender. The furan heterocycles were introduced through the diisocyanate, through the chain extender or through both, and took different configurations with respect to the main chains, i.e. as side groups or as an integral

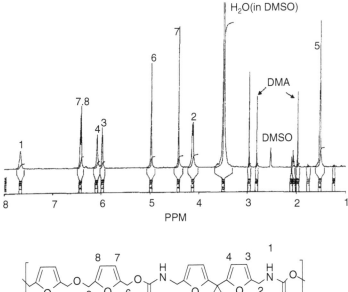

Figure 9.24 ¹H NMR of the polyurethane **48**.
SOURCE: Reproduced with the permission from reference: [107].

part of the backbone. Differential scanning calorimetry analyses of these materials provided information about glass transition, melting phenomena, and about microphase separation between soft and hard domains. With respect to commercial materials based on aromatic/aliphatic precursors, these furanic homologues tend to display lower moduli and thermal transitions, but can be thought to fulfill important applications specific to their properties. In particular, the polymer prepared by the combination of an aromatic diisocyanate with 2,5-bis(hydroxymethyl)furan, which is an industrial commodity, and appropriate macrodiols gives materials with good mechanical and thermal performances [108].

9.4.5 Furanic Polybenzoimidazoles

A preparation of a benzoimidazole unit containing thermally stable furanic polymers is also known. For example, a low temperature solution polycondensation of 2,5-furandicarbonyl dichloride (**13**) with *N1,N5*-diphenyl-1,2,4,5-tetraaminobenzene (**49**), followed by thermal cyclodehydration at 290–300°C gave furanic polybenzo-imidazoles [109], [110] of the type **50** as shown in Figure 9.25. These polymers were found to be very stable and did not show appreciable decomposition up to 400°C.

9.4.6   Furanic Polyoxadiazoles

The synthesis of a furanic oxadiazole polymer was described by Abid and co workers as shown in Figure 9.26. [111] In these experiments initial polycondensations were conducted using both solution and interfacial techniques. In the reactions conducted in solution, the amidic solvent insured itself the removal of HCl, whereas in the case of interfacial procedure, the presence of NaOH in the aqueous phase produced the corresponding neutralization. Progress of the reactions was followed by the evolution of the molecular weight, through the monitoring of the viscosities of the reaction solutions.

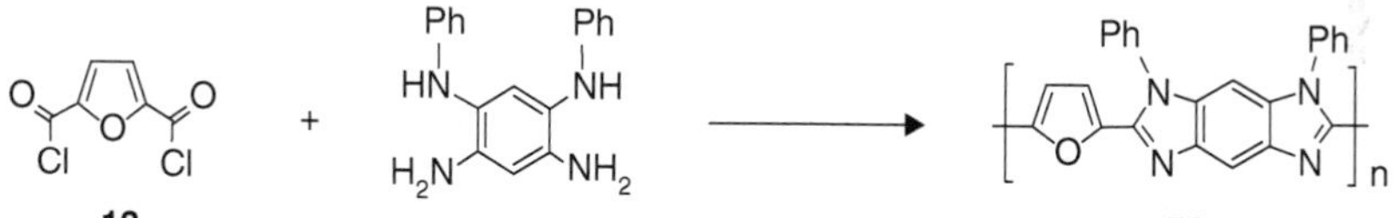

Figure 9.25 Synthesis of furanic polybenzoimidazole (**50**) from 2,5-furandicarbonyl dichloride (**13**) and *N1,N5*-diphenyl-1,2,4,5-tetraaminobenzene (**49**).

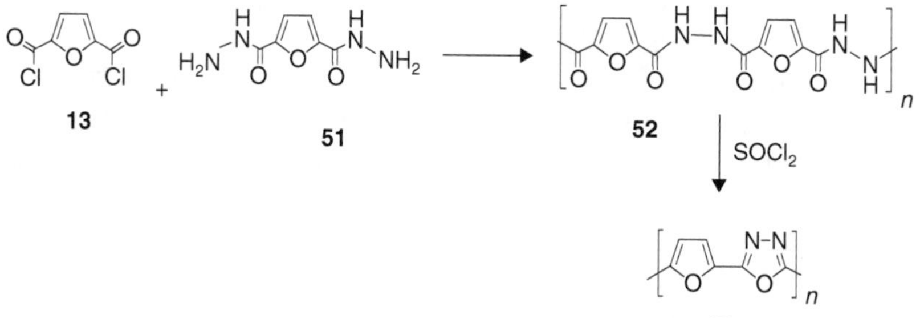

Figure 9.26 Synthesis of furanic polyoxadiazole (**53**) from 2,5-furandicarbonyl dichloride (**13**) and furanic bis hydrazide (**51**).

The interfacial procedure provided a more satisfactory per-formance compared to the use of homogeneous systems since only about 0.5 h of interfacial polycondensation was required for obtaining both the yields and viscosity attained after 12 hr by solution procedure. Polymers obtained after 3 hr by interfa-cial polycondensation were completely insoluble in DMSO and common solvents, and viscosity measurements could not be made. The polyhydrazides were then converted to polyoxadia-zoles, consisting of alternating furanic units and 1,3,4-oxadiazole rings,. Cyclodehydration of polyhydrazides were carried out by the chemical method, where polymers were heated in $SOCl_2$ con-taining a small aliquot of DMF in a nitrogen atmosphere at 80°C for 18 hr, and the corresponding polyoxadiazoles were then pre-cipitated in water and isolated by filtration. These polymers were characterized by FTIR and ^{1}H NMR spectroscopy. ^{1}H NMR spec-trums of polyhydrazide **52** and furanic oxadiazole polymer **53** are shown in the Figure 9.27.

9.4.7 Poly(furalidine bisamides)

In an application of HMF derived 2,5-bis(formyl)furan (BFF) (**6**), directly in the synthesis of polymeric materials, Amarasekara *et al.* [112] recently reported the synthesis and characterization of a poly(furalidine bisamides). This 2,5-diformylfuran–urea resin was prepared by melting a 1:2 ratio solid mixture of 2,5-bis(formyl) furan (BFF) (**6**) and urea at 110°C, as shown in Figure 9.28. The crystalline polymer resin was isolated in 90% yield after purifi-cation, and was characterized by elemental analysis, ^{1}H, ^{13}C and ^{1}H–^{1}H COSY NMR, IR, UV, TGA and DTA. The structural unit of this new material consisted of one BFF molecule condensed with two urea molecules at the aldehyde group of BFF. An analo-gous polymer was prepared by condensation of urea with furanic dialdehyde; 5,5′(oxy-bis(methylene))bis-2-furfural (OBMF) (**9**) as well.

9.4.8 Poly(furylenethylenediol)

The reductive coupling of aldehydes groups in 2,5-bis(formyl) furan (BFF) (**6**) is the technique used for the synthesis of

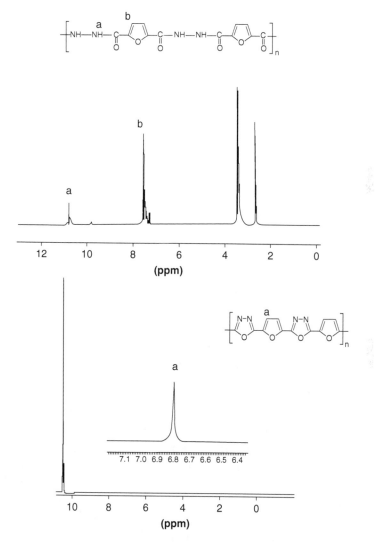

Figure 9.27 ¹H NMR spectrums of polyhydrazide **52** and furanic oxadiazole polymer **53**.

SOURCE: Reproduced with the permission from reference: [111].

poly(furylenethylenediol). This polymer was prepared by treatment of DFF with TiCl₄-Zn complex in dry THF at 0°C for 10 hr, and gave a furanic polymer with 1,2-diol units and DP of about 43 [113] as shown in Figure 9.29.

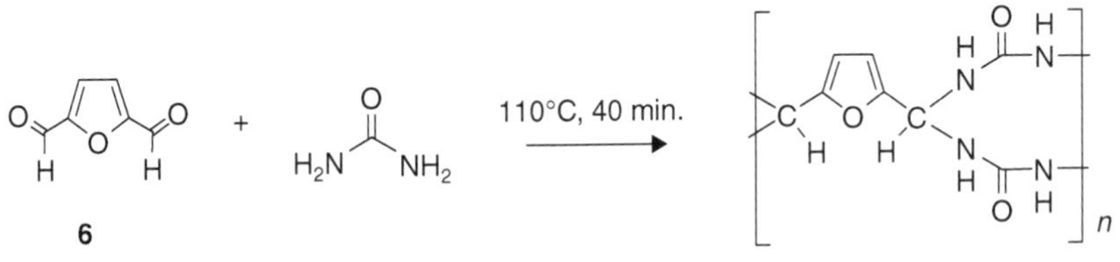

Figure 9.28 Synthesis of poly(furalidine bisamide) resin **54** by condensation of 2,5-bis(formyl)furan (BFF) (**6**) and urea.

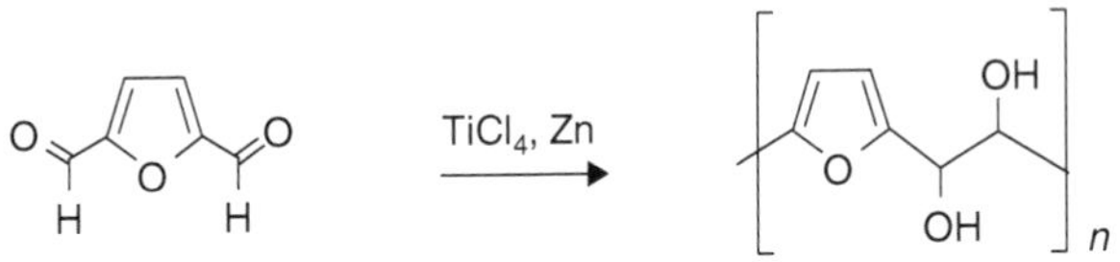

Figure 9.29 Synthesis of Poly(furylenethylenediol) (**55**) from reductive coupling of 2,5-bis(formyl)furan (BFF) (**6**).

9.4.9 Miscellaneous Furanic Polymers Derived from HMF

2,5-furandicarboxylic acid dichloride (**13**) has been used for the synthesis of various heat-resistant nitrogen containing furanic polymers like polyester urea, polysulfonamide, and polyamidourethane as well [114] [115]. Excellent thermal stability was seen in the benzoxazinone group containing furanic polymers. For instance, 3,3'-dicarboxybenzidine could be polymerized with 2,5-bis(chloroformyl) furan (**5**) in solutions or by interfacial polymerization to produce this new class of furanic polymers [116].

In another recent synthesis [117] of polymers from HMF, only the aldehyde function was modified to be used as the active polymerization terminal. In this method, HMF or its methylated derivative, 5-(methoxymethyl)furfural was efficiently converted to its vinyl derivative by the Wittig reaction in a solid-liquid phase transfer process. The vinylfuran monomer was subjected to a free radical polymerization in a bulk polymerization process. The structures of the vinyl monomers synthesized, and resulting polymers were confirmed by using NMR spectroscopy, including ^{1}H NMR, ^{13}C NMR, C-H COSY, and C-H COLOC. Moreover, the resulting polymers were subjected to GPC, thermogravimetric analysis (TGA), and differential scanning calorimetry (DSC), and they reported that the polymer from 5-methoxymethyl-2-vinylfuran exhibited much better heat resistance properties than the corresponding polymer without the methoxy group.

9.5 Conclusion

Recent progress in HMF chemistry driven by the search for new renewable resources based platform chemicals for the replacement of petroleum based feedstock has taken this virtually unknown furan derivative to the verge of becoming an industrial chemical. The main attraction in HMF is that it can be derived from the major fraction of the renewable carbon on earth. Still there are major challenges associated with large scale industrial production of HMF, like the choice of carbohydrate source, and problems associated in dissolution and efficient depolymerization of cellulose, in the case of direct use of lignocellulosic biomass as the raw material. The new developments in the use of ionic liquids as solvents and catalysts that can directly use raw biomass forms such as wood chips, and agricultural wastes is a major breakthrough, but there are enormous challenges like recycling the ionic liquid system. Instability of HMF is another concern, but HMF can be efficiently transformed into more stable and more useful derivatives like 2,5-bis(formyl)furan (BFF) or 2,5-furandicarboxylic acid (FDCA) before purification, storage, and transportation to the polymer industry. The recognition of FDCA as one of the high-value bio-based future platform chemicals by US Department of Energy National Renewable Energy Laboratory (NREL) is a significant development in the field. In a few examples like polyesters, and polyurethanes, furanic equivalents can be substituted to the currently available polymerization technologies to make substitutes to petroleum based polymeric materials. Nevertheless, there is more potential in the development of new polymeric materials with HMF derivatives. The renaissance in the HMF chemistry is reflected in the rapidly increasing number of publications in the last few years, and economically viable large scale industrial production of HMF from biomass is a possibility in the near future, as is the prerequisite for the HMF based furanic polymer industry.

References

1. A. Corma, S. Iborra, and A. Velty, *Chemical Routes for the Transformation of Biomass into Chemicals.* Chemical Reviews, Vol. 107(6), p. 2411, 2007.
2. G. Dull, Chem. Zeitg, Vol. 19, p. 216, 1895.
3. J. Kiermayer, Chem. Zeitg, Vol. 19, p. 1003, 1895.
4. L. Yan, W. Zuojun, C. Chuanjie, and L. Yingxin, *Preparation of 5-hydroxymethylfurfural by dehydration of carbohydrates.* Progress in Chemistry, Vol. 22(8), p. 1603, 2010.
5. X. Tong, Y. Ma, and Y. Li, *Biomass into chemicals: Conversion of sugars to furan derivatives by catalytic processes.* Applied Catalysis A: General, Vol. 385(1–2), p. 1, 2010.
6. M.E. Zakrzewska, E. Bogel-Łukasik, and R. Bogel-Łukasik, *Ionic Liquid-Mediated Formation of 5-Hydroxymethylfurfural—A Promising Biomass-Derived Building Block.* Chemical Reviews, Vol. 111(2), p. 397, 2011.
7. L.D. Cottier, and G. Trend, Heterocycl. Chem, Vol. 2, p. 223. 1991.
8. J. Lewkowski, *Synthesis, chemistry and applications of 5-hydroxymethyl-furfural and its derivatives.* Arkivoc, Vol. 2001(1), p. 17, 2001.
9. H. Shiyong, W. Fuli, and P. Lixia, *Synthesis of 5-hydroxymethylfurfural via dehydration of fructose.* Progress in Chemistry, Vol. 21(7–8), p. 1442, 2009.
10. L.R. Lynd, P.J. Weimer, W.H. van Zyl, and I.S. Pretorius, *Microbial cellulose utilization: fundamentals and biotechnology.* Microbiology and Molecular Biology Reviews, Vol. 66(3), p. 506, 2002.
11. H.E. Van Dam, A.P.G. Kieboom, and H. Van Bekkum, Starch-Starke, Vol. 38, p. 1995, 1986.
12. M.J. Antal, W. S. L. Mok, and G. N. Richards, Carbohydr Res, 199, p. 91. 1990.
13. B.F.M. Kuster, Starch-Starke, Vol. 42, p. 314, 1990.
14. B.F.M. Kuster, and L. M. Tebbens, Carbohydr Res, Vol. 54, p. 159, 1977.
15. Y. Román-Leshkov, J.N. Chheda, and J.A. Dumesic, *Phase modifiers promote efficient production of hydroxymethylfurfural from fructose.* Science, Vol. 312, p. 1933, 2006.
16. J.N. Chheda, Y. Román-Leshkov, and J.A. Dumesic, *Production of 5-hydroxymethylfurfural and furfural by dehydration of biomass-derived mono- and poly-saccharides.* Green Chemistry, Vol. 9(4), p. 342, 2007.
17. R.M. Musau, and R.M. Munavu, *The preparation of 5-hydroxymethyl-2-furaldehyde (HMF) from d-fructose in the presence of DMSO.* Biomass, Vol. 13(1), p. 67, 1987.
18. D.W. Brown, A.J. Floyd, R.G. Kinsman, and Y. Roshan-Ali, J. Chem.Technol. Biotechnol, Vol. 32, p. 920, 1982.
19. B.M. Smythe, and C.J. Moye, US Patent 3290263, 1965.
20. A.S. Amarasekara, L.D. Williams, and C.C. Ebede, *Mechanism of the dehydration of d-fructose to 5-hydroxymethylfurfural in dimethyl sulfoxide at 150°C: an NMR study.* Carbohydrate Research, Vol. 343(18), p. 3021, 2008.
21. C. Fayet, and J. Gelas, *Nouvelle méthode de préparation du 5-hydroxyméthyl-2-furaldéhyde par action de sels d'ammonium ou d'immonium sur les mono-, oligo- et poly-saccharides. Accès direct aux 5-halogénométhyl-2-furaldéhydes.* Carbohydrate Research, Vol. 122(1), p. 59, 1983.
22. C. Lansalot-Matras, and C. Moreau, *Dehydration of fructose into 5-hydroxymethylfurfural in the presence of ionic liquids.* Catalysis Communications, Vol. 4(10), p. 517, 2003.

23. C. Moreau, A. Finiels, and L. Vanoye, *Dehydration of fructose and sucrose into 5-hydroxymethylfurfural in the presence of 1-H-3-methyl imidazolium chloride acting both as solvent and catalyst.* Journal of Molecular Catalysis A: Chemical, Vol. 253(1–2), p. 165, 2006.

24. S. Hu, Z. Zhang, Y. Zhou, B. Han, H. Fan, W. Li, J. Song, and Y. Xie, *Conversion of fructose to 5-hydroxymethylfurfural using ionic liquids prepared from renewable materials.* Green Chemistry, Vol. 10(12), p. 1280, 2008.

25. L. Lai, and Y. Zhang, *The effect of imidazolium ionic liquid on the dehydration of fructose to 5-Hydroxymethylfurfural, and a room temperature catalytic system.* ChemSusChem, Vol. 3(11), p. 1257, 2010.

26. C. Li, Z. K. Zhao, A. Wang, M. Zheng, and T. Zhang, *Production of 5-hydroxymethylfurfural in ionic liquids under high fructose concentration conditions.* Carbohydrate Research, Vol. 345(13), p. 1846, 2010.

27. X. Qi, M. Watanabe, T.M. Aida, and R. L. Smith Jr., *Efficient process for conversion of fructose to 5-hydroxymethylfurfural with ionic liquids.* Green Chemistry, Vol. 11(9), p. 1327, 2009.

28. F. Du, X. H. Qi, Y. Z. Xu, and Y.Y. Zhuang, *Catalytic conversion of fructose to 5-hydroxymethylfurfural by ion-exchange resin in ionic liquid.* Gaodeng Xuexiao Huaxue Xuebao/Chemical Journal of Chinese Universities, Vol. 31(3), p. 548, 2010.

29. M. Chidambaram, and A.T. Bell, *A two-step approach for the catalytic conversion of glucose to 2,5-dimethylfuran in ionic liquids.* Green Chemistry, Vol. 12(7), p. 1253, 2010.

30. Y.Y. Zhang, J.H. Deng, Y.H. Shang, Y.S. Zhou, G.H. Wang, and L.M. Li, *Catalytic conversion of fructose into 5-hydroxymethylfurfural with sulfonic group bearing ionic liquid immobilized on micro-ball silica-gel.* Gao Xiao Hua Xue Gong Cheng Xue Bao/Journal of Chemical Engineering of Chinese Universities, Vol. 24(5), p. 824, 2010.

31. X. Tong, Y. Ma, and Y. Li, *An efficient catalytic dehydration of fructose and sucrose to 5-hydroxymethylfurfural with protic ionic liquids.* Carbohydrate Research, Vol. 345(12), p. 1698, 2010.

32. X. Tong, and Y. Li, *Efficient and selective dehydration of fructose to 5-hydroxymethylfurfural catalyzed by brønsted-acidic ionic liquids.* ChemSusChem, Vol. 3(3), p. 350, 2010.

33. S. Hu, Z. Zhang, Y. Zhou, J. Song, H. Fan, and B. Han, *Direct conversion of inulin to 5-hydroxymethylfurfural in biorenewable ionic liquids.* Green Chemistry, Vol. 11(6), p. 873, 2009.

34. X. Qi, M. Watanabe, T.M. Aida,and R.L. Smith Jr, , *Efficient one-pot production of 5-hydroxymethylfurfural from inulin in ionic liquids.* Green Chemistry, Vol. 12(10), p. 1855, 2010.

35. M. Benoit, Y. Bissonett, E. Guelou, K. De-Oliveira-Vigier, J. Barrault, and F. Jerome, *Acid-catalyzed dehydration of fructose and inulin with glycerol or glycerol carbonate as renewably sourced co-solvent.* ChemSusChem, Vol. 3(11), p. 1304, 2010.

36. H. Zhao, J. E. Holladay, H. Brown, and Z.C. Zhang, *Metal chlorides in ionic liquid solvents convert sugars to 5-hydroxymethylfurfural.* Science, Vol. 316(5831), p. 1597, 2007.

37. G. Yong, Y. Zhang, and J.Y. Ying, *Efficient catalytic system for the selective production of 5-hydroxymethylfurfural from glucose and fructose.* Angewandte Chemie - International Edition, Vol. 47(48), p. 9345, 2008.

38. S. Hu, Z. Zhang, J. Song, Y. Zhou, and B. Han, *Efficient conversion of glucose into 5-hydroxymethylfurfural catalyzed by a common Lewis acid $SnCl_4$ in an ionic liquid.* Green Chemistry, Vol. 11(11), p. 1746, 2009.

39. A.S. Amarasekara, and C.C. Ebede, *Zinc chloride mediated degradation of cellulose at 200°C and identification of the products.* Bioresource Technology, Vol. 100(21), p. 5301, 2009.

40. Y. Su, H.M. Brown, X. Hunag, X.D. Zhou, J.E. Amonette, and Z.C. Zhang, *Single-step conversion of cellulose to 5-hydroxymethylfurfural (HMF), a versatile platform chemical.* Applied Catalysis A: General, Vol. 361(1–2), p. 117, 2009.

41. C. Li, Z. Zhang, and Z.K. Zhao, *Direct conversion of glucose and cellulose to 5-hydroxymethylfurfural in ionic liquid under microwave irradiation.* Tetrahedron Letters, Vol. 50(38), p. 5403, 2009.

42. S. Wu, Z. Wang, Y. Gao, Y. Zhang, D. Ma, and Z. Zhao, *Production of 5-hydroxymethylfurfural from cellulose catalyzed by Lewis acid under microwave irradiation in ionic liquid.* Cuihua Xuebao/Chinese Journal of Catalysis, Vol. 31(9), p. 1157, 2010.

43. A.S. Amarasekara, and O.S. Owereh, *Hydrolysis and decomposition of cellulose in Brönsted acidic ionic liquids under mild conditions.* Industrial and Engineering Chemistry Research, Vol. 48(22), p. 10152, 2009.

44. F. Tao, H. Song, and L. Chou, *Hydrolysis of cellulose by using catalytic amounts of $FeCl_2$ in Ionic liquids.* ChemSusChem, Vol. 3(11), p. 1298, 2010.

45. C.V. McNeff, D.T. Nowlan, L.C. McNeff, B. Yan, and R.L. Fedie, *Continuous production of 5-hydroxymethylfurfural from simple and complex carbohydrates.* Applied Catalysis A: General, Vol. 384(1–2), p. 65, 2010.

46. Z. Zhang, and Z.K. Zhao, *Microwave-assisted conversion of lignocellulosic biomass into furans in ionic liquid.* Bioresource Technology, Vol. 101(3), p. 1111, 2010.

47. J.B. Binder, and R.T. Raines, *Simple chemical transformation of lignocellulosic biomass into furans for fuels and chemicals.* Journal of the American Chemical Society, Vol. 131(5), p. 1979, 2009.

48. H. Koch, and J. Pein, *Condensation reactions between phenol, formaldehyde and 5-hydroxymethylfurfural, formed as intermediate in the acid catalyzed dehydration of starchy products.* Polymer Bulletin, Vol. 13(6), p. 525, 1985.

49. A. Gandini, and M.N. Belgacem, *Furans in polymer chemistry.* Progress in Polymer Science (Oxford), Vol. 22(6), p. 1203, 1997.

50. A. Gandini, *Furans as offspring of sugars and polysaccharides and progenitors of a family of remarkable polymers: A review of recent progress.* Polymer Chemistry, Vol. 1(3), p. 245, 2010.

51. A. Gandini, and M.N. Belgacem, *Furan Derivatives and Furan Chemistry at the Service of Macromolecular Materials*, in *Monomers, Polymers and Composites from Renewable Resources*, B. Mohamed Naceur and G. Alessandro, Editors. Elsevier: Amsterdam. p. 115–152. 2008.

52. T.G. Utne, D. John, and R.E. Jones, *Hydrogenation of 5-hydroxymethylfurfural.* US Patent, US 3083236 19630326. 1963.

53. W.J. McKillip, and E. Sherman, Kirk-Othmer Encyclopedia of Chemical Technology, ed. M. Grayson. John Wiley & Sons, Inc., New York, Vol. 11, p. 499, 1981.

54. C. Carlini, P. Patrono, A.M.R. Galletti, G. Sbrana, and V. Zima, *Selective oxidation of 5-hydroxymethyl-2-furaldehyde to furan-2,5-dicarboxaldehyde by catalytic systems based on vanadyl phosphate.* Applied Catalysis A: General, Vol. 289(2), p. 197, 2005.

55. G.A. Halliday, R.J. Young Jr, and V.V. Grushin, *One-pot, two-step, practical catalytic synthesis of 2,5-diformylfuran from fructose.* Organic Letters, Vol. 5(11), p. 2003, 2003.

56. A.S. Amarasekara, D. Green, and E. McMillan, *Efficient oxidation of 5-hydroxymethylfurfural to 2,5-diformylfuran using Mn(III)-salen catalysts.* Catalysis Communications, Vol. 9(2), p. 286, 2008.

57. A. Gandini, A.J.D. Silvestre, C.P. Neto, A.F. Sousa, and M. Gomes, *The furan counterpart of polyethylene terephthalate: An alternative material based on renewable resources.* Journal of Polymer Science, Part A: Polymer Chemistry, Vol. 47(1), p. 295, 2009.

58. T. Werpy, and G. Peterson, *Top value added chemicals from biomass, No. DOE/ GO-102004–1992*, US Department of Energy, Office of Scientific and Technical Information, http://www.nrel.gov/docs/fy04osti/35523.pdf. 2004.

59. H. van Bekkum, Ed. F.W. Lichtenthaler, *Studies on Selective Carbohydrate Oxidation,.* VCH: Weinham, 1991.

60. P. Vinke, *Ph.D. Thesis, Technical University in Delft; Delft,* 1991.

61. P. Vinke, H.E. van Dam, and H. van Bekkum, *Platinum Catalyzed Oxidation of 5-Hydroxymethylfurfural.* Studies in Surface Science and Catalysis, Vol. 55, p. 147, 1990.

62. B.W. Lew, *(Atlas Chem.Ind.), US Patent, 3326944.* 1967.

63. P. Verdeguer, N. Merat, and A. Gaset, *Oxydation catalytique du HMF en acide 2,5-furane dicarboxylique.* Journal of Molecular Catalysis, Vol. 85(3), p. 327, 1993.

64. M. Kröger, U. Prüße, and K.D. Vorlop, *A new approach for the production of 2,5-furandicarboxylic acid by in situ oxidation of 5-hydroxymethylfurfural starting from fructose.* Topics in Catalysis, Vol. 13(3), p. 237, 2000.

65. W. Partenheimer, and V.V. Grushin, *Synthesis of 2,5-Diformylfuran and Furan-2,5-Dicarboxylic Acid by Catalytic Air-Oxidation of 5-Hydroxymethylfurfural. Unexpectedly Selective Aerobic Oxidation of Benzyl Alcohol to Benzaldehyde with Metal/Bromide Catalysts.* Advanced Synthesis and Catalysis, Vol. 343(1), p. 102, 2001.

66. O. Casanova, S. Iborra, and A. Corma, *Biomass into chemicals: Aerobic oxidation of 5-hydroxymethyl-2-furfural into 2,5-furandicarboxylic acid with gold nanoparticle catalysts.* ChemSusChem, Vol. 2(12), p. 1138, 2009.

67. Y.Y. Gorbanev, S.K. Klitgaard, J.M. Woodlay, C.H. Christensen, and A. Riisager, *Gold-catalyzed aerobic oxidation of 5-hydroxymethylfurfural in water at ambient temperature.* ChemSusChem, Vol. 2(7), p. 672, 2009.

68. M.A. Lilga, R.T. Hallen, and M. Gray, *Production of Oxidized Derivatives of 5-Hydroxymethylfurfural (HMF).* Topics in Catalysis, Vol. 53, p. 1264, 2010.

69. A. Boisen, C.H. Christensen, W. Fu, Y.Y. Gorbanev, T.S. Hansen, J.S. Jensen, S.K. Klitgaard, S. Pederson, T. Staahlberg, J.M. Woodlay, and A. Riisager, *Process integration for the conversion of glucose to 2,5-furandicarboxylic acid.* Chemical Engineering Research and Design, Vol. 87(9), p. 1318, 2009.

70. D. Chundury, and H.H. Szmant, *Preparation of polymeric building blocks from 5-hydroxymethyl- and 5-chloromethylfurfuraldehyde.* Industrial and Engineering Chemistry Product Research and Development, Vol. 20(1), p. 158, 1981.

71. O. Casanova, S. Iborra, and A. Corma, *Chemicals from biomass: Etherification of 5-hydroxymethyl-2-furfural (HMF) into 5,5'(oxy-bis(methylene))bis-2-furfural (OBMF) with solid catalysts.* Journal of Catalysis, Vol. 275(2), p. 236, 2010.

72. J. Quillerou, M.N. Belgachem, A. Gandini, J. Rivero, and G. Roux, *Urethanes and polyurethanes bearing furan moieties - 1. Synthesis and characterization of monourethanes.* Polymer Bulletin, Vol. 21(6), p. 555, 1989.

73. H. Takeda, *Preparation of furandicarboxylic chloride in high purity and with high efficiency.* Jpn. Kokai Tokkyo Koho, Japanese Patent. 2009298753 A 20091224. 2009.

74. Z. Hui, and A. Gandini, *Polymeric schiff bases bearing furan moieties.* European Polymer Journal, Vol. 28(12), p. 1461, 1992.

75. Méalares, C. and A. Gandini, *Polymeric Schiff bases bearing furan moieties 2. Polyazines and polyazomethines.* Polymer International, Vol. 40(1), p. 33, 1996.

76. J.G.N. Drewitt, and J. Lincoln, *US Patent.* US 2551731 19510508, 1951.

77. J.A. Moore, and J.E. Kelly, *Polyesters derived from furan and tetrahydrofuran nuclei.* Macromolecules, Vol. 11(3), p. 568, 1978.

78. O. Grosshardt, U. Fehernbacher, K. Kowollik, B. Tubke, N. Dingenouts, and M. Wilhelm, *Synthese und Charakterisierung von Polyestern und Polyamiden auf der Basis von Furan-2, 5-dicarbonsäure.* Chemie-Ingenieur-Technik, Vol. 81(11), p. 1829, 2009.

79. A. Khrouf, S. Boufi, R. El Gharbi, N.M. Belgacem, and A. Gandini, *Polyesters bearing furan moieties: 1. Polytransesterification involving difuranic diesters and aliphatic diols.* Polymer Bulletin, Vol. 37(5), p. 589, 1996.

80. A. Khrouf, M. Abid, S. Boufi, R. El Gharbi, and A. Gandini, *Polyesters bearing furan moieties, 2: A detailed investigation of the polytransesterification of difuranic diesters with different diols.* Macromolecular Chemistry and Physics, Vol. 199(12), p. 2755, 1998.

81. M. Okada, K. Tachikawa, and K. Aoi, *Biodegradable polymers based on renewable resources. II. Synthesis and biodegradability of polyesters containing furan rings.* Journal of Polymer Science, Part A: Polymer Chemistry, Vol. 35(13), p. 2729, 1997.

82. B.G. Girija, R.R.N. Sailaja, and G. Madras, *Thermal degradation and mechanical properties of PET blends.* Polymer Degradation and Stability, Vol. 90(1), p. 147, 2005.

83. M.P.J. Van Deurzen, F. Van Rantwijk, and R.A. Sheldon, *Chloroperoxidase-catalyzed oxidation of 5-hydroxymethylfurfural.* Journal of Carbohydrate Chemistry, Vol. 16(3), p. 299, 1997.

84. H. Hirai, *Oligomers from hydroxymethylfurancarboxylic acid.* Journal of macromolecular science. Chemistry, Vol. A21(8–9), p. 1165, 1983.

85. J.A. Moore, and J. E. Kelly, *Polyhydroxymethylfuroate [poly(2,5-furandiylcarbonyloxymethylene)]* Journal of Polymer Science: Polymer Chemistry Edition, Vol. 22(3), p. 863, 1984.

86. K. Matsuda, *Manufacture of poly(ethylene-2,5-furandicarboxylate) without staining using scandium triflate as polymn. catalyst.* Japanese Patent, Kokai Tokkyo Koho (2009), JP 2009215467 A 20090924. 2009.

87. W. Dong, C. Mingqing, D. Lan, Z. Ni, H. Yin, and J. Zhao, *Method for preparation of 2,5-furandicarboxylic acid-based polyester*. Chinese Patent, Faming Zhuanli Shenqing (2010), CN 101899145 A 20101201. 2010.

88. K. Yutaka, T. Miura, S. Eritachi, and T. Komuro, *Polyester resin, its manufacture, and molded article*. Japanese Patent, Jpn. Kokai Tokkyo Koho (2010), JP 2010280767 A 20101216. 2010.

89. S.Eritate, T. Miura., T. Komuro, and K. Yutaka, *Manufacture of polyester resins and molded products therewith*. Int Patent, PCT Int. Appl. (2010), WO 2010140599 A1 20101209. 2010.

90. R. Storbeck, and M. Ballauff, *Synthesis and properties of polyesters based on 2,5-furandicarboxylic acid and 1,4:3,6-dianhydrohexitols*. Polymer, Vol. 34(23), p. 5003, 1993.

91. R. Storbeck, M. Rehahn, and M. Ballauff, Makromol. Chem., Vol. 194, p. 53, 1993.

92. A. Ivanov, A.E. Primellese, and H.P. Aluija, *Synthesis and physical-chemical properties of poly(2,5-furandimethylene 2,5-furandicarboxylate)*. Revista sobre los Derivados de la Cana de Azucar, Vol. 9(2), p. 48, 1975.

93. A. A. Ivanov, A.E. Primelles, V.D. Quintana, G.J. Aleman, N.M. Exposito, D.C.L. Fernandez, S.M.D.P. Chaviano, and R.C. Hernandez, *Polyesters from 2,5-dihydroxymethylfuran and aliphatic acid chlorides*. Revista sobre los Derivados de la Cana de Azucar, Vol. 9(3), p. 15, 1975.

94. A.A. Ivanov, R. Delgado, and M. Fernandez, *Study of the acceptor-catalytic polycondensation of 2,5-dihydroxymethylfuran and chloroterephthalic acid*. CENTRO, Serie: Quimica y Tecnologia Quimica, Vol. 3(1), p. 85, 1975.

95. K. Hatanaka, D. Yoshida, K. Okuyama, A. Miyagawa, K. Tamura, N. Sato, K. Hashimoto, M. Sagehashi, and A. Sakoda, *Synthesis of polyesters from 5-hydroxymethyl-2-furfural as a starting material*. Kobunshi Ronbunshu, Vol. 62(7), p. 316, 2005.

96. K.A. Malyshevskaya, *Synthesis and study of polyamides from 2,5-furandicarboxylic acid and hexamethylenediamine*. Materialy Konf. po Itogam Nauchn.-Issled. Rabot za 1964 god, Sibirsk. Tekhnol. Inst., Krasnoyarsk, USSR 1964, p. 71–74. 1964.

97. K.A. Malyshevskaya, N. A. Tyukavniva, T.I. Aslapovskaya, A.I. Laletin, and A. F. Yuzepol'skii, *Synthesis of mixed polyamides based on 2,5-furandicarboxylic acid*. Fiz., Khim. Khim. Tekhnol., p. 237–239. 1969.

98. M.G. Grishkova, V. Trostnistkaya, K. A. Malyshevskaya, and N. A. Tyukavkina, *Synthesis of 2,5-furandicarboxylic acid based polyamides by equilibrium polycondensation*. Fiz., Khim. Khim. Tekhnol., Sb. Mater. Nauch.-Tekh. Konf. Rab. Nauki Proizvod, p. 240–244. 1969.

99. M. Russo, *New Heterocyclic Polyamides with High Heat Resistance and Mechanical Strength*. Kunststoffe, Vol. 65(6), p. 346, 1975.

100. N. Ogata, and K. Shimamura, *Active polycondensation of diesters having heterocyclic nuclei*. Polymer Journal, Vol. 7(1), p. 72, 1975.

101. A. Mitiakoudis, A. Gandini, and H. Cheradame, *Polyamides containing furanic moieties*. Polym. Commun, Vol. 26, p. 246, 1985.

102. A. Mitiakoudis, and A. Gandini, *Synthesis and characterization of furanic polyamides*. Macromolecules, Vol. 24(4), p. 830, 1991.

103. H.P. Benecke, A.W. Kawczak, and D.B. Garbark, *Furan-2,5-dicarboxylic acid-modified aromatic amine curing agents and chain extenders*. U.S. Pat. Appl. Publ, US 20080207847 A1 20080828, 2008.

104. H.P. Benecke, A. W. Kawczak, and D.B. Garbark, *Furanic-modified amine-based curatives, curative manufacture and control of cure time or potlife.* U.S. Pat. Appl. Publ, US 20100280186 A1 20101104. 2010.

105. M.N. Belgacem, J. Quillerou, A. Gandini, J. Rivero, and G. Roux, *Urethanes and polyurethanes bearing furan moieties-2. comparative kinetics and mechanism of the formation of furanic and other monourethanes.* European Polymer Journal, Vol. 25(11), p. 1125, 1989.

106. M.N. Belgacem, J. Quillerou, and A. Gandini, *Urethanes and polyurethanes bearing furan moieties-3. Synthesis, characterization and comparative kinetics of the formation of diurethanes.* European Polymer Journal, Vol. 29(9), p. 1217, 1993.

107. S. Boufi, M.N. Belgacem, J. Quillerou, and A. Gandini, *Urethanes and polyurethanes bearing furan moieties. 4. Synthesis, kinetics, and characterization of linear polymers.* Macromolecules, Vol. 26(25), p. 6706, 1993.

108. S. Boufi, A. Gandini, and M.N. Belgacem, *Urethanes and polyurethanes bearing furan moieties: 5. Thermoplastic elastomers based on sequenced structures.* Polymer, Vol. 36(8), p. 1689, 1995.

109. A. H. Kehayoglou, G. P. Karayannidis, and S-K. Irini, *Aromatic poly[(amino) amide]s and poly(imidazo[4,5-f]benzimidazole)s derived from poly[(tosylamino) amide]s. 1.* Makromolekulare Chemie, Vol. 193(2), p. 293, 1982.

110. G. P. Karayannidis, and A. H. Kehayoglou, *A two-stage polycondensation of 2,5-furandicarbonyl dichloride with N1,N5-diphenyl-1,2,4,5-tetraaminobenzene.* Chimika Chronika, Vol. 11(2), p. 149, 1982.

111. S.A. Aljia Afli, and R. El Gharbi, *All furanic polyhydrazides and corresponding-polyoxadiazoles.* e-Polymers 2008, p. 064, www.e-polymers.org, 2008.

112. A.S. Amarasekara, D. Green, and L.D. Williams, *Renewable resources based polymers: Synthesis and characterization of 2,5-diformylfuran-urea resin.* European Polymer Journal, Vol. 45(2), p. 595, 2009.

113. A. Wayne Cooke, and K.B. Wagener, *An investigation of polymerization via reductive coupling of carbonyls.* Macromolecules, Vol. 24(6), p. 1404, 1991.

114. A.A. Ivanov, A.E. Primelles, and J. Bango, *Synthesis and properties of some 2,5-furandicarboxylic acid polymers.* CENTRO, Serie: Quimica y Tecnologia Quimica, Vol. 3(1), p. 73, 1975.

115. A.A . Ivanov, A.E. Primelles, and F.A. Gil, *Synthesis and some characteristics of 2,5-furandicarboxylic acid polymers.* Revista sobre los Derivados de la Cana de Azucar, Vol. 10(2), p. 27, 1976.

116. A.H. Kehayoglou, *Polycondensation of 2,5-furandicarboxylic acid chloride with aromatic diaminodicarboxylic acids.* Chimika Chronika, Vol. 5(4), p. 303, 1976.

117. N. Yoshida, N. Kasuya, N. Haga, and K. Fukuda, *Brand-new biomass-based vinyl polymers from 5-hydroxymethylfurfural.* Polymer Journal, Vol. 40(12), p. 1164, 2008.

10

Natural Polymers—A Boon for Drug Delivery

Rajesh. N[1], Uma. N[2] and Valluru Ravi[3]

[1]*Department of Biochemistry, Holdsworth Memorial Hospital & College,*
Mysore, Karnataka, India
[2]*Department of Biochemistry, BGS International foundation for health science,*
Mysore, Karnataka, India
[3]*Department of Pharmaceutics, JSS College of Pharmacy, JSS University,*
Mysore, Karnataka, India

Abstract

Biopolymers like natural gums and polysaccharides are now extensively used for the development of dosage forms for delivery of drugs. Although natural polymers and their derivatives are used widely in pharmaceutical dosage forms, their use of biodegradable polymeric materials to deliver drugs has been hampered by the synthetic materials. These natural polysaccharides hold advantages over synthetic polymers, generally because they are nontoxic, less expensive and freely available. Biopolymers can be modified to have tailor-made materials for drug delivery systems and thus can compete with synthetic biodegradable materials available in the market. In this review article, recent efforts and approaches exploiting these natural materials in developing drug delivery systems are discussed.

Keywords: Natural gums, polysaccharides, pharmaceutical dosage, drug delivery systems

10.1   Introduction

Advances in polymer science have led to the development of several novel drug-delivery systems. A proper consideration of surface and bulk properties can aid in the designing of polymers for various drug-delivery applications. Natural polymers have been

Vikas Mittal (ed.) Renewable Polymers, (429–472) © Scrivener Publishing LLC

widely used in biomedical applications because of their known biocompatibility and biodegradability. These improvements contribute to make medical treatment more efficient and to minimize side effects and other types of inconveniences for patients.

The newer technological growth widely covers drug modification by chemical means, carrier based drug delivery and drug entrapment in polymeric matrices or within pumps that are placed in desired bodily compartments. The technical development in drug delivery/targeting approaches improve the efficacy of drug therapy thereby improve human health. Polymer chemists and chemical engineers, pharmaceutical scientists are engaged in bringing out design predictable, controlled delivery of bio active agents. Natural polysaccharides and their derivatives represent a group of polymers widely used in pharmaceutical dosage forms. Various kinds of natural gums are used in the food industry and are regarded as safe for human consumption. These polysaccharides are obtained usually as plant exudates containing various sugars other than glucose and having significant quantities of oxidized groups in adjunct to their normal polyhydroxy format. In many cases, water-soluble polysaccharides generally similar to the exudates are components of land and marine plants and their seeds. These materials result from normal metabolic processes, and many times, they represent the reserve carbohydrates in that system.

It is established that hydrophilic polymers release freely soluble drugs at a fairly constant rate [1]. Various synthetic polymers (e.g, cellulose ethers, polyalkyl methacrylates etc.,) used for this purpose have been reviewed [2–5]. These polymers, when come in contact with water, are hydrated and form a gel. Natural gums (like agar in the form of beads and konjac in the form of cylinders) have also been examined as matrices for the sustained release of drugs [6, 7]. When natural gums in the form of compressed tablets are placed in water from the medium and form a gel before they dissolve in the medium. If a drug is contained in the tablet it is expected to be released through the gel layer, and sustained release may be achieved.

Natural gums are often preferred to synthetic materials due to their non-toxicity, low cost and free availability. It should be noted that many "old" materials compete successfully today after almost a century of efforts to replace them. It is the usual balance of economics and performance that determines the commercial realities natural gums have been modified to overcome certain drawbacks like uncontrolled rate of hydration, thickening, and drop in viscosity on storage, microbial contamination etc., [8]. This review provides a comprehensive review of the area of natural and modified gums as well as polysaccharides used as carriers in the sustained release of drugs.

10.2   Acacia

Gum acacia or gum arabic is the dried gummy exudate obtained from the stems and branches of *Acacia senegal* and other related species of acacia (Leguminosae family). Acacia gum occurs as white or yellowish-white spheroidal tears of varying sizes or as angular fragments and is sometimes mixed with darker fragments. It has a molecular weight ranging between 2,40,000 and 5,80,000. It consists mainly of high molecular weight polysaccharides and their calcium, magnesium and potassium salts, which on hydrolysis yields L-arabinose, L-rhamnose, D-galactose and an aldobionic acid containing D-glucuronic acid and D-galactose. It is insoluble in alcohol, but completely soluble in twice its weight of water at room temperature [9, 10].

Acacia is mainly used in oral and topical phramacetical formulations as a suspending and emulsifying agent, often in combination with tragacanth. It is also used in the preparation of pastilles and lozenges and as a tablet binder [11]. Its use as sustained release carrier has been investigated by Baveja *et al.* [12]. Matrix tablet made using gum arabic and drug in a ratio of 4:1 completely dissolved within 2 hrs.

Ray *et al.* [13] prepared gum Arabic pellets from which sustained release of ferrous sulfate was achieved for 7 hrs. Release was further sustained by coating the pellets with polyvinyl acetate and ethylene vinyl acetate respectively. An increase in the amount of gum arabic in the pellets decreased the rate of release due to the gelling property of gum arabic. The gel layer acted as a barrier and retarded the rate of diffusion of ferrous sulfate through the pellet. In coated pellets, an increase in thickness of membrane helped to sustain the release of ferrous sulfate for a longer duration. Further a blend of synthetic polyvinyl alcohol and the natural macromolecule gum arabic was characterized [14]. Characterization of these blends by nuclear magnetic resonance (NMR), differential scanning calorimetry (DSC) and viscoelastic studies revealed a blend by composition with synergistic properties. This blend composition was used to release various antimicrobial drugs. The duration and release of the drug depended on the amount of drug loaded in the matrix, and the solubility of drug in the matrix and release medium. By this system, the release kinetics of the drug can be tailored by adjusting plasticizer, homopolymer and cross-linker composition depending on the type of drug to be released.

M. M. Meshali *et al.* [15] investigated the interpolymer complexes of chitosan with pectin and acacia using viscosity measurements. The binding ratios of the complexes were found to be 1:10 and 1:20 for chitosan-pectin and chitosan-acacia, respectively. The solid complexes were separated and dried to be used as tablet matrices. The release of the water-soluble chlorpromazine HCl from tablets showed that increase in the concentrations of the polymers, complexes or the physical mixtures in the tablets retarded drug release. Horomitsu Aoki *et al.* [16] examined the effect of gum concentration, molecular weight parameters and homogenization conditions on the droplet size and stability of emulsions prepared using a conventional *A. senegal* and *A. senegal* test gums prepared by controlled maturation process. The commercial product (Acacia (sen) SUPER GUM™ EM2) at a concentration below 5% w/w produced emulsions better than the emulsions produced by using 20% w/w of commercial grade.

10.3   Agar

Agar is an unbrranched polysaccharide obtained from the cell walls of some species of red algae or sea weed. Predominant agar-producing genera are *Gelidium, Gracilaria, Acanthopeltis, Ceramium, Pterocladia* found in Pacific and Indian oceans [10]. Chemically agar is a polymer made up of subunits of the sugar galactose (Figure 10.1). When dissolved in hot water and cooled, agar becomes gelatinous. Agar has been used as culture media in microbiology and agarose, which is a purified form of agar, has been employed extensively in separation and purification in biochemistry. Other uses are as a laxative, a thickener for soups, in jellies, ice cream, as a clarifying agent in brewing, and for paper sizing fabrics.

Possible use of agar for sustained release of sulfamethiazole as beads has been investigated [17]. The release profile from the beads showed that the agar beads exhibited sustained release of drug.

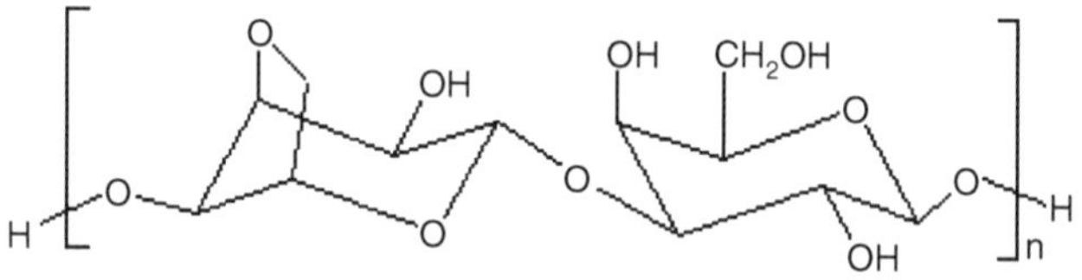

Figure 10.1 Chemical structure of agarose.

Desai *et al.* [18] prepared a novel floating controlled-release drug delivery system of theophylline using agar. The tablet maintained constant theophylline levels of about 2 mg/mL for 24 hrs, which may be attributable to the release from the agar gel matrix and the buoyancy of the tablet in the stomach. The use of agar as filler material in a hot melt extruded polymer matrix was studied [19]. Agar and microcrystalline cellulose were used as the filler materials in varying ratios, to examine the effect of percentage of filler content as well as filler type on the properties of the hot melt extruded matrix. Dissolution analysis showed that the presence of the fillers resulted in a slower release rate of API than for the matrix alone, indicating that agar is viable filler for extended release hot melt produced dosage forms.

10.4 Alginate

Alginic acid is a naturally occurring hydrophilic colloidal polysaccharide obtained from the various species of brown seaweed (*Phaeophyceae*). It is a linear copolymer consisting mainly of residues of β-1,4-linked D-mannuronic acid and α-1,4-linked L-glucuronic acid (Figure 10.2). These monomers are often arranged in homopolymeric blocks separated by regions approximating an alternating sequence of the two acid monomers. It occurs as white to yellowish brown filamentous, grainy, granular or powdered forms. Alginate is easily gelled in the presence of divalent cations, however Gelling depends on the ion binding ($Mg^{2+} < Ca^{2+} < Sr^{2+} < Ba^{2+}$). The gelation or cross-linking is due to the stacking of the guluronic acid blocks of alginate chains. Calcium alginate beads can be prepared by dropwise addition of the solution of sodium alginate into the solution of calcium chloride. Alginate is known to be nontoxic when taken orally and also to have a protective effect on the mucous membrane of the upper gastrointestinal tract. Since dried alginate beads reswell, they can act as controlled-release systems.

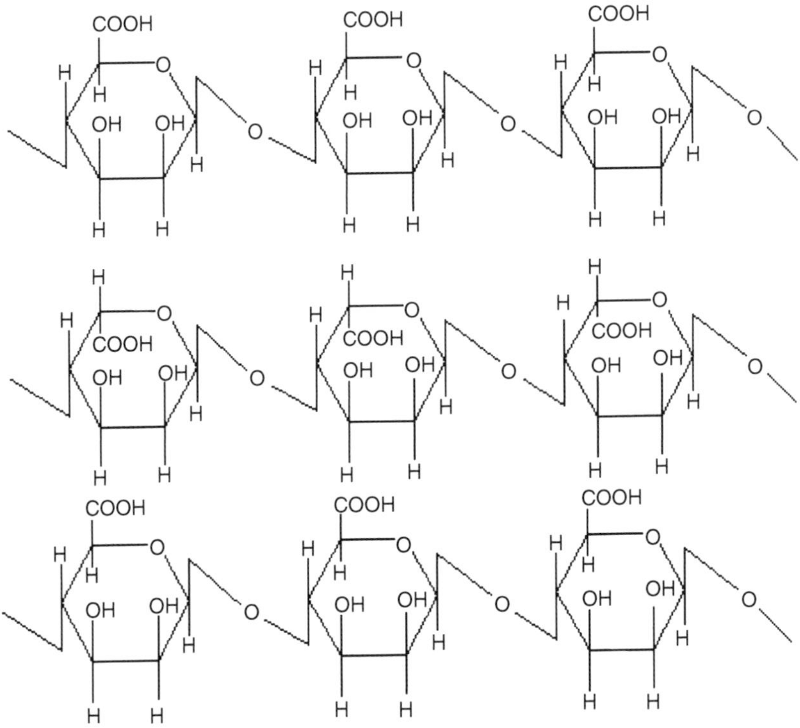

Figure 10.2 Structure of alginic acid.

However, porosity gives alginate beads not only a fast release pattern of incorporated drugs, but also very low efficiency of incorporation of low molecular weight drugs, except for sparingly soluble drugs [20]. Therefore it appears that alginate beads could be used for a controlled-release system of macromolecular drugs or low molecular weight drugs bound to macromolecules through covalent or non covalent bonds. The influence of erosion of calcium-induced alginate gel matrix on the release of brilliant blue has been reported [21]. Control of drug release from alginate gel beads by application of a complex formed between chondroitin sulfate and chitosan was investigated [22]. The complex suppressed the disintegration of the gel beads, and the release pattern of diclofenac incorporated within them was obviously changed.

Aerosol OT™ (AOT)-alginate nanoparticles were formulated using emulsion-crosslinking technology [23]. Nanoparticles, drug carriers in the sub-micron size range, can enhance the therapeutic efficacy of encapsulated drug by increasing and sustaining the delivery of the drug inside the cell. Rhodamine and doxorubicin were used as model water-soluble molecules. AOT-alginate nanoparticles demonstrated sustained release of doxorubicin over a 15-day period *in vitro*. Cell culture studies indicated that nanoparticles enhanced the cellular delivery of rhodamine by about two-tenfold compared to drug in solution.

Polymeric blend beads of acrylamide grafted polyvinyl alcohol (PVA-g-PAAm) and PVA with sodium alginate (NaAlg) were prepared by cross-linking with glutaraldehyde (GA), which was used to deliver model anti-inflammatory drug, diclofenac sodium (DS). Effects of variables such as PVA/NaAlg ratio, acrylamide content, exposure time to GA and drug/polymer ratio on the release of DS were discussed at three different pH values (1.2, 6.8, 7.4). It was found that DS release increased with increasing acrylamide content of the PVA-g-PAAm polymer. The highest DS release was obtained to be 92% for 1/1 PVA-g-PAAm/NaAlg ratio beads at high pH (6.8 and 7.4) than that at low pH (1.2) conditions [24].

Alginate was also used to crosslink with other polymers like chitosan, to deliver the drug effectively and efficiently. Alginate–chitosan (ALG–CS) blend gel beads were prepared based on Ca^{2+} or dual crosslinking with various proportions of alginate and chitosan [25]. The homogeneous solution of alginate and chitosan was dripped into the solution of calcium chloride and the resultant Ca^{2+} single crosslinked beads were dipped in the solution of sodium sulfate sequentially to prepare dual crosslinked beads. The cumulative release of BSA from ALG–CS mass ratios of 9:1, 7:3 and 5:5 were 2.35, 1.96 and 1.76% (in SGF 4 h), 82.86, 78.83 and 52.91% (in SIF 3 h) and 97.84, 96.81 and 87.26% (in SCF 3 h) respectively. This finding suggests that the dual crosslinked beads have potential small intestine or colon site-specific drug delivery.

Calcium alginate, another derivative of alginic acid was used in successfully developing oral floating tablet for the delivery of ibuprofen, niacinamide and metoclopramide HCl [26]. A pH sensitive alginate–guar gum hydrogel crosslinked with glutaraldehyde was prepared for the controlled delivery of proteins [27]. Guar gum was included in the alginate matrix along with a cross linking agent to ensure maximum encapsulation efficiency and controlled drug release. The glutaraldehyde concentration giving maximum (100%) encapsulation efficiency and the most appropriate swelling characteristics was found to be 0.5% (w/v). Protein release from test hydrogels was minimal at pH 1.2, and it was found to be significantly higher at pH 7.4. Presence of guar gum and glutaraldehyde crosslinking increases entrapment efficiency and prevents the rapid dissolution of alginate in higher pH of the intestine, ensuring a controlled release of the drug. Alginate was also used in developing some novel dosage forms such as microspheres [28], films [29], gels [30], *in situ* gels [31], microparticles [32], liposomes [33], implants [34], vaginal delivery [35], hydrogels [36, 37], colon targeting [38] etc.

10.5   Carrageenan

Carrageenans are marine hydrocolloids obtained by extraction from some members of the class Rhodophyceae. The most important members of this class are *Kappaphycus cottonii*, *Eucheuma spinosum* and *Gigartina stellata*. There are three different types of carrageenans viz., kappa (κ), iota (ι) and lambda (λ)-carrageenan. All consists chiefly of the sulfated esters of D-galactose and 3, 6-anhydro-D-galactose coplymers, linked α-1,3 and β-1,4 in the polymer (Figure 10.3). It appears as yellowish to colourless, coarse to fine powder which is practically odourless. The λ-carrageenan does not contain 3,6-anhydro galactose and is highly sulfated. It does not gel and is used as a thickeninig agent. The κ-ι-carrageenan are very similar, except ι-carrageenan is sulfated at carbon-2. Both the forms swell and form gels. The κ-carrageenan form strong, rigid and brittle gels. A very small amount of potassium ion is essential for this. The ι-carrageenan forms elastic gels that show thixotropy, mainly in the presence of calcium ions [39].

Carrageenans were mainly used as gelling and thickening agents. Only a few studies have dealt with carrageenans for controlled-release tablets (40, 41 and 42). These studies dealt only with drug delivery from tablets on a hydraulic press or from tablets that contain the carrageenans in a mixture with other excipients. In a study of four natural hydrophilic gums formulated as minimatrices in hard gelatin capsules, it was concluded that carrageenan used in the study did not produce sufficient sustained release [43]. In recent studies, the compaction and consolidation behavior of carrageenan were determined to prove their usefulness in tableting excipients for controlled-release tablets. The results indicated that the carrageenans were suitable as tableting excipients for controlled-release tablets. The compacts were formed easily and the material behaved

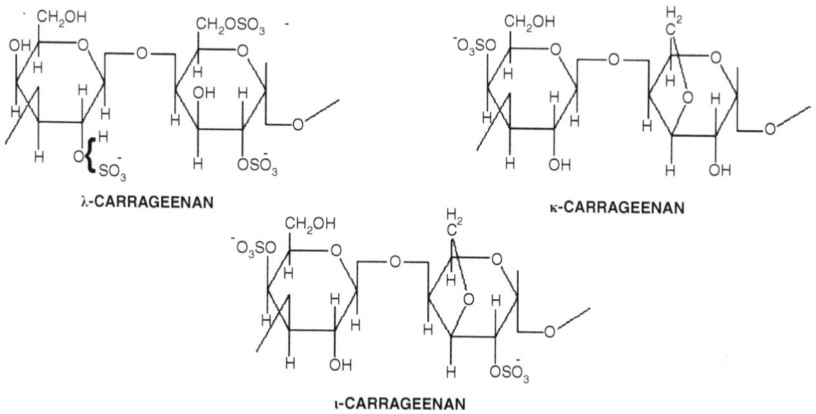

Figure 10.3 Chemical structures of types of carrageenans.

viscoelastically during compression. The resulting compacts were of high robustness and they showed good compactibility, indicated by a high tensile strength. The release behavior of model drugs diclofenac sodium and theophylline indicated that drug release was increased when water sorption and the extent of swelling decreased and viscosity increased [44–46].

Single unit, floating controlled drug delivery system was prepared using matrix-forming polymers like hydroxypropyl methylcellulose (HPMC), polyacrylates, sodium alginate, corn starch, carrageenan, gum guar and gum arabic and it was concluded that carrageenan can be used as a matrix material in formulating floating tablets [47]. Lambda-carrageenan, an anionic polymer was ionically interacted with alkaline drug, timolol maleate, resulting in a complex that releases the drug slowly. Carrageenan and gelatin in different ratios were combined for modulating the drug release profiles. A microsphere formulation was also tested *in vivo* in albino rabbits [48].

Sponge-like, *in situ* gelling inserts based on carrageenan using oxymetazoline HCl in was prepared [49]. The drug release, water uptake, mechanical properties, X-ray diffraction and bioadhesion potential of the nasal inserts were investigated. The drug release decreased with higher polymer content and increased drug loading of the insert. It was concluded that bioadhesive nasal inserts have a high potential as new nasal dosage form for extended drug delivery.

Novel, highly swelling hydrogels were prepared by grafting crosslinked polyacrylic acid-co-poly-2-acrylamido-2-methylpropanesulfonic acid (PAA-co-PAMPS) chains onto κ-carrageenan through free radical polymerization method [50]. The swelling of superabsorbent hydrogels was measured in various solutions with pH values ranging from 1 to 13. The hydrogels swelled to a range of 135–800 g/g indicating it to be a suitable candidate for drug delivery.

A novel approach of nanosizing a drug/polymeric complex to increase both solubility and dissolution rate of poorly water-soluble compounds was studied [51]. κ-carrageenan was complexed with a poorly water-soluble compound to increase the compound's aqueous solubility. The compound/carrageenan complex was further nanosized by wet-milling to enhance the dissolution rate, which increased the aqueous solubility of the compound from less than 1 μg/mL to 39 μg/mL.

10.6   Cellulose

Cellulose is the primary structural component of green plants. The primary cell wall of green plants is made up of cellulose. Cellulose is the major constituent of paper and textiles made from cotton, linen, and other plant fibers. Cellulose can be converted into cellophane, clear rolling papers made from viscous film, rayon, and more recently cellulose has been used to make modal, a bio-based textile derived from beechwood cellulose. Cellulose is used within the laboratory as the stationary phase for thin layer chromatography, and cotton linters, is used in the manufacture of nitrocellulose, historically used in smokeless gunpowder. Cellulose is derived from (β-glucose), which condense through β(1→4)-glycosidic-bonds.

The hydroxyl groups of cellulose can be partially or fully reacted with various reagents to afford derivatives with useful properties. Cellulose ethers are cellulose derivatives prepared by etherification of these available hydroxyl groups (e.g., ethyl cellulose, methyl cellulose, hydroxy propyl cellulose (HPC), cellulose acetate phthalate (CAP), hydroxyl propyl methyl cellulose (HPMC), hydroxyl propyl methyl cellulose phthalate (HPMCP), sodium carboxy methyl cellulose (Na CMC), and. Cellulose ethers are hydrophilic polymers that are quite popular in the design of controlled delivery dosage forms. They are biologically compatible and nontoxic. Apart from these advantages, their property of easy compression and the abilities to haydrate rapidly at body temperature and to accommodate a large percentage of the drug with negligible influence of the processing variables on the release rates are the main reasons for their popularity.

Transdermal release of primaquine from different matrix transdermal therapeutic systems (TTS) were prepared and the influence of polymer type (Eudragit® RL 100 or ethyl cellulose), adhesive layer and drug concentration in the polymeric matrix on the release profiles was studied [52]. Ethyl cellulose-based formulation with Mygliol® 840 as vehicle, showed a percutaneous flux of 180 µg cm^{-2} h^{-1}. Bilaminated films and bilayered tablets were prepared using chitosan, with or without an anionic crosslinking polymer (polycarbophil, sodium alginate, gellan gum), and the backing layer was made of ethylcellulose. Nifedipine and propranolol hydrochloride were used as model drugs and it was demonstrated that these new devices show promising potential for use in controlled delivery of drugs to the oral cavity [53]. Use of ethyl cellulose as a matrix material for periodontal treatment [54] and for preparation of microcapsules [55] was reported. Ethyl cellulose was also successively used in coating of hard gelatin capsules [56] and in preparation of microsponges [57].

A novel fast-gelling, non-cell adhesive, degradable, and biocompatible injectable gel as an injectable intrathecal drug delivery system was prepared [58]. The gel prepared was a blend of hyaluronan and methylcellulose (HA-MC). They suggested that HA-MC is a promising gel for localized delivery of therapeutic agents to the injured spinal cord.

A combination of hydroxypropyl cellulose (HPC) and microcrystalline cellulose (MCC) was used as base materials in order to develop nasal powder preparations with higher bioavailability for peptide delivery [59]. Significant enhanced absorption of leuprolide, calcitonin was attained by the addition of a small amount of HPC to MCC. It was suggested that MCC works as an absorption enhancer by causing a locally high concentration of drugs in the vicinity of the mucosa surface.

Cellulose acetate phthalate is widely used as film forming agent, which is used to protect the dosage form from acidic environment in the stomach [60–61]. Indomethacin was encapsulated in gelatin-CAP microcapsules by complex coacervation method and in CAP microcapsules by simple coacervation method were used to prepare sustained-release tablets [62]. The study showed that CAP can be used effectively for formulating controlled release drug delivery system.

Matrix tablets of sodium CMC and HPMC were prepared to improve controlled release performances of a water soluble drug [63]. *In vitro* release studies demonstrated that the mixture of the two cellulose derivatives enables a better control of the drug release profiles at pH 4.5 and at 6.8 both in term of rate and mechanism. Tablets containing 1:4, 1:1 and 4:1 weight ratios of pectin and HPMC were prepared for sustained release of diltiazem by sublingual administration [64]. An *in vitro* sustained release over 5 hrs was achieved with bilayer tablets composed of drug-free ethylcellulose layer in addition to the pectin/HPMC layer containing drug. However, HPMC is widely being used in developing gastroretentive drug delivery systems [65–67]. These systems release the drug in the stomach, providing local action of drugs and also avoid degradation of drugs which are sensitive to alkaline pH.

pH-sensitive cyclosporine A (CyA) nanoparticles were prepared by solvent displacement method using hydroxypropyl methylcellulose phthalate (HPMCP; including HP50 and HP55) [68]. The bioavailability of CyA-HP50 and CyA-HP55 nanoparticle colloids calculated by the AUC_{0-72} were 82.3% and 119.6%, similar to the reference of Neoral (marketed product), while the bioavailability of CyA-HP55 nanoparticle colloids was found to be higher than that of CyA-HP50 nanoparticle colloids. Researchers have extensively used HPMCP as coating material for enteric release formulations [69–71].

10.7   Chitosan

Chitosan is hydrophilic cationic polyelectrolyte obtained by alkaline N-deacetylation of chitin (Figure 10.4). Chitin is most abundant natural polymer next to cellulose and is obtained from lobsters, crab and shrimp shells. It appears as white to light yellow flakes having a viscosity of 20–2000 cps depending on the molecular weight. Chitosan is bioadhesive and readily binds to negatively charged surfaces such as mucosal membranes. Chitosan enhances the transport of polar drugs across epithelial surfaces, and is biocompatible and biodegradable. Purified qualities of chitosans are available for biomedical applications. Chitosan, in combination with bentonite,

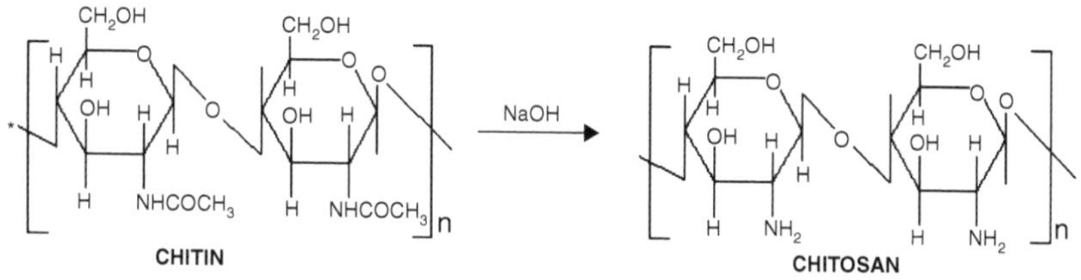

Figure 10.4 Conversion of chitin to chitosan.

gelatin, silicagel, isinglass, or other fining agents is used to clarify wine, mead, and beer. Chitosan's properties allow it to rapidly clot blood, and has recently gained approval in the USA for use in bandages and other hemostatic agents.

Chitosan has been found useful as a vehicle for sustained-release preparation of water-insoluble drugs like indomethacin, papavarine hydrochloride and water-soluble drugs such as propranolol hydrochloride [72–74]. It was observed that, when dissolution was performed in acidic medium, these dosage forms show an excellent sustained-release property. In another study, chitosan granules were prepared to achieve sustained release of indomethacin [75]. A unique characteristic of the chitosan granule was that they gradually swelled and floated in acid medium at pH1.2. This floating property of the granules on the acid medium can be applied to the formulation of sustained-release preparation of various drugs. The effect of the cross-linking procedure on the drug release patters from chitosan granules was also examined. The release of the drug from granules could be controlled by varying the cross-linking procedure. *In vivo* performance of indomethacin-chitosan granules was shown to be superior to conventional commercial capsules in terms of decrease in the peak of plasma concentration and maintaince of indomethacin concentration in plasma [76, 77].

The sustained release characteristics of chitosan have been investigated in the presence of citric acid or carbomer 934P in tablets containing theophylline as model drug [78]. When chitosan was used alone in a concentration greater than 50% of tablet weight, an insoluble non-erosion type of matrix was formed. Tablets prepared with chitosan concentration less than 33% were fast releasing and around 10% of chitosan acted as a disintegrant. Citric acid and carbomer 934P were used as coadjuvants in this study as acidifying agents, which gels the chitosan and thus imparted sustained-release property.

A matrix tablet for buccal drug delivery composed of chitosan and poloxamer 407 was prepared [79]. Different chitosan salts were prepared by reacting chitosan with acetic, citric, and lactic acid. Various proportions of poloxamer 407 were added to the aqueous solution of chitosan salt, and the residue obtained by lyophilisation was compressed into tablets. The drug release was in the order of chitosan acetate>chitosan citrate>chitosan lactate. Mucoadhesion was particularly favoured when poloxamer 407 was present at about 30% (w/w). Adusumilli and Bolton synthesized chitosan citrate complexes, which were found to be very effective in sustaining the release of theophylline and were directly compressible [80].

A chitosan/polyethylene vinyl acetate co-matrix has been developed for the controlled release of aspirin-heparin to prevent cardiovascular thrombosis. This amount of drug release initially was much higher, followed by a constant slow-release profile for a prolonged period. The released aspirin-heparin from the co-matrix system showed antiplatelet and anticoagulant functions. The results propose the possibility of delivering drug concentration with synergistic effects of therapeutic application [81].

A modified method to prepare chitosan–poly(acrylic acid) (CS–PAA) polymer magnetic microspheres was reported [82]. First, via self-assembly of positively charged CS and negatively charged Fe_3O_4 nanoparticles, magnetic CS cores with a large amount of Fe_3O_4 nanoparticles were successfully prepared. A continuous release of the entrapped ammonium glycyrrhizinate in such polymer magnetic microspheres occurred, which confirmed the potential applications of these microspheres for the targeted delivery of drugs.

Chitosan-Ca-alginate microparticles for colon-specific delivery and controlled release of 5-aminosalicylic acid after peroral administration were prepared using spray drying method followed by ionotropic gelation/polyelectrolyte complexation [83]. *In vitro* drug release studies carried out in simulated *in vivo* conditions in respect to pH, enzymatic and salt content confirmed the potential of the particles to release the drug in a controlled manner. Biodistribution studies of [I^{131}]-5-ASA loaded chitosan-Ca-alginate microparticles, carried out in Wistar male rats confirmed the localization of 5-ASA in the colon with low systemic bioavailability.

A combination of chitosan with sodium alginate for sustaining the release of theophylline has also been studied [84]. It was found that the release is independent of the pH of the medium. Daidzein-loaded chitosan microspheres wepe prepared by emulsification/chemical cross-linking technique [85]. *In vivo* pharmacokinetic characteristics were evaluated after intramuscular injection of the microspheres in rats and showed that the release of daidzein almost lasted for 35 days.

Gabr and Meshali [86] investigated and characterized the possible interaction between the natural cationic (chitosan) and anionic (pectin and acacia) polysaccharides. The solid complexes so formed were separated and dried to be used as tablet matrices. The release of water-soluble chlorpromazine hydrochloride from tablets containing various concentration of each of the polymers alone, the complexes or physical mixtures of chitosan and pectin and acacia in the same ratio as their respective complexes was evaluated. The physical mixture displayed the most efficient sustained release. A new highly porous, flexible sponge for buccal peptide administration by a very simple and mild casting/freeze-drying for insulin delivery was developed [87]. It consists of a mucoadhesive chitosan layer containing the peptide drug and an impermeable protective layer made of ethylcellulose.

Superporous hydrogels containing poly (acrylic acid-*co*-acrylamide)/*O*-carboxymethyl chitosan interpenetrating polymer networks (SPH-IPNs) were prepared by cross-linking *O*-carboxymethyl chitosan (*O*-CMC) with glutaraldehyde (GA) after superporous hydrogel (SPH) was synthesized. An enhanced loading capacity for insulin could be obtained by the SPH-IPNs as compared to non-porous hydrogel and more than 90% of the insulin was released within 1 hr [88].

A new type of nanoparticles made of chitosan (CS) and carboxymethyl-β-cyclodextrin (CM-β-CD) were prepared using ionotropic gelation technique and evaluated for their potential for association and delivery of macromolecular drugs [89]. The release profiles of the associated macromolecules were highly dependent on the type of molecule and its interaction with the nanomatrix. Insulin released very fast (84–97% insulin within 15 min) whereas heparin remained highly associated to the nanoparticles for several hours (8.3–9.1% heparin within 8 hrs). Fu Chen *et al.* prepared *N*-trimethyl chitosan chloride (TMC) nanoparticles by ionic crosslinking of TMC with tripolyphosphate (TPP) for the delivery of bovine serum albumin and bovine hemoglobin [90]. In another study, liposome–chitosan nanoparticle complexes (LCS-NP) were prepared as a complex between liposomes and chitosan nanoparticles (CS-NP). *In vivo* studies indicated that the prepared nanoparticles can be used effectively for ocular delivery of drugs [91].

10.8   Dextran

Dextran is a complex, branched polysaccharide made of many glucose molecules joined into chains of varying lengths consisting of α-1-6 glycosidic linkages between glucose molecules (Figure 10.5). Dextran is synthesized from sucrose by certain lactic-acid bacteria, the best-known being *Leuconostoc mesenteroides* and *Streptococcus mutans*. It is mainly used commonly by microsurgeons to decrease vascular thrombosis. The antithrombotic effect of dextran is mediated through its binding of erythrocytes, platelets, and vascular endothelium, increasing their electronegativity and thus reducing erythrocyte aggregation and platelet adhesiveness. Dextrans are available in multiple molecular weights ranging from 10,000 Da to 150,000 Da.

Degradation of dextran hydrogels, potential drug carriers for colon-specific drug delivery, was studied in simulated small intestinal juices as well as in a human colonic fermentation model [92]. Dextran hydrogels were found to be stable when incubated at 37°C with the small intestinal enzymes amyloglucosidase, invertase and

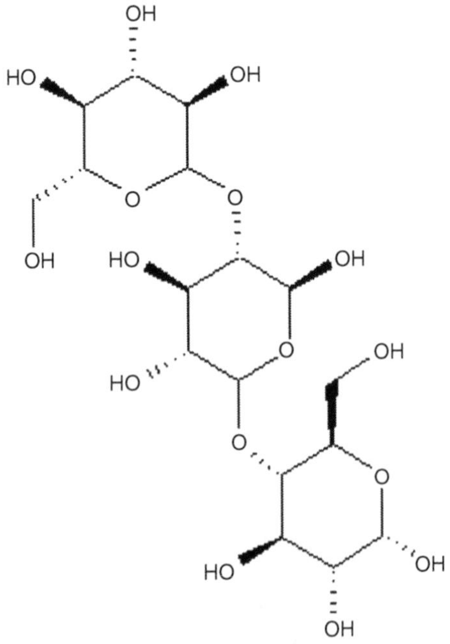

Figure 10.5 Chemical structure of dextran.

pancreatin as it showed release of less than 3.3% of free glucose after 24 hrs. The hydrogels were found to be completely degraded in the human colonic fermentation model. Rita Cortesi *et al.* used oxidized dextran as a cross-linker for the preparation of gelatin microspheres by thermal gelation method [93]. The obtained results indicated that oxidized dextran can form a cross-linked gelatin network which can reduce the dissolution of gelatin. The in vitro immunosuppressive activity of a conjugate of methylprednisolone (MP) with dextran 70 kDa (DEX-MPS) was tested using the lymphocyte proliferation assay after stimulation of lymphocytes with concanavalin A (Con-A) [94].

Eddy Castellanos Gil *et al.* developed a novel oral controlled delivery system for propranolol hydrochloride (PPL) [95]. The influence of matrix forming agents (native dextran, hydroxypropyl methylcellulose (HPMC), cetyl alcohol) and binary mixtures of them on PPL release in vitro was investigated. The sustained-release matrix tablets with good physical, mechanical and technological properties were obtained with a matrix excipient:PPL ratio of 60:40 (w/w), with a dextran: HPMC ratio of 4:1 (w/w) and with a cetyl alcohol amount of 15% (w/w). Nanoparticles combining a hydrophobically modified dextran core and a polysaccharide surface coverage were elaborated [96]. Dextran, was chemically modified by the covalent attachment of hydrocarbon groups (aliphatic or aromatic) via the formation of ether links.

10.9 Dextrin

Dextrins are a group of low-molecular-weight carbohydrates produced by the hydrolysis of starch. Dextrins are mixtures of linear α-(1,4)-linked D-glucose polymers (Figure 10.6). Dextrins are water soluble, white to slightly yellow solids which are optically active.

A solid dispersion system containing cyclosporin A (CsA) in order to improve the bioavailability of poorly water-soluble CsA was developed [97]. The CsA–microspheres at the CsA/SLS/dextrin ratio of 1/3/1 gave the highest dissolution rate of CsA. They demonstrated that CsA–microspheres prepared with SLS and dextrin, with improved bioavailability of CsA, would be useful to deliver a poorly water-soluble CsA and could be applicable to other poorly water-soluble drugs. The possible use of starch-agglutinant mixtures as principal excipients for extrusion–spheronization

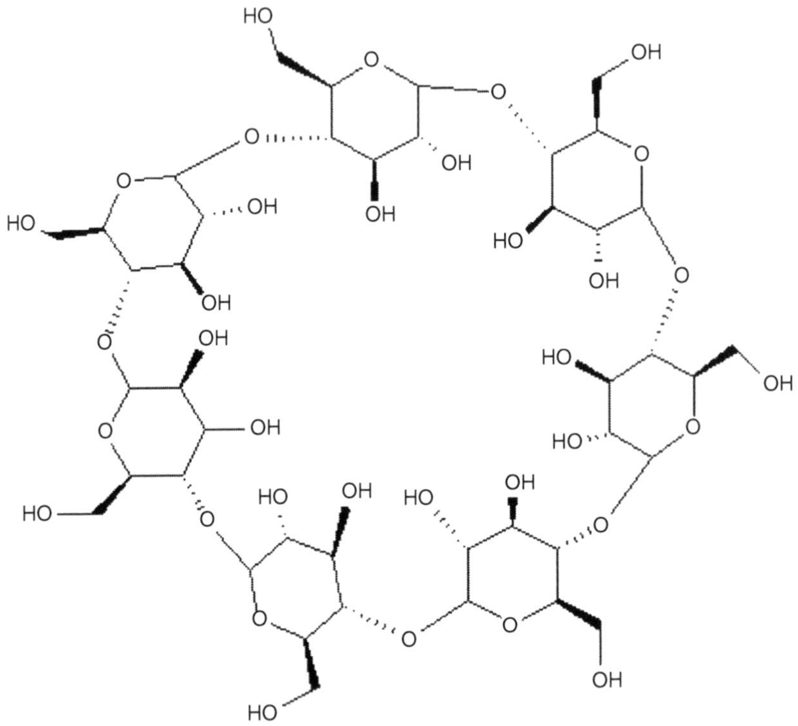

Figure 10.6 Chemical structure of β-cyclodextrin.

pellets was evaluated [98]. The results showed that some of the mixtures—notably starch (corn starch or wheat starch) with 20% white dextrin gave high-quality pellets with good size and shape distributions suggesting for use in drug delivery systems.

The cyclical dextrins are known as cyclodextrins. They are formed by enzymatic degradation of starch by certain bacteria. Cyclodextrins have toroidal structures formed by 6–8 glucose residues. Typical cyclodextrins contain a number of glucose monomers ranging from six to eight units in a ring, namely α-cyclodextrin (six sugar ring molecule), β-cyclodextrin (seven sugar ring molecule) and γ-cyclodextrin (eight sugar ring molecule).

The ability of cyclodextrins (α, β, γ-) to form non-covalent complexes with a number of drugs and altering their physico-chemical properties has been discussed [99]. The incorporation of various cyclodextrins into polymeric formulations and the mechanisms by which cyclodextrin/polymer formulations act has been reviewed.

Two types of cyclodextrin viz., γ cyclodextrin (GCD) and dimethyl- β -cyclodextrin (DMCD) as carriers in dry powder aerosol formulations was evaluated using salbutamol as model drug [100]. The drug release in both formulations containing GCD and DMCD was fast (over 70% was released in 5 min) and nearly all the drug was released within 30 min. It was concluded that GCD and DMCD are able to promote salbutamol delivery in dry powder inhaler compared to a formulation containing lactose. Brunella Cappello *et al.* developed a tablet for the buccal delivery of the poorly soluble drug, carvedilol (CAR) based on poly(ethyleneoxide) (PEO) as bioadhesive sustained-release platform and hydroxypropyl-β-cyclodextrin (HPβCD) as drug release enhancer [101]. Researchers showed that cyclodextrins act as permeation enhancers carrying the drug through the aqueous barrier, from the bulk solution towards the lipophilic surface of biological membranes, where the drug molecules partition from the complex into the lipophilic membrane [102].

10.10   Gellan Gum

Gellan gum is a polysaccharide produced by *Sphingomonas elodea,* a bacterium. It appears as off-white powder, soluble in water and used primarily as an alternative to agar as a gelling agent in microbiological culture. It can withstand to 120 degree Celsius heat, making it especially useful in culturing thermophilic organisms. As a food additive, gellan gum is used as a thickener, emulcifier, and stabilizer. The high molecular weight polysaccharide is principally composed of a tetrasaccharide repeating units of one rhamnose, one glucuronic acid, and two glucose units, and is substituted with acyl (glyceryl and acetyl) groups as the O-glycosidically-linked esters (Figure 10.7). The glucuronic acid is neutralized to a mixed potassium, sodium, calcium, and magnesium salt. It usually contains a small amount of nitrogen containing compounds resulting from the fermentation procedures.

The use of gellan gum in developing tablet dosage form in combination with ethyl cellulose is already discussed [53].

Gellan gum beads of propranolol hydrochloride, a hydrophilic model drug, were prepared by solubilising the drug in a dispersion of gellan gum and then dropping the dispersion into calcium chloride solution [103]. The droplets formed gelled beads

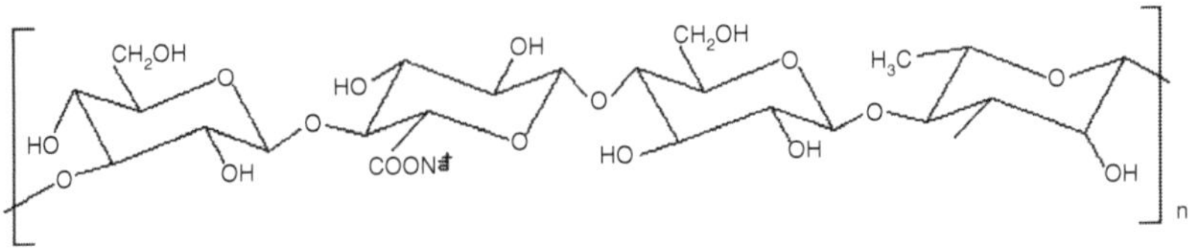

Figure 10.7 Chemical structure of gellan gum.

instantaneously by ionotropic gelation. Very high entrapment efficiencies were obtained (92%) after modifying the pH of both the gellan gum dispersion and the calcium chloride solution. Gellan gum could be a useful carrier for the encapsulation of fragile drugs and provides new opportunities in the field of bioencapsulation.

Gels formed *in situ* following oral administration of 1% (w/v) aqueous solutions of gellan to rats and rabbits were evaluated as sustained-release vehicles [104]. The formulation contained calcium ions in complexed form, the release of which in the acidic environment of the stomach caused gelation of the gellan gum. Bioavailability of theophylline from gellan gels formed by in situ gelation in the animal stomach was increased by four–fivefold in rats and threefold in rabbits compared with that from the commercial oral formulation.

The effect the different ions in tear fluid (Na^+, K^+, Ca^{2+}) on the gel strength and the consequences of dilution due to the ocular protective mechanisms were examined [105] and it was found that Na^+ was found to be the most important gel-promoting ion *in vivo*. *In situ* gels of gellan for the oral delivery of cimetidine [106], paracetamol [107] and amoxacillin [108] were also investigated. Research also proved that gellan can be used for targeting the drugs to colon [109].

10.11   Guar Gum

Guar gum is a gum obtained from the ground endosperms of *Cyamposis tetragonolobus* (Leguminosae family). It chiefly consists of high molecular weight hydrocolloidal polysaccharide, composed of galactan and mannan units combined through glycosidic linkages. The structure of guar gum is a linear chain of β-D-mannopyranosyl units linked (1→4) with single-member α-D-galactopyranosyl units occurring as side branches (Figure 10.8).

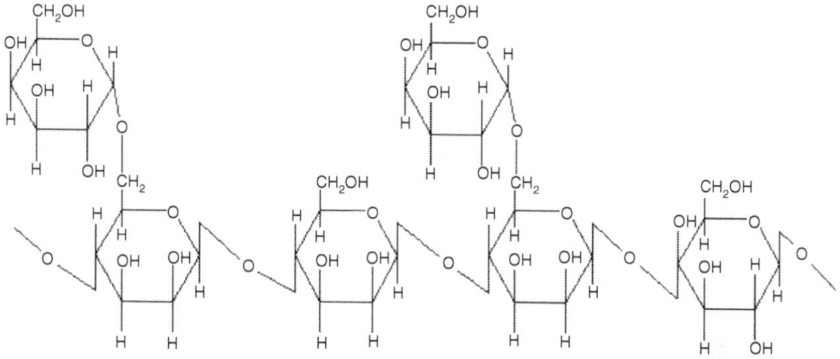

Figure 10.8 Chemical structure of guar gum.

Guar gum is an interesting polymer for the preparation of hydrophilic matrix tablets because of its high water swellability, nontoxicity, and low cost. Various groups of workers have used guar gum as a controlled-release carrier [110–112]. Baveja *et al.* [110] examined 12 natural gums and mucilages as sustaining materials in tablet dosage forms. Tablets prepared using guar gum as the sustaining material dissolved in 2 hrs due to the high erosion rate of the matrix. Bhalla and Gulati [111] evaluated sustained release theophylline tablets with Eudragits and guar gum. Formulations based on 5% guar gum gave an appropriate release pattern over a period of 12 hrs.

Alataf and co workers [112, 113] examined the use of guar gum to sustain the release of diltiazem. Results showed that varying the lot of guar gum as well as using guar gum from different suppliers had little effect on diltiazem dissultion. The stabilities of guar gum based formulations under stressed condition were established. All four formulations gave similar plasma concentration over time in pharmacokinetic studies of healthy volunteers. It was concluded that matrix tablets based on guar gum represent a simple and economic alternative to existing diltiazem sustained release dosage forms. Gebert and Friend [114] purified guar galactomannan and assessed certain pharmaceutical attributes. The viscosity of an aqueous 1% purifed galactomannan solution is typically 40–50% higher than its unpurified precursor. The hydration rate of 1% aqueous solution increases by 100% after purification. This data demonstrated the need for less guar gum to sustain the release of water-soluble drug.

In spite of the wide pharmaceutical application of guar gum, its use is limited by its uncontrolled rate of hydration, decreased viscosity on storage, and microbial contamination. Paranjothy and Thampi [115, 116] synthesized derivatives of guar, such as guar succinate, guar benzoate, polygalactomannan (borate reaction), oxidized guar gum, hydroxyl propyl guar and sodium carboxy methyl guar. The solubility studies showed that sodium carboxy methyl guar gave a transparent gel. A 2% sodium carboxy methyl guar solution in water poured over a mercury pool produced a very good flexible, clear, transparent film. Later, a transdermal patch of verapamil hydrochloride was prepared using sodium carboxy methyl guar as a polymer matrix [117]. *In vitro* release studies through mouse skin showed that sodium carboxy methyl guar was a suitable polymer.

The functioning of guar gum crosslinked products (GGP) as possible colon-specific drug carriers was analyzed by studying the release kinetics of pre-loaded hydrocortisone from GGP hydrogels into buffer solutions with, or without GG degrading enzymes (α-galactosidase and β-mannanase) [118]. It was concluded that crosslinked guar gum can be biodegraded enzymatically and is able to retard the release of a low water-soluble drug, which could potentially be used as a vehicle for colon-specific drug delivery.

Furthermore, guar gum and methylated guar gum were used to prepare hydrophilic matrix controlled release tablets using chlorpheniramine maleate as model drug [119]. Drug release profile from guar gum matrix tablets showed a high percentage of drug release (31.06 ± 2.56%) in the first half hour and then the rate of drug release decreased with time. In the case of methylated guar gum, significant reduction in the amount of drug released in the first half hour (18.77 ± 0.68%) was observed. The rate of hydration increased hence, the onset of the obstructive gel layer formation was faster compared to guar gum. The results also showed that the rate release increased with degree of methylation. Carboxymethyl guar (CMGS), an anionic semisynthetic guar gum derivative was evaluated for its suitability of use in transdermal drug-delivery systems using terbutaline sulfate (TS) as a model drug [120]. It was found that, diffusion of terbutaline sulfate from CMGS solution was relatively slower at pH 5 than at pH 10.

Cross-linked hydrogels of polyacrylamide-grafted guar gum were prepared by emulsification method by saponification of the -$CONH_2$ group to the -COOH group [121]. It was noted that the swelling of microgels increased when the pH of the medium changed from acidic to alkaline. The pH-sensitive microgels were loaded with diltiazem hydrochloride and nifedipine and their release studies indicated quicker release in pH 7.4 buffer than observed in 0.1 N HCl. M. George *et al.* prepared a pH sensitive alginate-guar gum hydrogel crosslinked with glutaraldehyde. The beads having an alginate to guar gum percentage combination of 3:1 showed better encapsulation efficiency, bead forming property and drug release.

Novel polyelectrolyte hydrogels (GA) based on cationic guar gum (CGG) and acrylic acid monomer were synthesized by pho-toinitiated free-radical polymerization [122]. Swelling experiments indicated that the ketoprofen loaded GA hydrogels were highly sensitive to pH environments. The results implied that the GA hydrogels can be exploited as potential carriers for colon-specific drug delivery.

10.12   Inulin

Inulins are a group of naturally occurring polysaccharides belonging to a class of carbohydrates known as fructans. Inulin is used by some plants as a means of storing energy and is typically found in roots or rhizomes. Inulins are polymers mainly comprised of fructose units and typically have a terminal glucose. The fructose units in inulins are joined by a beta-(2-1) glycosidic link (Figure 10.9). Inulin passes through the stomach and duodenum undigested and is highly broken down by the gut bacterial flora.

InulinHP (inulin with a high degree of polymerization) was formulated as a biodegradable colon-specific coating by suspending it in Eudragit RS films [123]. The *in vitro* degradability of the prepared isolated films was studied by incubating them in a faecal degradation medium and it was observed that inulin resisted hydrolysis and digestion in the upper gastro-intestinal tract, but fermented by the colonic microflora. Inulin hydrogels have been developed as carriers for colonic drug targeting [124]. The enzymatic digestibility of the prepared inulin hydrogels was assessed by performing an *in vitro* study using an inulinase preparation derived from *Aspergillus niger*. The data obtained suggested that inulinase enzymes are able to diffuse into the inulin hydrogel networks causing bulk degradation.

The azo-polysaccharide gels were synthesized by radical cross-linking of a mixture of methacrylated inulin or methacrylated

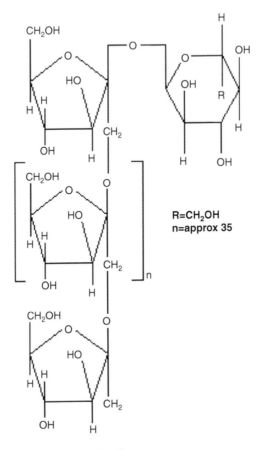

Figure 10.9 Chemical structure of inulin.

dextran and N,N'-bis (methacryloylamino) azobenzene (B(MA) AB). Increasing the amount of B(MA)AB resulted in denser azo-inulin and azo-dextran networks. However, the degradation of azo-dextran gels by dextranase seemed to be more pronounced than the degradation of the azo-inulin gels by inulinase [125].

10.13   Karaya Gum

Gum Karaya is an extract of *Sterculia urens* tree belonging to the family Sterculiaceae. It is used as a suspending or stabilizing agent, thickener, emulcifier and laxative in foods, and as a denture adhesive. Powdered Gum Karaya is white to greyish white in colour having a high molecular weight of about 9,500,000. On hydrolysis it yields galactose, rhamnose and galacturonic acid. Gum Karaya occurs as a partially acetylated derivative.

J. Sujja-areevath *et al.* evaluated the use of four natural hydrophilic gums, carrageenan (C), locust bean (LB), karaya (K) and xanthan gums (X) as mini-matrix formulations enclosed in a hard gelatin capsule [126]. The release profiles showed that sustained release of up to 77% of doclofenac drug was achieved from mini-matrices containing LB, X and K, while C did not produce sufficient sustained release. The amount of gum present played the dominant role in determining the drug release rate. Similar study was reported by J. Sujja-areevath *et al.* [127] mini matrix formulations. Directly compressed matrices containing xanthan gum and karaya gum were prepared to control the release of caffeine and diclofenac sodium [128]. Drug release from xanthan and karaya gum matrices depended on agitation speed, solubility and proportion of drug. Both xanthan and karaya gums produced near zero order drug release with the erosion mechanism playing a dominant role, especially in karaya gum matrices.

Modified gum karaya (MGK) was evaluated as carrier for dissolution enhancement of poorly soluble drug, nimodipine (NM) [129]. The advantages of MGK over the parent gum karaya (GK) were illustrated by differences in the *in vitro* dissolution profiles of respective solid mixtures prepared by co-grinding technique. The dissolution rate of NM was increased as the MGK concentration increased and optimum ratio was found to be 1:9 w/w ratio (NM:MGK).

10.14   Konjac Glucomannan

Konjac is a plant of the genus *Amorphophallus*. It is a perennial plant, growing from a large corm up to 25 cm in diameter. Konjac has almost no calories but is very high in fiber. Hence, it is often used as a diet food.

Kang Wang *et al.* prepared controlled release beads using alginate (ALG), konjac glucomannan (KGM) and chitosan (CHI). Bovine serum albumin and insulin were used as model proteins for *in vitro* assessments. After beads were treated by 0.1 N HCl for 4 h and put into pH 7.4 buffers, protein was released from ALG–CHI beads within 1 h, while it was lost from ALG–KGM–CHI beads for 3 h. However, the leaking of protein from ALG–KGM–CHI beads was also increased in 0.1 N HCl solution. The results indicated that the diffusion of protein was related to the viscosity and swelling properties of KGM [130].

Gels composed of konjac glucomannan (KGM), copolymerized with acrylic acid (AA) and cross-linked by N,N-methylene-bis-(acrylamide) (MBAAm) were evaluated for colon targeting [131]. The results of degradation test show that the hydrogels retain the enzymatic degradation character of KGM and they can be degraded for 52.5% in 5 days by Cellulase E0240. *In vitro* release of model drug 5-aminosalicylic acid (5-ASA) was studied in the presence of Cellulase E0240 in pH 7.4 phosphate buffer at 37°C and the drug release was found to be controlled by the swelling and degradation of the hydrogels. Novel polyelectrolyte beads based on carboxy-methyl Konjac glucomannan (CKGM) and chitosan (CS) were pre-pared via electrostatic interaction [132]. pH sensitivity tests showed that the swelling rate of test beads was larger (7.4) in the alkaline medium than in the acid medium (5.2) and at pH 5.3 swelling was the smallest (3.5). Release profiles at pH 1.2, 5.0 and 7.4 were carried out for bovine blood proteins (BSA).

10.15   Locust Bean Gum

Locust bean gum is a galactomannan vegetable gum extracted from the seeds of the Carob tree. It forms a food reserve for the seeds and helps to retain water under arid conditions. It is used as a thickener and gelling agent in food technology. It is also called Carob Gum or Carubin. It consists of a (1→4)-linked β-D-mannopyrannose backbone with branchpoints from their 6-positions linked to α-D-galactose (i.e. 1→6-linked α-D-galactopyranose).

Sujja-areevath *et al.* evaluated the use of four natural hydro-philic gums, carrageenan (C), locust bean (LB), karaya (K) and xanthan gums (X) as mini-matrix formulations enclosed in a hard gelatin capsule [126]. The release profiles showed that sustained release of up to 77% of doclofenac drug was achieved from mini-matrices containing LB, X and K, while C did not produce suf-ficient sustained release. The use of locust bean gum as a carrier for formulating colon targeted drug delivery systems has been investigated [133].

10.16   Locust Bean Gum

Honey locust gum (HLG) obtained from *Gleditsia triacanthos* (Figure 10.10), was investigated as a hydrophilic matrix material for theophylline drug release [134]. The matrix tablets containing hydroxyethylcellulose and hydroxypropyl methylcellulose as sustaining polymers at the same concentrations were prepared and a commercial sustained release (CSR) tablet containing 200 mg theophylline was examined for comparison of HLG performance. According to the results obtained from dissolution studies in distilled water, pH 1.2 HCl buffer and pH 7.2 phosphate buffer, no significant difference was found between CSR tablet and the matrix tablet containing 10% HLG in each medium ($P > 0.05$) and these tablets showed zero-order kinetic model in all the mediums.

Guar gum and Locust bean gum matrix tablets were compared with matrices obtained with scleroglucan [135]. Furthermore, the polymers were chemically crosslinked with glutaraldehyde to obtain a network suitable as a matrix for modified drug release. The delivery of the model molecules from the Guar gum and Locust bean gum gels, and from tablets prepared from the freeze-dried hydrogels of the three polymers was evaluated, and a comparison with the tablets prepared with the not-crosslinked polymers was carried out.

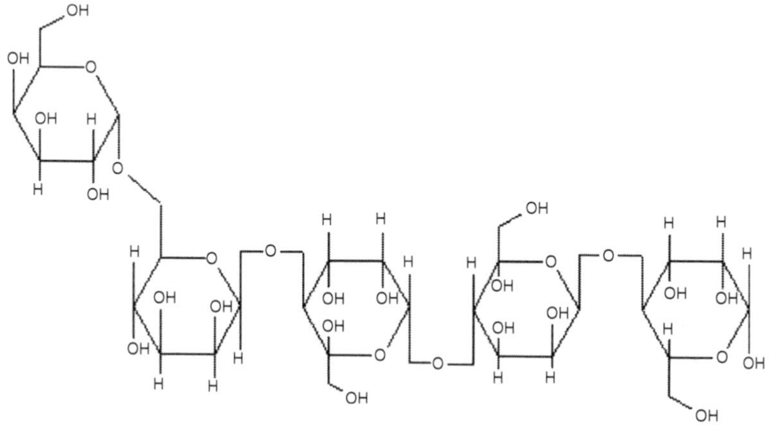

Figure 10.10 Chemical structure of locust bean gum.

10.17   Pectin

Pectins are important ionic polysaccharides found in plant cell walls. They consist mainly of linearly connected α-(1→ 4) D–galacturonic acid residues (Figure 10.11), which have carbonyl groups.

Recently, pectin beads prepared by the ionotropic gellation method have been investigated as a sustained-release drug delivery system. However, the use of pectin beads has some drawbacks due to their rapid *in vitro* release [136, 137]. In another study, calcium pectinate gel beads of indomethacin, a poorly soluble drug, were prepared [138]. Indomethacin was dispersed in a solution of pectin and then the dispersion was dropped in calcium chloride solution. The droplets instantaneously formed gelled spheres by ionotropic gellation. The effects of several factors (such as pectin type, the presence of hardening agent and the drug loading) on the percentage of drug entrapped, size distribution and drug release from calcium pectinate gel beads were investigated. Strong spherical beads with narrow size distribution, high yield and good entrapment efficiencies could be prepared. The mechanism of drug release from calcium pectinate gel beads followed the diffusion controlled model for an inert porous matrix.

A calcium-induced pectin gel bead (PB) containing pectin hydrolysate was prepared, and the drug release profiles and degradation properties were investigated [139]. It was found that the drug release rate of the PB in buffer solution decreased as the rate of gel erosion declined. It appeared that the PB gel matrix is an effective medium to control the release of drug within the gastrointestinal tract.

Tablets containing 1:4, 1:1 and 4:1 weight ratios of pectin and hydroxypropyl methylcellulose (HPMC) for the sustained release

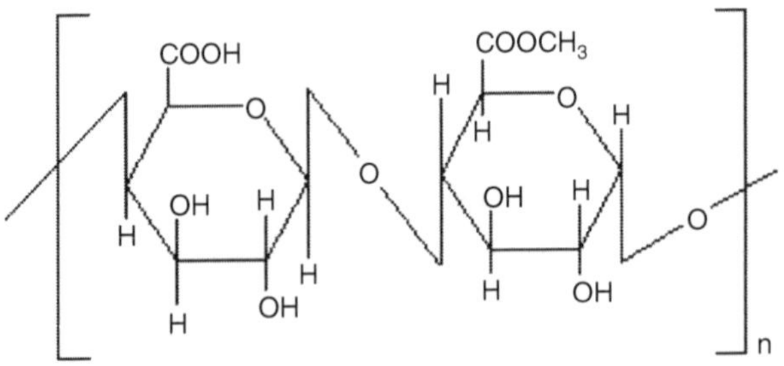

Figure 10.11 Chemical structure of pectin.

of diltiazem by sublingual administration have been investigated [140]. An *in vitro* sustained release of diltiazem over 5 hrs was achieved with bilayer tablets composed of a drug-free ethylcellulose layer in addition to the pectin/HPMC layer containing drug.

Pectin microspheres as ophthalmic delivery system of piroxicam (Px) has been investigated [141]. The microspheres were prepared by a spray-drying technique and *in vitro* release behavior was evaluated in pH 7.0 USP buffer using a flow-through apparatus. *In vivo* tests in rabbits of dispersions of Px-loaded microspheres also indicated a significant improvement of Px bioavailability in the aqueous humour (2.5-fold) when compared with commercial Px eye drops.

A new binary polymer matrix tablet has been developed for oral administration [142]. Highly methoxylated pectin and HPMC at different ratios were used as major formulation components and predinsolone was used as the drug model. The results indicated that, by increasing the ratio of pectin to HPMC, release rates were increased, but zero-order kinetics prevailed throught the dissolution period.

A novel colon specific drug delivery system containing flurbiprofen (FLB) microsponges has been developed [143]. The microsponges were prepared by quasi-emulsion solvent diffusion method. The colon specific formulations were prepared by compression coating and also pore plugging of microsponges with pectin: HPMC mixture followed by tabletting. *In vitro* studies exhibited that compression coated colon specific tablet formulations started to release the drug at the 8th hr corresponding to the proximal colon arrival time.

10.18   Psyllium Husk

Psyllium seed husks, also known as ispaghula or simply as psyllium, are portions of the seeds of the plant *Plantago psyllium* or *Plantago ovata*. Psyllium seed husks expand and become mucilaginous when wet and are soluble in water, but they are indigestible in human beings. The mucilage obtained from psyllium comes from the seed coat. Mucilage is obtained by mechanical milling/grinding of the outer layer of the seed. Mucilage yield amounts to approximately 25% or more (by weight) of the total seed yield.

Sustained release (SR)-gastroretentive dosage forms (GRDF) for ofloxacin preferably once daily has been prepared [144]. Different polymers, such as psyllium husk, HPMC K100M, crospovidone and its combinations were tried in order to get the desired sustained release profile over a period of 24 hrs. It was found that dimensional stability of the formulation increases with the increasing psyllium husk concentration. It was also found that *in vitro* drug release rate increased with increasing amount of crospovidone due to the increased water uptake, and hence increased driving force for drug release. Similar studies have been reported for ofloxacin gastroretentive dosage form [145]. Psyllium and acrylic acid based polymeric networks were prepared by using N,N'-methylenebisacrylamide (*N,N*-MBAAm) as crosslinker [146]. The swelling response indicated that the materials are potential candidates for use in colon specific drug delivery. Metal ion sorption shows that these polymeric networks can be used for removal, separation, and enrichment of hazardous metal ions from aqueous solutions and can play an important role for environmental remediation of municipal and industrial wastewater.

Baljit Singh *et al.* prepared psyllium and methacrylamide based polymeric hydrogels by using N,N'-methylenebisacrylamide (NN-MBAAm) as crosslinker and ammonium persulfate (APS) as initiator [147]. The effect of pH on the release pattern of tetracycline hydrochloride has been studied by varying the pH of the release medium. In each release medium, the values of the initial diffusion coefficient 'D_i' for the release of tetracycline hydrochloride was higher than the values of late time diffusion coefficient 'D_L' indicating that in the start, the diffusion of drug from the polymeric matrix was faster as compare to the latter stages.

10.19   Scleroglucan

Scleroglucan is a natural exocellur polysaccharide secreted by a fungus from the genus *Sclerotium rolfsii*. It is β (1→6) D-glucan with a single pendant glucose group attached through a β (1→3) linkage (Figure 10.12). It exhibits a gel-like structure in aqueous solution at low temperatures.

In recent years, particular attention has been focused on the possible utilization of scleroglucan because the physico-chemical properties of this polysaccharide suggest its suitability as a sustained release monolithic swellable matrix. The viscosity of scleroglucan solution and its stability over a range of temperature, pH and

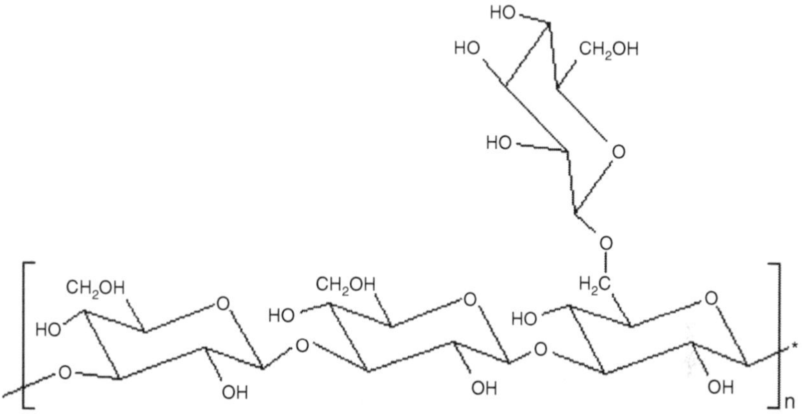

Figure 10.12 Chemical structure of scleroglucan.

salt concentration have been studied [148]. It was concluded that the high viscosity of scleroglucan solution at a low concentration (1–3% w/w), together with its stability make this polymer a subject of exploration for pharmaceutical preparations. In another study, physico-chemical and mechanical characteristics of scleroglucan were investigated. The physico-chemical properties of Actigum CS 11 (scleroglucan) suggested its suitability as a gelling polymer carrier matrix for slow release matrices [149].

Subsequently, pharmaceutical compatibility of this polymer with two common diluents, lactose Fast Flo® and Emcompress® was checked [150]. In another study, directly compressed hydrophilic matrices of theophylline were prepared with scleroglucan as the gelling agent [151]. It was concluded that, when a porous sclerogluican matrix is brought in contact with dissolution medium, the formation of the gel layer more or less quickly blocks the surface pores and prevents the ingress of the dissolution medium, assuming the control of drug transport.

Alhaique et al. [152–155] studied the release behaviour from tablets prepared with scleroglucan and concluded that the drug delivery is related directly to the penetration rate of the solvent into the polymer matrix and not to the outward diffusion of the drug through the gel layer that is formed around the tablet. Further, the behaviour of crosslinked scleroglucan indicated that the polymers could be used as monolithic swellable systems for sustained release of the drugs or as films capable of regulating the diffusion of the bioactive substance [156]. Recently, a cross linked polysaccharide hydrogel of scleroglucan and gellan gum was prepared [157]. For the characterization of the co-cross linked polysaccharide, diffusion experiments through the swelled hydrogels were carried out in different environmental conditions, and the release from the tablets prepared with co-cross linked polysaccharides and theophylline was evaluated. The addition of calcium chloride in the formulation of the dosage forms allowed a further marked reduction in delivery rate to be obtained. This effect is related to the free ionized carboxylic groups still present in the gellan moiety of co-cross linked polysaccharide.

New hydrogels prepared by a crosslinking reaction between the polycarboxylated derivative of scleroglucan (sclerox) and 1,6-hexanedibromide have been reported [158]. The diffusion of theophylline through the swelled crosslinked hydrogels behaved as swellable monolithic systems suitable for sustained drug release. Preparation of novel hydrogels with scleroglucan using borax as a crosslinker [159] and diffusion of Ca (II) ions into a solution of the carboxylated derivative of Scleroglucan [160] have been reported.

10.20   Starch

Starch is a mixture of two complex carbohydrates, amylose and amylopectin, both of which are polymers of glucose. It is used by plants as a way to store excess glucose. Usually, the amylose and amylopectin constituents of starch are found in a ratio of 30:70 or 20:80, with amylopectin found in larger amounts than amylose. Starch is often found in the fruits, seeds, rhizomes or tubers of plants. The major resources for starch production and consumption worldwide are rice, wheat, corn, and potatoes. Clothing starch or laundry starch is a liquid that is prepared by mixing a vegetable starch in water, and is used in the laundering of clothes.

Starch, in its native and modified forms, is used extensively throughout the pharmaceutical industry as a disintegrating agent, as a binder or as a diluent in the tableting process. Starches can be gelatinized to make them cold water swellable. When formulated as a tablet, these starches can form a hydrophilic gel matrix that prolongs the release of an active ingredient. Different processes are proposed for pregelatinization-spray-drying technique. The non-toxicity and low production costs of thermally modified starches make them of great interest for the formulation of controlled-release tablets.

For many years, a number of chemical modifications have been developed to improve the properties of starches. Because of the abundance of hydroxyl groups in the polymer, cross-linking occurs when starch is treated with a bifunctional or multifunctional reagent such as acid anhydride, aldehyde, ethylenic compound etc. Cross-linking reinforces hydrogen bonds holding the granules together. This produces considerable changes in the gelatinization properties of starch granules and leads to a restriction in swelling property.

Mohile [161] reported the formulation of an acetyl salicylic acid sustained release tablet based on modified starches (pregelatinzed starch). Acetyl salicylic acid was mixed with pregelatinzed starch and the blend was compressed into tablets. These tablets, when exposed to aqueous fluids, rapidly released a fraction of drug. However, due to hydration and gelation of the starch at the tablet liquid interface, remaining acetyl salicylic acid exhibited prolonged dissolution profile.

Aerde and Remon [162] investigated a variety of starches as possible hydrophilic matrices for controlled drug release using theophylline as model drug. The modifications were both physical and chemical, such as pre-gelatinization by drum drying or by extrusion, partial hydrolysis and cross-linking. The tablets made of native corn starch partially depolymerized by acid hydrolysis and purely cross-linked corn starch disintegrated completely within 10 min. Consequently, they can not be used as matrices for sustained release. The tablets made of pregelatinized starches presented prolonged drug release. Pregelatinization by extrusion seems to induce slower release in comparison to starches pregelatinzed by the drum drying process. Pregelatinization coupled with an increasing cross-linking degree reduced the delay of drug release.

In another study, Visavarungroj and coworkers [163] evaluated different types of cross-linked starches and pregelatinized starches for their use as hydrophilic matrices. The results indicated that cross-linked starches, either pregelatinized or not, are not suitable for sustained release agents.

Lenaerts and coworkers introduced amylose cross-linked by epichlohydrin as a matrix for controlled release of drugs [164, 165]. A linear release of theophylline from the cross linked amylose used for tablet was observed in all cases. Linear increase in the cross-linking degree of the cross-linked amylose used for tablet preparation generated nonlinear diminution of release time. Advantages of this material include the ease in tablet manufacturing, the possibility of achieving controlled release at high drug concentration and the relative independence of release kinetics from drug loading within certain limits.

A new starch product for controlled drug release was introduced [166]. It consists of linear glucose polymer with a mean degree of polymerization of 30 and was prepared by enzymatic degradation of gelatinized potato starch, followed by precipitation (retrogradation), filtration and washing with ethanol [167]. The last process created powder with specific surface area greater than 1.5 m^2/g. Tablets compressed from a physical mixture of this material with theophylline released the drug with a decreasing rate due to porous diffusion when the tablet porosity was more than 7%, but nearly constant drug release was observed for lower porosity. Release rates from retrograted pregelatinzed starch tablets can be enhanced or decreased to the desired profile desired profile by different parameters such as geometries of the tablet, compression force and the incorporation of additional excipients [168].

Crosslinked high amylose starch (CLHAS) as an implant matrix for the delivery of ciprofloxacin (CFX) was ionvestigated [169]. All CLHAS implants were subjected for 24 hrs dissolution tests to evaluate swelling, erosion, water uptake and CFX release studies. Of the different formulations tested, CLHAS implants with 1% hydrogenated vegetable oil and 7.5% CFX provided the longest period of drug delivery without any initial burst effect.

Novel hydrogels has been synthesized with the reaction of starch and succinic anhydride (SA), using 4-dimethylaminopyridine as esterification catalyst and dimethylsulfoxide (DMSO) as solvent, followed by NaOH neutralization. The degree of substitution (DS) was in the range of 0.4–1.4. The maximum water absorbency of the products was about 120 g-water/g-dry gel. The product degraded biologically up to 60–80% in a period of 20 days showing good biodegradability [170].

10.21   Xanthan Gum

Xanthan gum is a high molecular weight extracellular polysaccharide produced by the fermentation of gram negative bacteria *Xanthomonas campestris*. The primary structure of this naturally produced cellulose derivative contains a cellulosic backbone (β-D-glucose residues) and a trisaccharide side chain of β-D-mannose – β-D-glucuronic acid-α-D-mannose attached with alternate glucose residues of the main chain (Figure 10.13). The terminal D-mannose residue may carryu a pyruvate function, the distribution of which is dependant on the bacterial strain and the fermentation conditions. The non-terminal D-mannose unit in the side chain contains an acetyl function. The anionic character of this polymer is due to the presence of both glucuronic acid and pyruvic acid groups on the side chain. Xanthan gum offers potential utility as a drug carrier because of its inertness and bio-compatibility. Xanthan gum is compatible with virtually all salts, and solution pH temperature has very little effect on the viscosity of its gel. It has been reported by many group of workers that xanthan gum can be used as an effective exccipient for sustained release formulations.

Dhopeshwarkar and Zatz [171] evaluated xanthan gum as a matrix former for the preparation of sustained release tablets. It was very effective in prolonging the release of soluble (chlorpheniramine

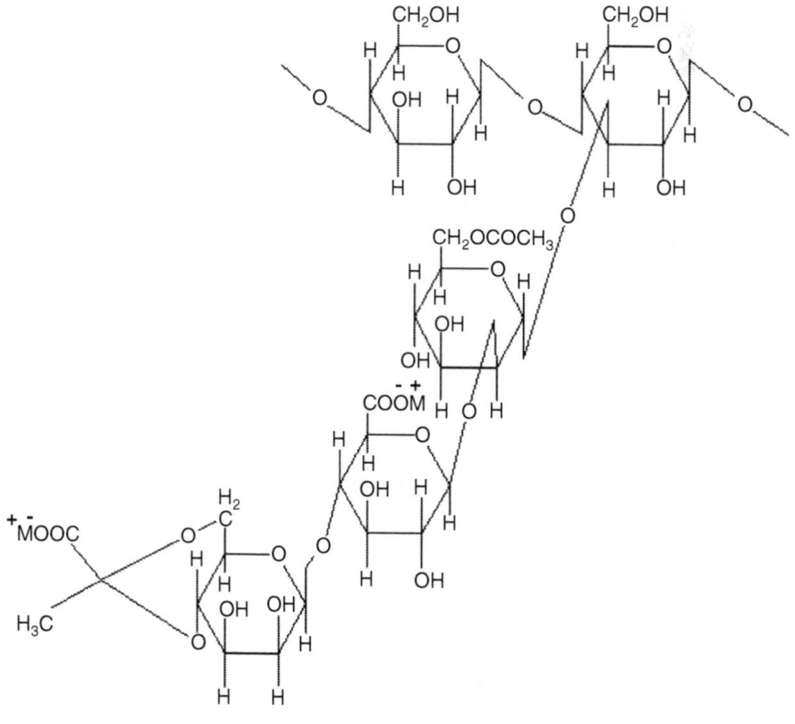

Figure 10.13 Chemical structure of xanthan gum.

maleate) and sparingly soluble (theophylline) drugs. The rate of release was slowed by decreasing the particle size of the gum or by increasing gum concentration. The release of soluble drugs was mainly via diffusion, whereas sparingly soluble or insoluble drugs were released principally via erosion. After hydration of the gum, drug release was essentially pH independent.

Talukdar *et al.* [172] undertook a comparative investigation to assess the performance of xanthan gum and hydroxy propyl methyl cellulose as hydrophilic matrix-forming agents in respect to compaction characteristics and *in vitro* drug release behavior. Considering the influence of ionic strength of the medium on drug release behavior, xanthan gum has the disadvantage that the drug release is influenced by the total salt concentration within the range of the gastrointestinal tract, whereas the drug release from HPMC matrices is independent of ionic strength. Compaction characteristics between the two polymers are quite similar, but the flowability of xanthan gum is better than that of HPMC.

Xanthan gum has been cross-linked with epichlorhydrin to obtain a hydrogel for use in drug retardation [173]. The obtained gel shows swelling degrees up to 100. Theophylline was inserted by diffusion into this gel. Controlled release of theophylline was estimated by elution in a closed recirculation system. The zero order kinetics was obtained in the diffusion process of theophylline into the eluent with in a 14 hr time interval.

Sinha *et al.* designed a formulation with a considerably reduced coat weight and gum concentration for colonic delivery of 5-fluorouracil for the treatment of colorectal cancer [174]. Rapidly disintegrating core tablets containing 50 mg of 5-fluorouracil were prepared and compression coating with 175 mg of granules containing a mixture of xanthan gum (XG) and guar gum (GG) in varying proportions was done. With this coat weight, a highly retarded drug release was observed. At the end of 24 h of dissolution the amount of drug released increased to $25\pm1.22\%$, $36.6\pm1.89\%$ and $42.6\pm2.22\%$, respectively in XG:GG 20:20, 20:10 and 10:20 tablets. Studies of XG:GG (10:20) tablets in presence of colonic contents showed an increased cumulative percent drug release of $67.2\pm5.23\%$ in presence of 2% cecal content and $80.34\pm3.89\%$ in presence of 4% cecal content after 19 hrs of incubation. Researchers have reported use of xanthan gum as carrier in developing drug loaded pellets [175, 176].

Xanthan-grafted copolymer of acrylamide (AAm) as a controlled release (CR) matrix for antihypertensive drugs such as atenolol

(ATL) and carvedilol (CDL) was reported [177]. Graft copolymerization of AAm onto xanthan gum (XG) was carried out by taking two different ratios of XG to AAm (1:5, 1:10) by free radical initiation polymerization using ceric ammonium nitrate (CAN). The *in vitro* release studies were also carried out on commercial tablet formulations of ATL and CDL along with grafted tablets. The release rate of CDL containing tablets with XG and grafted copolymer did not show significant difference, whereas CDL containing tablets with XG, grafted copolymer, and commercial formulations have shown significant differences.

References

1. S.K. Baveja, K.V. Ranga Rao, and K. Padmalatha Devi, *Intl. J. Pharm. Sci.*, Vol. 39, pp. 39–45,1987.
2. R.K. Khar, A. Ahuja and J. Ali, Controlled and Novel Drug Delivery (N.K. Jain, Ed.), CBS Publishers and Distributors, New Delhi, 1997.
3. P. Buri and E. Doelker, *Pharm. Acta Helv*, Vol. 55, pp. 189–197,1980.
4. A. Jayakrishnan and M.S. Latha, Controlled and novel Drug delivery (N.K. Jain, ed.), CBS Publishers and Distributors, New Delhi, 1997.
5. R.S.R. Murthy, Controlled and Novel Drug Delivery (N.K. Jain, Ed.), CBS Publishers and Distributors, New Delhi,1997.
6. M. Nakano, Y. Nakamura, K. Juni and T. Tomitsuka, *J. Pharm Dyn*. Vol.3, p. 1702, 1980.
7. M. Nakano, M. Kouketsu, Y. Nakamura, and K. Juni, *Chem. Pharm. Bull*, Vol. 28, p. 2905 1980.
8. D.F. Durso, in Handbook of water Soluble Gums and Resins (Davidson, Ed.), McGraw Hill, Kingsport Press, New York, 1980.
9. R. Alfonso Gennaro. In Remington: The science and practice of pharmacy. 20th Ed., Lippincott Williams and Wilkins, USA, 2000.
10. Merck Index. 4th Ed., Merck and Co., Inc. USA, 1989.
11. A. Wade and P. Jweller, in Handbook of Pharmaceutical Excipients, 2nd ed., American Pharmaceutical Association, Washington, DC, 1994.
12. S.K. Baveja, K.V. Ranga Rao and J. Arora, *Indian J. Pharm. Sci*, Vol. 50, pp. 89–92, 1988.
13. A. R. Ray, V. Batra, A. Bhowmick and B.K. Behra, *J. Pharm. Sci*, Vol. 83, pp. 632–635, 1994.
14. V. Kushwaha, A. Bhowmick, B.K. Behra, and A.R. Ray. *Artif. Cells Blood Substit. Immobil. Biotechnol*, Vol. 26, pp.159–172, 1998.
15. M. M. Meshali and K. E. Gabr, *Int. J. Pharm*, Vol. 89 (3), p. 177–181, 1993.
16. Hiromitsu Aoki, Tsuyoshi Katayama, Takeshi Ogasawara, Yasushi Sasaki, Saphwan Al-Assaf and Glyn O. Phillips, *Food Hydrocolloids*, Vol. 21 (3), pp. 353–358, 2007.
17. M. Nakano, Y. Nakamura, K. Takkiawa, M, Kousetsu, and T. Arita, *J. Pharm. Pharmacol*, Vol. 31, pp. 869–872, 1979.

18. S. Desai and S. Bolton, *Pharm. Res*, Vol. 10 (9), p. 1321–1325, 1993.

19. John G. Lyons, Declan M. Devine, James E. Kennedy, Luke M. Geever, Patrick O'Sullivan and Clement L. Higginbotham, *Eu. J. Pharm. Biopharm*, Vol. 64 (1), pp. 75–81, 2006.

20. G.P. Pfister, M. Bahader, and F. Korte, *J. Controlled Release*, Vol. 3, pp. 229–233, 1986.

21. Y. Murata, K. Nakada, E. Miyanito, S. Kawashima, and S.H. Seo, *J. Controlled Release*, Vol. 23, pp. 21–26, 1993.

22. Yoshifumi Murata, Etsuko Miyamoto and Susumu Kawashima, *J. Control. Rel*, Vol. 38 (2–3), 101–108, 1996.

23. M.D. Chavanpatil, A. Khadir and J. Panyam, *Pharm. Res*, Vol. 24 (4), pp. 803–810, 2007.

24. Oya Şanl, Nuran Ay and Nuran Işklan, *Eu. J. Pharm. Biopharm*, Vol. 65 (2), pp. 204–214, 2007.

25. Yongmei Xu, Changyou Zhan, Lihong Fan, Le Wang and Hua Zheng, *Int. J. Pharm*, Vol. 336 (2), pp. 329–337, 2007.

26. Yong-Dan Tang, Subbu S. Venkatraman, Freddy Y.C. Boey and Li-Wei Wang, *Int. J. Pharm*, Vol. 336 (1), pp. 159–165, 2007.

27. M. George and T.E. Abraham, *Int. J. Pharm*, Vol. 335 (1–2), pp. 123–129, 2007.

28. R. Rastogi, Y. Sultana, M. Aqil, A. Ali, S. Kumar, K. Chuttani and A.K. Mishra, *Int. J. Pharm*, Vol. 334 (1–2), pp. 71–77, 2007.

29. Zhanfeng Dong, Qun Wang and Yumin Du, *J. Membrane Sci*, *Int. J. Pharm*, Vol. 280 (1–2), pp. 37–44, 2006.

30. Pietro Matricardi, Ilenia Onorati, Tommasina Coviello and Franco Alhaique, *Int. J. Pharm*, Vol. 316 (1–2), pp. 21–28, 2006.

31. Zhidong Liu, Jiawei Li, Shufang Nie, Hui Liu, Pingtian Ding and Weisan Pan, *Int. J. Pharm*, Vol. 315 (1–2), pp. 12–17, 2006.

32. Sakchai Wittaya-areekul, Jittiporn Kruenate and Chureerat Prahsarn, *Int. J. Pharm*, Vol. 312 (1–2), pp. 113–118, 2006.

33. Chuanyun Dai, Bochu Wang, Hongwei Zhao, Biao Li and Jian Wang, *Colloids and surfaces B: Biointerfaces*, Vol. 47 (2), pp. 205–210, 2006.

34. Ulrike Bertram and Roland Bodmeier, *Eu. J. Pharm. Sci*, Vol. 27 (1), pp. 62–71, 2006.

35. Claudia Valenta, *Adv. Drug Del. Rev*, Vol. 57 (11), pp. 1692–1712, 2005.

36. Yu-Hsin Lin, Hsiang-Fa Liang, Ching-Kuang Chung, Mei-Chin Chen and Hsing-Wen Sung. *J. Biomaterials*, Vol. 26 (14), pp. 2105–2113, 2005.

37. Sung-Ching Chen, Yung-Chih Wu, Fwu-Long Mi, Yu-Hsin Lin, Lin-Chien Yu and Hsing-Wen Sung, *J. Control. Rel*, Vol. 96 (2), pp. 285–300, 2004.

38. Liu Xing, Chen Dawei, Xie Liping and Zhang Rongqing, *J. Control. Rel*, Vol. 93 (3), pp. 293–300, 2003.

39. K.B. Guiseley, N.F. Staneley, N.F. Stanley, and P.A. Whitehouse, Handbook of water Soluble Gums and Resins (R.L Davidson, Ed.), McGraw Hill, Kingsport Press, New York, 1980.

40. S. Takka, D. H. Dcak, and F. Acarturk, *Eur. J. Pharm. Sci*, Vol. 6, pp. 241–246, 1998.

41. M.C. Bonferoni, S. Rossi, M. Tamayo, J.L. Pedrez, G. Dominguez and C. Caramella, *J. Control. Rel*, Vol. 26, pp. 119–127, 1993.

42. M.C. Bonferoni, S. Rossi, M. Tamayo, J.L. Pedrez, G. Dominguez and C. Caramella, *J. Control. Rel*, Vol. 30, pp. 175–182, 1994.

43. D.L. Munday, J. Sujja-Areevath, P.J. Cox, and K.A. Khan, *Int. J. Pharm*, Vol. 139, pp. 53–62, 1996.

44. K.M. Picker and C. Gabelick, Proc, *Int, Syrup, Controlled Release Bioact. Mater*, Vol. 24, pp. 235–236, 1997.

45. K.M. Picker, *Drug Dev. Ind. Pharm*, Vol. 25, pp. 329–337, 1999.

46. K.M. Picker, *Drug Dev, Ind, Pharm*, Vol. 25, pp. 339–346, 1999.

47. A. Streubel, J. Siepmann and R. Bodmeier, *Eu. J. Pharm. Sci*, Vol. 18 (1), pp. 37–45, 2003.

48. Maria Cristina Bonferoni, Patrizia Chetoni, Paolo Giunchedi, Silvia Rossi, Franca Ferrari, Susi Burgalassi and Carla Caramella, *Eu. J. Pharm. Biopharm*, Vol. 57 (3), pp. 465–472, 2004.

49. Ulrike Bertram and Roland Bodmeier, *Eu. J. Pharm. Sci*, Vol. 27 (1), pp. 62–71, 2006.

50. A. Pourjavadi, Sh. Barzegar and F. Zeidabadi, *Reactive and Functional Polymers*, Vol. 67 (7), pp. 644–654, 2007.

51. Wei-Guo Dai, Liang C. Dong and Yan-Qiu Song, *Int. J. Pharm*, Vol. 342 (1–2), pp. 201–207, 2007.

52. P. Mayorga, F. Puisieux and G. Couarraze, *Int. J. Pharm*, Vol. 132 (1–2), pp. 71–79, 1996.

53. Carmen Remuñán-López, Ana Portero, José Luis Vila-Jato and María José Alonso, *J. Control. Rel*, Vol. 55 (2–3), pp. 143–152, 1998.

54. Lev E. Bromberg, Debra K. Buxton and Phillip M. Friden, *J. Control. Rel*, Vol. 71 (3), pp. 251–259, 2001.

55. S. S. Biju, S. Saisivam, N. S. Maria Gerald Rajan and P. R. Mishra, *Eu. J. Pharm. Biopharm*, Vol. 58 (1), pp. 61–67, 2004.

56. T. Bussemer and R. Bodmeier, *Int. J. Pharm*, Vol. 267 (1–2), pp. 59–68, 2003.

57. M. Jelvehgari, M.R. Siahi-Shadbad, S. Azarmi, Gary P. Martin and Ali Nokhodchi, *Int. J. Pharm*, Vol. 308 (1–2), pp. 124–132, 2006.

58. Dimpy Gupta, Charles H. Tator and Molly S. Shoichet, *J. Biomaterials*, Vol. 27 (11), pp. 2370–2379, 2006.

59. Y. Suzuki and Y. Makino, *J. Control. Rel*, Vol. 62 (1–2), pp. 101–107, 1999.

60. Jyrki Heinämäki, Martti Marvola, Irmeli Happonen and Elias Westermarck, *Int. J. Pharm*, 42 (1–3), pp. 105–115, 1988.

61. Marc S. Gordon, Anthony Fratis, Ronald Goldblum, Donald Jung, Kenneth E. Schwartz and Zak T. Chowhan, *Int. J. Pharm*, Vol. 115 (10), pp. 29–34, 1995.

62. Bin Lu, Rong Wen, Hong Yang and Yingju He, *Int. J. Pharm*, Vol. 333 (1–2), pp. 87–94, 2007.

63. S. Conti, L. Maggi, L. Segale, E. Ochoa Machiste, U. Conte, P. Grenier and G. Vergnault, *Int. J. pharm*, Vol. 333 (1–2), pp. 136–142, 2007.

64. S. Miyazaki, N. Kawasaki, T. Nakamura, M. Iwatsu, T. Hayashi, W.-M. Hou and D. Attwood, *Int. J. Pharm*, Vol. 204 (1–20), pp. 127–132, 2000.

65. Javed Ali, Shweta Arora, Alka Ahuja, Anil K. Babbar, Rakesh K. Sharma, Roop K. Khar and Sanjula Baboota, *Eu. J. Pharm. Biopharm*, Vol. 67 (1), pp. 196–201, 2007.

66. Xiaoqiang Xu, Minjie Sun, Feng Zhi and Yiqiao Hu, *Int. J. Pharm*, Vol. 310 (1–20), pp. 139–145, 2006.

67. Yasunori Sato, Y. Kawashima, H.Takeuchi and Hiromitsu Yamamoto, *Int. J. Pharm*, Vol. 275 (1–2), pp. 97–107, 2004.

68. Xue-qing Wang, Jun-dong Dai, Zhen Chen, Tao Zhang, Gui-min Xia, T. Nagai and Qiang Zhang, *J. Control. Rel*, Vol. 97 (3), pp. 421–429, 2004.

69. I. Hyuk Kim, Jung Hwan Park, In Woo Cheong and Jung Hyun Kim, *J. Control. Rel*, Vol. 89 (2), pp. 225–233, 2003.

70. Min-Soo Kim, Jeong-Soo Kim, Yeon-Hee You, Hee Jun Park, Sibeum Lee, Jeong-Sook Park, Jong-Soo Woo and Sung-Joo Hwang, *Int. J. Pharm*, Vol. 341 (1–2), pp. 97–104, 2007.

71. D. Torres, G. García-Encina, B. Seijo and J.L. Vila Jato, *Int. J. Pharm*, Vol. 121 (2), pp. 239–243, 1995.

72. S. Miyazaki, K. Ishi, and T. Nadai, *Chem. Pharm. Bull*, Vol. 29, pp. 3067–3069, 1981.

73. Y. Sawayanagi, N. Nambu, and T. Nagai, *Chem. Pharm. Bull*, Vol. 30, pp. 4213–4215, 1982.

74. Y. Sawayanagi, N. Nambu, and T. Nagai, *Chem. Pharm. Bull*, Vol. 30, pp. 4216–4218, 1982.

75. S. Miyazaki, W.M. Hou, M. Takada, and T. Komai, *Chem. Pharm. Bull*, Vol. 33, pp. 3986–3992, 1985.

76. S. Miyazaki. H. Yamaguchi, C. Yokouchi, M. Takada and W.M. Hou, Chem. J. *Pharm. Pharmacol*, Vol. 36, pp. 4033–4038, 1988.

77. S. Miyazaki., H. Yamaguchi, C. Yokouchi, M. Takada and W.M. Hou, *Chem. J. Pharm. Pharmacol*, Vol. 40, pp. 642–643, 1988.

78. A.G. Nigalaye, P. Adusumilli, and S. Botton, *Drug Dev. Ind. Pharm*, Vol. 16, pp. 449–467, 1990.

79. S. Cafaggi, R. Leardi, B. Parodi, G. Caviglioli, E. Russo and G. Bignardi, *J. Control. Rel*, Vol. 102 (1), pp. 159–169, 2005.

80. P.S. Adusumilli and S. M. Bolton, *Drug Dev, Ind. Pharm*, Vol. 17, pp. 1931–1945, 1991.

81. S.C. Vasudev, T. Chandy, and C.P. Sharma, *J. Biomaterial*, Vol. 18, pp. 375–381, 1997.

82. Yan Wu, Jia Guo, Wuli Yang, Changchun Wang and Shoukuan Fu, *Polymer*, Vol. 47 (15), pp. 5287–5294, 2006.

83. K. Mladenovska, R.S. Raicki, E.I. Janevik, T. Ristoski, M.J. Pavlova, Z. Kavrakovski, M.G. Dodov and K. Goracinova, *Int. J. Pharm*, Vol. 342 (1–2), pp. 124–136, 2007.

84. C. Yomota, T. Miyazaki and S. Okada, Yakugaku Zasshi, *Chem. Abstr*, Vol. 114, pp. 257–263, 1994.

85. Yue-bin Ge, Da-wei Chen, Li-ping Xie and Rong-qing Zhang, *Int. J. Pharm*, Vol. 338 (1–2), pp. 142–151, 2007.

86. K.E. Gabr and M.M. Meshali, *Ind, J, Pharm*, Vol. 89, pp. 177–183, 1993.

87. Ana Portero, Desirée Teijeiro-Osorio, María J. Alonso and Carmen Remuñán-López, *Carbohydrate Polymers*, Vol. 68 (4), pp. 617–625, 2007.

88. Lichen Yin, Likun Fei, Fuying Cui, Cui Tang and Chunhua Yin, *J. Biomaterials*, 28 (6), pp. 1258–1266, 2007.

89. Alexander H. Krauland and María José Alonso, *Int. J. Pharm*, Vol. 340 (1–2), pp. 134–142, 2007.

90. Fu Chen, Zhi-Rong Zhang and Yuan Huang, *Int. J. Pharm*, Vol. 336 (1), pp. 166–173, 2007.

91. Y. Diebold, M. Jarrín, Victoria Sáez, Edison L.S. Carvalho, María Orea, M. Calonge, B. Seijo and María J. Alonso, *J. Biomaterials*, Vol. 28 (8), pp. 1553–1564, 2007.

92. Lene Simonsen, Lars Hovgaard, Per Brøbech Mortensen and Helle Brøndsted, *Eu. J. Pharm. Sci*, Vol. 3 (6), pp. 329–337, 1995.

93. Rita Cortesi, Elisabetta Esposito, Maria Osti, Enea Menegatti, Giacomo Squarzoni, Stanley Spencer Davis and Claudio Nastruzzi, *Eu. J. Pharm. Sci*, Vol. 47 (2), pp. 153–160, 1999.

94. Katherine L. Rensberger, Dean A. Hoganson and Reza Mehvar, *Int. J. Pharm*, Vol. 207 (1–2), pp. 71–76, 2000.

95. Eddy Castellanos Gil, Antonio Iraizoz Colarte, Bernard Bataille, José Luis Pedraz, Fernand Rodríguez and Jyrki Heinämäki, *Int. J. Pharm*, Vol. 317 (1), pp. 32–39, 2006.

96. A. Aumelas, A. Serrero, A. Durand, E. Dellacherie and M. Leonard, *Colloids and Surfaces B: Biointerfaces*, Vol. 59 (1), pp. 74–80, 2007.

97. Eun-Jin Lee, Sa-Won Lee, Han-Gon Choi and Chong-Kook Kim, *Int. J. Pharm*, Vol. 218 (1–2), pp. 125–131, 2001.

98. S. Almeida Prieto, J. Blanco Méndez and F.J. Otero Espinar, *Eu. J. Pharm. Biopharm*, Vol. 59 (3), pp. 511–521, 2005.

99. David C. Bibby, Nigel M. Davies and Ian G. Tucker, *Int. J. Pharm*, Vol. 197 (1–2), pp. 1–11, 2000.

100. T. Srichana, R. Suedee and W. Reanmongkol, *Respiratory Medicine*, Vol. 95 (6), pp. 513–519, 2001.

101. Brunella Cappello, Giuseppe De Rosa, Lucia Giannini, Maria Immacolata La Rotonda, Giuseppe Mensitieri, Agnese Miro, Fabiana Quaglia and Roberto Russo, *Int. J. Pharm*, 319 (1–2), pp. 63–70, 2006.

102. Mar Másson, Thorsteinn Loftsson, Gisli Masson and Einar Stefansson, *J. Control. Rel*, Vol. 59 (1), pp. 107–118, 1999.

103. F. Kedzierewicz, C. Lombry, R. Rios, M. Hoffman and P. Maincent, *Int. J. Pharm*, Vol. 178 (1), pp. 129–136, 1999.

104. Shozo Miyazaki, Hirotatsu Aoyama, Naoko Kawasaki, Wataru Kubo and David Attwood, *J. Control. Rel*, Vol. 60 (2–3), pp. 287–295, 1999.

105. Mattias Paulsson, Helene Hägerström and Katarina Edsman, *Eu. J. Pharm. Sci*, Vol. 9 (1), pp. 99–105, 1999.

106. S. Miyazaki, N. Kawasaki, W. Kubo, K. Endo and D. Attwood, *Int. J. Pharm*, Vol. 220 (1–2), pp. 161–168, 2001.

107. Wataru Kubo, Shozo Miyazaki and David Attwood, *Int. J. Pharm*, Vol. 258 (1–2), pp. 55–64, 2003.

108. P.S. Rajinikanth, J. Balasubramaniam and B. Mishra, *Int. J. Pharm*, Vol. 335 (1–2), pp. 114–122, 2007.

109. Brahma N. Singh and Kwon H. Kim, *Int. J. Pharm*, Vol. 341 (1–2), pp. 143–151, 2007.

110. S.K. Baveja, K.V. Ranga Rao and J. Arora, *Ind. J. Pharm. Sci*, Vol. 50, pp. 89–92, 1988.

111. H.L. Bhalla and S.O. Gulati, *Ind. Drugs*, Vol. 24, pp. 338–342, 1987.

112. S. Alataf, K. Yu, J. Parasrampuria and D.R. Friend, *Proced. Int, Symp. Controlled Release Bioact, Mater*, Vol. 23, pp. 541–542, 1996.

113. S. Alataf, K. Yu, J. Parasrampuria and D.R. Friend, *Pharm. Res*, Vol. 15, pp. 1196–1201, 1998.
114. M.H. Gebert and D.R. Friend, *Pharm, Dev, Technol*, Vol. 3, pp. 315–320, 1998.
115. K.L.K. Paranjothy and P.P. Thampi, *Indian Drugs*, Vol. 29, pp. 84–87, 1997.
116. K.L.K. Paranjothy and P.P. Thampi, *Indian Drugs*, Vol. 29, pp. 404–407, 1992.
117. K.L.K. Paranjothy and P.P. Thampi, *Ind. J. Pharm. Sci*, Vol. 59, pp. 49–54, 1997.
118. Irit Gliko-Kabir, Boris Yagen, Muhammad Baluom and Abraham Rubinstein, *J. Control. Rel*, 63 (1–2), pp. 129–134, 2000.
119. A.N. Misra and J.N. Baweja, *Ind. J. Pharm. Sci*, Vol. 59, pp. 316–320, 1997.
120. S. Narasimha Murthy, Shobha Rani R. Hiremath and K.L.K. Paranjothy, *Int. J. Pharm*, 272 (1–2), pp. 11–18, 2004.
121. Kumaresh S. Soppimath, Anandrao R. Kulkarni and Tejraj M. Aminabhavi, *J. Control. Rel*, Vol. 75 (3), pp. 331–345, 2001.
122. M. George and T.E. Abraham, *Int. J. Pharm*, Vol. 335 (1–2), pp. 123–129, 2007.
123. L. Vervoort and R. Kinget, *Int. J. Pharm*, Vol. 129 (1–2), pp. 185–190, 1996.
124. Liesbeth Vervoort, Patrick Rombaut, Guy Van den Mooter, Patrick Augustijns and Renaat Kinget, *Int. J. Pharm*, Vol. 172 (1–2), pp. 137–145, 1998.
125. Barbara Stubbe, Bart Maris, Guy Van den Mooter, Stefaan C. De Smedt and Joseph Demeester, *J. Control. Rel*, Vol. 75 (1–2), pp. 103–114, 2001.
126. J. Sujja-areevath, D.L. Munday, P. J. Cox and K.A. Khan, *Int. J. Pharm*, Vol. 39 (1–2), pp. 53–62, 1996.
127. J. Sujja-areevath, D.L. Munday, P. J. Cox and K.A. Khan, *Eu. J. Pharm. Sci*, Vol. 6 (3), pp. 207–217, 1998.
128. Dale L. Munday and Philip J. Cox, *Int. J. Pharm*, Vol. 203 (1–2), pp. 179–192, 2000.
129. G. V. Murali Mohan Babu, Ch. D. S. Prasad and K.V. Ramana Murthy, *Int. J. Pharm*, Vol. 234 (1–2), pp. 1–17, 2002.
130. Kang Wang and Zhimin He, *Int. J. Pharm*, Vol. 244 (1–2), pp. 117–126, 2002.
131. Li-Gui Chen, Zhi-Lan Liu and Ren-Xi Zhuo, *Polymer*, Vol. 46 (16), pp. 6274–6281, 2005.
132. Jian Du, J. Dai, Jun-Long Liu and Theresa Dankovich, *Reactive and Functional Polymers*, Vol. 66 (10), pp.1055–1061, 2006.
133. V.R. Sinha and Rachna Kumria, *Int. J. Pharm*, Vol. 224 (1–2), pp. 19–38, 2001.
134. Melike Üner and Turan Altınkurt. I. Farmaco, 59 (7), 567–573 (2004).
135. Tommasina Coviello, Franco Alhaique, Antonello Dorigo, Pietro Matricardi and Mario Grassi, *Eu. J. Pharm. Biopharm*, Vol. 66 (2), pp. 200–209, 2007.
136. Z. Aydin and J. Akbuga, *Int, J. Pharm*, Vol. 137, pp. 133–136, 1996.
137. P. Sriamornsak, S. Puttipipatkhachorn, and S. Prekongpan, *Int. J. Pharm*, Vol. 156, pp. 189–194, 1997.
138. P. Sriamornsak and J. Nunthanid, *Int. J. Pharm*, Vol. 160, pp. 207–212, 1998.
139. Yoshifumi Murata, Michiko Miyashita, Kyouko Kofuji, Etsuko Miyamoto and Susumu Kawashima, *J. Control. Rel*, Vol. 95 (1), pp. 61–66, 2004.
140. S. Miyazaki, N. Kawasaki, T. Nakamura, M. Iwatsu, T. Hayashi, W.-M. Hou and D. Attwood, *Int. J. Pharm*, Vol. 204 (1–2), pp. 127–132, 2000.
141. P. Giunchedi, U. Conte, P. Chetoni and M.F. Saettone, *Eu. J. Pharm. Sci*, Vol. 9 (1), pp. 1–7, 1999.
142. R. Fassihi and H. Kim, *J. Pharm*, Vol. 160, pp. 207–212, 1997.

143. Mine Orlu, Erdal Cevher and Ahmet Araman, *Int. J. Pharm*, Vol. 318 (1–2), pp. 103–117, 2006.

144. Mahesh Chavanpatil, Paras Jain, Sachin Chaudhari, Rajesh Shear and Pradeep Vavia, *Int. J. Pharm*, Vol. 304 (1–2), pp. 178–184, 2005.

145. Mahesh D. Chavanpatil, Paras Jain, Sachin Chaudhari, Rajesh Shear and Pradeep R. Vavia, *Int. J. Pharm*, Vol. 316 (1–2), pp. 86–92, 2006.

146. Baljit Singh, G.S. Chauhan, S.S. Bhatt and Kiran Kumar, *Carbohyrate Polymers*, Vol. 64 (1), pp. 50–56, 2006.

147. Baljit Singh, Nisha Sharma and Nirmala Chauhan, *Carbohydrate Polymers*, Vol. 69 (4), pp. 631–643, 2006.

148. S. Rizk, C. Duru, L. Bardet, R. Furtunet, and M. Jacob, *J. Pharm. Belg*, Vol. 48, pp. 197–203, 1993.

149. S. Rizk, C. Duru, M. Bertoni, F. Ferrari, M. Jacob, D. Gaudy, and C. Carmella, *Int J. Pharm*, Vol. 112, pp. 125–131, 1994.

150. S. Rizk, C. Duru, R. Sabatier, and M. Jacob, *J. Pharm, Belg*, Vol. 48, p. 197, 1993.

151. S. Rizk, C. Duru, M. Jacob, D. Gaudy, *Drug Dev. Ind. Pharm*, Vol. 20, pp. 2563–2574, 1994.

152. F. Alhaique, G, Riccioni, F.M. Riccieri, E. Santucci and E. Touitou, *Drug Des, Delivary*, Vol. 5, pp. 141–148, 1989.

153. F. Alhaique, G. Riccioni, F.M. Riccieri, E. Santucci and E. Touitou, *Drug Des, Delivary*, Vol. 5, pp. 249–257, 1990.

154. F. Alhaique, E. Santucci, M. Carafa, T. Coviello, E. Murtus, and F.M. Riccieri, *J. Control. Rel*, Vol. 42, pp. 157–164, 1996.

155. F. Alhaique, M. Carafa, F.M. Riccieri, E. Santucci and E. Touitou, *Pharmazie*, Vol. 48, pp. 432–435, 1993.

156. F. Alhaique, T. Coviello, M. Carafa, E. Murtus, G. Rambone, F.M. Riccieri, E. Santucci and M. Paci, *Proc. Int. Symp. Controlled Release Society Bioactive mater*, Vol. 24, pp. 1043–1044, 1997.

157. F. Alhaique, T. Coviello, M. Dentine, M. Carafa, E. Murtus, G. Rambone, F.M. Riccieri, and P. Desidire, *J. Control. Rel*, Vol. 55, pp. 57–66, 1998.

158. Tommasina Coviello, Mario Grassi, Giuseppe Rambone, Eleonora Santucci, Maria Carafa, Evelina Murtas, Fulvio M. Riccieri and Franco Alhaique, *J. Control. Rel*, Vol. 60 (2–3), pp. 367–378, 1999.

159. T. Coviello, M. Grassi, R. Lapasin, A. Marino and F. Alhaique. *J. Biomaterials*, Vol. 24 (16), pp. 2789–2798, 2003.

160. Tommasina Coviello, Franco Alhaique, Chiara Parisi, Pietro Matricardi, Gianfranco Bocchinfuso and Mario Grassi, *J. Control. Rel*, Vol. 102 (3), pp. 643–656, 2005.

161. R.B. Mohile, *Ind. J. Pharm. Sci*, Vol. 48, pp. 150–156, 1986.

162. P.V. Aerde and J.P. Remon, *Int. J. Pharm. Sci*, Vol. 45, pp. 145–152, 1988.

163. N. Visavarungroj, J. Herman, and J.P. Remon, *Drug Dev. Ind. Pharm*, Vol. 16, pp. 1091–1108, 1990.

164. V. lenaerts, Y. Dumolin and M.A. Mateescu, *J. Control. Rel*, Vol. 15, pp. 39–46, 1991.

165. V. lenaerts, Y. Dumolin, L. Cartilier, and M.A. Mateescu, *Proc. Int. Symp. Controlled release Bioact. Mater*, Vol. 19, pp. 30–31, 1992.

166. G.H.P. Tewierik, A.C. Eissens, J. Bergsma, A.W. Arends and C.F. Lerk, *J. Control. Rel*, Vol. 45, pp. 25–33, 1997.

167. G.H.P. Tewierik, A.C. Eissens, J. Bergsma, T. Boersma, A.W. Arends and C.F. Lerk, *Int. J. Pharm*, Vol. 134, pp. 27–36, 1996.
168. G.H.P. Tewierik, A.C. Eissens, J. Bergsma A.W. Arends and G.K. Bolhuis, *Int, J. Pharm*, Vol. 157, pp. 181–187, 1997.
169. Cyril Désévaux, Pascal Dubreuil and Vincent Lenaerts, *J. Control. Rel*, Vol. 82 (1), pp. 83–93, 2002.
170. Toshio Yoshimura, Rumi Yoshimura, Chieko Seki and Rumiko Fujioka, *Carbohydrate Polymers*, Vol. 64 (2), pp. 345–349, 2006.
171. V. Dhopeshwarker and J.L. Zatz, *Drug Dev, Ind. Pharm*, Vol. 19, pp. 999–1002, 1993.
172. M.M. Talukdar, A. Michoel, P. Rombaut, and R. Kinget, *Int. J. Pharm*, Vol. 129, pp. 233–241, 1996.
173. S. Dumitriu and M. Dumitriu, *J. Biomaterials*, Vol. 12, pp. 821–826, 1991.
174. V.R. Sinha, B.R. Mittal, K.K. Bhutani and Rachna Kumria, *Int. J. Pharm*, Vol. 269 (1), pp. 101–108, 2004.
175. Helton Santos, Francisco Veiga, M Eugénia Pina and João José Sousa, *Eu. J. Pharm. Sci*, Vol. 21 (2–3), pp. 271–281, 2004.
176. Helton Santos, Francisco Veiga, M Eugénia Pina and João J. Sousa, *Int. J. Pharm*, Vol. 295 (1–2), pp. 15–27, 2005.
177. Raghavendra C. Mundargi, Sangamesh A. Patil and Tejraj M. Aminabhavi, Carbohydrate *Polymers*, Vol. 69 (1), pp. 130–141, 2007.

Index

1,2-Cyclohexanedicarboxylic anhydride, 40, 42–43
1,4:3,6-anhydohexitols, 405–407
1,4-butanediamine, 363–366
1,4-Diazobicyclo[2.2.2]octane, 40
1,4-Phenylenediisocyanate (PDI), 50, 53, 57–58
1,6-Diazidocyclohexane, 47, 49
1,6-Diazidohexane, 50
1,7-Octadiyne, 47
1-Methylvinyl isocyanate, 32–34
2,4-Diisocyanatotoluene (TDI), 44
2,5-Bis(formyl)furan
  polymer preparation, 396–397, 418
  synthesis, 392–393, 395
2,5-Bis(hydroxymethyl)furan
  polymer preparation, 407, 413
  synthesis, 392
2,5-Furandicarboxylic acid
  polymer preparation, 398, 400, 404, 408, 411, 419
  synthesis, 393–394
2,5-Furandicarboxylic acid dichloride
  polymer preparation, 405–408, 415
  synthesis, 396
2,5-Furandicarboxylic acid esters, 391, 398–400, 410
2-Hydroxyethyl methacrylate (HEMA), 44–45
2-Pyrrolidone, 357, 366–368
3-(Acryloxy)-2-hydroxypropyl methacrylate (AHPMA), 44–45

3-hydroxyacyl-acyl carrier protein CoA transferase, 284
4,4'-Methylenebis (phenylisocyanate), 47–48
4-coumarate-CoA ligase (4CL), 258, 259
5,5'(oxy-bis(methylene))bis-2-furfural, 392, 395, 397, 416
5-Norbornene-2-carboxylic acid, 159, 161

Acacia, 429
Acetic acid, 40
Acetic anhydride, 33
Acetoacetyl-CoA reductase (PhbB), 280
Acetyl coenzyme A 25
Acidic clays, 136
Acidogenic fermentation, 182–184
Acrylates, 223, 224, 232, 238
Acrylic acid, 36, 40, 43
Acrylonitrile, 235
Activated sludge, 179–182
Adhesive, 238, 242
Adipic acid, 362–363
Adipoyl chloride (AD), 58–61
Aerobic dynamic feeding (ADF), 180
Agar, 430
Agrobacterium-mediated transformation, 267, 269, 284, 288
Agro-industrial wastes, 178–184
AIBN 230

Alfalfa, 259, 260, 273, 281, 282
Alginate, 431
Aliphatic polyesters, 358, 370
Alkyd resins, 25, 27
alkyl halides
    bromide, 224, 230
    chloride, 224, 230
Alkylaluminum dihalides
    aqua complexes of, 79–80
    haloaluminates from, 79–80
    in butyl production, 79–81
    in conjunction with
        dialkylaluminum halides,
        81–82
Aluminoxanes
    modification of Parker system,
        130–131
    supported, 147–148
    with boron halides, 127, 129–131
    with Brønsted acids, 127–130
    with carbocation synthons,
        126–129
    with halogens, 127–130
    with silicenium ion synthons,
        127–130
Aluminum bromide
    in butyl production, 76–77
    self-ionization of, 78–79
Aluminum chloride
    drawbacks of in butyl
        production, 72–74
    initiation mechanisms, 76–79
    supported analogs, 135–136
    termination reactions, 72–74
Amflora, 247, 254, 289
Amylopectin, 247, 249–254
Amylose, 249–253
Anionic polymerization, 233
Aqueous cationic polymerization
        systems
    DBSA system, 153
    lanthanide triflates, 148–155
    PFLA based systems, 153–158
Arabidopsis thaliana, 273, 281–284

Arabinitol, 341
Architecture, 228
Arginine, 285–289
Aromatical monomers, 362
Aspect ratio, 19
Atom transfer radical
        polymerization, 6, 11
    activator concentration, 227
    active species, 225
    ambient conditions, 232, 234
    bulk, 233, 237
    controlled radical
        polymerization, 233
    conversion, 228, 233
    deactivator, 237
    dormant species, 225
    emulsion, 227
    equillibrium, 225
    functional initiators, 228, 229
    functionality, 221, 227, 228, 230
    heterogeneous, 227, 237
    homogeneous, 227, 237
    kinetics, 226, 233
    living radical polymerization,
        221, 222
    mechanistic steps, 226
    metal complex, 226, 236
    molecular weight, 221, 224,
        226, 230
    molecular weight distribution,
        223, 224, 226, 230, 241
    monomer concentration, 226
    persistent radical effect, 227
    polydispersity, 221, 233, 236, 239
    polymerization mechanism, 225
    polymerization rate, 226, 227
    prepolymers, 236
    propagation, 223, 226
    rate constant, 226
    reaction medium, 227
    solid supports, 227
    synthesis, 221, 224
    termination, 224, 226
Atomic force microscopy, 6

Azaleic acid, 344
Azelaoyl chloride (AZ), 58–62
Azobis(2-aminopropane) dihydrochloride, 36

Bacteria, 354–356
Bacterial fermentation, 354
Benzoyl peroxide, 29, 33
Beta vulgaris, 282
Biobased polymers, 353, 358
Bisphenol A (BA), 27, 29
Bisphenol-A 330, 341
Block copolymers, 5, 6
Brassica napus, 282
Butanediol, 2,3- 306
Butyl rubber
  production methods and limitations, 72–75
  structure of, 72
Butyrolactones, 4

Caffeic acid 3-O-Methyltransferase (COMT), 257, 258
Calcium oxide, 29, 33
Carbamoylation, 32
Carbon dioxide, 331
Carbon tetrabromide, 50
Carrageenan, 434
Cassava, 249, 250, 253
Castor oil, 308, 329, 343
Catalyst, 223, 224, 230, 237, 241
Cationic olefin polymerization
  heterogeneous initiator systems for, 134–158
  homogeneous initiator systems for, 75–134
  of renewable monomers, 161–163
Cationic polymerization, 39–41
Catylysts, 358
Cellulose, 247, 248, 254–257, 259, 261, 436, 306, 387–390
Chain transfer agent, 17

Chain-extender
  hexamethylene diisocyanate, 358–359
Chemical composition, 227
Chemical transformations, 231
Chitosan, 438
Cinnamoyl-CoA reductase (CCR), 258, 260
Cinnamyl alcohol dehydrogenase (CAD), 258–260
cis-prenyltransferase (CTP), 262, 263, 268, 269
Citric acid, 306, 321–322
Click chemistry, 229
Climate protection, 222
Coating, 236, 242
Coatings
  curing agents, 324
  performance, 324–325, 341
  UV stability, 324, 337
Cobalt naphthenate, 36, 39, 43
Collagen, 247, 248, 269, 274–276
Composites, 239
Composition, 221, 227
Continuous flow stirred tank reactors, 74
Copolyester
  poly(butylene succinate-co-adipate), 359
  poly(butylene succinate-co-butylene adipate), 361
  poly(butylene-co-hexylene succinate), 359
  poly(L-lactic acid-co-succinic acid-co-1,4-butanediol), 361
Copolymers
  block, 227, 230, 233, 234, 240
  brush, 228
  graft, 241
  star, 227, 228
  triblock, 233, 234, 241
Copper sulfate, 47, 50
Corn, 249, 250, 252, 253, 256, 257, 259, 281, 282, 290

Cotton, 255, 259, 281, 282, 284

Coumarate, 3-hydroxylase (C3H), 258, 260

Crosslink density, 19

Cyanophycin, 247, 248, 269, 285–290

Cyanophycin synthetase, 286, 288, 289

Cyclic polymer chains, 318–320, 336

Cyclohexane-1,4-dicarbonyl dichloride (CH), 58–59

Dextran, 442

Dextrin, 443

Dialkylaluminum halides
aqua complexes of, 81
haloaluminates from, 81
overview of compatible initiators, 88–89
with alkali/alkaline metal and actinide salts, 94–95
with Brønsted acids, 89–90
with carbocation synthons, 90–91
with electron acceptors, 93–94
with halogens, 91–92
with metal halides, alkoxides, alkoxy halides, 94–95

Dianhydrohexitol, 1,4:3,6- 306, 311–328, 331–341, 345

Dibutyltin dilaurate, 33

Dicarboxylic acid

Diester, 358

Differential scanning calorimetry, 6

Diol, 358
1,4-butanediol, 354, 356–357, 362–363
ethylen glycol, 359, 362–363

Diphenyl carbonate, 330–331, 336–337

Divinylbenzene, 39, 41

Drug delivery, 241

Elastin, 247, 248, 269, 276–278

Elastine like polypeptide (ELP), 277, 278

Electron transfer initiator systems, 96–99

Environment friendly, 222

Epoxidation, 35, 37, 40

Eugenol, 160–163

Fatty acid
calendic acid, 232
eleostearic acid, 232
erucic acid, 232
lauric acid, 232
linoleic acid, 232
linolenic acid, 232
oleic acid, 232
petroselinic acid, 232
ricinoleic acid, 232
stearic acid, 232
vernolic acid, 232

Fatty acids
linoleic, 24, 26, 36
linolenic, 24, 26, 36
oleic, 24, 26, 35–37, 39, 51, 66
palmitic, 24, 26
ricinoleic, 29, 33, 35
sebacic acid, 33, 35
stearic, 24, 26

Feast and Famine, 180–181, 185–187

Ferulate, 5-hydroxylase (F5H), 258–260

Festuca arundinacea, 259

Fibrous Proteins, 269, 270–279

Fig tree, 265

Fillers, 234

Flame retardants, 235

Flax, 259, 282

Formic acid, 35

Fossil fuels, 2

Fossil reserves, 222

Free-radical polymerization, 39

Fructose, 383–385

Functional materials, 227

Functional polymers, 224

Functionality, 221, 227, 228, 230

Furandicarboxylic acid, 2,5- 307, 328, 347
Furanic poly esters, 398–408
Furanic poly Schiff bases, 396–398
Furanic polyamides, 408–412
Furanic polybenzoimidazoles, 414
Furanic polyoxadiazoles, 415
Furanic polyurethanes, 412–414

Gamma radiation, 71, 132–134
Gamma-butyrolactone, 356
Gellan Gum, 445
Glass transition temperature, 10
Glucose, 306, 386–387
Glutamic acid, 306
Glycerol, 24, 27, 29–30, 32–33, 36, 50–51 , 307, 329, 339
Glycerolysis, 25, 27, 29, 32
Glyceryl trioleate (Triolein), 50
Glycogen accumulating organisms, 179
Gossypium hirsutum, 282
Granule bound starch synthase (GBSS), 249, 250, 252–254
Green chemistry attributes of, 70
Gross energy requirement, 222, 241
Group transfer polymerization, 11
Guar Gum, 446
Guayule, 264–269

Heteropolyacids, 138–139
Hevea brasiliensis, 261–268
High tyrosine-content peptide, 261
Honey Locust Bean Gum, 453
Human procollagen, 276
Hydrofluorocarbons, 158
Hydrogen peroxide, 35, 40
Hydrogenation, 356
Hydroquinone, 27–31, 36
Hydroxycinnamoyl-CoA:shikimate hydroxycinnamoyl transferase (HCT), 258, 260

Hydroxylevulinic acid, 5- 309
Hydroxymethylfurfural, 5- 307

Initiators
  aliquat-336 234
  alkyne, 229
  amine, 229, 225
  azide, 229
  difunctional, 229
  ethyl-2-bromoisobutyrate, 234
  functional groups, 223, 228
  hydroxy, 229
  initiation, 223, 225, 226
  initiator concentration, 226
Ink, 238
Intercalated Lewis acids, 139
Inulin, 449, 386
Ionic liquids, 384–390
Isohexide bis(alkyl / aryl carbonate), 337–339
Itaconic acid, 306, 311

Karaya Gum, 450
Konjac Glucomannan, 451

Lactuca serriola, 265
Latex coagulation, 268
Lettuce, 264, 265
Ligand
  bipyridine, 240
  pentamethyl diethylene triamine (PMDETA), 233, 237
  tris[2-(dimethyl amino)ethyl] amine (Me6Tren), 239
Light scattering, 16
Lignin, 247, 248, 254–261, 309, 311
Limonene, 13
Linum usitatissimum, 259, 282
Living polymerization, 4
Locust Bean Gum, 452
Lysine, 347
Lysyl hydroxylase (LH3), 276

Magnesium chloride, 137–138

Major ampullate spidroin (MaSp), 271–273, 278

MALDI-ToF-MS 318–320, 333–335, 339

Maleic acid / anhydride, 311, 321, 323

Maleic anhydride, 27, 29, 40, 43–44, 46, 40, 43–44, 46

Manihot esculenta, 253

Material properties
    flexibility, 238, 239
    glass transition temperature, 233, 238
    gloss properties, 236
    hardness, 236
    mecanical properties, 222
    morphology, 234, 241
    optical properties, 240
    physical crosslinks, 234
    physical properties, 227, 233, 234
    side chain crystallinity, 233, 239
    solvent resistance, 240
    steric effect, 238
    thermal decomposition, 236
    thermal stability, 242
    thermoplastic elastomers, 233, 234, 239, 240, 241

Medicago sativa, 259, 282

Medium-chain-length PHA (mclPHA), 279, 282–285

Metal cation WCA initiator systems
    bis-cyclopentadienyl WCA systems, 111–117
    mono-cyclopentadienyl WCA systems, 117–119
    non-cyclopentadienyl Group IV WCA systems, 119–121
    transition metal nitrile WCA systems, 121–125

Metathesis, 309, 345–346

Metathesis reaction, 10

Methacrylates, 232, 233

Methyl chloride, 75

Methyl ethyl ketone peroxide, 43

Methyl methacrylate, 233, 235, 237, 240

Methyl methacrylate (MMA), 29, 36, 44–45

Methyl oleate, 35–37

Micelles, 227

Microdomains, 234

Miscanthus, 290

Mixed Lewis acid initiator systems
    aluminum based (Parker system), 81–82
    heterogeneous strong/weak Lewis acids, 142–144
    Marek complex solid system, 83, 139–140
    Mitsubishi system, 83
    modes of initiation for, 86–88
    Nippon Oil Company system, 83–84
    Nippon Petrochemical system, 83–84
    Priola system, 84–85
    Strohmayer system, 83
    Sumitomo Chemical Company systems, 84–86
    Young and Kellog system, 82

Modified processes
    activator regenerated by electron transfer (ARGET), 230
    ascorbic acid, 230
    continuous activator regeneration (ICAR), 230

Molecular architecture, 227, 230

Molecular sieves, 136–137

Molecular weight distribution, 5

Monomers
    from saccharides, 306–307
    from vegetable oils, 307–309

Montmorillonite, 19

N, N-dimethylaniline, 40

N-bromosuccinimide (NBS), 36–37, 43–45

Neopentyl glycol (NPG), 24, 27, 29
Nicotiana tabacum, 259, 268, 282, 288
Nutrient limitation, 188–189

Oilseed rape, 281, 282
Olive mill effluents
    Biological treatment, 203–206
    Composition, 198–200
    Disposal, 201–203
    Production, 197–198
    Properties, 198–200
One-pot polymerization, 6
Organic load rate, 186
Organozinc halides with carbocation synthons, 131, 133
Oryza sativa, 259
Oxypropylation, 12
Ozonolysis, 47

Panicum virgatum, 282
Parthenium argentatum, 264, 265
Pectin, 454
Pentaerythriol, 29, 32
Perchlorate salt initiators, 140–142
Perfluoroarylated Lewis acids
    salts of with adventitious moisture, 103–104
    salts of with carbocation synthons, 107–108
    salts of with silicenium ion synthons, 108–111
    with adventitious moisture, 99–103
    with carbocation synthons, 104–109
PHB synthase (PhbC), 280, 281
Phosgene (derivatives), 330–336, 342
Photoperoxidation, 7
Pinus abies, 259
Pinus radiata, 259
Poly(1,3-propylene succinate), 309, 342

Poly(3-hydroxybutyrate) (PHB), 279–285, 288
Poly(3-hydroxybutyrate-co-3-hydroxyvalerate (P(HB-co-HV)), 282, 283
Poly(alkylene succinate), 358
    poly(butylene succinate), 358–363
    poly(ethylene succinate), 358–362
    poly(hexamethylene succinate), 360
    poly(propylene succinate), 358–359
Poly(butylene terephthalate), 326–327
Poly(ester amide), 357, 368–370
Poly(ethylene-2,5-furandicarboxylate), 400–402
Poly(furalidine bisamides), 416
Poly(furylenethylenediol), 416
Poly(L-lactic acid), 361
Poly(propylene terephthalate), 309
Polyamides , 357, 363–368
    aliphatic, 342–346
    applications, 343–344
    fatty acid-based, 344
    PA 10.10 344
    PA 11 343
    PA 4.10 343, 345
    PA 44 363–364, 366
    PA 5.10 344
    PA 6.10 343
    PA 6.9 344
    poly(2-pyrrolidone), Polyamide 4, PA 4 366, 368
Polyaspartate, 286
Polycarbonates
    aliphatic, 331–341
    aromatic, 341–342
    properties, 332, 334
    rheology, 339–340
    thermal stbility, 339–340
Polycondensation, 358, 360
Polyester, 357

Polyesters
  aliphatic, 309–323, 329–330
  aromatic, 325–328
  from furan monomers, 328–329
  from saccharide-derivatives, 309
  from vegetable oils, 329–330
  functionality, 316–321
  liquid crystalline, 328
  morphology, 312–316
  terpolyesters, 314–317
  thermal properties, 312–316
  unsaturated, 312, 321–323
Polyhydroxyalkanoate, 248, 269,
      282–285
Polyhydroxyalkanoates
  Applications, 177
  By mixed microbial cultures,
      178–182
  By pure microbial cultures,
      177–178
  Cellular content, 189–195
  Chemical and physical
      properties, 176–177
  From olive mill effluents, 206–211
  Hydroxybutyrate/
      Hydroxyvalerate ratio,
      190–196
  Molecular weight, 176–177
  Storage yield, 197
  Structure, 176
Polymer supported Lewis acids,
      143–146
Polyphenoloxidase (PPO), 268
Polyphosphate accumulating
      organisms, 179
Polysaccharides, 2
Poplar, 255, 258–261
Populus, 258
Potato, 247, 249, 250, 252–254, 272,
      273, 277, 282, 284, 286–290
Pressure-sensitive adhesives
      (PSA), 35
Proline-4-hydroxylase (P4H), 276
Propanediol, 1,3- 307, 337

Protein targeting to
  cell wall, 261
  endoplasmic reticulum, 273,
      278, 289
  peroxisome, 284
  plastids, 281, 283, 284, 288
  vacuole, 276
Psyllium Husk, 455
Pyrophosphate synthase, 267

Reactor fouling, 74, 158–159
Renewable monomers , 71–72,
      161–164
  acrylated methyl oleate, 235
  biomass, 240
  bromoacrylated methyl oleate,
      235
  butyrolactone, 240
  cardanol, 235, 236
  cardanyl acrylate, 236, 237
  cashew nut shell liquid, 235
  cellulosic, 240
  dehydroabietic acid, 238
  dehydroabietic ethyl acrylate,
      239
  epoxidized methyl oleate, 235
  glycopolymers, 241
  hydrophilic, 241
  hydrophobic monomers, 227
  lauryl methacraylate, 232, 237
  levulinic acid, 240
  methyl oleate acrylamide, 235
  methylene butyrolactone, 240
  phenanthrene, 238
  phenolic lipid, 235
  plant oils, 222, 232, 235
  polylauryl methacrylate, 233
  polysacharides, 222
  rosin gum, 238, 239
  stearyl acrylate, 237
  stearyl methacrylate, 237
  sugar acrylates, 241
  sugars, 222
  triglycerides, 231

tulipalin A 240
unsaturation, 236
vegetable oils, 222, 231, 232, 237
Rice, 249, 252, 256, 257, 259
Ring-opening polymerization, 14
Rubber, 247, 248, 261–269
Rubber tree, 261–265, 269
Russian dandelion, 264, 267–269

Scleroglucan, 456
Sebacic acid, 343
Sequencing batch reactor, 185
Silk, 247, 248, 269, 270–274
Silk-like protein (DP1B), 273
Simultaneous interpenetrating
    networks (SINs), 44
Slurry process
    improvements to, 158–161
    problems with, 74–75, 158–161
Sodium ascorbate, 49–50
Sodium azide, 50
Sodium cyanoborohydride, 40, 50
Solanum tuberosum, 282
Solid state polymerization, 327
Sorbitol, 311
South American leaf blight (SALB),
    263, 264, 269
Soybean oil, 308
Spider silk-elastin fusion protein,
    277, 278
ß-ketoacetyl-CoA thiolase (PhbA),
    280, 281
Starch, 247–254, 269, 289, 290, 458
Starch branching enzyme (SBE),
    250, 252, 253
Sterically hindered pyridines
    adducts of titanium cations, 104,
        110, 113
    decomposition of carbocations
        by, 103–104, 106, 110, 113
    structures, 79
Strain at break, 19
Styrene, 223, 224, 27–34, 36–37,
    39–45, 60

Suberic acid, 313
Suberin, 309
Succinic acid, 306, 312, 354–356,
    358, 361–364, 369
Succinic anhydride, 33
Sucrose, 306
Sugar beet, 281, 282, 290
Sugar cane, 256, 281, 282, 290
Supported perchlorates, 141–142
Supported triflates, 141–142
Sustainable chemical Industry, 222
Switchgrass, 281, 282, 290
Synthetic spider silk protein (SO1),
    277, 278

Tall fescue, 259
Taraxacum koksaghyz, 264,
    265, 268
t-Butyl perbenzoate, 64
Terephthalic acid, 326
Tetrahydrofuran, 357
Thale cress, 282
Thermal degradation resistance, 9
Thermal polymerization, 50
Thermogravimetric analysis, 10
Thermoresponsive behavior, 16
Threonine deaminase, 283
Tobacco, 259, 272, 273, 275–278,
    281–284, 288, 290
Toluene, 33, 35, 44, 46, 53–56
Topology, 221, 227
Transgenic production/
    modification in plants
    collagen, 275–276
    cyanophycin, 288–289
    elastin, 277–279
    lignin, 258–261
    polyhydroxyalkanoate, 281–285
    rubber, 267–269
    silk, 272–274
    starch, 253–254
Trialkylaluminums
    overview of compatible
        initiators, 88–89

with carbocation synthons, 90–91
with halogens, 91–92
Triazabicyclo[4.4.0]dec-5-ene, 1,5,7- 345
Triethylamine, 40
Triflate salt initiators, 140–142
Triflic acid, 159–162
Triglyceride oils, 307
Triphenylantimony, 27
Triphenylphosphine, 50

Varnish, 238
Vegetable oils
    castor, 23–24, 29, 32, 43–45, 62–63
    cotton, 24

lesquerella, 24
licania, 24
linseed, 24, 26, 29–30, 33
soybean, 24, 26–27, 29, 32, 39–44, 46–50, 62–63
sunflower, 24, 30
vernonia, 24
Volatile fatty acids (VFAs), 180, 182–184, 206

Xanthan Gum, 460
Xylene, 44, 46
Xylitol, 341

Yield strain, 19

Zea mays, 259, 282

Also of Interest

Check out these published and forthcoming related titles from Scrivener Publishing

Handbook of Bioplastics and Biocomposites Engineering Applications
Edited by Srikanth Pilla
Published 2011. ISBN 978-0-470-62607-8

Biopolymers: Biomedical and Environmental Applications
Edited by Susheel Kalia and Luc Avérous
Published 2011. ISBN 978-0-470-63923-8

Renewable Polymers: Synthesis, Processing, and Technology
Edited by Vikas Mittal
Forthcoming September 2011. ISBN 978-0-470-93877-5

Plastics Sustainability
Towards a Peaceful Coexistence between Bio-based and Fossil fuel-based Plastics
Michael Tolinski
Forthcoming Octoberr 2011. ISBN 978-0-470-93878-2

Polymers from Renewable Resources
Ram Nagarajan
Forthcoming Spring 2012. ISBN 9780470626092

Green Chemistry for Environmental Remediation
Edited by Rashmi Sanghi and Vandana Singh
Forthcoming September 2011. ISBN 978-0-470-94308-3

High Performance Polymers and Engineering Plastics
Edited by Vikas Mittal
Published 2011. ISBN 978-1-1180-1669-5

Polymer Nanotube Nanocomposites: Synthesis, Properties, and Applications
Edited by Vikas Mittal.
Published 2010. ISBN 978-0-470-62592-7

Handbook of Engineering and Specialty Thermoplastics
Part 1: Polyolefins and Styrenics by Johannes Karl Fink
Published 2010. ISBN 978-0-470-62483-5
Part 2: Water Soluble Polymers by Johannes Karl Fink
Published 2011. ISBN 978-1-118-06275-3
Part 3: Polyethers and Polyesters edited by Sabu Thomas and Visakh P.M.
Published 2011. ISBN 978-0-470-63926-9
Part 4: Nylons edited by Sabu Thomas and Visakh P.M.
Forthcoming October 2011. ISBN 978-0-470-63925-2

A Concise Introduction to Additives for Thermoplastic Polymers
By Johannes Karl Fink.
Published 2010. ISBN 978-0-470-60955-2

Introduction to Industrial Polyethylene: Properties, Catalysts, Processes by Dennis P. Malpass.
Published 2010. ISBN 978-0-470-62598-9

The Basics of Troubleshooting in Plastics Processing
By Muralisrinivasan Subramanian
Published 2011. ISBN 978-0-470-62606-1

Miniemulsion Polymerization Technology edited by Vikas Mittal
Published 2010. ISBN 978-0-470-62596-5